Advanced Electrical and Electronics Materials

Scrivener Publishing
100 Cummings Center, Suite 541J
Beverly, MA 01915-6106

Advanced Materials Series

The Advanced Materials Series provides recent advancements of the fascinating field of advanced materials science and technology, particularly in the area of structure, synthesis and processing, characterization, advanced-state properties, and applications. The volumes will cover theoretical and experimental approaches of molecular device materials, biomimetic materials, hybrid-type composite materials, functionalized polymers, supramolecular systems, information- and energy-transfer materials, biobased and biodegradable or environmental friendly materials. Each volume will be devoted to one broad subject and the multi-disciplinary aspects will be drawn out in full.

Series Editor: Dr. Ashutosh Tiwari
Biosensors and Bioelectronics Centre
Linköping University
SE-581 83 Linköping
Sweden
E-mail: ashutosh.tiwari@liu.se

Publishers at Scrivener
Martin Scrivener(martin@scrivenerpublishing.com)
Phillip Carmical (pcarmical@scrivenerpublishing.com)

Advanced Electrical and Electronics Materials

Processes and Applications

K.M. Gupta and Nishu Gupta

Scrivener Publishing
WILEY

Co-published by John Wiley & Sons, Inc. Hoboken, New Jersey, and Scrivener Publishing LLC, Salem, Massachusetts.
Published simultaneously in Canada.

For general information on our other products and services or for technical support, please contact our Customer Care Department within the United States at (800) 762-2974, outside the United States at (317) 572-3993 or fax (317) 572-4002.

Wiley also publishes its books in a variety of electronic formats. Some content that appears in print may not be available in electronic formats. For more information about Wiley products, visit our web site at www.wiley.com.

For more information about Scrivener products please visit www.scrivenerpublishing.com.

Cover design by Russell Richardson

Library of Congress Cataloging-in-Publication Data:

ISBN 978-1-118-99835-9

Printed in the United States of America

10 9 8 7 6 5 4 3 2 1

Dedicated to
my respected mother **BELA,**
father (Late) **Ram Nath,**
Godfather (Late) **Lakhan Lal,**
father-in-law (Late) **Kishori Lal,**
brother-in-law and sister: **Jawahar and Savitri,**
nephew (Late) **Jayant (Babul)**
and
all forefathers and foremothers
whose
blessings have always been a boon in my life

Contents

Preface

This book is intended to cover the vast and fast growing field of electrical and electronic materials and their science in accordance with the modern trends. The level of the book covers the syllabi being taught at graduate and undergraduate standard of Engineering Colleges in many countries around the world including the United States and India. It also covers the syllabi of various Indian national level examinations. The contents of the book will also be very helpful to the students of postgraduate studies.

Basic and pre-requisite information has been included for easy transition to newer topics. Latest developments in various fields of materials; their sciences, processes and applications have been accommodated. Latest topics like spintronics, high energy hard magnetic materials, ferrites, plzt, vacuum as insulator, fibre-optics, high temperature superconductors, ferroelectrics, smart materials, ferromagnetic semiconductors etc. have been included.

Illustrations, examples, and details of sciences are such that they include different disciplines of engineering e.g. robotics, electrical, mechanical, electronics, instrumentation and control, computer, and inter-disciplinary branches. Topics like electron theories, magnetostriction, high voltage engineering, brillouin zone theory, gas-filled conductors etc. have been explained. A variety of materials ranging from iridium to garnets, resistors, capacitors, printed circuit boards to microelectronics, micro alloys to memory devices, left-handed materials, advance and futuristic materials are described.

Review and Objective Type Questions are based on concepts, design, construction, applications and practical orientations. Objective questions of all kinds viz.; 'Multiple Choice' type, 'Assertion and Reasoning' type, 'Matching' type, 'Fill in the Blanks' type, 'Pick up the Correct Choice', 'True-False' type are included. A large number of numerical examples have been worked out. Numerical exercises for practice and self evaluation are also given with their answers. SI units, in general, have been incorporated throughout the text but for familiarity the MKS units have also been used.

The book is substantiated by a large number of diagrams, tables, equations, review questions, objective questions, solved numericals, unsolved numerical exercises, question-answers, and terms in quick revision summary. I gratefully acknowledge the authors and publishers of the books and other literature quoted in references which have have been consulted in preparing this book.

I acknowledge the inspiration and blessings of my respected mother Smt. BELA DEVI, brother-in-law Sri JAWAHAR LAL, sister Smt. SAVITRI LAL, elder brother Sri GOPAL DAS GUPTA and other family members. I am full of gratitude to my daughter NIDHI, son NISHU, wife RITA, son-in-law RITESH, grandson AKARSH (RAM) and granddaughter GAURI for the patience shown and encouragement given to complete this venture.

I am highly obliged to my postgraduate student Mr. Kishor Kalauni, M.Tech. (Materials Science) without whose help and support it was not possible to publish this book. He took great pains in formatting and typing of a large number of pages, typesetting, making figures etc. I also acknowledge my student Mr. Saurabh Kumar Singh, M.Tech. (Materials Science) for some typing work. I extend my heartfelt gratitude to my friends Er. Ranjeet Singh Virmani, Er. K.R.D. Tewari, and Er. Satish Chandra Srivastava for their technical support in several ways. Last but not the least, I owe a lot to Sri S.C. Sant, Mr. Anurag Sant and Mrs. Shilpa Sant for their continued support, guidance and cooperation in preparing this book.

Enormous effort has been made to avoid errors and mistakes; however, their presence cannot be ruled out. Any suggestion to improve the standard of this book, indication towards errors, omissions and mistakes will be highly appreciated.

January 2015
Allahabad, India

Dr. K.M. GUPTA
kmgupta1948@gmail.com
and
NISHU GUPTA
dce.nishu@gmail.com

Acknowledgement

The Author of this book acknowledges with heartfelt gratitude to Mr. S.C.Sant and Mr. Anurag Sant, the publishers of the Umesh Publications (4230/1 Ansari Road, Daryaganj, Delhi-110002) India, for being kind enough to provide some literature support and valuable information that proved very useful in preparing this book. Author (Dr. K. M. Gupta) thanks especially Mr. Anurag Sant for his courteous gesture.

Dr. K.M. Gupta

About the Authors

Dr. K.M. Gupta is a Professor in the Department of Applied Mechanics, Motilal Nehru National Institute of Technology, Allahabad. He has over 38 years of teaching, research and consultancy experience. He obtained Diploma (with Honours) in Mechanical Engineering, Bachelor of Engineering (Gr. I.E., AMIE) in Mechanical Engineering, Postgraduation (M.E. with Honours) in 1977, and completed his Doctorate (Ph.D.) degree from University of Allahabad. Although a Mechanical Engineer but he has also specialised in Automobile Engineering discipline. He has authored 29 books and edited 2 books on Engineering subjects, and a chapter in Scrivener Wiley published 'Handbook of Bioplastics and Biocomposites Engineering Applications'. He has also authored 120 research papers in reputed International and National Journals and Conferences to his credit. Professor K.M.Gupta has presented his research papers in 16 International conferences abroad at USA, UK, Japan, China, France, Muscat, Bangkok, South Africa, Hongkong etc. He has also chaired 8 International Conferences in China, Singapore, Dubai, Bangkok etc. He has acted as Editor-in-Chief of *The International Journal of Materials, Mechanics and Manufacturing (IJMMM)* Singapore.

Recipient of many Gold Medals and Prizes for his outstanding career from Diploma to Doctorate (a rare achievement in this India); he has served as Head of Automobile Engineering Department at the Institute of Engineering and Rural Technology, Allahabad. He developed several laboratories viz. Automobile related Labs, Materials Science Lab., Strength of Materials Lab., Hydraulics Lab. etc. at different Institutes/colleges.

Currently, Dr. Gupta is teaching materials science, engineering mechanics, thermodynamics of materials, electrical and electronic materials etc. his research interests are in the fields of materials science, composite materials, stress analysis, solid mechanics etc.

Nishu Gupta is a Research Scholar in the department of Electronics and Communication Engineering, Motilal Nehru National Institute of Technology, Allahabad, India. He received his B.Tech. Degree in Electronics and Communication Engineering from U.P. Technical University, M.Tech. Degree in Nanoscience and Technology from Delhi Technological University (formerly Delhi College of Engineering), Delhi. He served as Visiting Faculty in the department of Electronics and Communication Engineering, Motilal Nehru National Institute of Technology, Allahabad; as Adhoc Faculty in the department of Electronics and Communication Engineering at National Institute of Technology, Sikkim. He also served as Deputed Faculty at NIT Calicut.

Prior to coming to academic profession, he has served as Software Engineer at Infosys Technologies Limited, and as Senior Software Engineer at Tech Mahindra Limited. He has 18 Research papers to his credit which were published in various International and National Conferences in India and abroad, and out of these has published 5 papers in International Journals. He has attended 5 International conferences abroad out of which he delivered presentations in 3 conferences, all as a student, which is an incredible accomplishment. His honours and awards include the recommendations for "Top 2000 Intellectuals of the 21st century by International Biographical Centre, Cambridge, England", and for inclusion in the list of "2009 edition of Marquis Who's Who in the World". He has been shortlisted by the Ministry of HRD, Government of India for the Commonwealth Scholarship/Fellowship Plan, 2012 offered by the Government of United Kingdom.

His research interests are in the field of Semiconductor Devices, Hybrid Solar Cells, Photonics, Nano-optics and related fields.

Abbreviations

ACSR	Aluminum Conductor Steel Reinforced
AR	Anti Reflective
BCT	Body Centred Tetragonal
BJT	Bipolar Junction Transistor
BMG	Bulk Metallic Glass
CCD	Charge Coupled Device
CIC	Cable-In-Conduit
CNC	Computerized Numerically Controlled
DC	Diamond Cubic
EGS	Electronic Grade Silicon
EHP	Electron-Hole Pair
FCC	Face Centred Cube
FFC	Flexible-Film-Circuit
FPC	Flexible Printed Circuit
GGG	Gadolium Gallium Garnet
GMR	Giant Magneto-Resistance
HCP	Hexagonally Closed Packed
HMM	Hard Magnetic Materials
HTS	High Temperature Superconductor
IGFET	Insulated Gate Field-Effect Transistor
IFF	Ionic Ferrofluid
KDP	Potassium Dihydrogen Phosphate (a piezoelectric material)
LBL	Layer-By-Layer
LDR	Light Dependent Resistor
LED	Light Emitting Diode
LHM	Left-Handed Material
LTS	Low Temperature Superconductor
MBD	Magnetic Bipolar Diode
MBT	Magnetic Bipolar Transistor

MCE	Magneto–Caloric Effect
MEE	Magneto–Electro–Elastic
MGS	Metallurgical Grade Silicon
MHD	Magneto–Hydro-Dynamic
MOSFET	Metallic Oxide Semiconductor Field- effect Transistor
MRAM	Magnetic Random Access Memory
NLO	Non–Linearly Optical
NTC	Negative Temperature Coefficient (Thermistor)
OFHC	Oxygen Free High Conductivity
OMR	Optical Magnetic Reader
PCB	Printed Circuit Board
PLZT	Lead Lanthanum Zirconate Titanate
PTC	Positive Temperature Coefficient (Thermistor)
PZT	Lead Zirconate Titanate
RHM	Right Handed Material
SET	Single Electron Transistor
SFF	Surfated Ferrofluid
SMC	Surface Mounted Device
SOI	Silicon-On-Insulator
SQUID	Superconducting Quantum Interference Device
TEC	Thermo–Electric Cooler
TEE	Thermo–Electric Effect
UPT	Unipolar Transistor
USB	Universal Serial Bus (interface)
VDR	Voltage- Dependent Resistor
VLSI	Very Large Scale Integration
VVR	Voltage–Variable Resistor
WAN	Wide Area Network
WC	Tungsten Carbide
XRD	X-ray Diffraction

1

General Introduction to Electrical and Electronic Materials

1.1 Importance of Materials

Use of materials is an indispensible requirement for the development of engineering and technology. They provide the basis for manufacturing, fabrication, operations, and constructions etc. It may be the construction of a building, manufacturing of a machine, generation of electricity, transmission of message from one place to another, or control instruments; they all make use of some materials. These materials are of different natures viz., R.C.C. (reinforced cement concrete), steel and iron, copper and aluminium, mica and rubber, alloy and glass etc.

Depending upon the areas in which they are used, the materials may be known as

1. Civil engineering materials,
2. Electrical engineering materials,
3. Mechanical engineering materials,
4. Electronics engineering materials,

5. Nuclear engineering materials, and
6. Computer engineering materials etc.

These days, most materials find inter-disciplinary uses. It, therefore, becomes difficult to conceive as to which material belongs to which category. As an illustration, we take the example of aluminium. It is generally known as a mechanical engineering material but it finds use in electrical engineering also as a conductor; is used in electronics engineering for doping of intrinsic semiconductors; is used in metallurgical engineering as an alloying element; is used in civil engineering for decorative items. These are inter-disciplinary uses. However, conventionally we call RCC as civil engineering material, insulator (e.g. mica) as electrical engineering material, metals as mechanical engineering materials, semiconductors as electronics engineering materials, uranium as nuclear engineering materials etc,

1.2 Importance of Electrical and Electronic Materials

Advancement of any engineering discipline is not possible without the development of materials suitable for appropriate uses. The development necessitates progress in the science: physics and chemistry, engineering and technology of the materials. Rapid advancement in electron-based computers, revolutionary changes in electronics engineering from vacuum valves to very large scale Integration (VLSI); developments of conducting polymers, ferroelectrics as a modern breed of dielectrics, and ferrites as a superb magnetic material (in addition to several other versatilities) are some illustrations which are the outcome of developments in electrical materials technology.

Further advancement in electrical, electronics, computers, and instrumentation fields are likely to be in the form of whisker-based fibre optics, light-based computers, high temperature superconductors etc. But dreams of these futuristic advances will become a reality only after achieving a breakthrough in certain materials properties. Most likely the 21st century will see the high voltage transmission through hair- sized conducting wires, the hybrid magnets of more than 100 T (tesla) capacities, the magnetic refrigerators operating well above 100 K (kelvin), and hybrid crystals (e.g. Hg-Cd-Te) serving as sensor elements.

Smart materials, ferrites, ferroelectrics, garnets, whiskers and ceramic superconductors are the latest kind of materials of today and tomorrow. More materials in this list are ruby laser, superalloys, crystalline carbon 60 (C_{60}) etc. Details of these topics are described in appropriate chapters in this book.

1.3 Classification of Electrical and Electronic Materials

Electrical engineering materials can be classified into following types:

1. Conductors
   i. high voltage and low voltage conductors
   ii. high temperature and low temperature conductors
   iii. bared and insulated conductors
2. Semiconductors
   i. intrinsic (or element) type
   ii. extrinsic (compound and alloy) type
      - n type
      - *p* type
3. Dielectrics (or Insulators)
   i. solid type
   ii. liquid type
   iii. gaseous type
   iv. ceramic type
   v. polymeric type
   vi. fibrous type
4. Superconductors
   i. metallic type
   ii. ceramic type
   iii. ideal and hard types
   iv. low and high temperature types
   v. magnetic and non-magnetic types
5. Magnetic materials
   i. diamagnetic
   ii. paramagnetic
   iii. ferromagnetic
   iv. antiferromagnetic
   v. ferrimagnetic (or ferrites)
6. Ferroelectrics
   i. zirconates
   ii. hafnates

   iii. titanates
   iv. PLZT

7. Piezoelectrics
   i. natural (as rochelle salt)
   ii. artificial (as tourmaline, metaniobate)
8. Perovskites (or mixed oxides)
9. Spinels, Garnets, and Magnetoplumbites
   i. normal spinel (as $ZnFe_2O_4$)
   ii. inverse spinel (as magnetite)
   iii. metallic garnet
   iv. rare earth garnet

An introductory description of these materials is given below. They have been described at length and detail in forthcoming chapters.

1.3.1 Conductors

These are the materials having plenty of free electrons in their valence orbit which are responsible for electrical and heat conduction. The valence and conduction band energies overlap in them (see Fig. 1.1). The mechanism of electron conduction through conductors occurs in accordance with free electron theory as described in chapter 4. Gold (Au), silver (Ag), copper (Cu), aluminium (Al) are good electrical conductors. Considering the cost factor, copper and aluminium are widely used as transmission wires and cables, and windings of rotating machines (motors, alternators etc.). Oxygen-free high conductivity (OFHC) copper conductor is very suitable for low temperature applications.

Electrical resistivity of conductors is of the order of 10^{-9} ohm-meter. It is influenced by temperature, impurities, alloying elements, and plastic deformation. Effects of these factors are determined by Nordheim equation and Matthilssen rule whose complete details are given in chapter 4.

1.3.2 Semiconductors

These are the solids having energy gap E_g lying in between the conductor and insulator (see art. 1.7 for classification of solids based on energy gap). Their conductivity is more than that of the dielectrics but less than that of conductors. They are basically electronic materials viz. silicon (Si) and germanium (Ge) in element form. The element form remains in pure state and is called

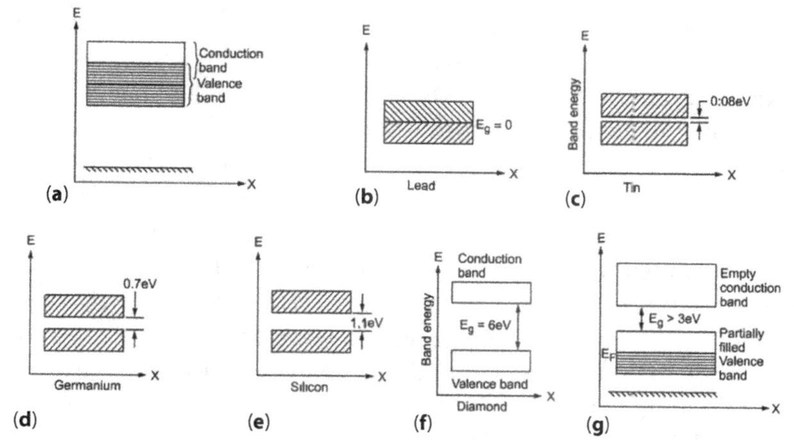

Figure 1.1 Classification of solids based on energy gap (*a*) conductor having overlapping energy level, (*b*) conductor having $E_g = 0$, (*c*) conductor, (*d*) and (*e*) semiconductors, and (*f*) and (*g*) insulators

intrinsic semiconductor. Intrinsic semiconductors are of no practical utility since they desire an extremely high voltage 10^8 volts) for conduction across the energy gap. Hence, they are doped to make extrinsic semiconductors.

Extrinsic semiconductors are primarily of *n*-type and *p*-type. They may be in compound form such as GaAs, CdTe etc. and in alloy form such as $HgCd_xTe_{1-x}$. A number of semiconducting compounds are available in oxide, halide and sulphide forms also. The *n* and *p*-types are used to make *n*-*p* junction for diode, *n*-*p*-*n* and *p*-*n*-*p* for transistor, etc.

Semiconductors are generally hard and brittle, and possess negative coefficient of temperature resistance. In contrast with conductors whose conductivity increases with purification, the conductivity of semiconductors decreases with purification. Semiconductors are widely used as rectifiers, amplifiers, photocells etc. Their properties are of greater importance in telecommunication, power electronics, computer hardware etc. Complete details about them are given in chapters 6, 7 and 8.

1.3.3 Dielectrics

These are the insulating materials in which the valence electrons are tightly bound to their parent atoms. Therefore, they require a very large electric field to remove these electrons from the atoms. Dielectrics are characterized by (i) a full valence band (ii) an empty conduction band, and (iii) a large energy gap ($E_g > 3$ eV).

Dielectrics are available in solid, liquid, and gaseous states. They may be of natural or synthetic types, polymeric or ceramic in nature, and fibrous or flaked in structure. Mica, Bakelite, elastomeric fibre, and paper are some examples of solid dielectrics; Transformer oil, silicon oil, varnishes are liquid dielectrics; while the air, nitrogen, ozone are gaseous dielectrics. Dielectrics are used as capacitors, as insulation, in strain gauges and sonar devices.

Main properties of dielectrics are their dielectric constant, dielectric strength, and dielectric loss. Dielectric constant is influenced by temperature and frequency. A good dielectric should have high dielectric constant and high dielectric strength but a lower dielectric loss. The dielectric loss is the least in transformer oil and fused silica ($\tan \delta = 0.0001$). Dielectrics are subjected to different kinds of polarization under the influence of applied electric field. These polarization processes are affected by time, frequency and temperature. The time effect is expressed in terms of relaxation time. More details are given in chapters 9 and 10.

1.3.4 Superconductors

These are those materials which exhibit abrupt and sudden changes in their resistance and other behaviour, at extremely low temperatures. It is because of a phenomenon called superconductivity which is noticed below a certain critical temperature. Superconductors show extraordinary properties like (i) resistance R of the material ≈ 0, (ii) magnetic flux density B ≈ 0, and (iii) power (copper) loss $I^2R \approx 0$.

Superconducting materials mainly fall in the category of metals and ceramics, and may be in element, compound, or alloy forms. They may be of magnetic or non-magnetic nature, type I (ideal) or type II (hard), and low temperature or high temperature superconductors. Since the peculiar behaviour of superconductivity in materials is presently available at low temperatures only, there is a need to develop these properties at high temperatures (say room temperature).

Ceramic superconductors are more promising development in this field. Nb_3Sn, $La_{80}Au_{20}$, Nb-Zr-Ti are some examples of superconducting compounds and alloys. Yattrium barium copper oxide ($YBa_2Cu_3O_{7-x}$ bismuth strontium calcium copper oxide (Bi Sr Ca CuO) and thallium barium calcium copper oxide (Th Ba Ca CuO) are some modern ceramic (i.e. oxide) superconductors.

Superconductors are used in leviated trains (wheel-less train moving on magnetic field), as powerful magnets for nuclear fusion reactors, in medical imaging machines, in cryogenic engines, for high speed electronic signal processing, and magnetohydrodynamic (MHD) power generation etc. Its existence at room temperature in future is likely to bring a revolution in life style of mankind. More details on them are given in chapter 13.

1.3.5 Magnetic Materials

These are those materials in which the magnetization can be easily induced to make them magnets. The generation of induced dipole is attributed to electron spins, orbital motion of electrons and the interaction between them. Depending upon the arrangement of magnetic moments in them, the magnetic materials may be of diamagnetic, paramagnetic, ferromagnetic, antiferromagnetic and ferrimagnetic (or ferrites) types.

Diamagnetic materials develop no magnetism due to the absence of magnetic moment. Gold, bismuth, copper are such examples.

Paramagnetic materials form weak magnets. Aluminium, platinum, $NiSO_4$ etc. are examples of this kind.

Ferromagnetic materials develop very strong magnetism. Steel and iron (Fe), nickel (Ni), cobalt (Co) and gadolinium (Gd) are such examples. Depending upon the area under B-H curve, ferromagnetic materials may be either soft magnetic materials or hard magnetic materials. Soft magnetic materials are used as transformer core because hystersis loss is least in them. Hard magnetic materials are used to make permanent magnets for electric motors, alternators etc. Ferromagnetic materials find use as magnetic screens and magnetic waves also.

Antiferromagnetic materials are those kind of materials in which the quantum mechanical exchange forces produce a tendency for antiparallel alignment of electron spins of neighbouring atoms. Oxides and sulphides of metals are some examples of this kind. Their susceptibility is of the order of $+10^3$ which varies with change in temperature. Maximum susceptibility in them occurs at Neel temperature.

1.3.6 Ferrites

These are those kind of magnetic materials in which the magnitude of magnetic moment is more in one direction than the other. They are compounds of two metallic oxides of which one is invariably an iron oxide. $FeO.Fe_2O_3$, $BaO.6Fe_2O_3$ are some examples of ferrites. Basically they are ceramics having high electrical resistivity and low power loss, and are similar to semiconductors in conduction behaviour.

Ferrites may be soft ferrite or hard ferrite. Lithium ferrite is a soft ferrite and strontium ferrite is a hard ferrite. A soft ferrite possesses square hysteresis loop while the hard ferrite shows semiconductor behaviour. Mg-Mn and Zn-Mn ferrites are used as memory cores in computers. Ferrites are made by powder compaction technique. Their relaxation time is of the order of 10^{-9} to 10^{-10} second, resistivity of 10^7 ohm-meter, and relative permeability of the order of 10^4.

Complete details on magnetic materials are given in chapters 11 and 12.

1.3.7 Ferroelectrics

These are advanced kind of dielectrics having $\varepsilon_r > 2000$. They are much superior to conventional dielectrics such as mica whose ε_r is 8 only. Such a high value of ε_r in ferroelectrics is due to alignment of all the electric dipoles in the direction of applied field. It, therefore, causes large and spontaneous polarization. Thus the ferroelectrics are analogous to ferromagnetic materials. Rochelle salt, $SrTiO_3$ (strontium titanate), $PbZrO_3$ (lead-zirconium titanate) are some examples of ferroelectrics. They are used in miniature capacitors, as electro-optic materials, and in sonar devices etc.

1.3.8 Piezoelectrics

These are those kinds of dielectrics which exhibit piezoelectric effect. According to this effect, an electric field induces a mechanical strain in piezoelectric crystals and vice-versa. Quartz, CdS (calcium sulphide), ZnO (zinc oxide) are some examples. These are used as transducers in various devices and instruments for changing electrical signal into mechanical signal and vice-versa. A quartz crystal is used to produce ultrasonic waves.

1.3.9 Perovskites (Titanates, Zirconates, Hafnates)

These are the mixed oxides which are obtained on fusing the oxides of titanium (Ti), zirconium (Zr), and hafnium (Hf) with oxides of other metals. The other metals are quite different in size from Ti, Zr or Hf. Thus the fusion of the two metals results in formation of titanates, zirconates, and hafnates. Perovskites (named after its inventor Perovsky) are used to make high temperature ceramic superconductors.

1.3.10 Spinels, Garnets, and Magnetoplumbite

These are basically magnetic materials. They belong to ferrite family and are different crystal forms of it. Spinels and garnets have cubic structures while the magnetoplumbite has hexagonal structure. Typical example of a spinel is $MnFe_2O_4$, of a garnet is $Er_3Fe_2O_{12}$, and of magnetoplumbite is $BaFe_{12}O_{19}$. They behave as soft and hard ferrites both, possess square hysteresis loop, and serve as low mobility semiconductors.

Since they belong to ferrite's family, they are used as memory cores in computers, in microwave frequency applications etc.

1.4 Scope of Electrical and Electronic Materials

Scope of application of electrical materials is very vast. These materials find utilities in not only electrical engineering machines, equipments, devices etc. but are also used as components, circuits and other auxiliaries related to electronics, computers, and instrumentation fields. Their importance is also realized in cable networking, wireless networking, satellites, optical devices etc. They find very useful applications even in medical, mechanical, nuclear, biotechnological fields. The details given below focuses upon the scope of electrical materials in numerous applications.

In Consumer Items

- Bulb filaments
- Heaters
- Remote control devices
- Telephone
- Domestic wiring
- Tape record
- Switches
- Iron press
- Television
- Invertors
- Radio
- Microwave ovens, etc.

In Electrical Engineering

- Contacts
- Cables
- Magnets
- Alternators
- Motors
- Voltage Dividers
- Conductors
- Dielectrics/Insulators
- Piezoelectrics
- Transformers
- Capacitors
- Bus Bars, etc.

In Electronics

- Amplifiers
- Integrated Circuits
- Antenna
- Broadcasting Systems
- Printed Circuits
- Rectifiers
- Filters
- Regulators
- Satellite
- Photoconductive Cell
- Transistors
- Modulators, etc.

In Robotics

- Sensors
- Controls
- Manipulators
- Grippers
- Actuators
- Processors
- Encoders
- Pendants, etc.

In Computer Engineering

- Hardware
- Monitors
- Peripherals
- Floppy
- Printers
- Memory devices
- Ports
- Display devices
- Hard disk
- CDs, etc.

In Instrumentation

- Transducers
- Signal generators
- Microprocessors
- Strain gauges
- Cathode ray oscilloscopes
- Recorders
- Thermistors
- Energy meters, etc.

In Information Technology

- Networking cables
- Web camera
- Cam corder
- Multimedia devices
- Routers
- Optical fibres, etc.

In Mechanical Engineering

- Furnace
- Thermocouples
- Arc welding set
- Electrically heated ovens
- X-ray systems, etc.

In Biotechnology and Medical Sciences

- Electro-cardiography
- Fibre-optic endoscopy
- Electronic photography
- Magnetic resonance imaging, etc.

Most of the above systems/devices are inter-disciplinary, and are indicative only. Besides above, the electrical materials are also used in the fields of aeronautics, marine, defence, chemical, metallurgical, civil and automobile engineering applications.

1.5 Requirements of Engineering Materials

Materials technology does not mean just knowing the physics and chemistry of materials, their behaviour and properties. It is also essential to know as to how a material can be suitably and economically put to practical uses under wide range of conditions. These conditions may relate to operation, to fabrication, or to stability of materials. An electrical engineering material is used in one or all of the following areas.

- *Machines* (as motor, alternator, robots etc.)
- *Structures* (as transformer, cathode ray tube, antenna etc.)
- *Devices* (as strain gauge, integrated circuit, control switch, thermistor, bimetal gauge etc.)
- *Instruments* (as multimeter, transducers, thermocouples etc.)

Each material possesses several properties. Some important properties are listed as follows.

1. Electrical	resistivity, conductivity, dielectric constant, dielectric strength,relaxation time, loss angle, power factor.
2. Magnetic	hysteresis, retentivity, permeability, susceptibility, coercive force,reluctivity.
3. Electronic	semi-conduction, drift, diffusion, concentration, energy gap, Fermi energy,mobility, carrier density, ionization energy, effective mass, density of state, rectifying action.
4. Optical	reflection, refraction, transmission, fluorescence, lustre, luminescence.
5. Physical	density, melting point, colour, shape, size, finish, porosity.
6. Thermal	expansion, conductivity, specific heat, thermal fatigue, thermal stress, thermal shock, latent heat of fusion.

7. Mechanical — creep, fatigue, toughness, hardness, impact, ductility, malleability,resilience, brittleness.
8. Chemical — corrosion resistance, passivity, atomic number, molecular weight, acidity, alkalinity, oxidation.
9. Cryogenic — ductile-brittle behaviour, low temperature impact behaviour, very low temperature phase changes, superconductivity.
10. Acoustical — sound reflection, absorption, damping, transmission.
11. Structural — strength, stiffness, elasticity, plasticity.
12. Surface — friction, abrasion, wear, erosion.
13. Metallurgical — phase rule, solid solution, crystallization rate, diffusion.
14. Technological — weldability, machinability, formability, castability, fabricationability, hardenability.
15. Aesthetic — feel, texture, appearance, lustre.
16. Nuclear — half-life period, decay constant, radiation absorptivity.

Amongst these the magnetic, electrical, electronical and optical properties are of greater importance in selection of electrical materials for specific purposes. Therefore, they have been dealt in detail in different chapters ahead. However, other properties are also of utmost consideration. Hence, important among them are briefly explained as follows.

Porosity is a desired requirement in intrinsic semiconductor for doping, but is undesired in insulators exposed to moist air.

Thermal expansion of unequal values are desired in bimetallic strips of dissimilar materials for use in thermostatic controls and thermocouples.

Fatigue resistance is a desired property in lamp filaments against fluctuation of voltage, otherwise the life of filaments will decrease.

Corrosion resistance is desired in all the materials against environmental effects.

Damping is a desired property for various components of an electrical instrument to prevent against mechanical shocks and vibrations, and against the effects of external magnetic and electrical fields.

Strength is very essential for the casing of alternators and motors.

Elasticity and ductility are also needed in cables, wires and coils.

Abrasion resistance is an important requirement for contact materials which are subjected to sparking of arc, such as in electrodes.

Wear resistance is desired in contacts which are subjected to physical/ mechanical rubbings.

Diffusion is desired for doping of silicon and germanium elements. It is also needed in making of binary (metal) alloy systems such as CuZn, Cu-Ni etc.

Weldability is desired in the construction of laminated core of transformer.

Appearance of any machine, equipment, or device provides elegancy and aesthetics to the product.

1.6 Operational Requirements of Electrical and Electronic Materials

Electrical materials have to satisfy widely varying needs of different operational parameters. These parameters are voltage, current, temperature, frequency, polarization, remanance, resistivity, emission etc. Accordingly, the materials have to be suitable for meeting the following requirements.

- high and low voltage applications
- high and low conduction applications
- high and low temperature services
- high and low frequency services
- high and low resistivity devices
- high and low emission applications, etc.

The meaning of high and low in above applications/services may vary between extreme values of parameters. For example, a high voltage conductor may have to carry a current at 1000 kV as in overhead transmission but a low voltage conductor may be subjected to 5 mV only as in audio signal generator. Similarly, a dielectric may be subjected to a low of 50/60 Hz frequency as in electrical power supply and a high frequency of 10^6 Hz as in infrared systems. Use of a magnetic material at about 1200°C (as in booster rocket motor), semiconductor at room temperature (as in television cameras), and of a superconductor below cryogenic temperature (−157.5°C) are some other operational needs to which a material has to meet during uses.

1.6.1 High and Low Temperature (Service) Materials

Temperature measurement of furnaces and ovens in industrial applications is essentially required for a proper process control. The measurement may have to be done well above the melting point of the metal in many instances. It may be a high value of 3410°C for tungsten (W) and 2454°C for iridium (Ir), a medium range value of 1769°C for platinum (Pt) and 1083°C for copper, or a low value of 30°C for gallium (Ga) and -220°C for fluorine (F). In case of electrically heated ovens, the heating coils of the heater are normally subjected to above said temperatures.

Normally a thermocouple is used for temperature measurement. It mainly employs nichrome as the material. Nichrome is a heat resisting alloy of 80% Ni and 20% Cr composition. Constantan (60% Cu + 40% Ni), kanthal, platinum-rhodium etc. are other materials used for this purpose. For low-temperature measurement, various nickel based cryogenic steels are used. These are

Percentage of nickel	3.5	5.0	9.0
Service temperature (°C)	−100	−120	−190

1.6.2 High Voltage (Service) Materials

With increasing demand of electricity, the generation and transmission requirements of voltages are on increase. It is because at high voltage, the line losses are relatively reduced and the power system becomes economical. Whereas the 400 kV transmission lines are very much in use, 800 kV transmission lines in India are also in operation now (1000 kV line is operational in Russia). Because of increased power demand, the distribution voltages are also growing in value (110 kV, 220 kV etc.).

Since the transmission system, switchgear and other electrical systems made of conventional materials are bulky and expensive; therefore, newer materials have to be used to cut-down the cost and space requirements. Such newer materials are known as 'high voltage materials' and include the materials for the purpose of live conductors, insulators etc. Table 1.1 lists some important high voltage materials in this regard.

Table 1.1 High voltage materials

Equipments	Materials
• Overhead live conductors	ACSR (Aluminium conductor reinforced with steel)
• Underground cables	Nitrogen gas filled pressure cables in conjunction with oil-treated paper insulation
• Insulators	Highly purified benzene, hexane, silicone oil; ferroelectrics, mica, vacuum
• High frequency coils	Powdered metals
• Circuit breakers	Pd-Rh and W-Pt contacts, Petroleum oil as insulation
• Low-loss capacitors	Ferroelectrics, vacuum as insulation

1.7 Classification of Solids on the Basis of Energy Gap

Solids (or materials) are broadly classified into following categories on the basis of their energy gap which is an important electronic behaviour.

1. Conductors,
2. Semiconductors, and
3. Dielectrics or insulators.

- Conductors are mainly metals and alloys. Their electrical resistivity ρ (reciprocal of electrical conductivity) is the least.
- Semiconductors have properties in-between the conductors and dielectrics. They are widely used for making solid state devices.
- Dielectrics have the highest values of resistivity. They are used as insulators and capacitors etc.

The order of electrical resistivity of solids vary between 10^{-9} for good conductors to 10^{+17} for best quality insulators. The classification of solids on the basis of their electrical resistivities are shown in Table 1.2.

Table 1.2 The electrical resistivity of solids at room temperature

Solid	**Electrical resistivity (ohm m)**	**Solid**	**Electrical resistivity (ohm m)**
Metals		*Semiconductors*	
Silver	10^{-9}	Doped germanium	10^{-3}
Gold	10^{-9}	Pure germanium	10^{-1}
Copper	10^{-9}	Pure silicon	10^{-2} to 10^{-3}
Aluminium	10^{-8}	*Insulators*	
Nickel	10^{-7}	Glass	10^{5}
Iron	10^{-6}	Fireclay	10^{8}
Antimony	10^{-5}	Bakelite	10^{9}
Bismuth	10^{-4}	Alumina, diamond, rubber, polyethylene	10^{11}
Alloys			
Brass	10^{-7}	Mica, PTFE	10^{13}
Constantan	10^{-3}	PVC	10^{15}
Nichrome	10^{-2}	Pure silica	10^{17}

1.7.1 Energy Gap for Different Solids

Valence and conduction band. The outermost filled or partially filled energy band is called *valence band.* The energy band above the valence band which is empty at 0 K is known as *conduction band.* The electrons available in the valence band are to be pushed into the conduction band for conduction. Thermal energy or electrical voltage is given to the solid for this purpose. The electrons conduct from the top energy level of the valence band to the bottom energy level of the conduction band.

There may or may not be an energy gap between these two bands. Based on the magnitude of energy gap, the solids are classified as conductors, semiconductors and insulators. The energy gap E_g for different solids are shown in Figs. 1.1*a*-*g*. Based on the magnitude of E_g, various solids are classified as follows:

i. Conductors if energy levels overlap, Fig.1.1*a*,
ii. Conductors if $E_g = 0$, Fig. 1.1*b*,
iii. Conductors if $E_g < 1$ eV Fig. 1.1*c*,
iv. Semiconductors if 3 eV $> E_g >$ 1 eV, Fig. 1.1*d* and e, and
v. Insulators if $E_g > 3$ eV, Fig. 1.1 *f* and g.

Here eV stands for electron-volts. The symbol x in Fig. 1.1 indicates an arbitrary length of the energy band.

1.7.2 Comparison among Conductors, Semiconductors and Insulators

The properties and characteristics of conductors, semiconductors, and insulators are compared below for a quick remembrance.

Description	Conductors	Semiconductors	Insulators
• Conductivity	High to very high	Low to medium	Nil to very low
• Resistivity	Low to very low	Medium to low	Very high to high
• Temperature coefficient of resistance	Always positive	Always negative	May be positive or negative
• Energy band	Overlapping or unfilled	Filled	Filled
• Forbidden zone	Nil or very small	Small	Large
• Current carriers			
– without external energy	Free electrons	Nil	Nil
– with external energy	Free electrons	Electrons and holes	Electrons
• Behaviour at low temperatures	Near absolute zero, they become super-conductors	In lower range, they become dielectrics	ε_r drops sharply
• Effect of increasing temperature on conductivity	Conductivity decreases	Conductivity increases	Conductivity increases
• Effect of high electrical voltage on breakdown	No breakdown	They breakdown	They generally breakdown
• Bonding nature of valence electrons with their parent atom	Very loosely (feebly) bound	Loosely bound	Very tightly bound
• Ionization energy	Much less	Less	Very large

1.8 Glimpse of Some Electronic Products, Their Working Principles and Choicest Materials

Some products and their choicest materials are given below along with the working principle incorporated in them. The services rendered by these products are also mentioned.

Product	Working principle	Materials
• Sound recording in films	Magnetism	Metallic glass
• Accelerometer	Piezoelectric	Quartz
• Stroboscope disk	Flashing light called Strobotron (electro optics)	Opto-electronic polymers
• Solar cells	Photovoltaic action	Compound semiconductors
• Cinematography	Photocell effect	Se, CdS, $PbSO_4$ semiconductors
• Automatic door opener	Photoconductivity	CdSe, CdS, CdTe semiconductors
• Computer memory devices	Soft magnetism	Magnetic bubbles
• Wheel-less train magnets	Superconductivity	Ceramic superconductors
• Oscillographic recorder	Electromagnetic	FeCoNi alloy, $Gd_3Fe_5O_{12}$
• Capacitive transducer	Capacitance	Saline impregnated paper
• Cold cathode glow transfer tube	Gas ionization	Neon (lamp)
• LVDT (linear variable differential transformer) core	High permeability magnetic effect	Supermalloy, permalloy, Si-iron
• Photo tube pulse pick-up microphone	Piezoelectricity	Carbon, quartz, Rochelle salt, lithium sulphate
• Antenna rod	Conductivity	Al, Cu

1.9 Different Types of Engineering Materials

The materials are broadly classified in different categories as shown in Fig 1.2:

1.9.1 Metals

Metals are elemental substances capable of changing their shape permanently. They are good conductors of heat and electricity. These may be of ferrous or non-ferrous type. The behaviour and properties of ferrous metals depend on the percentage and the form (phase and constituents) of carbon present in them. The difference between steel and iron, and their specific names according to percentage of carbon are given in Table 1.3.

1.9.2 Non-Ferrous Metals

Non-ferrous metals do not contain Fe and C as their constituents. Aluminium, copper, silver, nickel, zinc, tin, chromium etc. are some examples. Al, Cu, Ag and Au are good conductors of electricity; Ag is most malleable, Au is most ductile, and chromium is corrosion resistant. Zinc is used in the metals plating, tin is used to make bushes, and nickel imparts strength and creep resistance.

1.9.3 Ceramics

Ceramics are generally metallic or non-metallic oxides. Physically separable and chemically homogeneous constituents of materials consisting different

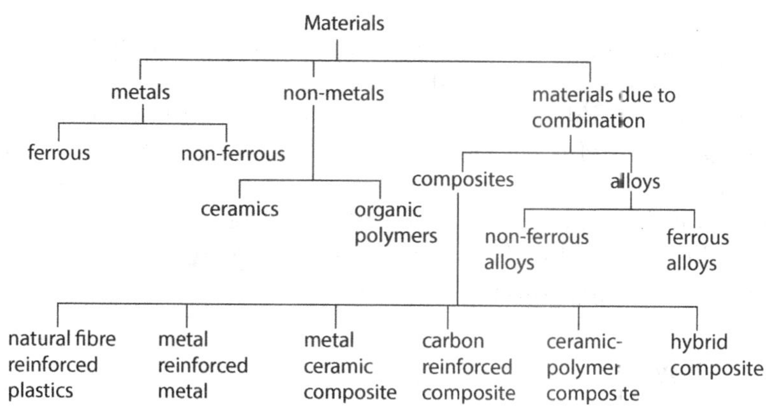

Figure 1.2 Classification of different engineering materials

Table 1.3 Steel and iron on the basis of percentage of carbon*

Percentage of carbon	Specific names	Abbreviated name	Steel or iron
Purest to 0.05	Wrought iron	WI	Plain carbon steel
0.05 to 0.10	Dead mild steel	DMS	
0.10 to 0.30	Mild steel	MS	
0.30 to 0.70	Medium carbon steel	MCS	
0.70 to 1.50	High carbon steel	HCS	
1.50 to 2.00	Semi-steel or semi-iron	—	
2.00 to 4.50	Cast iron (grey, white, mottled, spongy, malleable, chilled, spheroidal, nodular)	CI	Iron
4.50 to 6.67	Pig iron	PI	
6.67 or more	Ore	—	

* Data is averaged and reasonably approximated

phases are also ceramics. Rocks, glasses, fireclay and firebricks, cements and limes are ceramics. Ferrites, garnets, ferroelectrics and ceramic superconductors are the latest developments in this area.

1.9.4 Organic Polymers

Organic polymers are relatively inert and light, and generally have a high degree of plasticity. These are derived mainly from the hydrocarbons. These consist of covalent bonds formed by carbon, chemically combined with oxygen and hydrogen. The word 'mer' in Greek means a unit, 'mono' means one and 'poly' means many. Thus, polymers are obtained from monomers bonded by a chemical reaction (a process called polymerization). In this process, long molecular chain having high molecular weight is generated. Bakelite, polyethylene, nylon, teflon are some examples.

1.9.5 Alloys

An *alloy* is a combination of two or more metals. They possess properties quite different from those of their constituent metals. An alloy is prepared for a specific purpose to meet the particular requirement of an application. Alloys may be ferrous-alloy or non-ferrous depending on the base metal used.

1.9.6 Composites

Composites may be inorganic or organic. They have two or more constituents of dissimilar properties. The two major constituents may be metals and ceramics, or metals and polymers, or ceramics and polymers or other combinations. Alloys may also be used instead of metals to make composites. One of the constituent (called reinforcing constituent) may be in particulate form, fibrous form, or flake form. Fibrous composites are more common in present day applications. Whisker reinforced composites are likely to be the future material.

1.10 Different Levels of Materials Structure

It is possible to study the *internal structure* of a material at different *levels of observation.* We can look at a material with naked eye and see some details of its structure. To see the finer details, we take help of a microscope. Depending on whether we observe the material with naked eye, or under a low magnification, or under a microscope of high magnification, there are two broad levels of structure of a material viz.

1. Macro-structure, and
2. Micro-structure

The macro-structure can be observed with naked eye or with a low magnification using an optical microscope. Etched and polished crystals of cast brass can be directly seen with naked eyes.

1.10.1 Micro-Structure Levels

The human eye is capable of distinguishing two lines as separate lines only when their separation is more than 0.1 mm. This is known as *limit of resolution* of the human eye. It means that if the separation between two lines is less than 0.1 mm, the human eye will see them as one line. The micro-structural levels may further be classified as below.

1. Sub-structure
2. Crystal structure
3. Electron structure
4. Nuclear structure

Table 1.4 Structure levels, dimensional range, examples and magnification for visualization

Structure level	Dimensional range, (order in metre)	Example	Magnification (×)
Macro	10^{-1} to 10^{-3}	Fractured pieces of metals, internal symmetry of quartz	Naked eye to 10 ×
Micro	10^{-4} to 10^{-6}	Crystals	200 × to 1500 ×
Sub	10^{-6} to 10^{-8}	Crystal imperfections	10 000 × to 100 000 ×
Crystal	10^{-8} to 10^{-10}	Unit cells	X-ray diffraction technique
Electronic	10^{-10}	Electrons in the outermost shell of the atoms	Spectroscopic techniques
Nuclear	Still smaller	Proton and neutron	Nuclear magnetic resonance (NMR) techniques

Their dimension levels vary from nanometre (10^{-9} m) to micrometre (10^{-6} m). These structures are also called as nano-level, Angstrom level (1 Å = 10^{-10} m) and micro-level structures.

1.10.2 Dimensional Range and Examples

Various levels of structures, their dimensional range, examples and the magnification needed to visualize them are summarized in Table 1.4.

RECENT DEVELOPMENT IN ELECTRONIC MATERIALS

1.11 Spintronics (The Electronics of Tomorrow) and Spintronic Materials

The development of electronics has spread in many specialized branches such as bio-electronics, molecular electronics, spintronics, rubber electronics, polymer electronics, nano electronics etc. Amongst these, the spintronics (i.e. spin electronics) is one of the fast emerging fields. It relates to spin-dependent phenomena which are manipulated to achieve a desired electronic outcome, such as *quantum-computing*. The spintronics devices store information into spins of up and down orientations, which are then attached to the mobile electrons to carry them along a wire, to be read at a terminal.

Several kinds of spintronic devices are in the stage of development and in concept. They are/may be bipolar devices such as spin diodes, spin transistors, spin-polarized solar cells, magnetic diodes etc. These devices are attractive for the purpose of magnetic sensor and memory storage applications. Other aspects of spintronics are in the use of NSOM (Near Field Scanning Optical Microscopy) to detect electrons in semiconductor quantum dots, in quantum computing, and in SET (Single Electron Transistor) as a single electron spin detector.

1.11.1 Major Fields of Spintronic Research

Modern researches on spintronics concentrate on development of advanced materials. Almost all material systems are tried for it. Magnetic metal multilayers, $Ga_{1-x}Mn_xAs$ type ferromagnetic semiconductor, GaN etc. are prominent among them. However, the ferromagnetic semiconductors are the most favourable material system, which combine the characteristics of ferromagnetism and semiconductor both. Some major fields of current spintronic research are the following.

1. Development of spin-based devices such as p-n junctions and amplifiers.
2. Spin relaxation behaviour in metals and semiconductors.
3. Spin polarized transportation through the semiconductor/ semiconductor interfaces.
4. Spin-based quantum computation.
5. Quantum computer hardware.

1.11.2 Operational Mechanisms of Spintronic Devices

Spin orientation of conduction electrons survives for nanoseconds (10^{-9}s), whereas the time elapsed in decay of electron momentum is tens of femto seconds (10^{-15}s). Consequently, the spintronic devices are especially useful for quantum computing. Here, the electron spin represents a bit of information, which is known as 'qubit'. The understanding of spin relaxation and efforts needed to enhance the spin memory of conduction electrons in metals and semiconductors are essential for study of spintronic devices.

1.11.3 Working Principle of Spintronic Devices

On application of magnetic field in normal metals, there is a change in their resistance. This change is normally of the order of 1% for the magnetic field of 1 tesla (T). In some ferromagnetic metals, the direction of magnetization can be reversed when a magnetic field of the order of 0.0001 tesla is applied. This results in a phenomenon called *Magneto-resistance (MR),* which is utilized in commercial production of small magnetic 'read heads'. They can sense very small magnetic fields in written information on hard discs. This causes considerable decrease in the space required for storing the bits of information. Thus the storage capacity of hard discs increases.

1.11.4 Emerging and Futuristic Spintronic Materials

The utility of semiconductor spintronics devices depends on the availability of suitable materials which retain ferromagnetic properties above room temperature. Some recent and important materials in this list are the following.

- Half-metallic materials such as Fe_3O_4,. CrO_2 etc.
- Colossal magneto-resistive materials such as Sr-doped $LaMnO_3$
- Double perovskites
- Heusler alloys having composition of X_2YZ, where X and Y are transition elements and Z is an element of group III, IV or V.
- Half-Heuslar alloys having a composition of XYZ.
- Ferromagnetic semiconductors

1.12 Ferromagnetic Semiconductor

Magnetic bipolar diodes (MBD) such as p-type GaMnAs and n-type GaAs are salient applications of ferromagnetic semiconductor. In this diode, the p-region is a magnetic semiconductor while the n- region is a conventional semiconductor compound. Similarly, a magnetic bipolar transistor (MBT) may be of great use, if produced. Conceptually, it may consist of a conventional semiconductor emitter and collector, and a base of p-type magnetic semiconductor. With the advent of such materials, presently existing memory devices e.g. SRAM (semiconductor random access memory) and DRAM (dynamic random access memory) can be combined together to produce MRAM (magnetic random access memory), which combines the advantages of high speed access, high density information storage and non-volatility.

Ferromagnetic semiconductors such as $CrBr_3$ are highly spin polarized, Co-doped TiO_2 are optically transparent, and Mn-doped III-V compounds have small solubility of magnetic dopant. Thus the ferromagnetic semiconductors having a wider range of spin polarization are desirable materials. In order to be ferromagnetic at room temperature, they are required to have Curie temperature above room temperature.

1.12.1 Emerging Wide Bandgap Semiconductors

Other popular/emerging and futuristic materials having wide bandgap semiconductors that retain ferromagnetic properties above room temperature are given below.

InMnAs. It is a metal-insulator semiconductor which is suitable for working at low temperature. It may be used as magnetic sensors.

GaMnAs. It is suitable for high spin-injection. It may be used for infrared LED and lasers.

$CoTiO_2$. It has high coherent length.

NiGaAs. It is suitable for efficient carrier injection. It may be used for high speed digital electronic applications

$ZnGeP_2$. It is a chalcopyrite material that exhibits unusual nonlinear optical properties. It can be used in optical oscillations.

$ZnSnAs_2$. This is a chalcopyrite material. It shows promise for frequency converters and IR-generation.

$ZnSiN_2$. It is a wide bandgap chalcopyrite having lattice parameter close to GaN and SiC. Achievement of ferromagnetism in it will enable its usefulness to make ultravoilet solar-blind detector and microwave power electronic devices.

EuS. It is a dilute magnetic semiconductor (DMS) material. It can be used as a tunneling barrier, which can effectively function as spin filter.

1.13 Left-Handed (LH) Materials

This is a novel metamaterial characterized by simultaneously negative permittivity ε and negative permeability μ, i.e.

$$\varepsilon < 0 \qquad \text{and} \qquad \mu < 0$$

LH materials are recently developed interesting materials having novel concepts and applications. The unusual properties of LHMs will lead to

1. reversal of Doppler effect, and
2. reversal of Snell's law.

Consequences of these unusual properties are that

i. the light rays incident on a convex lens will diverge instead of converging, as happens in conventional i.e. right handed (RH) materials.
ii. it will support waves with antiparallel group and phase velocities, known as backward waves.
iii. in a left-handed medium, the light propagates in opposite direction to the direction of energy flow.

A left-handed material has negative index of refraction.

The name left-handed comes from the fact that the properties of these materials are just opposite to those having $\varepsilon > 0$ and $\mu > 0$ i.e. right- handed materials (RHMs).

Applications. The unusual behaviour of LHMs has lead to novel applications such as

- microwave components
- miniaturized antennas
- probes
- waveguides
- interconnections between nanodevices and external terminations
- Novel filters in cellular phones
- micro-lenses

1.14 Solved Examples

Example 1.1 Suggest materials which are suitable for *(a)* polishing of glass insulators, *(b)* pivots for instruments, and (c) giving colours to discharge tubes.

Solution. (a) Cerium Ce is suitable for glass polishing. Its specific gravity is 6.77 and melting point is 804°C. (b) Iridium Ir is suitable for pivots of instruments. Its specific gravity is 22.5 and melting point 2454°C. (c) Krypton Kr is suitable for giving colours to discharge tubes. Its melting point is –157.2°C.

Example 1.2 Which materials are used (a) as power source in satellite, (b) in producing electricity for 'pace-maker'?

Solution. (a) Neptunium Np is used as power source in satellite. Its melting point is 637°C. (b) Plutonium Pu is used for producing electricity for 'pace-maker'. Its specific gravity is 19.5 and melting point is 640°C.

Example 1.3 Name the major applications of Rhenium Re, Samarium Sm, Selenium Se, Tantalum Ta, and Thorium Th elements.

Solution. Major applications and properties of these elements are as follows.

Element	Major use	Specific gravity	Melting point
Re	Electric furnace winding	21.04	3180°C
Sm	High temperature superconductors	7.49	1072°C
Se	Xerox-type photocopiers	4.79	217°C
Ta	Capacitors for electronics industries	16.6	2996°C
Th	Incandescent gas mantles	11.66	1750°C

Example 1.4 Enumerate major uses of Yttrium in electrical and other fields.

Solution. It is used as 'red phospher for TV tubes', for synthetic garnet in radar, as gemstones, for making high temperature ceramic superconductors in conjunction with barium and copper etc. Its specific gravity is 4.47 and melting point is 1509°C.

Example 1.5 Differentiate between 'mylar' and 'kevlar' as electrical materials, quoting their uses.

Solution. Mylar is aluminized plastic film and kevlar is polyamide fibre. Mylar is used as reflective insulation for the tankers containing liquid O_2, liquid N_2 and liquid H_2. Kevlar is used as reinforcing fibre to produce polymeric composites for use as solid dielectric. Mylar is generally used in sheet form while the kevlar is used in fibre form. Specific gravity of kevlar is 1.45.

Example 1.6 Which materials are generally used in mobile phones?

Solution. A mobile phone is one of the most intricate electronic devices. It is a sophisticated radio that communicates both ways: in sending and in receiving the signals. It operates under a very low power. In past few years, the size of mobile (cell) phones has diminished considerably from small-brick size to miniature pocket size. It has become more reliable and less expensive too. Such dramatic change has become possible due to various factors amongst which, one important factor is the use of specific materials possessing the unique communication properties. These materials are as follows.

- Tantalum
- Palladium
- Gold plating that covers the surface of the circuit board
- Copper
- Nickel

Example 1.7 What is the importance of tantalum as cell phone material and in passive capacitors?

Solution. Tantalum is a silver-gray coloured precious metal, which is very important for a mobile phone. It helps to run the heart of the cell phone. In powder compacted form, the tantalum is used to produce 'passive capacitors', which is a key factor in reducing the size of cell phones. These capacitors are tiny but sophisticated components, which regulate the voltage at high temperatures. Tantalum capacitors are crucial items that are used as

i. storage vessels
ii. storing energy, and
iii. when there is a big surge of energy to a cellular phone.

Various properties of tantalum are the following.

- Density = 16600 kg/m^3
- Young's modulus = 181 GPa
- Melting point = 2996°C
- Atomic number = 73

Example 1.8 What are the other uses of tantalum as electronic material?

Solution. Besides use in mobile phones, the tantalum also finds use in the following applications.

- Portable computers
- Game consoles
- Electronic devices where smaller size is the main requirement
- As ingredient of superalloys, mainly meant for spacecraft and aircraft engines.

Review Questions

1. Discuss the importance of materials in development of electrical machines; electronic devices, components and instruments.
2. Classify the engineering materials from various viewpoints. Give a brief account of their especialities and applications.
3. Enumerate the requirement of materials suitable for various electronic items. Make special mention of electrical, magnetic and electronic devices.
4. Classify the electrical engineering materials on the basis of their energy gaps and briefly explain each of them. How do the conductors, semiconductors, and superconductors differ from each other?
5. What are ferrite, zirconates, spinels and garnets? Explain their features and applications.
6. Discuss the scope of electrical materials. Where do these find use in consumer items, robotics, information sector and biotechnology?
7. Which factor should be considered while selecting a suitable material for a particular application? How is a right type of material selected for specific applications in (a) high

temperature, (b) high voltage, and (c) high frequency situations.

8. What are different levels of materials structure? Discuss them with the help of their dimensional range, examples and the magnification required for viewing them.
9. What are spintronic devices? State the principle of their operation. Discuss the various fields of spintronic research.
10. Enumerate different wide bandgap semiconductors. State their suitability and applications.
11. Suggest suitable materials for the following applications, and write the logic of such selections.
    a. Commutator of an alternator
    b. Transformer core
    c. Compact disc
    d. Thermistor
    e. C.R.O. screen
    f. Memory devices in computers
    g. Thermocouples
    h. Hard magnets
12. Write notes on the following:
    i. Emerging Spintronic materials
    ii. Futuristic electronic materials
    iii. Ferromagnetic semiconductors
    iv. Synthetic diamond
13. Explain various materials used in (a) mobile phones, (b) coloured discharge tubes (c) CPU of a computer, and (d) electrical contacts on keyboard.
14. Discuss the role of following materials as futuristic electronic materials.
    a. Left-handed materials
    b. Wide bandgap semiconductors

Objective Questions

1. Match List I (Materials) with List II (Applications) and select the correct answer using the codes given below the lists.

List I	*List II*
A. Ferroelectrics	1. Memory devices

B. Iron garnet	2. Microphone
C. Refractory metal	3. IC-technology
D. Polysilicon	4. Furnace

Codes:

	A B C D		A B C D
a.	2 3 1 4	b.	1 4 2 3
c.	1 3 2 4	d.	2 1 4 3

2. In a Photostat machine, the drum is made of

a. selenium coating b. silicon coating
c. cadmium coating d. zircon coating

3. Match the items of list I with their applications given in list II, and choose the correct answer from the codes given below the lists.

List I (Items)	*List II (Applications)*
A. Tungsten	1. Commutator
B. Lead-tin alloy	2. Transformer
C. Mineral oil	3. Fuse
D. Mica	4. Lamp filament
	5. Heating element

Codes:

	A B C D		A B C D
a.	1 2 3 4	b.	5 4 2 3
c.	4 3 2 1	d.	4 5 3 2

4. Match list I with list II, and select the correct answer using the codes given below the lists.

List I (Materials)	*List II (Applications)*
A. Platinum	1. Computer memory core
B. Ferrites	2. Ferroelectric
C. Selenium	3. Contact material
D. Rochelle salt	4. Solar battery

Codes:

	A B C D		A B C D
a.	1 2 4 3	b.	2 4 3 1
c.	3 1 4 2	d.	4 3 1 2

5. *Assertion* A. Innovation in engineering materials means clever and careful use of new materials but not necessarily a 'latest material'.
   Reasoning R. Paper clips made of plastics do better than previously used metals for them.
   Of these

   a. A is correct but *R* is wrong
   b. *A* is correct and *R* is also correct
   c. *A* and *R* both are wrong
   d. *A* is wrong but *R* is correct

2

Atomic Models, Bonding in Solids, Crystal Geometry, and Miller Indices

2.1 Atomic Models

The history of development of atomic models may be enumerated as under:

1. Thomson's plum pudding atomic model
2. Rutherford's nuclear atomic model
3. Bohr's quantum atomic model
4. Sommerfeld's relativistic atomic model
5. Wave mechanical or de Broglie's atomic model, or modern concept of atomic model.

2.2 Bohr's Quantum Atomic Model

Bohr conceived-off a new atomic model employing the principles of *quantum theory* suggested by Planck. This model provided adequate explanation for stability of the atom. The model also accounted for the origin of spectral lines in the hydrogen atom.

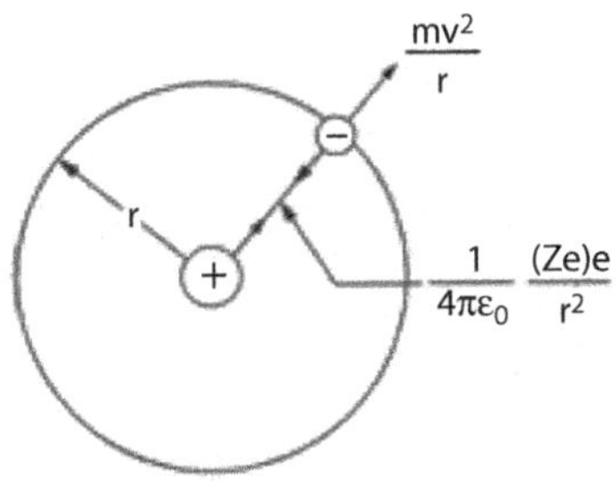

Figure 2.1 Balance forces keeps the electron in orbital motion

Bohr postulates. He proposed new ideas which are now known as Bohr's postulates. These are:

i. Electrons revolve in non-radiating stationary orbits.
Centripetal force provided by Coulomb's force of attraction between the electron and the nucleus keeps the electron in orbital motion, Fig. 2.1.

Thus

$$\frac{mv^2}{r} = \frac{1}{4\pi \in_0} \frac{(Ze)(e)}{r^2} \tag{2.1}$$

where, Z = atomic number of nucleus, m = mass of the electron, v = velocity of electron in the orbit, r = radius of the orbit, e = charge of electron.

ii. Angular momentum of the moving electron is an integral multiple of $\frac{h}{2\pi}$ where h is *Planck's constant*. Thus

$$mvr = \frac{nh}{2\pi} \tag{2.2}$$

where, $n = 1, 2, 3, \ldots \infty$, and is called *principal quantum number.*

iii. The electron does not radiate energy while moving in stationary orbit. Energy is emitted when the electron falls from higher energy orbit to lower energy orbit. If the electron jumps-up to higher energy orbit from lower energy orbit, absorption of energy takes place. The energy absorbed or emitted is expressed by Bohr's frequency condition given as

$$\Delta E = E_f - E_i = hf \tag{2.3}$$

where f is the frequency of emitted radiation, E_i and E_f are the energies of initial and final orbits respectively.

2.2.1 Radii of Orbits, Velocity and Frequency of Electrons

Radius of *n*th orbit. The radius of *n*th stationary orbit is obtained from Eqs. 2.4 and 2.5 on eliminating v from them. It is, then, given as

$$r_n = \frac{n^2 \epsilon_0 h^2}{\pi m Z e^2} \tag{2.4}$$

Velocity of *n*th orbit. Velocity of *n*th orbit electrons is determined by

$$v_n = \frac{Ze^2}{2n \epsilon_0 h} \tag{2.5}$$

Frequency of *n*th orbit. The orbital frequency of an electron in *n*th orbit is given as

$$f_n = \frac{v_n}{2\pi r} \tag{2.6}$$

Substituting the values of r_n and v_n from Eqs. 2.4 and 2.5 in Eq. 2.6, we get

$$f_n = \frac{mZ^2 e^4}{4 \epsilon_0^2 n^3 h^3} \tag{2.7}$$

As ε_0, h, m and e are constants, hence for $Z = 1$, Eq. 2.4 shows that $r_n \propto n^2$. Thus for $n = 1, 2, 3, 4, \ldots$; the radii of orbits are proportional to $1^2, 2^2, 3^2, 4^2$, ... *i.e.* in the ratio of 1 : 4 : 9 : 16, ..., etc.

Conclusions. Equation 2.5 reveals $v_n \propto \frac{1}{n}$ It means that

- v the velocity of electrons in outer orbits is lower than those in the inner orbits.

Similarly Eq. 2.7 interprets that $f_n \propto \frac{1}{n^3}$. Therefore

- orbital frequency of electrons, for $Z = 1$, is in the ratio of $\frac{1}{1^3} : \frac{1}{2^3} : \frac{1}{3^3} : \frac{1}{4^3} : \ldots$ etc.

2.2.2 Normal, Excited and Ionized Atoms

Hydrogen atom has a single electron. When this electron lies in the innermost K orbit ($n = 1$), the atom is said to be in *normal state.* Normal state is also called *ground state.* The electron in normal state remains stable. The atom is said to be *excited* if its electron is forced into an outer orbit. The electron absorbs energy in this case. The atom is said to be *ionized* when its electron is removed out from the atom completely. The electron emits energy in ionized state.

The radius r_o, velocity v_o and frequency f_o of hydrogen atom in ground state ($n = 1$) are obtained using Eqs. 2.4, 2.5 and 2.7 These values are

$$r_o = 0.53\ \text{Å},\ v_o = 2.2 \times 10^6\ \text{m/s and } f_o = 0.66 \times 10\ \text{m/s}^2$$

2.2.3 Kinetic and Potential Energy of Electron

Energy of an electron is the sum of kinetic energy E_k due to its motion and potential energy E_p due to its position in the orbit. The value of kinetic energy is given as

$$E_k = \frac{1}{2} m v_n^2 = \frac{mZ^2 e^4}{8 \in_0^2 n^2 h^2} \tag{2.8}$$

and the value of potential energy may be obtained from

$$E_p = \int_{\infty}^{r} F dr = \frac{Ze^2}{4\pi \in_0} \int_{\infty}^{r} \frac{1}{r^2} dr$$

$$= -\frac{Ze^2}{4\pi \in_0 r}$$

In above expression, right hand side term of Eq. 2.4 has been taken for F, which is Coulombian force. On substituting the value of r from Eq. 2.7, we get

$$E_p = -\frac{mZ^2 e^4}{4 \in_0^2 n^2 h^2} \tag{2.9}$$

The potential energy of an electron is equal to the work done in bringing the electron from infinity to the present position. This work is required to be done due to Coulombian force produced by the nucleus. Since this force is attractive, the work done and hence the potential energy E_p has negative value.

Energy level of electrons in *n*th level. On summing-up Eqs. 2.8 and 2.9, the energy level of electron in *n*th orbit is obtained as

$$E_n = -\frac{mZ^2e^4}{8\epsilon_0^2 n^2h^2} \tag{2.10}$$

2.3 Modern Concept of Atomic Model

2.3.1 De Broglie Wave

The modern concept of atom also considers the concept of wave nature of electron. According to this concept a particle (say electron) of mass m moving with velocity v is associated with a wave propagating in the direction of moving particle. This wave is known as de *Broglie wave or matter wave.* Wavelength λ of such a wave associated with the particle is given by

$$\lambda = \frac{h}{mv} \tag{2.11}$$

Velocity of electron accelerated by a potential difference of V volts can be obtained from

$$\mathrm{eV} = \frac{1}{2}\mathrm{mv}^2, \mathrm{or\ v} = \frac{\sqrt{2eV}}{m}$$

$$\mathrm{mv} = \sqrt{2meV} \tag{2.12}$$

2.3.2 Wavelength of Electron Wave

Therefore, wavelength of electron wave in Eq. 2.11 becomes

$$\lambda = \frac{h}{\sqrt{2meV}}$$

On substituting $h = 6.62 \times 10^{-34}$ Js, $m = 9.1 \times 10^{-31}$ kg, and $e = 1.602 \times 10^{-19}$ C, we get

$$\lambda = \frac{12.25}{\sqrt{V}}\mathring{A} \tag{2.13}$$

2.3.3 Concept of Standing Wave

Modern concept of atomic model, also known as wave- mechanical model, assumes behaviour of electron as a *standing* or *stationary wave*. Such a wave will establish when the length of orbit is a whole number multiple of electron wavelength as shown in Fig. 2.2.

Hence $$2\pi r = n\lambda = \frac{nh}{mv}$$

or $$mvr = \frac{nh}{2\pi} \quad \text{for } n = 1, 2, 3, \ldots \text{ etc.} \tag{2.14}$$

Here r is radius of the circular orbit, mvr is angular momentum of electron as a particle.

Characteristics of standing waves.

- The stationary waves have their nodes and antinodes.
- Motion is practically zero at nodes, and there are almost no charge on them at these points.
- The amount of charge is maximum at antinodes.
- Whole of electron charge and mass is uniformly distributed around nucleus of the atom.
- The wavelength of electron is of the order of interatomic spacing in the crystals.
- A beam of electron incident on a crystal shows diffraction pattern thus depicting concept of wave nature of electron.

Example 2.1 Calculate the minimum uncertainty in determining the position of a particle when the uncertainty in its momentum does not exceed 10^{-27} kg ms^{-1}.

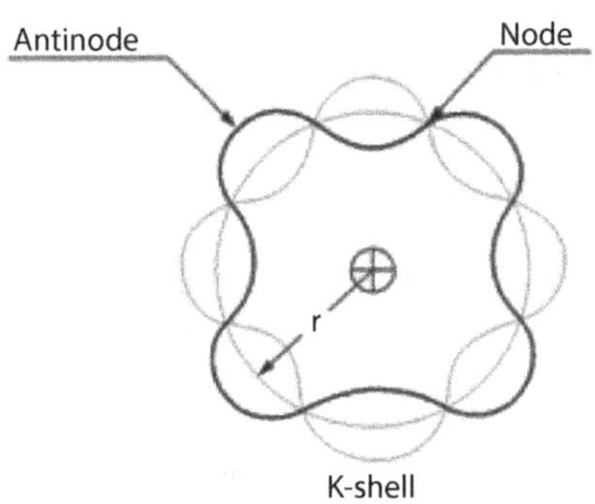

Figure 2.2 Wave nature of electron in K-shell of an atom showing stationary wave

Solution: From Eq. 2.11,

$$(\Delta p)(\Delta x) > \frac{h}{2\pi}$$

Given is the value of $(\Delta p) \leq 10^{-27}$ kg ms^{-1}, and taking $h = 6.626 \times 10^{-34}$ Js, the uncertainty in position (displacement) of the particle will be obtained from

$$\begin{aligned}\Delta x &\geq \frac{h}{2\pi.(\Delta p)} \\ &\geq \frac{6.626 \times 10^{-34}}{2\pi(10^{-27})} \\ &\geq 1.054 \times 10^{-7} \text{ metre}\end{aligned}$$

Thus the minimum uncertainty = 1.054×10^{-7} m.

2.4 Electron Configuration

The atomic structure of elements may be shown by their orbital model. Showing the arrangement of electrons in different orbitals is termed as electronic configuration. Electronic configurations for some elements are given below.

Element	Symbol	Configuration	Atomic Number
Hydrogen	H	$1s^1$	1
Nitrogen	N	$1s^2\ 2s^2\ 2p^3$	7
Silicon	Si	$1s^2\ 2s^2\ 2p^6\ 3s^2\ 3p^2$	14
Titanium	Ti	$1s^2\ 2s^2\ 2p^6\ 3s^2\ 3p^6\ 3d^2\ 4s^2$	22
Copper	Cu	$1s^2\ 2s^2\ 2p^6\ 3s^2\ 3p^6\ 3d^{10}\ 4s^1$	29
Molybdenum	Mo	$1s^2\ 2s^2\ 2p^6\ 3s^2\ 3p^6\ 3d^{10}\ 4s^2\ 4p^6\ 4d^5\ 5s^1$	42
Antimony	Sb	$1s^2\ 2s^2\ 2p^6\ 3s^2\ 3p^6\ 3d^{10}\ 4s^2\ 4p^6\ 4d^{10}\ 5s^2\ 5p^3$	51
Neutral iron	Fe	$1s^2\ 2s^2\ 2p^6\ 3s^2\ 3p^6\ 3d^6\ 4s^2$	26
Ferrous ion	Fe^{2+}	$1s^2\ 2s^2\ 2p^6\ 3s^2\ 3p^6\ 3d^6$	24
Ferric ion	Fe^{3+}	$1s^2\ 2s^2\ 2p^6\ 3s^2\ 3p^6\ 3d^5$	23

BONDING IN SOLIDS

2.5 Meaning of Chemical (or Atomic) Bonding

If we try to break a material, it requires application of some breaking force. The magnitude of applied force varies widely for different materials. A *chalk* may break by applying a small force; timber may require application of medium magnitude of force, but steel necessitates application of substantial external force. One may think as to why is it so? In fact, atoms in the solids are held together by internal forces. These forces are known as *bonding forces* or atomic bonding forces or chemical bonding forces.

2.6 Classification of Chemical Bonds

Atomic bonds in solids may be classified into two main types as shown in Fig. 2.3. These are:

1. Primary or fundamental bonds, and
2. Secondary or molecular bonds.

Primary bonds are stronger and more stable than the secondary bonds. They are further classified into following types.

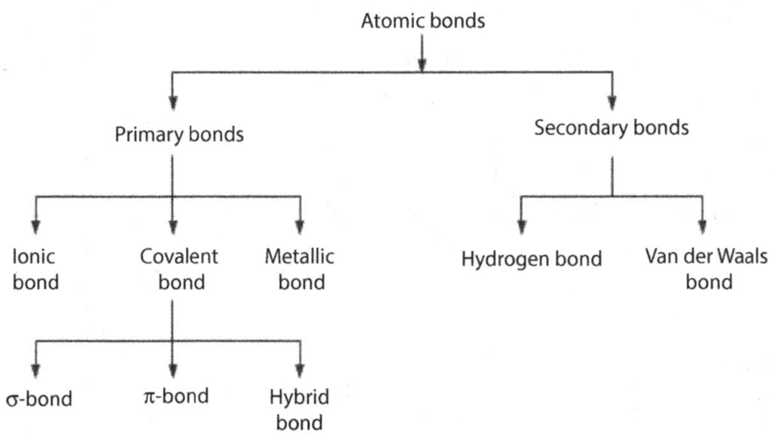

Figure 2.3 Different kinds of bonds

a. Ionic bond
b. Covalent bond
c. Metallic bond

Out of these three bonds, the ionic bonds are strongest and the metallic bonds are weakest. The hardness, strength and other properties are imparted to the solids due to primary bonds. Mostly solids have primary bond of one kind or the other. Examples of primary bonded materials are ceramics (refractories) such as alumina (Al_2O_3) and magnesia (MgO); hydrogen, diamond, silicon, plastics etc; and ferrous and non-ferrous metals. Primary bonds are interatomic in nature.

The ***secondary bonds*** are further classified into two types.

a. Hydrogen bond
b. Van der Waals bond

Secondary bonds are intermolecular. Gases and liquids, generally, form such bonds. They are feeble and less stable. Van der Waals bonds are the weakest amongst all known bonds. Secondary bonds may also be classified as follows.

a. Dipole bond
b. Dispersion bond

Both the hydrogen and Van der Waals bond, develop dipole in their formation. Hence they are also referred to as dipole bonds. Van der Waals bond is known as dispersion bond due to the dispersing nature of its dipoles.

2.7 Ionic Bond

Ionic bonds are formed between two oppositely charged ions. These are produced by transfer of electrons from one atom to another. An attractive force between a positive and a negative ion develops when they are brought in close proximity. These ions are formed when the atoms involved lose or gain electrons to stabilize their outer shell configurations.

Let us consider the case of sodium chloride (NaCl) which is ionically bonded.

- The sodium atom has a single electron in its outer orbit.
- This electron transfers to join the seven electrons in the outer orbit of the chlorine atom.

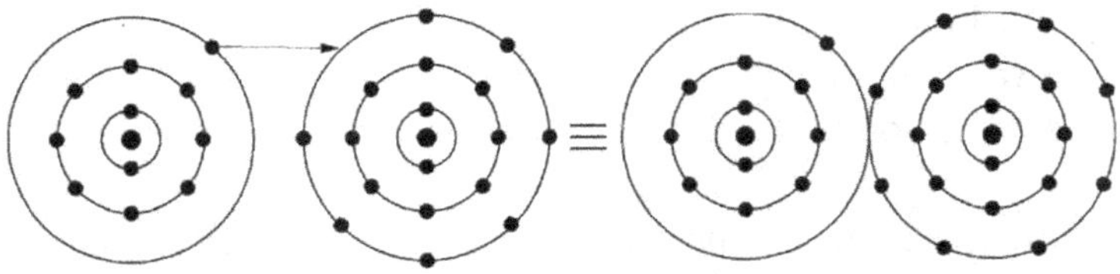

Figure 2.4 Arrangement of ionic bond.

- This atomic interaction involving transfer of electron from one atom to another leads to formation of ions which are held together by Coulomb's electrostatic attraction.

The bonding process is shown in Fig. 2.4.

2.8 Covalent Bonds

A covalent bond is formed when a pair of electrons are shared by several atoms. As a result of this sharing, their energies are lowered. Stable covalent bonds are formed between many non-metallic elements as the atoms of these elements usually posses half filled outer electron orbit. Such elements are hydrogen, carbon, nitrogen, oxygen and chlorine etc.

Examples. Elements like silicon, germanium, arsenic and selenium etc. form partly covalent and partly metallic bond. Covalent bonds are also found in organic compounds such as alcohol, benzene, chloroform and turpentine etc. Polymers, rubbers, elastomers, glasses, potteries, brick, Mo, Ta and high strength fibres are also covalent bonded.

2.8.1 Types of Covalent Bonds

Covalent bonds may be further classified into following three kinds.

1. Sigma (σ) bond
2. Pi (π) bond
3. Hybrid bond

End to end overlap of p suborbits gives rise to σ-*bond.* Bonding in sulphur, selenium and tellurium are the examples. π-*bonds* are formed when overlap of p suborbits are *lateral.* Bond in oxygen molecule is the example of

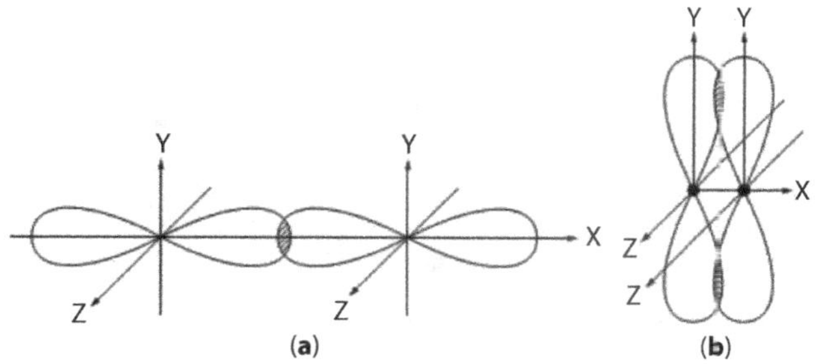

Figure 2.5 Covalent bond of type (a) σ-bond, and (b) π -bond.

this kind. Schematic diagrams are shown in Fig. 2.5 to explain σ-bond and π-bond. A comparison between them is summarized as follows:

Comparison Between Sigma (σ) Bond and Pi (π) Bond

S.No.	Characteristics	Sigma (σ) bond	Pi (π) bond
1.	Formation due to	Axial overlapping of orbitals	Lateral overlapping of orbitals
2.	Extent of overlapping	Greater	Smaller
3.	Orbitals involved and their axes	*p* along the same axis	*p* only at an angle
4.	Strength	Stronger	Comparatively weaker
5.	Rotation of atoms	Freely around the bond	No

2.8.2 Bond Angle

Covalent bonds are easily formed between the atoms having unfilled *p* suborbits. The *p* suborbits are directional in nature. They allow overlapping of suborbits in the direction of densest electron probability cloud. Hence the covalent bonds are directional. The bond is formed at specific angles in different elements. This angle is called *bond angle* θ.

Bond angle is 90° between the unbonded *p* suborbits. If *p* suborbits are bonded without showing any distortion then also the bond angle is 90°, and also in the case when distortion of this suborbit does not take place during formation of bond. Bond angle in water is 104° owing to hybridization of *s* and *p* suborbits of oxygen. Diamond develops tetrahedral bond angle of 109.5°. Figure 2.6 shows the concept of bond angle. Bond angles in some elements are given in Table 2.1.

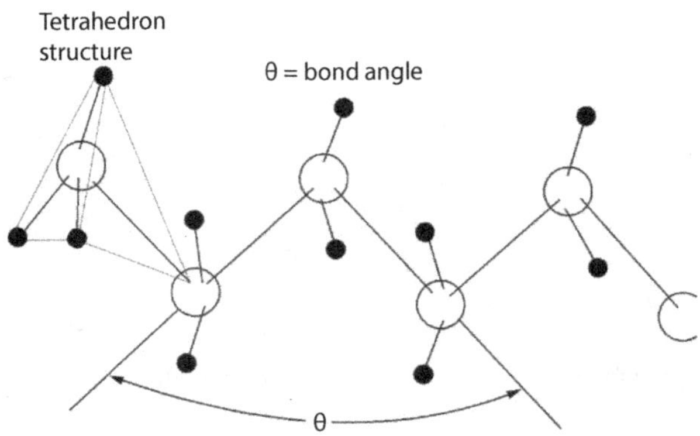

Figure 2.6 Tetrahedral bond angle.

Table 2.1 Bond angle of some elements and molecules

Element	Bond angle	Molecule and other	Bond angle
Antimony	96°	Water	104°
Phosphorus	99°	Ammonia	107°
Telurium	104°	Methane	109°
Selenium	104°	BF_3	120°
Sulphur	107°	$BeCl_2$	180°

2.8.3 Directional and Non-Directional Bonds

Bonds, whether primary or secondary, may have properties which are either

i. directional
ii. non-directional

Depending on this behaviour, the bonds are termed as either 'directional bond' or 'non-directional bond'. The bond strength is unequal in different directions in case of *directional bonds.* It is, thus, *direction dependent.* The non-directional bonds possess equal strength in all directions. *Hence they are direction independent.* The ionic, metallic and Van der Waals bonds are non-directional while the covalent and hydrogen bonds are directional.

2.8.4 Mixed bonds

Mixed bonds are more common in materials. Such bonds are formed due to a combination of two or more pure bonds (primary or secondary). All these bonds and their properties etc. have been discussed in subsequent articles.

CRYSTAL GEOMETRY

2.9 Monocrystalline and Polycrystalline Crystal Structures

Materials, on the basis of their structure, may be classified into two groups viz.

1. Crystalline materials, and
2. Non-crystalline (or amorphous) materials

Generally metals are crystalline, and non-metals are non-crystalline. But this is not a rule. Plastics, a non-metal may be obtained in almost crystalline form. Crystalline solids have periodically repeating arrangement of atoms. Such solids can be further sub-classified as follows.

i. Monocrystalline, and
ii. Polycrystalline.

Most of materials in engineering applications are polycrystalline. Monocrystalline material has a *single crystal.* It finds use in specific applications. As an example, a single crystal quartz is employed in generating ultrasonic waves. Examples of some monocrystalline and polycrystalline materials are given below.

Monocrystalline materials	Quartz, Single crystal Ti, Single crystal garnet, Single crystal Si, etc.
Polycrystalline materials	Steel and iron, Nickel, Copper, Magnesium, Zinc, Tungsten, Gold, etc.

We are concerned here with the geometry of crystalline materials only. Non-crystalline materials will be taken up in later articles.

2.9.1 Construction of a solid

The smallest visible part of a material is made up of a large number of *crystals*, Fig. 2.7*a*. These crystals may be of different shapes and sizes. They generally have random orientation. Each crystal is further composed of basic

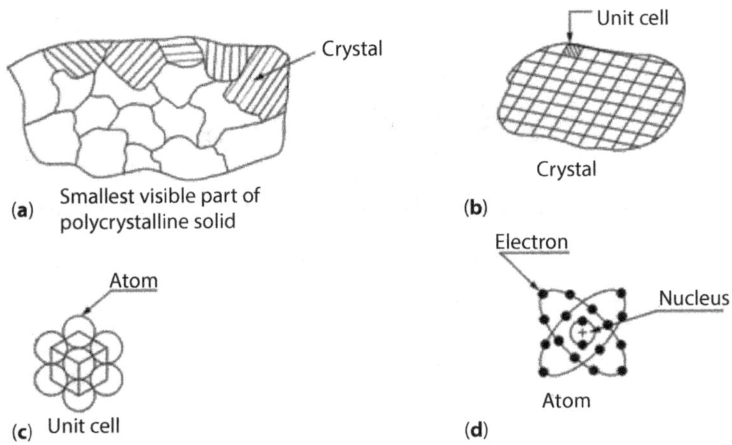

Figure 2.7 Constructional details of a solid material.

structural item called *unit cell*, Fig. 2.7*b*. Unit cells are of different types. These unit cells contain *atoms* arranged in a very systematic pattern. The details of an atom has already been discussed earlier. Thus the complete system of solid materials may be summarily shown as in Fig. 2.7.

2.10 Space Lattice

A *space lattice is defined as an infinite array of points in three-dimensional space in which each point is identically located with respect to the other.* Concept of space lattice is helpful in understanding the crystal structure of existing materials, and also those materials which are likely to be developed in future.

2.11 Basis

The way of filling-up of points in a space lattice by the atoms is known as *Basis*. Each point may be occupied by one, two or many atoms in different solids. The space lattice when combines with the *basis* generates a *unit cell*. Thus

$$\textit{space lattice} + \textit{basis} = \textit{unit cell}$$

The unit cell will be called *monoatomic* if one atom occupies a lattice point. When two atoms occupy a lattice point, it will make a *diatomic* unit

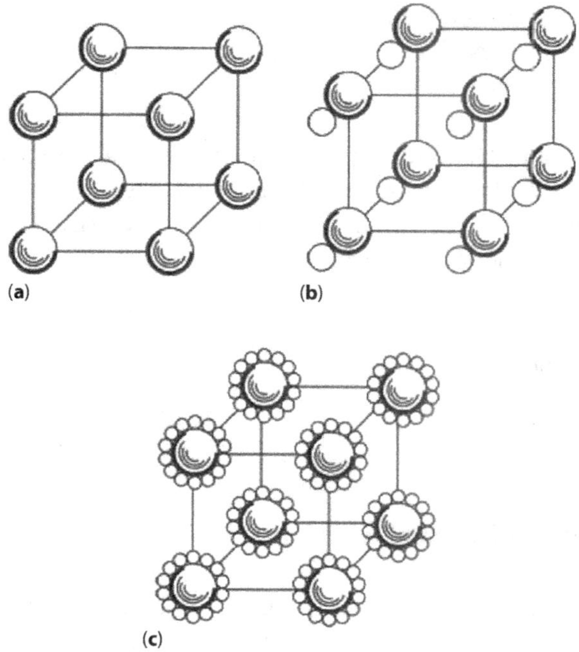

Arrangement of atoms are different on different unit cells

Figure 2.8 Types of unit cells according to the basis (a) monoatomic, (b) diatomic, and (c) multiatomic

cell. Similarly the unit cell will be known as *multiatomic* when too many atoms occupy a lattice point. These types of unit cells are shown in Figs. 2.8*a*-*b*-*c*. Here the atoms are shown separated from each other for clarity, which in actual materials are not separated. In diatomic (Fig. 2.8*b*) and multiatomic (Fig. 2.8*c*) unit cells, the centre of larger atom coincides with the lattice point.

In manganese, there are 29 atoms at each corner and 29 atoms at centre of the cube. Thus its unit cell contains 58 atoms.

2.12 Unit Cell and Crystal

Definition. After developing the concept of unit cell, we now proceed to define it. A *unit cell is defined as the basic structural part in the composition of materials.* It is analogous to a brick used in the building construction.

When many unit cells repeat in a three-dimensional space, a *crystal* is obtained. The structure of a crystal is same as that of a repeating unit cell. Hence crystal structure may be classified as

1. monoatomic crystal,
2. diatomic crystal, and
3. multiatomic crystal.

Molecular crystal. Diatomic and multiatomic crystals are also known as *molecular crystals.* There exist thousands of varieties of crystal structures. Many of them have *complex nature.* Complex crystal structure has two or more types mixed together, such as in

- Sulphur
- Phosphorous
- Gallium
- Uranium, etc.

2.13 Bravais Crystal System

Crystals have unit cells of various geometries. The geometries are defined in terms of their linear dimensions **a**, **b** and **c**, and angular dimensions α, β and γ. Dimensions *a, b* and *c* are along the *x, y* and *z* axes respectively; and angles α, β and γ between *xy*, *yz* and *zx* axes respectively. These are shown in Fig. 2.9.

Bravais has classified 14 space lattices into 7 crystal systems. These are displayed in Table 2.2.

Crystals have inherent symmetry of different types. Different crystals exhibit vivid varieties of symmetry. The *cubic* crystal is most symmetric while the *triclinic* is least symmetric. Symmetry decreases as we move from cubic crystal system towards triclinic crystal system in Table 2.2.

Example 2.2 Why is it not possible to have a face centred tetragonal (FCT) space lattice?

> ***Solution***: Tetragonal crystal system has only two space lattices viz. simple tetragonal (ST) and body centred tetragonal (BCT). Geometries of BCT and FCT both can be described by the same array of points. In such cases, the unit cell is represented by those lattices which have smaller number of lattice points. As BCT has lesser number of lattice points than FCT, hence FCT lattice is not included in the list of Bravais system.
>
> Figure 2.10 is self-illustrative in this regard. There are two FCT unit cells showing corner atoms marked *C* and face atoms marked *F*. A BCT unit cell fits well into it, and is shown by dotted lines.

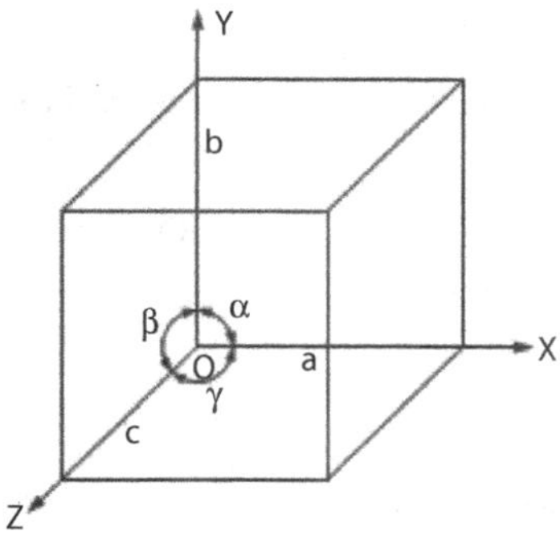

Figure 2.9 Geometry of unit cell

Table 2.2 Bravais crystal system

Unit cell (geometry)	Space lattice	Abbreviation	Example
1. Cubic $a = b = c$ $\alpha = \beta = \gamma = 90^\circ$ 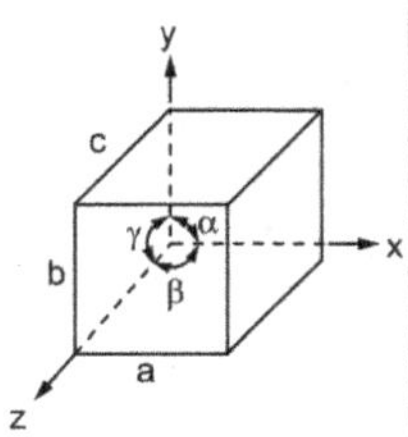	1. Simple (Points at the eight corners of the unit cell) 2. Body Centred (Points at the eight corners and at the body centre) 3. Face Centred (Points at the eight corners and at the six face centres)	SC BCC FCC*	Po Li, Na, V, Ta, Cr, Mo, W, Fe, CsCl Ni, Pt, Cu, Ag, Au, Al, Pb, Ne, Ar, Kr, Xe, NaCl
2. Tetragonal $a = b \neq c$ $\alpha = \beta = \gamma = 90^\circ$ 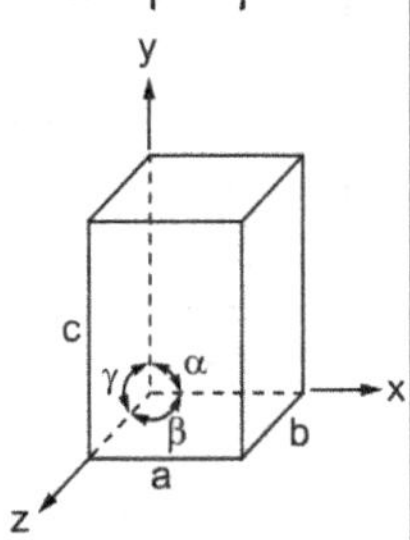	4. Simple (Points at the eight corners of the unit cell) 5. Body Centered (Points at the eight corners and at the body centre)	ST BCT	Pa, In White Sn, KH_2PO_4, U between 668 °C–774 °C

(*Continued*)

Table 2.2 (Cont.)

Unit cell (geometry)	**Space lattice**	**Abbreviation**	**Example**
3. Orthorhombic $a \neq b \neq c$ $\alpha = \beta = \gamma = 90°$ 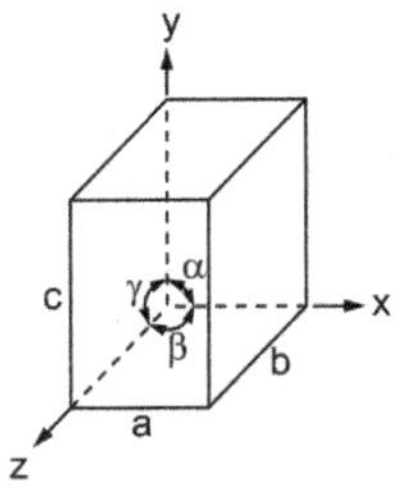	6. Simple (Points at the eight corners of the unit cell)	SO	As, Sb, Bi
	7. End Centred. (Also called side Centred or base Centred) (Points at the eight corners and at the face centres opposite to each other)	ECO	$BaSO_4$, KNO_3, $MgSO_4$
	8. Body Centred (Points at the eight corners and at the body centre)	BCO	–
	9. Face Centred (Points at the eight corners and at the six face centres)	FCO	S, Ga
4. Rhombohedral (or trigonal) $a = b = c$ $\alpha = \beta = \gamma \neq 90°$ 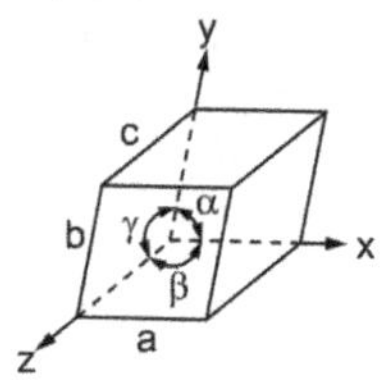	10. Simple (Points at the eight corners of the unit cell)	SR	B, SiO_2 $CaCO_3$, Lithium niobate ($LiNbO_3$), Lithium tantalate ($LiTaO_3$)
5. Hexagonal $a = b \neq c$ $\alpha = \beta = 90°$, $\gamma = 120°$	11. Simple [(*i*) Points at the eight corners of the unit cell outlined by thick lines OR	SH	La, Pr, Nd, Am, Se, Te, graphite, AgCl

(*Continued*)

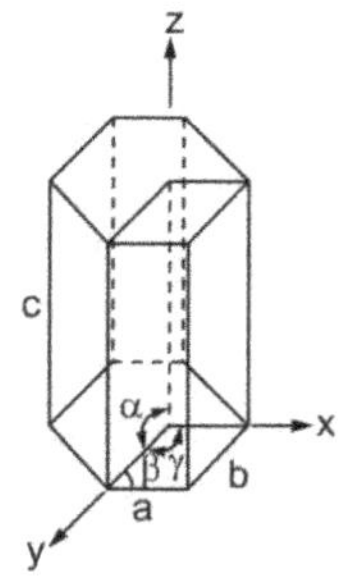	(*ii*) Points at the twelve corners of the hexagonal prism and at the centres of the two hexagonal faces]		
6. Monoclinic $a \neq b \neq c$ $\alpha = \beta = 90° \neq \gamma$ 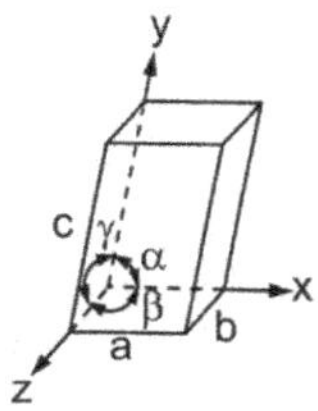	12. Simple (Points at the eight corners of the unit cell) 13. End Centred (Points at the eight corners and at two face centres opposite to each other)	SM ECM	$NaSO_4$, Li_2SO_4, $CaSO_4$, Rochelle salt, $FeSO_4$ S
7. Triclinic $a \neq b \neq c$ $\alpha \neq \beta \neq \gamma \neq 90°$ 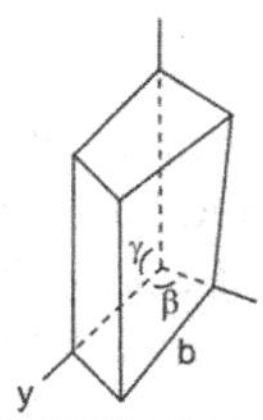	14. Simple (Points at the eight corners of the unit cell)	STC	$K_2Cr_2O_7$, $CuSO_4$

* FCC is also known as CCP (Cubic Closed Packed)

2.14 Primitive and Non-Primitive Unit Cells

Primitive cells are those unit cells which contain atoms at corner lattice points only. So these cells have least number of total atoms and the least volume of atoms per unit cell. All unit cells of Table 2.2 namely simple cube (SC), simple tetragon (ST), simple orthorhombic (SO), simple rhombohedral (SR) etc. are primitive cells.

All those unit cells which do not fall under this category are non-primitive cells.

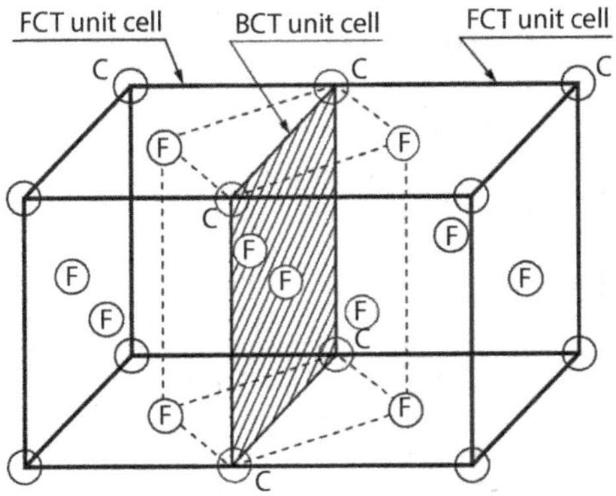

Figure 2.10 A BCT unit cell shown by dotted lines within two FCT unit cells. C and F indicate corner and face atoms respectively.

2.15 Coordination Number

Definition. Each atom in a crystal is surrounded by a number of atoms. The surrounding atoms are located at different distances. *The coordination number is defined as the number of nearest and equidistant atoms with respect to any other atom in a unit cell.*

Effective number of atoms per unit cell N_e is different from total number of atoms per unit cell. The atom at the corner of a cubical unit cell has only 1/8 of it inside the boundary of that unit cell. The remaining 7/8 of it lies in the surrounding unit cells of the crystal. Similarly the atom at the face in FCC is shared 1/2 by that atom and 1/2 by the neighbouring atom. In BCC, the atom at the centroid is wholly occupied by that unit cell in which it lies. Thus the effective number of atoms are 1, 2 and 4 in SC, BCC and FCC respectively. Figures 2.11 *a-b-c* explain these details. Effective number of atoms for different unit cells are summarized in Table 2.3.

2.16 Atomic Packing Fraction

The atomic packing fraction is defined as the ratio of total volume of atoms per unit cell, v, to the total volume of unit cell, V. It is also known as *relative density of packing* or *Atomic Packing Factor* (APF) or *Atomic Packing Efficiency* (APE). This may be expressed by

$$\text{APF} = \frac{v}{V} \tag{2.15a}$$

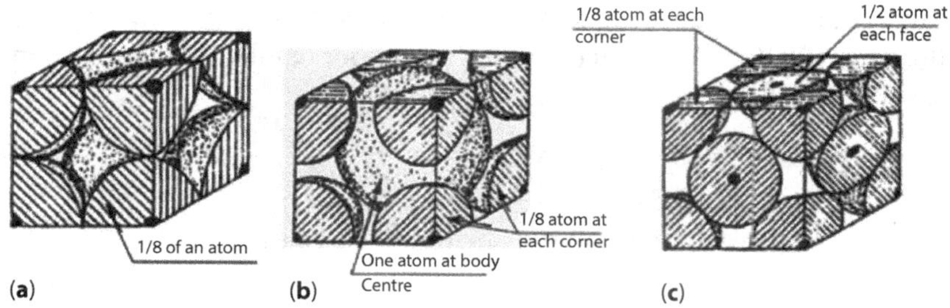

Figure 2.11 Effective number of atoms in (a) SC, (b) BCC, and (c) FCC

Table 2.3 Details of crystal geometry

Types of crystal	Effective number of atoms	Coordination number	APF	Relation between 'a' and 'r'
SC	1	6	0.52	$2r = a$
FCC	4	12	0.74	$2r = \frac{a}{\sqrt{2}}$
BCC	2	8	0.68	$2r = \frac{a\sqrt{3}}{2}$
ST	1	4	$\pi a/6c$	$2r = a = b < c$
BCT	2	8	$\frac{\pi(2a^2 + c^2)^{3/2}}{24a^2c}$	$r = \sqrt{(2a^2 + c^2)/4}$
SO	1	2	$\pi a^2 / 6bc$	$2r = a$ for $a < b$, $a < c$
ECO	2	2	$(b^2 \quad c^2)/24abc$	$r = \sqrt{(b^2 + c^2)}/4$
BCO	2	8	$\pi\left(\frac{a^2 + b^2 + c^2}{24abc}\right)^{3/2}$	$4r = \sqrt{a^2 + b^2 + c^2}$
FCO	4	2	—	—
SH	3	—	$\pi a / 3\sqrt{3c}$	$2r = a$

Volume of an atom, supposed to be spherical, is equal to $4\pi r^3/3$ where r is the radius of atom. If the unit cell contains N_e number of effective atoms, then

$$v = \frac{4\pi r^3}{3} \times N_e \tag{2.15b}$$

Hence atomic packing fraction may be determined by

$$\text{APF} = \frac{4\pi r^3}{3V} \times N_e \tag{2.15c}$$

where V may be found for different geometries of unit cells from the following relations.

Type of unit cell	Volume of unit cell V
Cubic	a^3
Tetragonal	a^2c
Orthorhombic	abc
Rhombohedral	$a^3\sqrt{1-3\cos^2 a - 2\cos^3 a}$
Hexagonal	$\frac{3\sqrt{3}z^2c}{2}$
Monoclinic	$abc\sin\beta$
Triclinic	$abc\sqrt{1-\cos^2 ga - \cos^2\beta - \cos^2\gamma + 2\cos a\cos\beta\cos\gamma}$

Example 2.3 Calculate the ionic packing fraction of NaCl having FCC structure.

Solution: Ionic radius of Na = 0.98 Å, Ionic radius of Cl = 1.8l Å
In FCC NaCl, there are 4 sodium ions and 4 chloride ions.

$$\text{APF} = \frac{\text{Total volume of ions per unit cell}}{\text{Volume of the unit cell}}$$

$$= \frac{\text{Total volume of Na ions} + \text{Total volume of Cl ions}}{\text{Volume of the unit cell}}$$

$$= \frac{4\times\frac{4}{3}\pi(0.98)^3 + 4\times\frac{4}{3}\pi(1.81)^3}{a^3}$$

The lattice constant $a = (2 \times 0.98 + 2 \times 1.81)$

$$\therefore \quad \text{APF} = \frac{16\pi\{(0.98)^3 + (1.81)^3\}}{3\times\{2\times 0.98 + 2\times 1.81\}^3} = 0.67$$

2.17 Calculation of Density (or Bulk Density)

The density ρ of a material is expressed as

$$\rho = \frac{\text{mass of unit cell}}{\text{volume of unit cell}} = \frac{M}{V} \tag{2.16a}$$

where, $M = m \,.\, N_e$, and m = mass of one atom.

The mass of one atom m may be obtained by

$$\mathrm{m} = \frac{\text{Atomic weight}}{\text{Avogadro's number}} = \frac{A_W}{N_A} \tag{2.16b}$$

The volume of cubical unit cell = a^3. Thus the density may be calculated by the given formula

$$\rho = \frac{\mathrm{A_W . N_e}}{\mathrm{N_A . a^3}} \tag{2.16c}$$

Example 2.4 The atomic radius of copper is 1.278 Å. Find the density of copper.

Solution: Copper has FCC structure for which

$$\mathrm{a} = \frac{4r}{\sqrt{2}},$$

atomic weight $A_W = 63.54$

$$\therefore \qquad \mathrm{a} = \frac{4 \times 1.278}{\sqrt{2}} = 3.61\ \text{Å}$$

Using Eq. 2.16c we get

$$\rho = \frac{A_W \times N_e}{N_A \times a^3} = \frac{63.54 \times 4}{6.023 \times 10^{23} \times (3.61 \times 10^{-8})^3}$$
$$= 8.98\ \mathrm{gm/cm^3} = 8930\ \mathrm{kg/m^3}$$

2.18 Miller Indices

Miller is the name of a material scientist. Indices is plural of index. Miller indices are the styles to designate the planes and directions in the unit cells and crystals. Such designations become essential for investigations of various properties in different elements. In the most general way, Miller indices are expressed by *(hkl)* and no comma is used between them. Here, symbols *h, k* and *l* indicate unknown integers. The kinds of brackets have special meanings which are elaborated as below.

(*hkl*) denotes a plane.
[*hkl*] denotes a direction.
{ *hkl*} denotes family of planes.
< *hkl* > denotes family of directions.

Miller indices (*hkl*) *are* expressed as a reciprocal of intercepts *p*, *q* and *r* made by the plane on the three rectangular axes *x*, *y* and *z* respectively. The intercepts are measured in terms of the dimensions of unit cell. These are the unit distances from the origin along the three axes. Thus

$$h = \frac{1}{p}, \quad k = \frac{1}{q} \quad \text{and} \quad l = \frac{1}{r}$$

where, p = intercept of the plane on x–axis,
q = intercept of the plane on y–axis, and
r = intercept of the plane on z–axis.

Reciprocal of these intercepts are then converted into whole numbers. This can be done by multiplying each reciprocal by a number obtained after taking LCM (lowest common multiplier) of denominator. This gives the Miller indices of required plane. The Miller indices are expressed by three smallest integers.

2.18.1 Determining the Miller Indices of a Given Plane

Refer Fig. 2.12 showing a plane *ABC* in a cubical unit cell. We have to determine the Miller indices of plane *ABC*. Origin *O* and coordinate axes *x*, *y* and *z* are shown in the unit cell in which *OP* = *OQ* = *OR* = *a*. The plane in question intercepts at *A*, *B* and *C* on *x*, *y* and *z* axes in such a way that

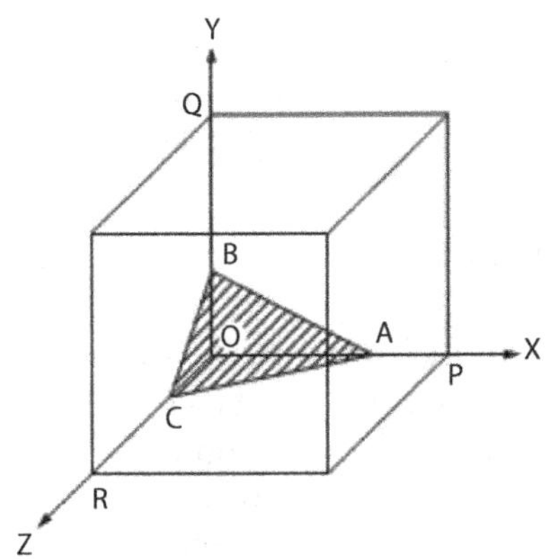

Figure 2.12 Determination of Miller indices of unknown plane

$OA = (2a/3)$ along x-axis,
$OB = (2a/5)$ along y-axis, and
$OC = (1a/3)$ along z-axis.

To obtain the required Miller indices, we proceed as below :

i. The intercepts are $c_1 = OA$, $c_2 = OB$ and $c_3 = OC$, therefore

$$p \text{ (along } x\text{-axis)} = \frac{OA}{OP} = \frac{(2/3)a}{a} = \frac{2}{3},$$

$$q \text{ (along } y\text{-axis)} = \frac{OB}{OQ} = \frac{(2/5)a}{a} = \frac{2}{5}, \text{ and}$$

$$r \text{ (along } z\text{-axis)} = \frac{OC}{OR} = \frac{(1/3)a}{a} = \frac{1}{3}.$$

ii. As $h = \frac{1}{p}$, $k = \frac{1}{q}$ and $l = \frac{1}{r}$. therefore

$$h = \frac{1}{(2/3)} = \frac{3}{2},$$

$$k = \frac{1}{(2/5)} = \frac{5}{2} \text{ and}$$

$$l = \frac{1}{(1/3)} = \frac{3}{1}$$

$$\therefore \qquad (hkl) = \left(\frac{3\,5\,3}{2\,2\,1}\right)$$

iii. As (hkl) should have smallest possible integers, hence we can write

$$(hkl) = \frac{1}{2}(3\;5\;6)$$

Here the factor 1/2 is the result of conversion of reciprocals to integers, and is usually omitted. We, therefore, conclude that Miller indices of the plane ABC is

$$(hkl) = (356)$$

Miller indices of a plane in non-cubical unit cells may be determined in the same manner. Method to determine Miller indices in an orthorhombic crystal has been given in example 2.5.

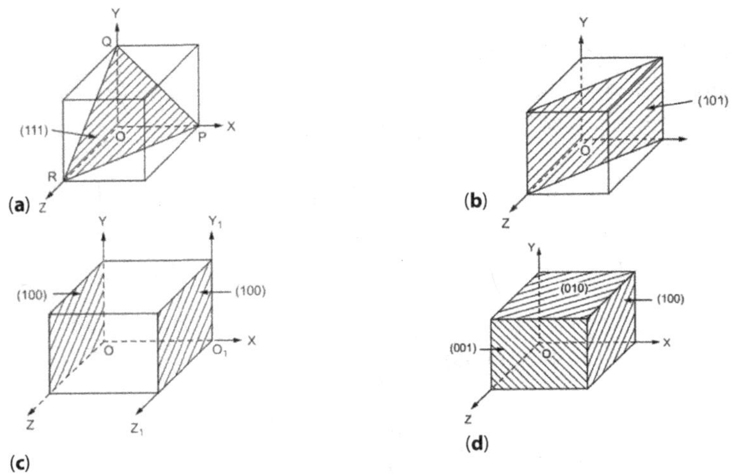

Figure 2.13 Planes showing Miller indices *(a)* (111), *(b)* (101), *(c)* $(\bar{1}00)$ and (100), and *(d)* {100} in a cubic unit cell.

2.18.2 Drawing a Plane Whose Miller Indices are Given

In the previous section, we have determined the Miller indices of an unknown plane. Now, we shall draw a plane (in an unit cell) whose Miller indices are given. We take the case of a plane whose Miller indices are (111).

As $h = 1$, $k = 1$, and $l = 1$; therefore p = 1/h = 1/1, q = 1/k = 1/1, and r = 1/1 = 1/1. Hence intercepts are

$$c_1 \text{ (on } x\text{-axis)} = pa = 1 \times a = a,$$
$$c_2 \text{ (on } y\text{-axis)} = qb = 1 \times b = b, \text{ and}$$
$$c_3 \text{ (on } z\text{-axis)} = rc = 1 \times c = c.$$

If the plane (111) is to be drawn in cubical unit cell, then $a = b = c$. Therefore intercepts $c_1 = c_2 = c_3 = a$.

Now a cubical unit cell is drawn showing origin O and coordinate axes x, y and z (Fig. 2.13*a*). Intercepts $c_1 = a = OP$, $c_2 = a = OQ$ and $c_3 = a = OR$ are located. Points P, Q and R are joined. The hatched plane PQR represents the required Miller indices (111).

2.18.3 Drawing a Plane which is Parallel to an Axis

Refer Fig. 2.13 *b* showing a plane whose Miller indices are (101). This plane intercepts x and z axes, and is parallel to y axis. A plane parallel to an axis is supposed to have its intercept at infinity. In this case $h = 1$, $k = 0$ and $l = 1$. Therefore, $p = \frac{1}{1}$, $q = \frac{1}{0} \rightarrow \infty$, and $r = \frac{1}{1}$. Hence intercepts $c_1 = 1$, $c_2 \rightarrow \infty$ and $c_3 = 1$.

2.18.4 Planes with Negative Indices

Refer Fig. 2.13*c* showing origin at O and the coordinate axes x, y and z. Miller indices (100) of a plane is depicted in accordance with the above discussions. If we have to draw a plane whose Miller indices are $(\bar{1}00)$, then the origin O has to be shifted to O_1. Now the coordinate system x, y_1 and z_1 is under consideration. The direction O_1O indicates negative side of x. Plane with Miller indices $(\bar{1}00)$ will be one as shown in above figure. To indicate a minus sign, the symbol is marked $\bar{1}$, and is spoken as 'bar 1'.

2.18.5 Family of Planes

Different planes in a cubical unit cell are shown in Fig. 2.13*d*. Their Miller indices are (100), (010) and (001). Miller indices of a plane is (100) in Fig. 2.13*c*. Similarly other planes in a cubical unit cell may be designated as $(0\ \bar{1}0)$ and $(00\ \bar{1})$. Digits in all these planes are 1, 0 and 0 at different sequence. They belong to the same family, and are called family of planes. These planes can be designated by {100} only using a curly bracket. As an another illustration, the planes having Miller indices (123), (231), (321), (312), (123), (123), (213) etc. form a family of planes. They can be designated by {123} which are the lowest integers.

Example 2.5 Obtain the Miller indices of a plane whose intercepts are a, $b/2$, and $3c$ on x, y and z axes respectively in a simple cubic unit cell.

Solution: The intercepts c_1 (along x-axis) = a, c_2 (along y-axis) = $b/2$, and c_3 (along z-axis) = $3c$.

$$\text{Therefore,}\quad p = \frac{c_1}{a} = \frac{a}{a} = 1,$$

$$q = \frac{c_2}{b} = \frac{b/2}{b} = \frac{1}{2} \text{ and}$$

$$r = \frac{c_3}{c} = \frac{3c}{c} = 3$$

$$\therefore \quad h = \frac{1}{p} = \frac{1}{1} = 1,$$

$$k = \frac{1}{q} = \frac{1}{1/2} = 2, \text{ and}$$

$$l = \frac{1}{r} = \frac{1}{3}$$

Hence $(hkl) = (1\ 2\ \frac{1}{3}) = \frac{1}{3}$ (361) or 361.

Example 2.6 Determine the Miller indices of a plane that makes intercepts of 2Å, 3Å and 4Å on the coordinate axes of an orthorhombic crystal with $a : b : c = 4 : 3 : 2$.

Solution: Given are $a : b : c = 4 : 3 : 2$
The intercepts in terms of unit axial dimensions are p, q and r. So

$$p:q:r=\frac{2\text{Å}}{4}:\frac{3\text{Å}}{3}:\frac{4\text{Å}}{2}=\frac{1}{2}:1:2$$

Reciprocal of these numbers are found as

$$h=\frac{1}{p}=2,\ k=\frac{1}{q}=1,\ l=\frac{1}{r}=\frac{1}{2}$$

Reducing the value of (*hkl*) obtained above into smallest set of whole number, we obtain the Miller indices as

$$(hkl)=(2\,1\,\frac{1}{2})=\frac{1}{2}(421)\text{ i.e.}(421).$$

2.18.6 Miller Indices: Crystallographic Notation of Atomic Crystal Directions

Miller indices are also used to specify directions within a unit cell or a crystal. An unknown direction is designated by [*hkl*]. Here a square bracket is used to designate direction. To illustrate the procedure of determining the crystal direction, let us take a line passing through the origin and parallel to a given direction. A vector r passing through the origin O of the unit cell to a lattice point may be given by

$$r = r_1 x + r_2 y + r_3 z \tag{2.17}$$

where r_1, r_2 and r_3 are integers. The length of the projections of this line on the coordinate axes x, y and z are noted. These intercepts are reduced to smallest integers to obtain [*hkl*].

Consider vector $r = OB$ in Fig. 2.14*a* in which B is a lattice point. Here $r_1 = 1$, $r_2 = 1$ and $r_3 = 1$ from the geometry of the cubical unit cell along x, y and z axes. Hence this direction is [111].

Illustrations. As an other case, consider point C located at a distance (2/3) of the dimension of cube along x-axis. The direction vector $r = OC$ is

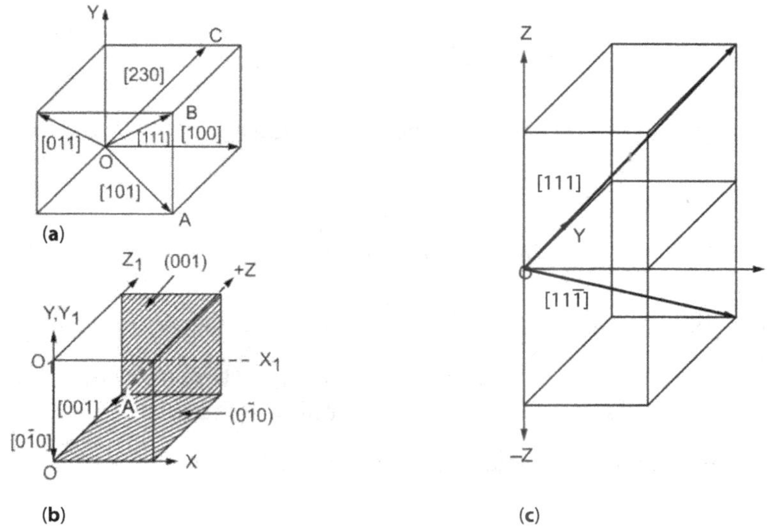

Figure 2.14 Miller indices of directions *(a)* [111], [230], [100], [101], [011] *(b)* [001], [010] and *(c)* [111], [11$\bar{1}$]

parallel to direction *z*. In this case $r_1 = 2/3$, $r_2 = 1$ and $r_3 = 0$. Therefore, its direction will be found as

$$[hkl] = [230].$$

Miller indices of directions [100], [110] and [011] are also shown in Fig. 2.14 a-b-c.

2.19 Interplaner Spacing

The spacing between a plane (*hkl*) and the other parallel plane passing through the origin is called *interplaner* spacing. It is denoted by $d_{[hkl]}$. It is measured at right angles to the planes. As an example, the interplaner spacing of (100) planes in cubic crystal is equal to the lattice constant *a*. For (200) planes, this value is (*a*/2). Interplaner spacing may be obtained from the following relations in cubic and tetragonal crystals.

For cubic unit cell,

$$d_{[hkl]} = \frac{a}{\sqrt{h^2 + k^2 + l^2}} \tag{2.18a}$$

for tetragonal unit cell

$$d_{[hkl]} = \frac{a}{\sqrt{h^2 + k^2 + l^2(a^2/c^2)}} \tag{2.18b}$$

and for orthogonal unit cell, it can be generalized as

$$\frac{1}{d_{hkl}^2} = \left(\frac{h}{a}\right)^2 + \left(\frac{k}{b}\right)^2 + \left(\frac{l}{c}\right)^2 \tag{2.18c}$$

Members of a family of planes $\{hkl\}$ have the same interplaner spacing.

Example 2.6 Determine the interplaner spacing between (200), (220) and (111) planes in a FCC crystal. The atomic radius is 1.246 Å.

Solution : In a FCC crystal, the interatomic distance is

$$a = \frac{4r}{\sqrt{2}}$$

If atomic radius r = 1.246 Å, then

$$a = 2\sqrt{2} \times 1.246\,\text{Å} = 2.492\sqrt{2}\,\text{Å}$$

$$\because \quad d_{[hkl]} = \frac{a}{\sqrt{h^2 + k^2 + l^2}}$$

$$\therefore \quad d_{(200)} = \frac{2.492\sqrt{2}}{\sqrt{2^2 + 0^2 + 0^2}} = 1.762\,\text{Å}$$

$$d_{(220)} = \frac{2.492\sqrt{2}}{\sqrt{2^2 + 2^2 + 0^2}} = 1.24\,\text{Å}$$

and

$$d_{(111)} = \frac{2.492\sqrt{2}}{\sqrt{1^2 + 1^2 + 1^2}} = 2.034\,\text{Å}$$

2.20 Linear Density

Definition. Linear density ρ_L may be defined as the number of effective atoms N_{eL} per unit length on certain length L along any direction in a unit cell. It can be expressed as

$$\rho_L = \frac{N_{eL}}{L} \tag{2.19}$$

2.21 Planer Density

Definition. The number of atoms per unit area of a crystal plane is known as *planer density.* This density infact expresses the packing of atoms on a plane. The rate of plastic deformation is significantly influenced by it. Hence calculation of the number of atoms per unit area (generally per square millimetre) becomes essential. The study on common planes (100), (110) and (111) in cubic system will follow in the text below.

The planer density ρ_p can be expressed as

$$\rho_p = \frac{N_e}{A} \tag{2.20}$$

where N_e is the effective number of atoms on the plane whose area is A.

2.21.1 Planer Density in Face Centred Cube (FCC) on (100) Plane

This plane as shown in Fig. 2.15*a* contains

$$N_e = (1/4 \times 4) + 1 = 2$$

$$A = a \times a = a^2$$

$$\rho_p = \frac{N_e}{A} = \frac{2}{a^2} \tag{2.21}$$

2.21.2 Planer Density in FCC on (110) Plane

This plane contains one-fourth atom on each of the four corners in addition to half atom in each of *AB* and *CD* edges. The arrangement shown in Fig. 2.15*b* has

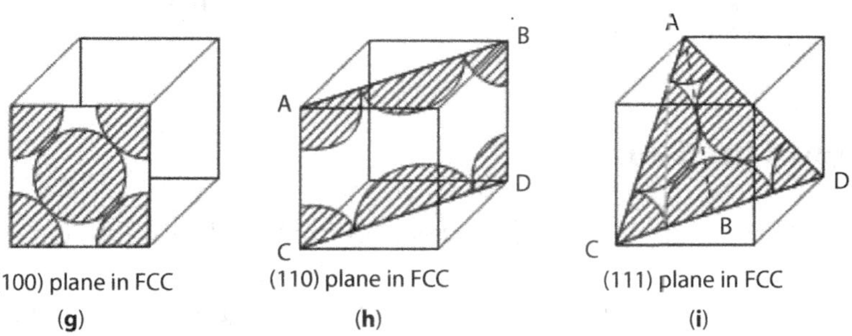

Figure 2.15 Planer density in face centred cube on (a) (100) plane, (b) (110) plane, (c) (111) plane.

$$N_e = (1/4 \text{ x } 4) + (1/2 \text{ x } 2) = 2$$
$$\text{length } AB = CD = \sqrt{2}a$$

$$A = a.\sqrt{2}a = \sqrt{2}a^2$$

Therefore

$$\rho_p = \frac{N_e}{A} = \frac{2}{\sqrt{2}a^2} = \frac{1.414}{a^2} \quad (2.22)$$

2.21.3 Planer Density in FCC on (111) Plane

The arrangement on this plane is shown in Fig. 2.15*c*. It contains one-sixth atom in each of the three corners in addition to half atom on each of the edges *AC*, *CD* and *AD*. Thus

$$N_e = (1/6 \text{ x } 3) + (1/2 \text{ x } 3) = 2$$
$$\text{length } AC = CD = AD = \sqrt{2}a$$

$$\text{perpendicular height } AB = \sqrt{2}a.\sin 60°$$

$$\therefore \quad A = \frac{CD \times AB}{2} = 0.866a^2$$

Therefore

$$\rho_p = \frac{N_e}{A} = \frac{2}{0.866a^2} = \frac{2.31}{a^2} \quad (2.23)$$

Quick Revision Summary

Photons are discrete radiations emitted from the materials in the form of small energy packets. These packets are also called as '*quanta*'. They possess energy, mass and momentum.

Review Questions

1. Discuss in brief, different stages in the development of atomic model concept.
2. State and explain the Bohr's postulates regarding structure of hydrogen atom. Interpret the meaning of negative sign in the expression of electron energy.

3. Show that the radii of the Bohr's atomic orbits are in the ratio of $1^2 : 2^2 : 3^2 : 4^2$ ... etc.
4. Enumerate the limitations of Bohr's atomic theory. In this regard, explain the improvement in Sommerfeld's model.
5. Describe the modern concept of atom. Discuss the arrangement of subshells in different shells?
6. Write notes on the following.
   a. Heisenberg's uncertainty principle
   b. Probability density cloud
   c. de Broglie wave
   d. Quantum numbers and their significance
7. What do you mean by electronic configuration? Write electron configuration of silicon, copper, chromium and zinc.
8. What do you mean by bonding in solids? Why are some solids strongly bonded than the others?
9. Classify the various types of bonds, and illustrate their examples.
10. Differentiate between the following:
    a. Directional and non-directional bond
    b. Primary bond and secondary bond
    c. Ionic bond and covalent bond
    d. Metallic bond and hydrogen bond
11. Why is the covalent bond directional while the ionic bond non-directional?
12. Compare the characteristics of σ-bonds, π-bonds and hybrid bonds.
13. Discuss the effects of elements of different columns on various properties and characteristics of hybridized bonds.
14. Explain the composition of a solid illustrating the crystals, unit cells, atoms and electrons in it.
15. Are the triangular or heptagonal space lattices possible? Justify your answers with logic and diagrams.
16. There are three space lattices in cubic crystal system (SC, BCC and FCC) but four in orthorhombic crystal system (SO, ECO, BCO and FCO). Why is it so?
17. There is no end-centred tetragonal lattice in the Bravais list, but there is an end-centered orthorhombic lattice. Explain why this is so?
18. Write notes on the following:
    a. Basis
    b. Primitive unit cell

   c. Molecular crystal
   d. Coordination number
19. Obtain the coordination number for (*a*) BCT (*b*) ECO (*c*) SR, and (*d*) ECM; unit cells.
20. Determine the *basis* in magnesium, diamond and silicon.
21. Derive the expression for relation between atomic radius and lattice constant in case of (*a*) BCC (*b*) FCC, and (*c*) ST.
22. Define the atomic packing factor. Obtain its expression for SC, FCC and BCC.
23. How much is the volume of atoms in 1 cm^3 of (*a*) vanadium (*b*) polonium, and (*c*) strontium? Justify your answers.
24. Discuss different types of crystal symmetries with the help of sketches.
25. What is the importance of Miller indices? How does it help in the study of crystallography?
26. Detail out the procedure to obtain Miller-Bravais index in a hexagonal crystal. [see any related book]
27. Sketch the following planes and directions in cubical and hexagonal unit cells or their crystals as required.

    $(221), (\overline{10}\,1), <100>, [234], \{\bar{1}00\}, (2020)$.
28. Derive the expression which relates interplaner spacing, Miller indices and dimension of the (*a*) cubic unit cell, and (*b*) tetragonal unit cell.
29. Prove that the plane (*hkl*) is perpendicular to the direction [*hkl*].

    Hint: A plane passing through the origin is $hx + ky + lz = 0$, and a plane parallel to above plane and nearest to the origin is $hx + ky + lz = a$. Scalar product of an arbitrary vector in a plane through origin and [*hkl*] will be zero.
30. Find the Miller indices of a plane that makes an intercept of 1 on the *a*-axis and 2 on the *b*-axis and is parallel to the *c*-axis. Draw it.

Numerical Questions

1. For a hydrogen atom in the ground state, determine in electron volts: (*a*) the kinetic energy, (*b*) the potential energy, (*c*) the total energy, (*d*) the energy required to remove the electron completely.

2. At what speed must the electron revolve around the nucleus of a hydrogen atom in order that it may not be pulled into the nucleus by electrostatic attraction? Also calculate its orbital energy and frequency
3. Find the number of electrons in the atoms which have the following levels filled in the ground state
(*a*) *K* and *L* shells, the 3*s* subshell and one half of the 3*p* subshell
(*b*) The *K*, *L*, and *M* shells and the 4*s*, 4*p* and 4*d* subshells.
4. Calculate for the hydrogen atom (*a*) the velocity of electron in the ground state (*b*) radius of the orbit in the ground state, and (*c*) time taken by the electron to traverse the first Bohr's orbit.
5. The ultraviolet wavelength of 800 Å and 700 Å is allowed to fall on hydrogen atom in the ground state. It is found to liberate electrons with kinetic energy 1.8 eV and 4.0 eV respectively. Find the value of Planck's constant.
6. The energy of an electron in the second and third Bohr's orbits of the hydrogen atom is -5.42×10^{-19} joule and -2.41×10^{-19} joule respectively. Calculate the wavelength of the emitted radiation when the electron drops from the third to the second orbit.
7. How many revolutions does an electron make in the $n = 2$ state of hydrogen atom before dropping to $n = 1$ state. The average life time of an excited state is about 10^{-8} *s*. Assume the required constants.
8. Calculate the frequency of radiation and the wave number when the electron jumps from third orbit to the second orbit in a hydrogen atom.
9. Calculate the lattice constant and atomic radius of aluminium whose specific gravity is 2.7. Assume any suitable data, if required.
10. Determine the density of CsCl crystal from radii of ions givens $Cs^+ = 1.65$ Å, $Cl^- = 1.81$ Å. Assume any data suitably if needed.
11. Volume of a FCC unit cell is 67.42×10^{-30} m^3. Calculate the atomic diameter of its atom. Guess as which metal it can be. Determine the number of unit cells in 2 mm^3 volume of this metal.
12. Calculate the number of atoms in the unit cells of (*a*) calcium, (*b*) vanadium, and (*c*) manganese. Their densities are 1550,

6100 and 7440 kg/m^3; and lattice constant a = 6.04, 3.03 and 8.92 Å respectively. Assume any data if required. Comment on the values obtained, and the possible space lattices.

13. The atomic packing efficiency of a crystal is 68%, and the closest distance of approach between its neighbouring atoms is 2.62 Å. Calculate the density of the crystal if the mass of its one atom is 50.04 amu.
14. Calculate the specific gravity of gallium arsenide (GaAs) having FCC structure and lattice constant 5.65 Å. The atomic weights of Ga and As are 69.7 and 74.9 respectively.
15. Determine the density of BCC iron (Atomic mass = 55.847 g/mol), which has a lattice parameter of 0.2866 nm.
16. The centre distance between the atoms in the basal plane of zinc unit cell is 2.66 Å. Calculate (*a*) the effective number of atoms per unit cell, (*b*) the volume of the unit cell, and (*c*) the density of zinc. The height of the hexagonal unit cell is 4.935 Å.
17. The dimensions of an orthorhombic unit cell are 2.3 Å, 4.6 Å and 6.9 Å. Obtain the intercepts of a plane of Miller indices (120).
18. Determine the Miller indices of a plane that makes an intercept of 2 Å, 3 Å and 4 Å on the coordinate axes of an orthorhombic crystal with $a : b : c = 4 : 3 : 2$. Draw it.
19. Calculate the Miller indices of a plane passing through the orthorhombic gallium crystal, and intersecting the three coordinate axes along *a, b* and *c* at 1.763, 8.040 and 1.915 Å respectively. The lattice constants for gallium are a = 3.526 Å, b = 4.020 Å, and c = 7.660 Å.
20. Show that (*a*) the ratio of $d^{(111)}/\, d^{(200)}$ in lead (atomic radius 1.746 Å) is 1.156, (*b*) the ratio of number of atoms/mm^2 in (111) plane over (100) plane is 1.158.
21. Calculate the linear atomic density in an aluminium unit cell along direction [110].The lattice constant is 0.4049 nm.
22. There are 3.82×10^{12} atoms/mm^2 on certain plane in a strontium (Sr) FCC unit cell. Its lattice constant is 6.08 Å. Determine the Miller indices of the plane in question. Consider (100), (110) and (111) planes only.
23. Calculate the planer density of atoms on (111) plane in the above question.
24. Calculate the number of atoms along <110> directions, and per unit area of {111} planes, in a DC germanium unit cell

of lattice constant 5.66 Å. Which of the planes (110) or (111) will contain more atoms per square centimetre?

25. The number of atoms per square millimetre on (110) plane of a BCC crystal is 9.8×10^{12}. Calculate the atomic radius of the atom of this crystal. Show that the number of atoms on (100) plane is 63% of that on (110) plane.
26. Calculate the spacing between (*a*) (111) (*b*) (110) and (*c*) (100) plane, in a crystal of NaCl whose lattice parameter is 5.64 Å.
27. Obtain the number of atoms per unit area on (100), (110) and (111) planes of copper. Take its lattice constant as 3.61 Å.

Objective Questions

1. An unit cell is
   a. the smallest group of atoms which when regularly repeated forms the crystal.
   b. the basic building block of a crystal.
   c. a cube containing the largest number of atoms.
   d. an agglomerated structure.
2. In a simple cubic structure, are the direction [100] and plane (100) parallel?
   a. no b. yes
   c. may be yes d. can't be predicted
3. The crystal structure of most of the common metals are
   a. cubic b. rhombohedral
   c. tetragonal d. triclinic
4. The Miller indices of a plane are proportional to
   a. the reciprocal of numerical parameters of the intercepts.
   b. the square of unit cell dimensions.
   c. the intercepts of the planes on the coordinate axes.
   d. interplaner spacing.
5. A crystalline solid has all similar atoms relative to neighbouring atoms in
   a. periodical positions b. reversed positions
   c. random fluctuating position d. similar positions
6. The [110] direction in a cubic unit cell is parallel to
   a. face diagonal of unit cell b. edge of the cube
   c. body diagonal of the cube d. none of the above

7. Miller indices of the diagonal plane of a cube are
   a. (200) b. (111) c. (010) d. (110)
8. If Z is the atomic number of a multi-electron atom, the energy of an electron in the Bohr's atomic model is proportional to
   a. Z^2 b. $1/Z$ (c) Z d. $1/Z^2$
9. Consider the spectral line resulting from the transition $n = 2 \rightarrow n = 1$ in the atoms and ions given below. The shortest wavelength is produced by
   a. hydrogen atom b. deuterium atom
   c. singly ionized helium d. doubly ionized lithium
10. In Bohr model of hydrogen atom,
    a. the radius of the nth orbit is proportional to n
    b. the total energy of electron in nth orbit is proportional to n
    c. the angular momentum of the electron in an orbit is an integral multiple of Planck's constant
    d. the magnitude of the potential energy of an electron in any orbit is greater than its kinetic energy
11. The maximum number of electrons in the nth main energy level of an atom is
    a. $\sqrt{2}n$ b. $2n^2$ c. $2n$ d. $2/n$
12. If l is the number of subsell in the main shell, the maximum number of electrons in this subshell is
    a. $2l(2l + 1)$ b. $2l$ c. $2(2l + 1)$ d. $4l$
13. Which of the following atomic orbital do not follow quantum theory?
    a. $4f$ b. $3f$ c. $6s$ d. $3p$.

3

Solid Structures, Characterization of Materials, Crystal Imperfections, and Mechanical Properties of Materials

3.1 Crystallography

The study of geometry and structure of solids is a science known as *crystallography*. X-rays and microscopes have substantially contributed to this area. The atoms and molecules are held together by chemical bonds in the solids. The orientation, periodicity, and symmetries of atoms and molecules vary widely. The arrangement of atoms may be regular as in metals or irregular as in rubber; closely packed as in metals or loosely packed as in non-metals. In this way, the solids may be grouped as under :

1. Crystalline solids
2. Non-crystalline or amorphous solids.

Structure of metals are generally crystalline. Non-metals such as plastics, rubber, ceramic, wood, organic fibres and glass etc. have non-crystalline structure in normal conditions.

- A metal may behave as non-crystalline, and a non-metal can exhibit crystalline nature under specific conditions.
- Silica, a non-metal, takes on the crystalline form as *quartz* and non-crystalline form such as *silica glass.*
- The high density polyethylene (HDPE), another non-metal, is almost crystalline due to its long aligned molecular chains.

3.2 Crystalline and Non-Crystalline Structures

As already discussed, a crystalline form of solid has periodically repeated arrangement of atoms. But the solids in non-crystalline form do not have long range periodic repetition. However, in both of the above forms, the coordination number is almost the same. Formation of non-crystalline structure is characterized by several factors enumerated below:

1. Non-formation of three-dimensional primary bond.
2. Formation of one-dimensional chain molecule.
3. Formation of two-dimensional sheet molecule.
4. Absence of primary bonds in all the directions.
5. Weak secondary bond.
6. Non-parallel, entangled chain configuration.
7. Open network of the atomic packing.

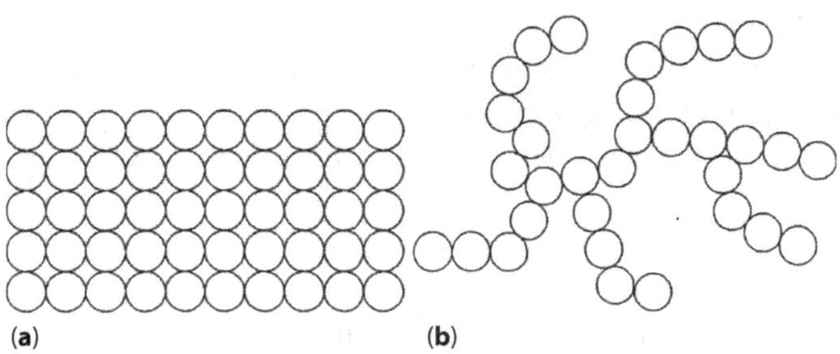

Figure 3.1 Structure of (a) crystalline solid, and (b) non-crystalline solid.

Figure 3.1 shows the difference in structures of crystalline and non-crystalline solids.

3.3 Hexagonally Closed Packed Structure (HCP)

This is also known as closed packed hexagonal (CPH) structure. Arrangement of atomic packing and sequential piling of atoms in HCP is of ...*ABAB*... form as shown in Fig. 3.2. We will study more about the geometry of HCP crystal now.

HCP structure is different from hexagonal structure. It is more denser than the hexagonal one. A HCP unit cell is shown in Fig. 3.2. There are total seventeen atoms in it. Six atoms are placed on each of the bottom and top hexagonal corners, one each on bottom and top hexagonal faces, and three atoms on vertical alternate planes. The atoms are closed packed, but are shown separated for clarity. The atoms in planes *A* are placed at lattice points but the atoms in plane *B* are not.

The geometrical details of HCP unit cell are given as follows :

• Effective number of lattice points	3
• Effective number of atom	6
• Basis (effective number of atoms/effective number of lattice points)	6/3=2
• Coordination number	12
• c/a ratio for a perfect crystal	1.633
• Atomic packing factor	0.74

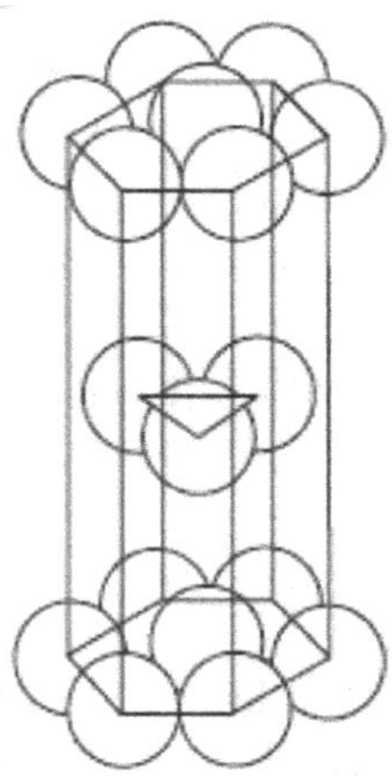

Figure 3.2 A hexagonally closed packed (HCP) structure.

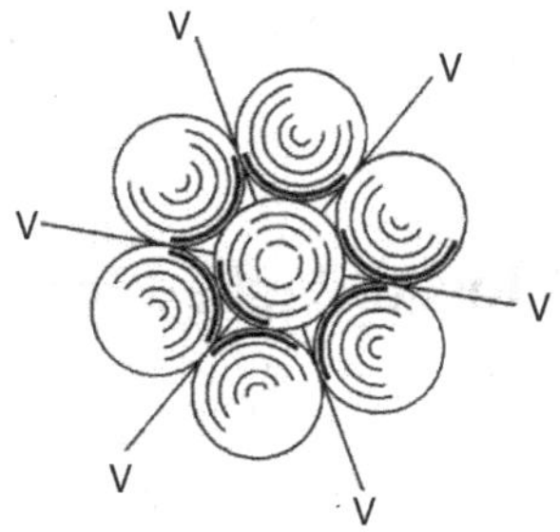

Figure 3.3 Interstitial voids.

3.4 VOIDS

The closed packing of atoms are redrawn in Fig. 3.3 showing the plan (top view). The space enclosed by atoms marked *V* are vacant. The vacant space between atoms is called *void.* These voids are also known as *interstitial voids.* Voids are of two different types.

1. Tetrahedral voids, and
2. Octahedral voids.

Depending on the geometry of above two kinds of voids, the space available in them can be computed. The atoms of other elements fit into these spaces during formation of an alloy. We shall now discuss details of these voids.

3.4.1 Tetrahedral Voids

A tetrahedral void will form in a four atoms system. Of these, three atoms *A*, *B* and *C*, Fig. 3.4*a* are placed on a closed packed plane while the fourth atom *D* on a plane above or below it. The fourth atom occupies cavity space formed between atoms *A*, *B* and *C*. If the centres of all four atoms are joined, a regular tetrahedral geometry, Fig. 3.4*b* will appear. There are eight tetrahedral voids in FCC. This makes two tetrahedral voids for each atom.

3.4.2 Octahedral Void

Formation of such voids requires at least six atoms of which three atoms *A*, *B* and *C* are on a closed packed plane, and other three atoms *D*, *E* and

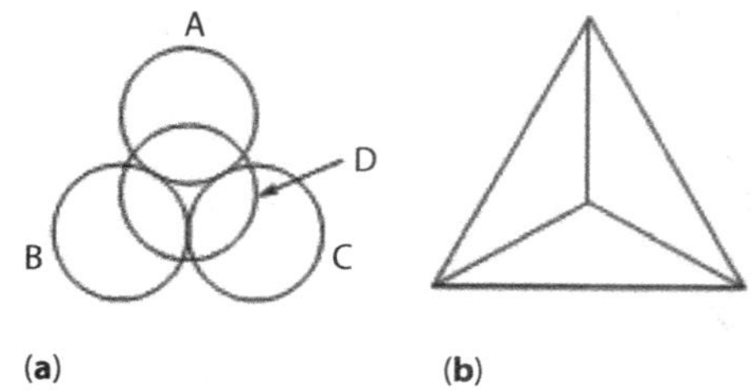

Figure 3.4 Tetrahedral geometry (a) arrangement of atoms in plan, and (b) in elevation

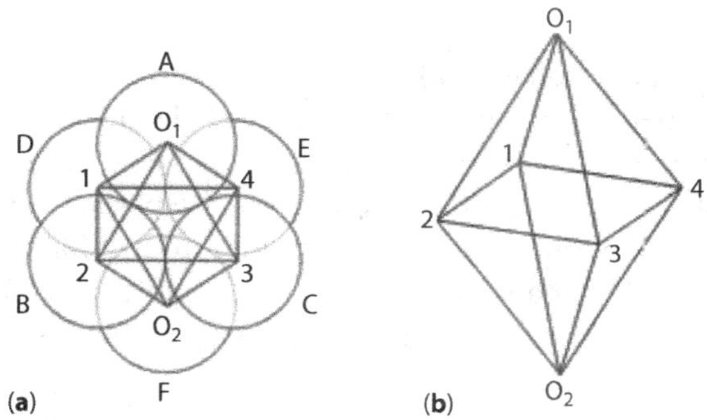

Figure 3.5 (a) Octahedral void in a closed packed structure, and (b) actual shape of octahedral void

F are above or below these atoms as shown in, Fig. 3.5*a*. When centres of all six atoms are joined, a regular octahedron shape emerges. This shape is marked O_1 1234 on one side and O_2 1234 on the other side. Figure 3.5*b* showing octahedron shape is drawn for more clarity. Octahedron is a type of polyhedron having eight equal and equilateral triangles with six corners. In FCC, effective number of octahedral voids per unit cell is 4.

3.5 Covalent Solids

3.5.1 Diamond Cubic (DC) Structure

Carbon exists in two forms viz. diamond and graphite. Both have quite different characteristics and properties. Diamond has (sp^3) hybrid covalent bond. Each of its atoms has four bonds. The bonds are directional in nature. The bonds are primary in nature and extend in a three-dimensional network.

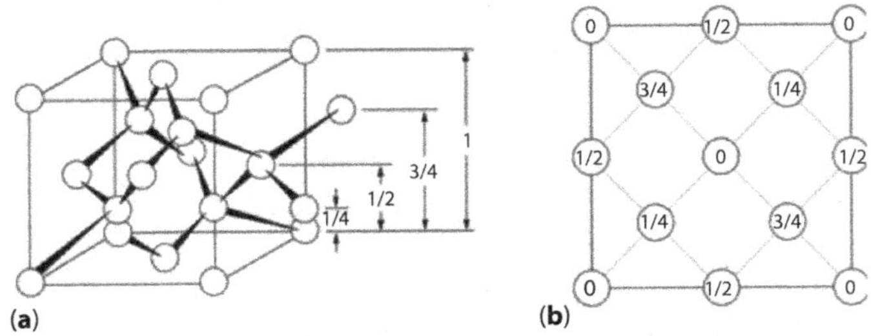

Figure 3.6 Structure of diamond (a) 3-dimensional view, and (b) top view.

Geometry. The directional bond angle in diamond is 109.5°. The structure of diamond is better known as *diamond cubic* (DC). Unit cell of a DC is shown in Fig. 3.6(*a*). It contains a total of 18 atoms. Of these, 8 are placed on corners of the cube, one on each of the six faces and four completely inside. The two inside atoms are placed at three-fourth (3/4) and other two at one-fourth (1/4) distance above the base when the height of unit cell is unity i.e. 1. These four atoms are placed on body diagonal, and are not the lattice points. An alternative way to depict the DC structure is shown in Fig. 3.6(*b*). It shows the top view in which the location of each atom is marked by the numerical value written thereupon.

Diamond is the hardest known solid. Its hardness on Moh's scale is 10. Other geometrical details are given below:

• Number of atoms per lattice point	2
• Basis	2
• Distance of separation between two atoms	$a\sqrt{3}/4$
• Atomic packing factor	0.34
• Effective number of atoms per unit cell	8
• Specific gravity	3.5

Besides diamond, the DC structure also exists in silicon, germanium, grey tin etc.

CHARACTERIZATION

3.6 Bragg's Law of X-Rays Diffraction

Father and son, W.H. Bragg and L. Bragg, devised a X-ray spectrometer, and used a crystal as reflecting grating for its study. The crystal was mounted on a turntable. A beam of X-rays was made incident on this crystal. Beam of incident X-rays, at a glancing angle θ, were scattered from the crystal planes which were rich in atoms. The X-rays were scattered by each individual atom lying on a parallel plane. This occurrence is known as *X-rays diffraction*. The Bragg's law of X-rays diffraction is explained as follows:

3.6.1 Bragg's Equation

A crystal contains millions of atoms. They produce combined scattering i.e. reflection of X-rays from these planes. The system of incident X-rays, glancing angle and the planes in a crystal is shown in Fig. 3.7. Here *A*, *B*, *C*, etc., are the atoms. Ray number 1 reflects from atom *A*, ray number 2 from atom *B*, and so on. Path difference between the two rays will decide whether the reflected rays are in phase or not. The phase difference may be found by drawing perpendiculars *AD* and *AE* as shown. It is obvious that the extra distance travelled by ray number 2 is equal to

$$= DB + BE = d \sin \theta + d \sin \theta = 2d \sin \theta$$

where *d* is the interplaner spacing.

If the path difference between two reflected rays is equal to an integral multiple of wavelength λ, the rays will be in phase. Hence condition of Bragg's diffraction gives

$$2d \sin\theta = n\lambda \tag{3.1}$$

where $n = 1, 2, 3 \ldots$ is an integer. Order of reflections will be 1, 2 and 3 depending on the value of *n*. If $n = 1$, the reflection will be known as

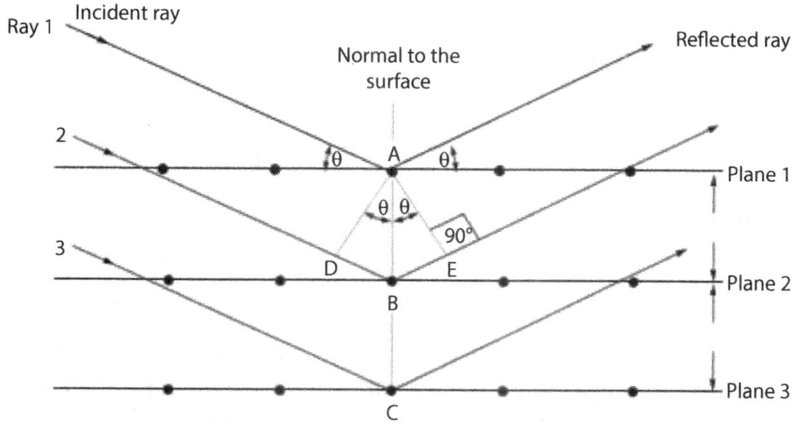

Figure 3.7 Bragg's law : X-rays diffraction.

first order; for $n = 2$, it will be called second order reflection; and so on. Equation 3.1 is known as Bragg's equation.

3.6.2 Reflections from Various Sets

X-rays of different wavelengths can be used to get reflections from various set of planes. As sin θ cannot be more than unity i.e. 1, hence λ is governed by $2d = \lambda$ for $n = 1$. So interplaner spacing of 3 Å gives upper limit of λ as 6 Å. If $\lambda > 6$ Å, there will be no reflection. Normally K_α radiation of molybdenum target is used in diffraction studies as its wavelength 0.71 Å lies within a reasonable range of interatomic spacing in solids.

3.7 Structure Determination

Lattice constants of a cubic crystal can be obtained using Bragg's Eq. 3.1. On considering (100) plane and $n = 2$ and 3, we get

$$2\lambda = 2d_{100} \sin \theta$$

$$\therefore \qquad \lambda = d_{100} \sin \theta \tag{3.2a}$$

and

$$3\lambda = 2d_{100} \sin \theta \tag{3.2b}$$

Now consider (200) plane and $n = 1$. We get

$$\lambda = 2d_{200} \sin \theta \tag{3.3}$$

Similarly for plane (300) and $n = 1$, we have

$$\lambda = 2d_{300} \sin\theta \tag{3.4}$$

On comparing Eqs. 3.2*a* and 3.3, we find that

$$d_{100} \sin \theta = 2d_{200} \sin \theta \tag{3.5}$$

Thus, from Eq. 3.5 it can be shown that $d_{100} = 2d_{200}$. Therefore, its substitution in Eq. 3.5 suggests that Eqs. 3.2*a* and 3.3 are identical. Infact, sets of above two planes are parallel, therefore their reflections superimpose each other. Similarly Eqs. 3.2*b* and 3.4 are identical as $d_{100} = 3d_{300}$. Hence third order reflections from (100) planes and first order reflections from (300) planes superimpose each other.

Example 3.1. X-rays beam of 0.824 Å wavelength is incident on a crystal at glancing angle of 8°35′ when the first order diffraction occurs. Determine the glancing angle for third order diffraction, and the interplaner spacing of the crystal.

Solution : Given are: $\lambda = 0.824$ Å, $\theta_1 = 8°35'$, $n_1 = 1$, and $n_3 = 3$.

Since $$2d \sin \theta = n\lambda$$

∴ $$2d \sin \theta_1 = 1\lambda,$$

and $$2d \sin \theta_3 = 3\lambda,$$

Therefore $$\frac{\sin \theta_3}{\sin \theta_1} = \frac{3}{1}$$

∴ $$\sin \theta_3 = 3 \sin \theta_1 = 3 \sin 8°35'$$

or $$\theta_3 = 26°36'$$

From Eq. (*i*), we get

$$d = \frac{\lambda}{2 \sin \theta_1} = \frac{0.824}{2 \sin 8°35'}$$

$$= 2.761 \text{ Å}$$

MICROSCOPES AND MICROSCOPIC METHODS

3.8 Microscopy

The closest distance of separation visible to the naked human eyes is 0.1 mm. Inside details of the solids, crystals, unit cells, atoms, electrons, and imperfections demand for the visibility of much smaller dimensions. The smaller dimensions may be as small as 1 Å or less. These situations need a magnified vision. This is possible through a microscope. Microscopes of various magnification ratio are used according to the need. Their magnification generally varies between 5 to 1 000 000. Microscopes are broadly classified into following main categories.

Table 3.1 Magnification and Applications of Different Types of Microscopes

Type	Range of magnification	Used in study of
• Optical microscope	10 to 2000 × linear	Microstructure
• Electron microscope	100 000 × linear	Finer particles, dislocations
• Field ion microscope	upto 1 000 000 ×	Imperfections
• Scanning tunneling microscope	more than 1 000 000 ×	Atomic image

1. Optical microscope
2. Electron microscope
3. Field ion microscope
4. Scanning tunneling microscope

Table 3.1 indicates their magnification and areas of application.

Before conducting the metallographic studies, some pre-requisites are needed to be fulfilled. In this regard, preparation of test specimen is very essential and important.

3.8.1 Microscopic Principle

The microscopic examination is based on optical principle. In it the rays from a light source is passed-on to a glass reflector through a diffusing disk and an Iris diaphragm as shown in Fig. 3.8. The diffusing disk helps in diffusing the light, the Iris diaphragm controls the width of light beam, and the reflector kept at 45° partially reflects the light rays onto the sample (or object).

After illuminating the polished sample; the rays return by reflection, pass through the objective and glass reflector, and then form the image. This image can be seen through an eye-piece lens to get the view of the sample surface.

The eye-piece is carried by a 'draw tube' at its top end. The draw tube can slide within the 'body tube' of microscope through a rack and pinion mechanism, on rotating the coarse and fine adjusting knobs. By doing so the distance between objective and the eye-piece can be varied for focussing the object. The coarse adjustment is done for initial focussing and the fine adjustment for final focussing.

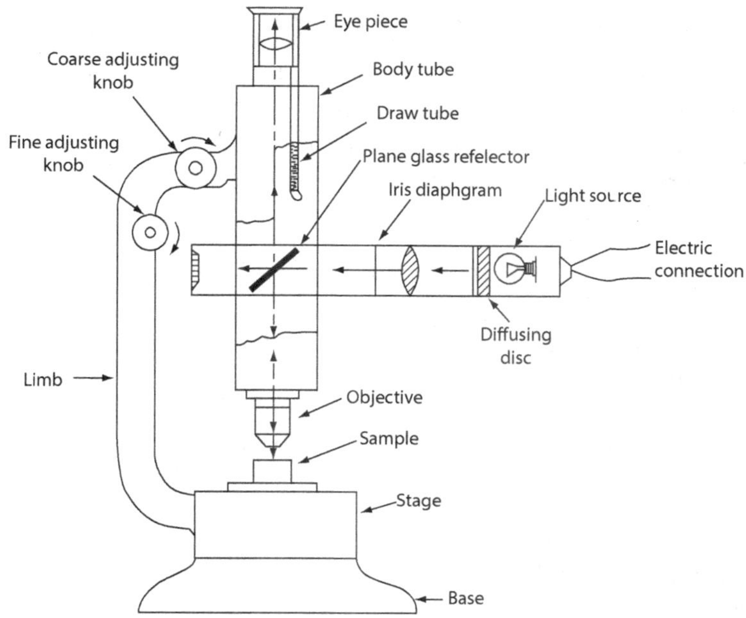

Figure 3.8 A metallurgical microscope illustrating the principle of microscopy.

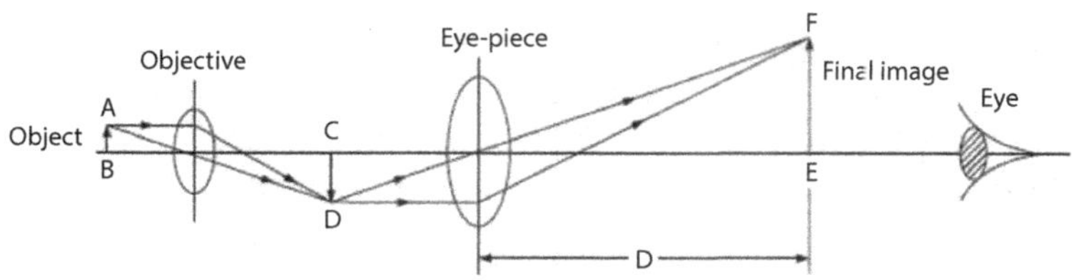

Figure 3.9 Schematic arrangement of a (optical) metallurgical microscope to explain the principle of magnification

3.8.2 Ray Diagram and Principle of Magnification

In a metallurgical microscope the *objective* is provided for resolving the sample structure whereas, the *eye-piece* enlarges the image formed by 'objective'. Figure 3.9 depicts a schematic arrangement of an objective and eye-piece to explain the formation of magnified image in a metallurgical microscope. The travel path of light rays is also shown in this diagram which is self-explanatory. Its optical combinations may bring-out a magnification of 50 × to 1500 ×.

3.8.3 Magnifying Power of Microscope

Magnifying power of a microscope is its ability to enhance the size of real object many times. In an optical microscope, the objective and the eye piece both magnify the real object. Therefore, magnifying power of objective is different from the magnifying power of eye piece. Objectives are generally available in magnifying powers of

$$5\times,\ 10\times,\ 40\times,\ \text{and}\ 1000\times$$

where × denotes a linear magnification. Eye pieces are generally available in magnifying powers of

$$5\times,\ 6\times,\ 7.5\times,\ 10\times,\ 15\times,\ 20\times,\ \text{and}\ 25\times$$

In Fig. 3.9, *AB* is an object whose magnified image is *CD* due to objective lens. The image of *CD* is further magnified to *EF* by eye piece. The total magnification M_t is obtained by combining the magnifications of the objective and the eye piece. This is expressed by

$$M_t = M_o M_e D \times \tag{3.6}$$

where $M_o = CD/AB$ and $M_e = EF/CD$ are magnifications of objective and the eye piece respectively, and D is the projection distance. The projection distance is measured between the eye lens of eye piece and the screen on which the image is projected. Generally, the objectives are designed for use at a definite length D. Hence Eq. 3.6 modifies to

$$M_t = \frac{M_o M_e}{D} \times \tag{3.7}$$

The value of D = 250mm is very common.

3.9 Different Types of Metallurgical Microscopes and Their Features

Several kinds of metallurgical microscopes are employed to examine the prepared samples. Depending upon the layout of viewing head, type of construction, purpose served and other versatile features; these may be classified into following main types:

1. Monocular, Binocular, or Trinocular head type
2. Straight, Inclined, or Inverted head type

3. Stationary or Portable type
4. Dissecting or Polarising type
5. Stereoscopic or stereoscopic zoom type
6. Modular or Profile projection type
7. Research or Fluorescence type

Amongst these a *binocular* microscope provides better vision through both the eyes.

- A *trinocular* microscope provides the facility of extra tube for simultaneous microphotography.
- The *inclined head* type microscopes are convenient for prolonged observations.
- *Portable microscopes* are suitable for field work.
- The *zoom stereoscopic* microscope enables to view small things with both eyes at high magnifications, which can be easily and quickly changed also by zoom optical system.
- A *polarising* microscope is used to view the objects in polarised light.
- A *research* microscope is equipped with most advanced systems and is used for highly minute observations.
- A *fluorescence* microscope is used for study in transmitted light.

All these microscopes are equipped with various attachments and accessories for versatility and additional features. Some of the normal features incorporated in them are the following:

i. Eye-piece
ii. Viewing head
iii. Objectives
iv. Objective changer
v. Stages for *x-y* movement
vi. Illuminator
vii. Focussing movement, etc.

More sophisticated microscopes include the following facilities also.

- Phase contrast and dark field system
- B & W and CC (closed circuit) TV system
- Microphotographic system
- Computerised inspection system.

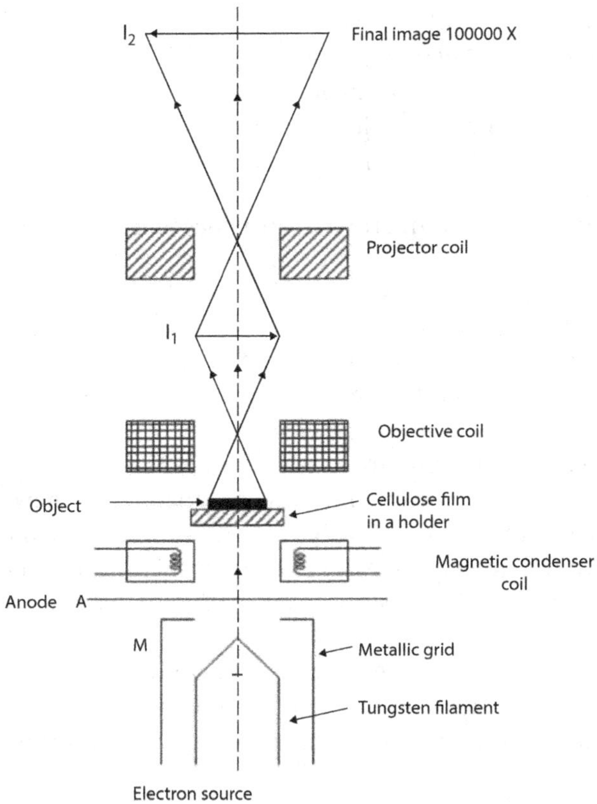

Figure 3.10 Layout of an electron microscope.

3.10 Working Principle of Electron Microscope

Schematic layout. This microscope is a very useful tool in crystallographic research. It can produce photographic images and diffraction patterns. It employs much shorter wavelength ($\approx 1/20$ of X-rays). Therefore, it results in better resolution of photographic image than those through an optical microscope. Schematic layout of an electron microscope is shown in Fig. 3.10. It has much higher power of resolution (PR). The object is illuminated by a beam of electrons. It employs magnetic focussing.

3.10.1 Formation of Magnified Image

In its operation, the tungsten filament T is thermionically heated. Due to this the electrons are emitted which are collimated by metallic grid M. The collimated electron beam is accelerated by anode A to a potential of about

5000 volts. Accelerated beam is then focussed on the object by magnetic condenser coil. The object is placed on a cellulose film held in a holder. The incident electron beam on the object scans it, and then the objective coil produces a magnified image I_1 of the object.

This image I_1 acts as an object for projector coil which magnifies it to image I_2. Final image I_2 may be seen on a fluorescent screen or can be photographed on a photographic plate. Whole system is placed in a metal casing which is evacuated to produce vacuum. Arrangements of keeping and removing the object and its adjustment are incorporated in the microscope.

Uses. Electron microscope may be used in the fields of medicine and biology to study bacteria and virus, in colloidal solutions to examine minute particles, in textile industry to study structure of fibres, and in industries like paper, paints, plastics, lubricants and metals.

CRYSTAL IMPERFECTIONS

3.11 Ideal and Real Crystals, and Imperfections

Till now, we have discussed crystal structures and bonding. We have also described crystalline and non-crystalline solids. Crystals discussed upto now are perfect (ideal) crystals. They are perfectly ordered arrangement of atoms at the points of infinite space lattice. Any departure from this idealized arrangement makes the crystals imperfect. In general, crystalline materials have *real crystals.* They are of finite dimensions. They have broken bonds at the boundaries where bonding forces remain unbalanced. Thus the boundary of a crystal is a defect in itself.

3.11.1 Disadvantageous Effects of Imperfections

These imperfections lead to several deficiencies in the solids. Solids are generally imperfect. A perfect single crystal is rarely available. Polycrystalline solids are, inevitably, imperfect. We shall deal with all such imperfections (defects) of crystalline solids in this chapter.

- Presence of imperfections in crystalline materials is generally undesired.
- They decrease the mechanical strength of materials.

- We have already discussed that a material does not attain its theoretical strength due to presence of imperfections.

3.11.2 Advantageous Effect of Imperfection

Imperfections affect structure-sensitive properties of crystals. Their presence is advantageous in certain applications.

- Parts per million (ppm) doping of phosphorus in silicon changes the behaviour of intrinsic semiconductor and makes it suitable for various applications.

3.12 Classification of Imperfections

Imperfections in crystalline solids are classified as follows.

S. No.	Main types	Sub-types
1.	Point imperfections (zero-dimensional defects)	Vacancy, substitutional impurity, interstitial impurity, frenkel's defect, schottky's defect.
2.	Line imperfections (1-dimensional defects)	Edge dislocation, screw dislocation, mixed dislocation.
3.	Surface or planer imperfections (2-dimensional defects)	Grain boundary, twin or twinning, low angle boundary, high angle boundary, twist boundary, stacking fault, interphase.
4.	Volume imperfections (3-dimensional defects)	Pores, foreign particle inclusions, non-compatibility regions, dissimilar natured regions.

Point imperfections, line and surface imperfections may occur together in crystals. These defects are not visible to the naked eye. They can be visualized by using X-rays diffraction techniques and microscopes, already described in chapter 1. Imperfections are also classified on the basis of their dimensions, such as given below.

1. Nano-level (10^{-9} m) imperfections
2. Angstrom level (10^{-10} m) imperfections
3. Micro-level (10^{-6} m) imperfections

We shall now deal with each type of imperfection one by one.

3.13 Point Imperfections

As the name suggests, they are imperfect point-like regions in the crystal. These defects are of one or two atomic diameters only. Hence, these are known as zero-dimensional defects. Various kinds of point imperfections are discussed below.

3.13.1 Vacancy

Vacancy refers to a vacant atomic site in a crystal. At these sites the atoms are missing. One or more atoms may remain absent from their respective locations. The missing of atoms is random and not according to any rule. As shown in Fig. 3.11*a*, the atoms *A* and *B* are missing from a FCC unit cell. It is not necessary that atoms from *A* and *B* sites only will abstain when the unit cell repeats to form a crystal. Figure 3.11*b* shows a simple cubic (SC) crystal. Sites marked *V* are the vacant atomic sites. In above figures, the atoms are shown separated for clarity.

3.13.2 Substitutional Impurity

This defect refers to a foreign atom that substitutes a parent atom at its site in the crystal. Atoms marked *A* in Figs. 3.12 *a-b* are the foreign atoms. The substituting foreign atoms are called *solute* and the substituted (or dislodged) parent atoms are known as the *solvent*. Solute and solvent of comparable sizes mix randomly to form an alloy. For example, copper and zinc mix together to form alpha-brass. Boron or antimony doped in germanium is another example of substituted impurities in a crystal.

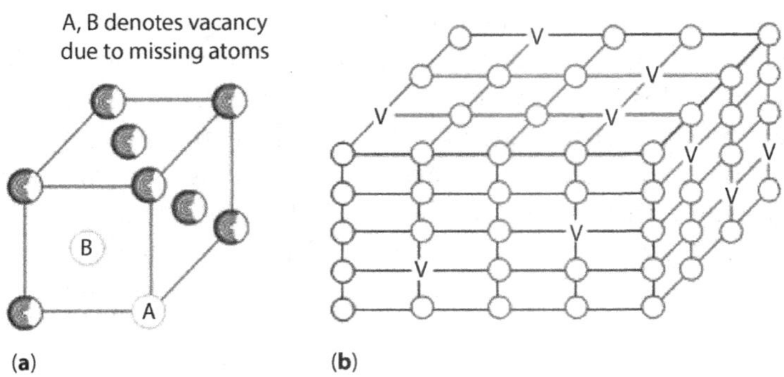

Figure 3.11 Point imperfection (a) vacancy in FCC unit cell, and (b) vacancy in SC crystal

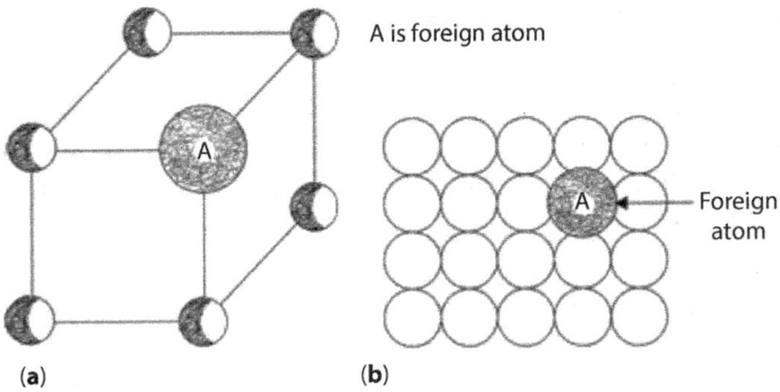

Figure 3.12 Substitutional impurity (a) in SC unit cell, and (b) on front face of a crystal.

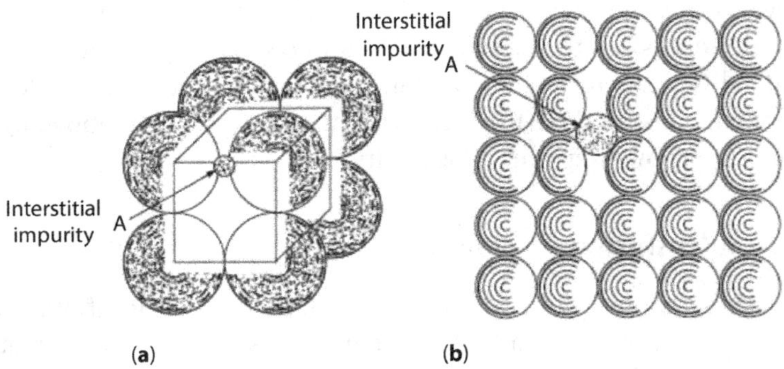

Figure 3.13 Interstitial impurity (a) in a SC unit cell, and (b) on front face of a crystal.

3.13.3 Interstitial Impurity

When a small sized foreign atom occupies a void space in the parent crystal (or its unit cell), the defect is known as interstitial impurity. Atoms marked *A* in Figs. 3.13 *a-b* are the interstitial atoms. Dislodging of parent atoms from their sites does not occur in this case. However, they can squeeze due to forced entry of a foreign atom or even the parent atom.

An atom can enter into the interstitial void when it is quite smaller in size than the parent atom. We have already discussed in chapter 2 that the largest atom that can fit into the tetrahedral and octahedral voids has radius of 0.225 *r* and 0.414 *r* respectively. Carbon, an interstitial solute in FCC iron (between 910°C and 1410°C), is an example of this kind.

3.13.4 Frenkel's Defect

An ion, displaced from a regular location to an interstitial location, in an ionic solid is called Frenkel's defect, Fig. 3.14 *a*. The ions of two different

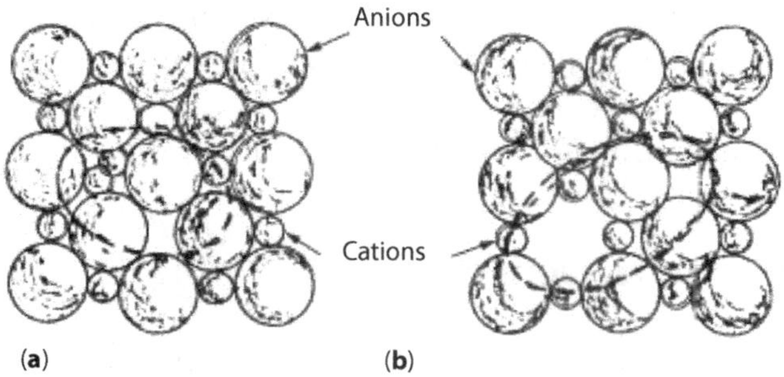

Figure 3.14 (a) Frenkel's defect in an ionic crystal shows presence of a cation elsewhere within the crystal, and (b) Schottky's defect in an ionic crystal shows a missing pair of anion and cation

kinds are known as *cations* and *anions*. Cations are the smaller ions while the anions are the larger ones. Cations may easily get displaced into the void. Anions, on account of their larger size, do not displace in small sized voids. The presence of this defect does not change the overall electrical neutrality of the crystal. Imperfections in CaF_2 and silver halides are the examples of this kind.

3.13.5 Schottky's Defect

When a pair of one cation and one anion are absent from an ionic crystal, Fig. 3.14*b*, the defect is called Schottky's defect. The valancy of missing pair of ions maintains electrical neutrality in the crystal. Such imperfections are dominant in alkali halides such as LiCl, LiBr, LiI etc.

3.14 Effects of Point Imperfections

Presence of point imperfections induce distortions in their surroundings. Consequently, various point imperfections have the following effects on crystals and their properties.

1. When the imperfection is a vacancy, the bonds with its neighbouring atoms do not exist.
2. In case of substitutional impurity, elastic strains develop in the surrounding region due to size difference of parent and foreign atoms. A larger foreign atom induces compressive stress and strain, while a smaller atom produces stress and strain field of tensile nature.
3. An interstitial atom creates strains around its surrounding.

4. Point imperfections of different types interact with each other and in doing so lower the total energy. Consequently, the stability of crystals is affected.
5. These are thermodynamically stable.

3.15 Line Imperfections

Line imperfections are called *dislocations.* The two primary types of dislocations are:

1. Edge dislocation, and
2. Screw dislocation.

3.16 Features of Edge Dislocation

To understand an edge dislocation, let us first consider a perfect crystal as shown in Fig. 3.15*a*. Atoms (not shown in diagram) are located on each cross-point in it. A perfect crystal is free from dislocations. Now consider the above crystal with an extra or incomplete plane of atoms *ABCD* as shown in Fig. 3.15*b*. The spacings of the atoms are slightly different in the portions above and below the plane S_1S_2 containing atom at B. This is because of the presence of dislocations. The line *BC* is called *dislocation edge.*

3.17 Screw Dislocation

Features of screw dislocation. Contrary to edge dislocation, there is no extra plane in the case of screw dislocation in a crystal. Screw dislocation forms when a part of the crystal displaces angularly over the remaining part. The plane of atoms converts into a helical surface, or a screw. The angular displacement is similar to the movement of a screw when turned. The designation 'screw' is derived from this feature. Screw dislocations are symbolically represented by ↶· and ·↷, and are referred to as 'clockwise' and 'anticlockwise', or 'positive' and 'negative' screw dislocations.

3.17.1 Stress-Strain Field in Screw Dislocation

Model of a crystal having screw dislocation is shown in Fig. 3.16. Part of it *MNB′C* has displaced over part *FGHJ* from right to left direction. Displacement has occurred along the slip plane *CDOH.* Vertical edge

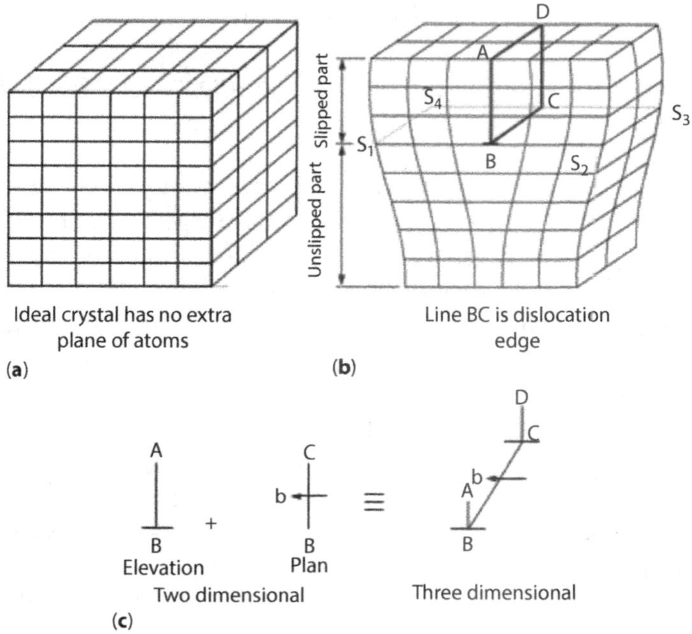

Figure 3.15 (a) Perfect crystal model, (b) crystal model with edge dislocation, and (c) representation of a dislocation

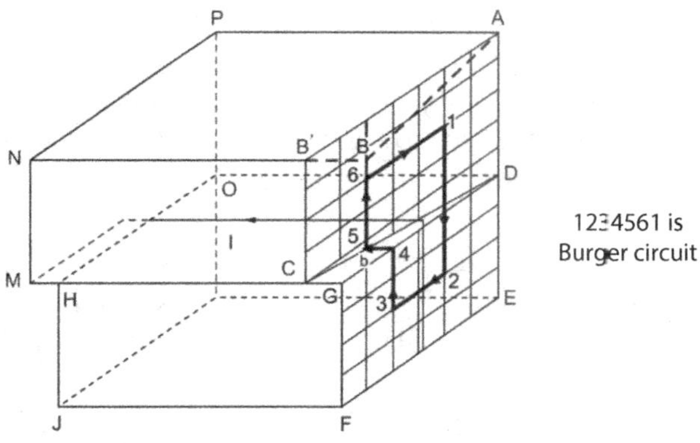

Figure 3.16 Screw dislocation model.

ADE is uneffected by this shifting. Originally the right hand side face *ADEFGB* has angularly deformed. Line *DG* has angularly displaced to *DC*, and edge *AB* has shifted to *AB′*. The screw dislocation line is marked in the direction of arrow *l*. The atomic bonds in the vicinity of dislocation line undergo through shear deformation. This gives rise to shear stress-shear strain field.

3.18 Characteristics of Dislocations

The properties and behaviour of dislocations are characterized by certain geometries. These are:

1. A crystal normally incorporates large number of dislocations. Hence, there exists numerous Burgers vectors. The sum of these Burgers vectors meeting at a point, called nodal point, inside the crystal remains zero.
2. A dislocation does not end abruptly within the crystal. It vanishes either at a nodal point or on the surface of the crystal.
3. A dislocation under the influence of stress-field may close on as a loop. The profile of the loop may be a circle or otherwise.
4. The distortional energy associated with dislocations may be the source of crystals instability. Distortional energy is produced due to tensile and compressive stress-strain field in edge dislocations, and due to shear stress-strain field in screw dislocations.
5. The elastic strain energy U per unit length of a dislocation is directly proportional to the square of Burgers vector **b**. It is

   given by $$U = \frac{\pi}{8} G b^2 \approx \frac{G b^2}{2} \tag{3.8}$$

   where G is shear modulus or modulus of rigidity of the crystal.
6. Dislocations may have Burgers vector of *full lattice translation* or partial lattice translation *i.e.*, the Burgers vector may be of b, $2b$, $3b$, ... and $b/2$, $b/3$, ... magnitudes.
7. Dislocations have inherent tendency to keep smallest possible Burgers vector. By doing so an enhanced stability is developed in the crystals. It can be understood by the following equations. A dislocation of Burgers vector of 3**b** magnitude tries to break into three dislocations each of **b** magnitude.

   $$3\text{b} \rightarrow \mathbf{b} + \mathbf{b} + \mathbf{b}$$

   By doing so it possesses elastic strain energy as

   $$\frac{G(3b)^2}{2} \neq \frac{G(b)^2}{2} + \frac{G(b)^2}{2} + \frac{G(b)^2}{2} \tag{3.9}$$

The amount of energy given by right hand side of above expression is much lower than that of left hand side term. Hence stability is increased.

8. Two edge dislocations of opposite sign $\perp$ and $\top$, of equal Burgers vector, and on the same slip plane cancel-out. This is because the distortional strain energy field superimpose and annihilate each other.
9. Edge dislocations travel much faster ($\approx$ 50 times) than screw dislocations.

3.18.1 Burgers Vectors of Dislocations in Cubic Crystals

Probable Burgers vectors of full dislocations in some cubic crystals are given in Table 3.2.

In NaCl crystal, the Burgers vector cannot be from the centre of a chlorine ion to the centre of a sodium ion. Similarly in CsCl crystal, the Burgers vector cannot be from a chlorine ion at body corner to cesium ion at body centre. It is because the Burgers vector in above cases will not give a full lattice translation. Burgers vector in ionic crystals are larger than those in metallic crystals.

3.19 Sources of Dislocations, Their Effects and Remedies

There are two main sources of dislocations in the solids. These are:

1. Mishandling during grain growth, and
2. Mechanical deformation.

Table 3.2 Burgers vectors of dislocations

• Monoatomic SC	<100>
• Monoatomic BCC	1/2 <111>
• Monoatomic FCC	1/2 <110>
• Diamond Cubic DC	1/2 <110>
• Ionic solid: NaCl structure : CsCl structure	1/2 <110> <110>

Crystals are obtained by the process of crystallization. In doing so, the molten metal is solidified. Recovery, recrystallization and grain growth are the essential features of material's manufacturing and processes. It is almost impossible to achieve perfection in the process of grain growth, and control on the size and orientation of grains. Process mishandling cannot be ruled out. That is why the dislocations invariably crop-in the solids.

Mechanical deformation is another source of dislocations. Crystals are subjected to various kinds of situations during manufacturing and fabrication in which they get deformed. Deformation may be geometrically linear, angular or complex. Due to these, the dislocations enter into the crystals.

3.19.1 Effects of Dislocations

Effects of dislocations are detrimental to the properties of crystalline materials.

- Their mechanical strengths lower down. Due to this, the structural and machine components become uneconomical.
- Dislocations are responsible for reduced electrical conduction.
- They also influence the surface-sensitive properties of materials negatively.
- Infact, almost all the surface related properties listed in art. 1.5 deteriorate mildly to drastically.

However, the density of the solid is negligibly affected.

3.19.2 Remedies to Minimize the Dislocations

Following remedies can be suggested to avoid or minimize the occurrence of dislocations.

1. Use of thermal energy.
2. More careful control on crystallization.
3. Prevention of undesired mechanical deformation.
4. Use of *whisker* form of material.

Dislocations can be minimized by use of thermal energy. On heating the crystals to a high temperature, many dislocations either annihilate among themselves or drive out of the crystal surfaces. Other way to remove dislocations will be to change the form of crystalline material. Normally the materials are in bulk form. If they could be produced as whiskers, dislocation density will definitely decrease immensely.

3.20 Grain Boundary

We have already described in earlier chapters that the polycrystalline solids consist of several crystals of different sizes oriented randomly with respect to each other. They grow during the process of recrystallization. Growth of crystals is a direct consequence of addition of atoms. If the number of atoms are more in a crystal, the size of crystal will be large. As shown in Fig. 3.17, the crystal marked *A* is smaller than the crystal marked *B* because of lesser number of atoms joining the crystal *A*.

3.20.1 Mechanism of grain boundary formation

The crystals grow randomly and in doing so impinge upon each other. When adjoining crystals impinge together, some atoms marked *C* are caught in between them. These atoms are forced to join one or the other crystal. But they do not join any crystal due to their opposing forces. These atoms occupy positions at the junction of adjoining crystals. Junction or the boundary region is distorted and behaves as noncrystalline material. This boundary region is the defect called *grain boundary.*

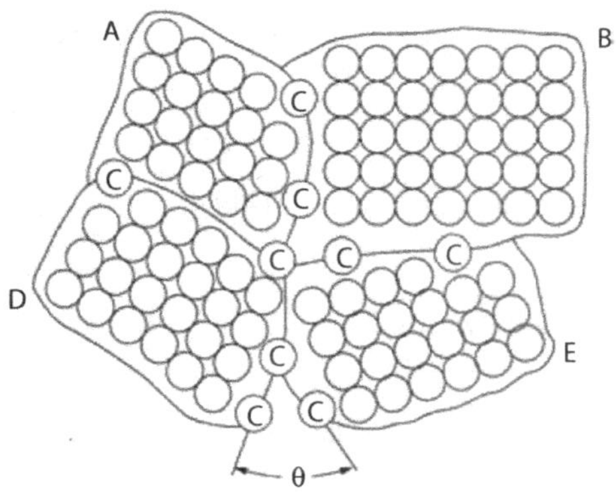

Figure 3.17 Grain boundary in polycrystalline material.

Lines joining various atoms *C* are grain boundaries, and θ is the grain orientation angle.

3.21 Twin or Twinning

This defect is also called *twin boundary.* As the name implies, twin boundaries occur in pairs. The arrangement of atoms is such that one side of twin boundary is a mirror replica of the other side. As shown in Fig. 3.18, side *A* is a mirror image of side *B*. The zone *CDEF* is known as *twinned zone. DE* and *FC* are twin boundaries.

3.21.1 Annealing Twin and Deformation Twin

Twins can form during the process of recrystallization or during plastic deformation of materials. In the first case it is called *annealing twin* and in the other as *deformation twin.* Occurrence of twins is common in brass and metallic sheets. Twins in crystalline solids can be visualized by an optical microscope.

The twin planes and twin directions in different crystal structures are given in Table 3.3.

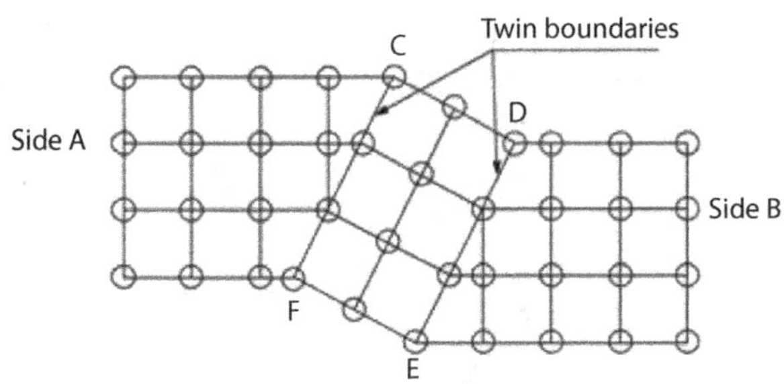

Figure 3.18 Twin and twin boundaries in a crystalline solid.

Table 3.3 Twin planes and twin directions in some crystals

Structure	Twin plane	Twin direction	Example
BCC	(112)	[111]	Fe, W, V
FCC	(111)	[112]	Cu, Al, Ag
HOP	$(10\bar{1}2)$	$[10\bar{1}\bar{1}]$	Mg, Ti, Zn

MECHANICAL PROPERTIES OF METALS MEANT FOR ELECTRICAL APPLICATIONS

3.22 Mechanical Properties of Metals

Various mechanical properties of metals are the following.

- Ductility
- Stiffness
- Strength
- Plasticity
- Deformability
- Creep
- Malleability
- Elasticity
- Resilience
- Weldability
- Castability
- Fatigue
- Hardness
- Toughness
- Formability
- Damping ability
- Proof stress
- Brittleness, etc.

3.22.1 Isotropic, Anisotropic and Orthotropic Materials

Isotropic material. In an *isotropic* material, all the planes are *planes of symmetry* and thus the properties are independent of direction. The number of independent elastic constants are only two in this case. Isotropic materials obey Hooke's law. All metals are isotropic.

Anisotropic material. In an *anisotropic* material, no plane is a plane of symmetry and thus the properties are direction dependent. The number of elastic constants in most general anisotropic case is 36, out of which 21 are independent. Anisotropic materials such as composites *obey generalized Hooke's law* which is different from Hooke's law.

Orthotropic material. *Orthotropic* material is a special kind of anisotropic material. The number of elastic constants are 12 out of which 9 are independent. An unidirectional (U/D) fibre composite is the example. It has three orthogonal planes of symmetry.

Isotropy Orthotropy, and Anisotropy. The nature of identical properties in all the directions is called *isotropy* while the dissimilar properties in all the directions is known as *anisotropy.* When properties are identical in some directions and dissimilar in others, the nature is called *orthotropy.*

3.22.2 Homogeneity and Heterogeneity

Definition of homogeneity. When the material properties do not change from point to point in a certain direction, the nature is termed as

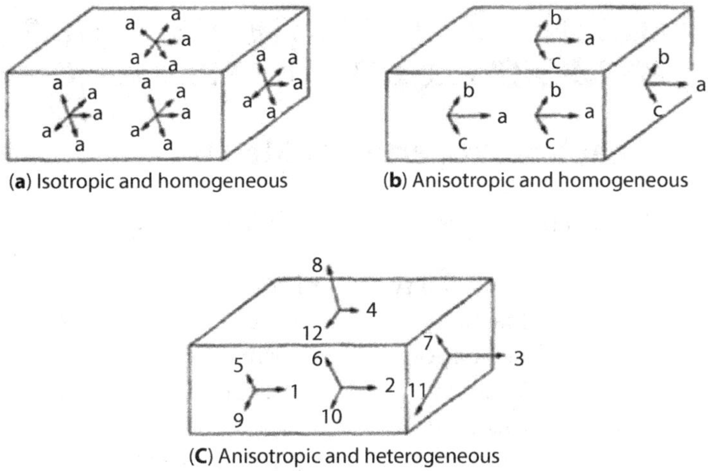

Figure 3.19 Explanation for homogeneous and heterogeneous natures.

homogeneity. It has already been discussed that when properties do not change with direction, it is isotropy.

Definition of heterogeneity. If the properties change from point to point in a certain direction, the nature is called *heterogeneity.* Figures 3.19a-c illustrate these materials. In these diagrams; markings a, *b* and c; and 1, 2, ... 12 indicate the elastic properties.

- Metals are isotropic and homogeneous.
- Composites of most general nature are anisotropic and heterogeneous
- Orthotropic materials with controlled anisotropy are homogeneous.

3.22.3 Strain Energy Absorbed by the Materials

A material subjected to a gradual tensile load, starting from 0 upto P elongates by δl and stores strain energy U in the form of stress σ and strain ε due to work done W. As the work done is equal to the strain energy absorbed,

$$W = U$$

therefore,

$$\left(\frac{0+P}{2}\right)\times \delta l = \frac{\sigma\varepsilon}{2}$$

where (0 + P)/2 is the average force.

$$U = \frac{\sigma^2}{2E} \times V \tag{3.10}$$

where V is volume of the material and E its Young's modulus.

Modulus of resilience. It is defined as the strain energy absorbed per unit volume when the material is stressed to its proportional limit. It is given by

$$U_r = \frac{\sigma_p \varepsilon_p}{2} = \frac{\sigma_p^2}{2E} \times V \tag{3.11}$$

where σ_p and ε_p are the stress and strain at limit of proportionality. Modulus of resilience decreases with increase in stiffness as $U_r \propto (1/E)$.

3.22.4 Strength

The strength of a material is its ability to sustain failure under the action of applied load. Depending on the type of loading, the strengths are termed as tensile strength, compressive strength, shear strength, flexural strength, or torsional strength of a material. The strength may be expressed as follows.

1. Elastic strength,
2. Ultimate strength,
3. Breaking strength, and
4. Specific strength.

Elastic strength is determined from the highest stress at which the behaviour of materials remains elastic. It is generally measured by the stress at the end of proportionality limit.

Ultimate strength is the maximum stress sustained by the material before fracture. Breaking strength is the highest stress attained by the material just at the time of breaking.

Specific strength is the strength per unit density of the material. It can be elastic specific strength (σ_A/ρ) or (σ_y/ρ), and ultimate specific strength (σ_u/ρ) where ρ is the density of material. A material with high specific strength is desired as it will result in light-weight construction in addition to considerable strength. Composite materials posses higher specific strength than the conventional materials, and hence are widely employed in aircrafts, satellites and space-going-vehicles.

3.22.5 Stiffness

The ability of a material to resist elastic deformation or deflection is called *stiffness*. It depends on the shape of the structural or machine member. For identical geometries of components, their stiffness is proportional to the elastic modulus of materials. In that case, elastic modulus may be termed as *modulus of stiffness*. However, stiffness should not be confused with elastic modulus. Stiffness of a member under different types of loading is obtained as below.

1. In axial loading, it is EA,
2. In flexural loading, it is EI, and
3. In torsional loading, it is GI_p

where A is area of cross-section, I the area moment of inertia, I_p polar moment of inertia, E the Young's modulus and G the shear modulus.

3.22.6 Resilience, Proof Resilience and Toughness

Resilience or *strain energy* is the capacity of a material to absorb energy within elastic limit. The maximum energy stored in a body upto elastic limit is termed *as proof resilience.* These are shown in Figs. 3.20a and *b* respectively in which point *B* indicates elastic limit. Toughness is different from resilience and proof resilience. Consideration made is beyond the elastic limit in its determination. *Toughness* is the strain energy absorbing capacity of a material in both elastic and plastic deformation upto fracture. It is shown in Fig.3.*20c*.

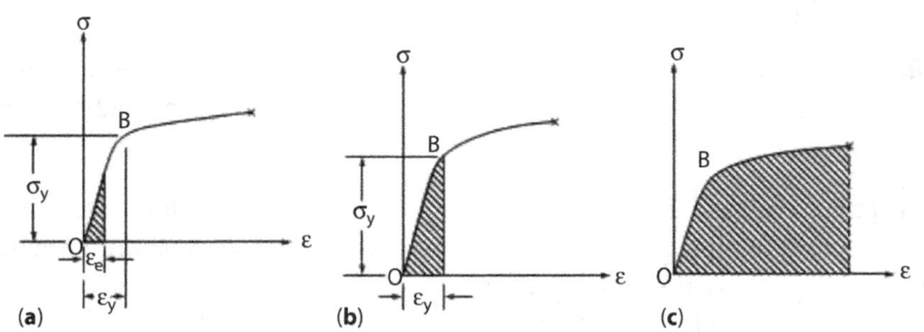

Figure 3.20 Hatched area indicates (a) resilience, (b) proof resilience, and (c) toughness.

3.22.7 Elasticity and Plasticity

Definition. Behaviour of a material by virtue of which the strains (deformations) disappear on removal of load, is known as *elasticity.* Ability of a material to undergo permanent deformation before rupture is called *plasticity.* Elasticity is characterized by elastic action, and plasticity by yielding, strain hardening and neck formation.

Salient features. Design of machine components or structures based on elastic theory employ elastic properties already discussed above. Plastic properties are considered in plastic design of structures and in mechanical parts at elevated temperatures.

Brittle materials show negligible plasticity while it is considerable in ductile materials.

- Plasticity increases with increase in temperature.
- Materials do not obey Hooke's law in plastic region. They follow the stress-strain relation given as

$$\sigma = E\,\varepsilon^{n} \tag{3.12}$$

where power index $n \neq 1$.

- It is different from the value of $n = 1$ which holds true for elastic region.

3.22.8 Ductility and Brittleness

Definitions. A material is said to be ductile if it elongates considerably under tension in plastic range. Total elongation is the sum of elastic and plastic elongations. Property of a material that enables it to be elongated is known as ductility. Ductility is associated with tensile loading.

A brittle material fractures under tension with negligible plastic elongation. Thus brittleness is the property that restricts elongation generally beyond the elastic limit.

Demarcating factor between ductile and brittle materials. The stress-strain diagrams of perfectly brittle, more and less ductile materials are shown in Figs. 3.21a, 3.21b and 3.21c respectively. According to a general classification, the ductile and brittle materials may be categorized as below on the basis of percentage plastic strain (% ε_p). Here $\varepsilon_e << \varepsilon_p$ and is recoverable.

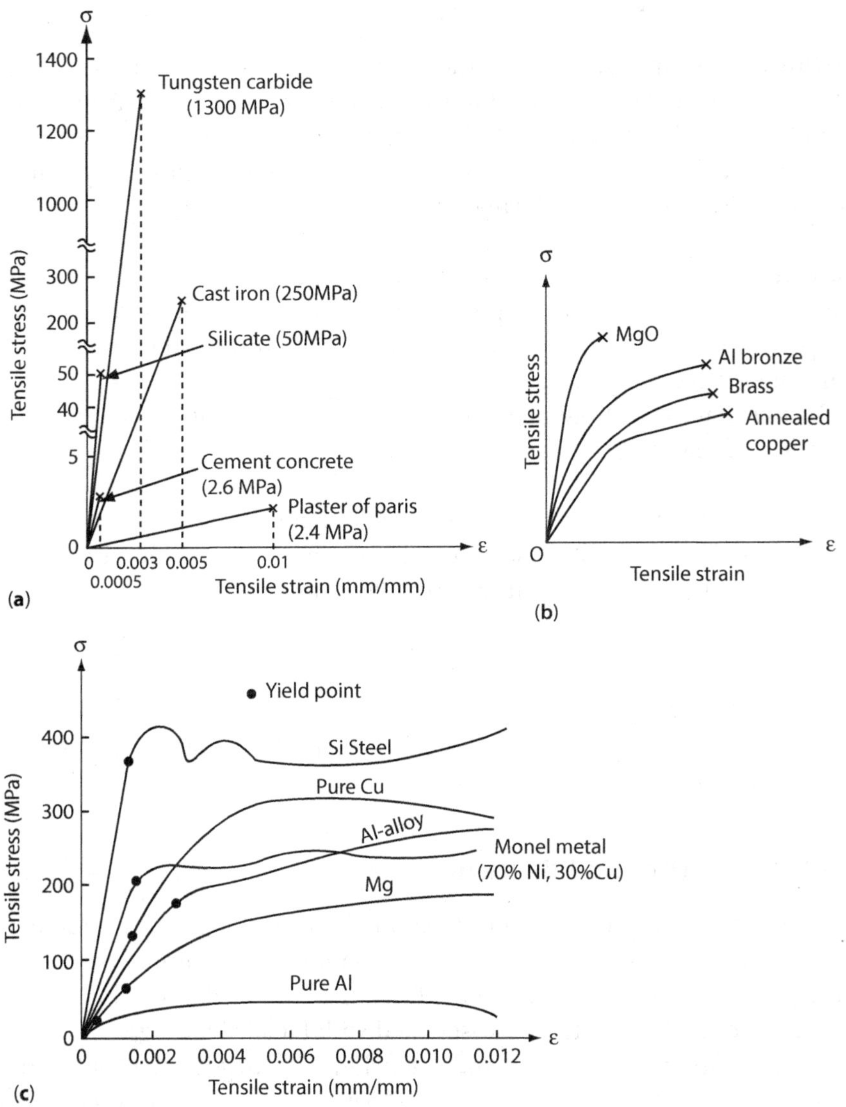

Figure 3.21 a-b-c Stress–strain curves of different materials under tensile loading.

Illustration

- Gold is the most ductile metal. Wires many metres long can be drawn from 1 gram of gold.
- Other ductile metals in the order of decreasing ductility are silver, aluminium and copper.
- Silicate is the most brittle material.

3.22.9 Malleability

A material that can be pressed into the form of a sheet under compressive load is said to be malleable. Thus malleability is the property that enables a material to be converted into flat sheet. Silver is most malleable metal followed by gold.

3.22.10 Fatigue

Definition. The behaviour of materials under fluctuating and reversing loads (or stresses) is known as fatigue. This behaviour is different from that under the steady load. Fatigue is, however, not a dynamic effect. The rate of loading is usually not a factor in fatigue behaviour. Fatigue behaviour is experienced by all materials whether metals, plastics, concretes, or composites.

Main effects of fatigue. The main effects of fatigue on the properties of materials are the following.

1. Loss of ductility,
2. Loss of strength, and
3. Enhanced uncertainty in strength and the service life of materials.

3.22.11 Creep

Definition. The permanent deformation (strain) of a material under steady load as a function of time is called *creep.* A very common observation in which the length of our waist belt increases after some duration, is due to creep effect. It is thermally actuated process, and hence is influenced by temperature. It is, however, appreciable at temperatures above 0.4 T_m where T_m *is* melting point of material in degree kelvin.

Creep occurs at room temperature in many materials such as lead, zinc, solder wire (an alloy of Pb and Sn), white metals, rubber, plastics and leather etc.

Critical applications for creep consideration. Consideration of creep is very important in applications such as given below.

1. Industrial belts,
2. Blades of gas turbines,
3. Blades of steam turbines,
4. Pistons of I.C. engines,
5. Rockets and missiles,

6. Nuclear reactors,
7. Tubes of heat exchangers etc. and
8. Polymeric and elastomeric constructions

3.22.12 Need of Different Properties for Different Applications

Engineering applications of materials cannot be counted on fingers. Different properties are desired for different utilities. For example,

- *Elasticity* is desired in precision machines,
- *Stiffness* is desired in beams and columns,
- *Resilience* is desired in springs,
- *Ductility is* desired in cold wire drawing,
- *Hardness* and *toughness* are desired in cutting tools etc.

3.22.13 Hardness

Definition. Hardness is a *surface property.* It is defined as the resistance of a material against permanent deformation of the surface in the form of scratch, cutting, indentation, or mechanical wear. Diamond is the hardest known material. In various hardness tests, the indenters are used to introduce indentation on the surface. The shape of indenters may be a spherical ball, a cone, or a pyramid. Various hardness test methods are Brinnel test, Rockwell test, Vicker's test, etc.

3.22.14 Impact

In manufacturing locomotive wheels, coins, connecting rods etc., the components are subjected to impact loads (or shock loads). These loads are applied suddenly. The stress induced in components are many times more than the stress produced by gradual loading. Hence materials should be able to sustain such loads. Therefore, impact tests are performed to assess shock absorbing capability of materials subjected to suddenly applied shock loads. These capabilities are expressed as *(i)* rupture energy, *(ii)* modulus of rupture, and *(iii)* notch impact strength.

3.22.15 Factors Affecting Mechanical Properties

Mechanical properties of materials are influenced by various factors. Main among these factors are the following.

1. Grain size
2. High temperature
3. Low temperature
4. Radiation exposure
5. Heat treatment
6. Rate of loading, and
7. Moisture absorption etc.

The advantageous or disadvantageous effects depend on proper control of these factors, for example,

- Micro-alloyed steels made of very fine grains need not to be heat treated, hence they are economical in use.
- The impact strength of iron and steel improves at higher temperatures.
- A ductile metal exhibits brittle nature at much lower temperatures.
- Strength of a material increases while ductility decreases due to radiation.

Review Questions

1. Differentiate between the crystalline and non-crystalline solids. Which different factors promote the crystalline solid structures?
2. Name those non-metals which behave as crystalline materials. What are the causes of such a behaviour?
3. Classify the solids on the basis of their structure. How do the metallic, ionic and covalent solids differ from each other?
4. Astatine (At) has melting point of 302°C and cesium (Cs) of 28°C. A polymer has melting point in the range of 135°C to 145°C. Why are these so?
5. Describe the geometries of HCP crystal structure. Explain them in consideration of lattice point, effective number of atoms, coordination number and atomic packing factor.
6. What do you mean by void? Sketch and explain the tetrahedral and octahedral voids and differentiate between them.
7. Is the presence of voids in solids desirable? Discuss the effects of voids on density of solids.

8. Gold ornaments are made by mixing copper. Can't we mix aluminium or iron in gold? Support your reply with logical reasons.
9. Enumerate the main examples of covalent bonded solids. Explain their structures.
10. Describe the geometries of DC, mentioning the number of atoms per lattice point, basis, distance of separation between two atoms, and the atomic packing efficiency.
11. Diamond is the hardest known mineral – comment why? Derive its atomic packing factor. Can't we enhance its *APF*?
12. Illustrate the examples of ionic, covalent and metallic solids. Why do they behave like themselves?
13. What do you mean by material characterization? Name the various techniques employed for it.
14. What are several techniques employed in determining the crystal structure of solids? Compare their advantages and disadvantages.
15. What do you mean by X-rays diffraction? How is Bragg's diffraction different from Laue's concept of X-ray diffraction?
16. Derive Bragg's equation. Explain the meaning of first order, second order and third order reflections.
17. Elaborate the procedure to determine Miller indices of planes by X-ray technique. Also mention as how the crystal structure is determined.
18. Explain the working principle of a microscope. Draw ray diagram and show as to how a magnified image is obtained in a metallurgical microscope.
19. Enlist various kinds of metallurgical microscopes used for material characterization. Write features of each of them.
20. Sketch and explain the construction and working of any high power metallurgical microscope. State its features also.
21. Explain various microscopic methods. What are different kinds of microscopes used in crystallographic investigations? Write their special features.
22. What is etching? Why is it necessary before microscopic studies? Name different etching agents.
23. Sketch and explain the construction and working of an electron microscope.
24. Define the following in one or two sentences.
    a. Ideal crystal
    b. Real crystal
    c. Solids with nano-level imperfections.

25. Classify the imperfections as exhaustively as you can.
26. Discuss different types of point imperfections.
27. Why doesn't electrical neutrality change in cases of Frenkel's defect and Schottky's defect?
28. What is a slip plane? How is it related to dislocation?
29. Why is the name 'screw dislocation'?
30. Discuss different characteristics of dislocations.
31. Why does a dislocation of larger Burgers vector try to break itself into smaller dislocations?
32. A dislocation reaction in a BCC solid is given by $b_1 + b_2 = b_3$ where $b_1 = a/2\ [111]$, $b_2 = a/2\ [1\ 1\ 1]$, and $b_3 = a[100]$. Show that the reaction is vectorially correct and energetically favourable.
33. What are various sources of dislocations? How do they affect the properties of materials? Suggest ways and means to remove or minimize them.
34. Screw dislocation cannot glide whereas edge dislocation can. Explain why?
35. Enumerate different types of surface imperfections with the help of suitable diagrams.
36. Why does grain boundary defect develop in materials?
37. What are the effects of imperfections on the geometry of unit cells and crystals?
38. Discuss different types of twins. Where does the defect of twin occur?
39. How are mechanical properties important to a design engineer and a material scientist?
40. Differentiate among the following:
    a. Isotropy, orthotropy and anisotropy.
    b. Homogeneous and heterogeneous materials, and
    c. Isotropic-homogeneous and anisotropic-homogeneous materials.
41. Bring out the differences between the following.
    a. Elastic action and plastic action
    b. Toughness and resilience,
    c. Ductility and brittleness, and
    d. Modulus of resilience and proof stress.
42. The resilience of strong metals increases when they are used in slender shapes such as leaf springs in auto vehicles, or as spiral springs in watches. Why?
43. The thin gaskets of lead is used at compressive stresses much higher than its yield strength when not compressed - Why?

44. What is fatigue? What are its effect on properties of materials? Describe fatigue limit and its criticality in aeroplanes.
45. Why do the blades of gas turbine creep? What are the different stages in the creep curve?
46. Why is the hardness of materials of interest to (*a*) an engineer testing a reciprocating pump, (*b*) the mineralogist, (*c*) the design engineer designing a journal and bush bearing assembly, and (*d*) the botanist?
47. Name some specific engineering applications where various hardness tests are essential.

Numerical Problems

1. The incident angle of X-rays for the first reflection from silicon carbide (SiC) is 17.2°. Obtain the incident angle for second reflection.
2. The dimension of nickel unit cell is being determined by calculating $d_{(200)}$ using the X-rays of 0.58 Å wavelength. If the reflection angle is 9.5°, what is the size of unit cell?
3. An X-ray analysis of a crystal is made with monochromatic X-rays of wavelength 0.58 Å. Bragg's reflections are obtained at angles of (*a*) 6.45° (*b*) 9.15°, and (c) 13°. Calculate the interplaner spacings of the crystal.
4. In determining the crystal structure of an element by Bragg's spectrometer, the following observations are noticed: Reflection $(h^2 + k^2 + l^2) = 4$, Glancing angle = 4.06°, Wavelength of X-rays = 1.54 Å. Calculate lattice constant of the crystal.
5. Write specifications of an optical metallurgical microscope. A microscope has eyepieces of 10 × and 15 ×; and objectives of 5 ×, 10 ×, 20 × and 45 ×. Find the range of its magnifications.
6. Is magnifications of 10 ×, 225 × and 300 × possible in case of above problem?

4

Conductive Materials: Electron Theories, Properties and Behaviour

4.1 Electrons and Their Role in Conductivity

Electron is the most important particle of an atom. The electrical nature of atoms is attributed to a highly systematic configuration of electrons in them. They are universal constituent since all the metals contain them in their structure. Electrons have similar behaviour in conductors, semiconductors, dielectrics, magnetic or superconducting materials. They also have same behaviour in all applications whether a computer, television, radiography, or electrical conduction.

4.1.1 Valence and Free Electrons

Electrons play vital role in determining electrical properties of metals. Valence electrons in the outermost orbit of an atom decide the manner in which they respond to external effects. For example, due to their free movement within the metal they provide conduction, but on actuation by thermal energy they jump over the energy gap in semiconductors. As a whole the arrangement of electrons in an atom, behaviour of valence electrons, and inter-atomic interactions govern the electrical properties (conduction as well) of materials.

ELECTRON THEORIES

4.2 Electron Theories of Solids

The electrical properties of solids are determined by the electrons in the outermost orbit of an atom. The valence electrons in an atom decide the manner in which they will respond to external effects. The cohesive and repulsive forces, and the potential energy in the formation of atomic bond are not sufficient to describe many properties of solids. Various electron theories have been propagated to study the behaviour of solids. The classical free electron theory and other acceptable theories in this regard are

1. Drude-Lorentz (classical) theory,
2. Free electron theory,
3. Energy band theory, and
4. Brillouin zone theory.

Whereas the Drude-Lorentz and Free electron theories explain the mechanism of conduction in solids, the Energy band and Brillouin zone theories explain the mechanism of semiconduction. Here, we shall discuss the modern theories of conduction in solids.

4.3 Free Electron Theory

Assumption. We have already studied in chapter 2 that the electrons in the outermost orbit are not bound to its atom, and are free to move throughout the solid. These free electrons are known as *Fermi gas* or *electron cloud,* and their potential field remains uniform throughout the solid due to the ion-cores. The free electron theory is based on the assumption that the average potential energy E_p is a constant throughout the solid, and its energy difference $dE_p = 0$. So the total energy E is equal to the kinetic energy E_k only. This energy is substantially lower than the energy of bound electrons in an isolated atom.

de-Broglie's wavelength. Due to dual nature of particle (*i.e.* electron), the de-Broglie's wavelength is given by

$$\lambda = \frac{h}{mv} \tag{4.1}$$

where h is Planck's constant, v is velocity of the free electron and m its mass. Defining the wave number k by

$$k = \frac{2\pi}{\lambda} \tag{4.2}$$

and neglecting relativistic effect* as the velocity of electrons is much lower than the velocity of light ($v_{electron} << v_{light}$), the total energy may be expressed by

$$E = E_k = \frac{1}{2}mv^2 \tag{4.3}$$

4.3.1 Kinetic Energy in Terms of Wave Number

Substitution of Eqs. 4.1 and 4.2, in Eq. 4.3, yields a parabolic relation of continuous energy. It is shown in Fig. 4.1*a* as kinetic energy versus wave number curve. The kinetic energy is given as follows.

By putting Eq. 4.1 in Eq. 4.2, we get

$$k = \frac{2\pi}{\lambda} = \frac{2\pi}{(h/mv)} = \frac{2\pi mv}{h}$$

$\therefore$

$$v = \frac{hk}{2\pi m}$$

$\therefore$

$$E = E_k = \frac{1}{2}m\left(\frac{hk}{2\pi m}\right)^2$$

or

$$E = \frac{h^2k^2}{8\pi^2 m} \tag{4.4}$$

---

* Einstein has given that mass at velocity v will be $m = m_0\sqrt{(1-(v/c)^2)}$ where m_0 is rest mass and c is velocity of light.

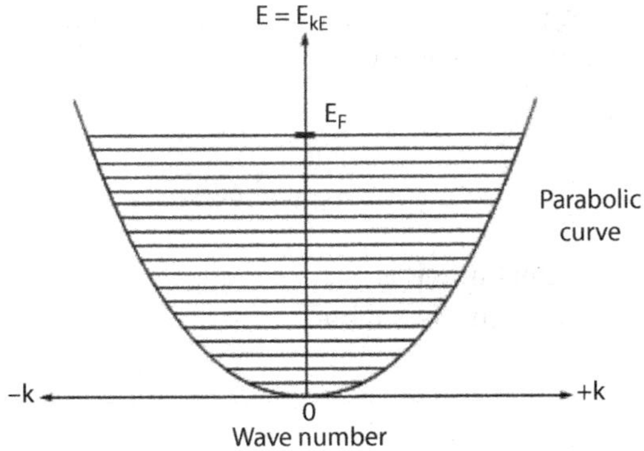

Figure 4.1(a) Relation between the kinetic energy $E_k = E$ and wave number k of a free electron, in free electron theory where potential energy is assumed constant.

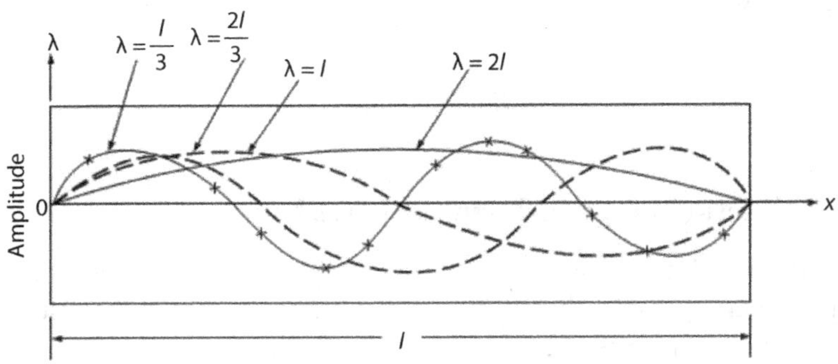

Figure 4.1(b) de Broglie wavelengths λ of some unidirectionally moving electrons.

It implies that the increase in kinetic energy E is proportional to the square of the wave number. And also the electron having largest λ possesses lowest kinetic energy as λ is inversely proportional to k.

4.3.2 Kinetic Energy in Terms of Length of the Solid

We consider a solid of length l in which the electrons move in unidirection x only. The permissible de Broglie wavelengths of free electrons are shown in Fig. 4.1(b). The longest wavelength $\lambda = 2l$ and the others are $\lambda = l, \lambda = \frac{2l}{3}, \lambda = \frac{l}{2}, \ldots$ etc. Consequently the corresponding values of k are $\frac{\pi}{l}, \frac{2\pi}{l}, \ldots$ and so on.

As the electrons in an atom move in opposite directions at same speed, so the wave number may take positive and negative values. In critical conditions, these values are $k = \pm \pi/l, \pm 2\pi/l$, ... etc. and in general $k = \pm n\pi/l$ where n is the principal quantum number and l is length of solid. Substituting this value in Eq. 4.4, we get

$$E = \frac{h^2\left(\pm\frac{n\pi}{l}\right)^2}{8\pi^2 m} = \frac{h^2 n^2 \pi^2}{8\pi^2 m l^2}$$

or

$$E = \frac{h^2 n^2}{8ml^2} \tag{4.5}$$

4.3.3 Energy Equation for 3-Dimensional Solid

This equation is true for the case of unidirectional flow of electrons. The free electrons can move in any direction in a three-dimensional solid. Hence we can write $n^2 = n_x^2 + n_y^2 + n_z^2$, and now Eq. 4.5 gets modified to

$$E = h^2\left(\frac{n_x^2 + n_y^2 + n_z^2}{8ml^2}\right) \tag{4.6}$$

The electron waves are the travelling waves by nature, and have constant velocity throughout the solid. Hence the time averaged probability of finding an electron is constant throughout the solid.

Example 4.1 Calculate the energy levels between $n_x = n_y = n_z = 1$ and next higher energy level $n_x = 2, n_y = n_z = 1$ for free electrons in a solid cube of 20 mm side, and show that the successive energy levels are so close to one another that assuming E to be varying continuously with k is justified. Take standard values of h and m.

Solution. Given are: l = 20 mm = 2×10^{-2} m, $h = 6.626 \times 10^{-34}$ Js, $m = 9.109 \times 10^{-31}$ kg

From Eq. 4.6, we get

$$E = h^2\left(\frac{n_x^2 + n_y^2 + n_z^2}{8ml^2}\right)$$

$$= h^2 \frac{\left(6.626\times10^{-34}\right)^2\left(1^2+1^2+1^2\right)}{8\times9.109\times10^{-31}\times\left(2\times10^{-2}\right)^2}$$

$$= 0.452 \times 10^{-33} \text{ J} \tag{i}$$

Calculating again for next higher level, we get

$$E = h^2 \frac{\left(6.626\times10^{-34}\right)^2\left(2^2+1^2+1^2\right)}{8\times9.109\times10^{-31}\times\left(2\times10^{-2}\right)^2}$$

$$= 0.905 \times 10^{-33}\ \text{J} \qquad (ii)$$

From the values of E obtained from Eqs. (*i*) and (*ii*), we notice that the energy difference is 0.453×10^{-33} J which is extremely small. **Hence the assumption that *E* varies continuously with *k*, is justified.**

4.3.4 Mechanism of Conduction by Free Electrons

The electrical and thermal conduction through a metal bar is a common feature noticed by us. The electrical conductivity in an aluminium rod or in a copper wire, and the thermal conductivity through a steel bar is possible due to movement of free electrons. For every electron moving in a direction with some speed, there is another electron moving in the opposite direction with the same speed. So the net velocity becomes unidirectional on application of an external field, and hence the solid conducts.

Redistribution of energy field. As shown in Fig. 4.2, the negatively charged electrons accelerate towards the positive end of the applied field. In this action, some electrons acquire additional velocity while the electrons moving in opposite direction lose some velocity. Due to this, *E vs* k *curve* of

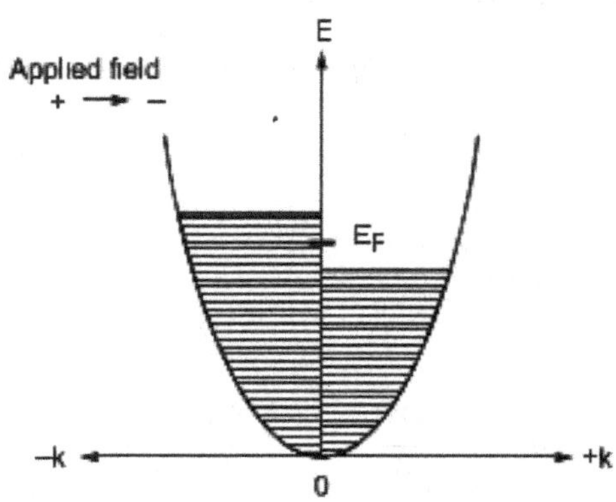

Figure 4.2 Redistributed *E* vs *k* profile on application of external field. Some electrons gain extra velocity at the cost of electrons moving in opposite direction.

Fig. 4.1 gets modified. Such occurrence is the characteristic of *conductors*, and is possible only when empty energy levels are available above the Fermi energy level E_F. In this figure, it is shown that some electrons acquire $E > E_F$ while others have $E < E_F$.

4.3.5 Drift Velocity and Collision Time

The accelerated movement of electrons results in a gain of extra velocity which is above their normal velocity v_n. This gained velocity is called the *drift velocity* v_d. When accelerated, these electrons collide with obstacles present within the solid. The drift velocity in a real situation is shown in Fig. 4.3*a* whose idealized representation is depicted in Fig. 4.3*b*. If an electron of mass *m* and charge *e* undergoes an acceleration *a* due to the applied field gradient ζ, then

$$\zeta e = ma \tag{4.7}$$

The accelerated electrons collide with obstacles at successive interval of time *t* called collision time. As $a = \frac{v_d - 0}{t}$, hence Eq. 4.7 may be rewritten as

$$\zeta e = \frac{mv_d}{t} \tag{4.8}$$

or

$$v_d = \frac{\zeta et}{m} \tag{4.9}$$

Current density. The current density I_d is the flux caused due to the flow of *n* number of free electrons, and is expressed by

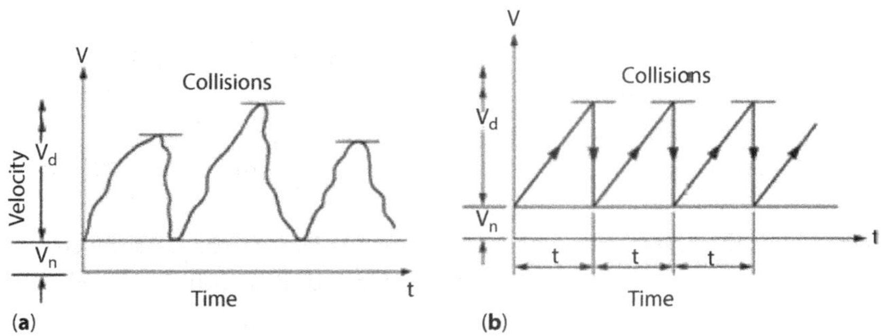

Figure 4.3 Accelerated movement of electrons under an applied field shows normal and drift velocities v_n and v_s in (*a*) real situation, and (*b*) an idealized case.

$$I_d = nev_d = \frac{ne^2t\zeta}{m} \tag{4.10}$$

Conductivity. The flux per unit potential gradient is called conductivity σ which is determined from Eq. (4.10) as

$$\sigma = \frac{I_d}{\zeta} = \frac{ne^2t}{m} \tag{4.11}$$

Resistivity. The electrical resistivity ρ of a conductor is given by

$$\rho = \frac{1}{\sigma} = \frac{1}{\frac{ne^2t}{m}} = \frac{m}{ne^2t} \tag{4.12}$$

Example 4.2 How much is the electric field between a pair of conducting plates, 5 mm apart, when the potential difference between them is 230 V?

Solution. The electric field E between two plates is equal to the potential gradient, which may be obtained as follows.

As $$E = -\frac{dV}{dx} = -\frac{V}{d}$$

∴ $$E = -\frac{230\,\text{volt}}{5\,\text{mm}} = -\frac{230\,\text{V}}{0.005\,\text{m}} = -4.6 \times 10^4\,\text{V/m}$$

Example 4.3 The carrier concentration in a material of conductivity 0.018/ohm m is 10^{19} electrons/m^3. A voltage of 0.16 volts is applied across the 0.29 mm thick material. Determine the drift velocity of the carriers. Take standard values of mass of electron and electronic charge.

Solution. The given data and the standard values for electron are

$n = 10^{19}$ electrons/m^3, $e = 1.602 \times 10^{-19}$ C,
$\sigma = 0.018$/ohm m, $m = 9.1 \times 10^{-31}$ kg,
$V = 0.16$ volt, and $t = 0.29$ mm

where V is the voltage, and t is the thickness of the material. As the electric field gradient ζ is equal to the voltage per unit thickness, therefore

$$\zeta = \frac{V}{t} = \frac{0.16}{0.29 \times 10^{-3}} = 551.72\,\text{V/m}$$

We can write Eq. 4.10 as

$$\frac{v_d}{\zeta}=\frac{et}{m} \qquad (i)$$

Substitution of Eq. (*i*) in Eq. 4.12, yields

$$\sigma=\frac{nev_d}{\zeta} \qquad (ii)$$

Therefore,

$$0.018=\frac{10^{19}\times 1.602\times 10^{-19}\times v_d}{551.72}=0.0029\ v_d$$

$$\therefore \qquad v_d=\frac{0.018}{0.0029}=6.19\ \text{m/s}$$

4.3.6 Mean Free Path (or Mean Free Length)

The average distance travelled by an electron between two successive collisions is called *mean free path l*. It is expressed by

$$l=\frac{l_1+l_2+\ldots+l_p}{p} \qquad (4.13)$$

where $l_1, l_2 \ldots, l_p$ are the distances travelled between any two random collisions, and p is the number of collisions.

- More the mean free path in a crystal, more will be its conductivity.
- *In an ideal crystal at* 0 K, $l \to \infty$ *therefore* $\sigma \to \infty$,
- *but in a real crystal* $l \neq \infty$, *therefore* σ *is finite.*

It happens because there are no collisions in an ideal crystal as it is free from imperfections, whereas the real crystals have too many imperfections.

4.3.7 Effect of Temperature on Mean Free Path

The effect of high temperature is to decrease the mean free path and conductivity of the material. This effect varies *cubically* at low temperatures

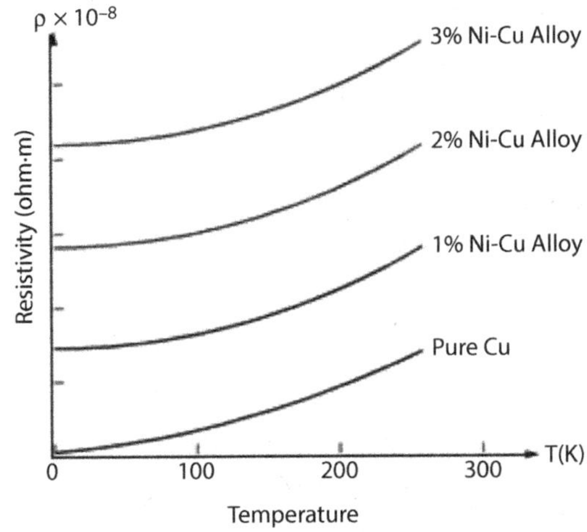

Figure 4.4 Effect of temperature on the conductivity (reciprocal of resistivity) of pure copper and its alloy with nickel.

and *linearly* at higher temperatures for resistivity. Conductivity of a pure metal is higher than the conductivity of its alloy. Effect of temperature on conductivity (reciprocal of resistivity) of pure copper and its alloy with nickel is shown in Fig. 4.4. It clearly shows that conductivity decreases with increasing percentage of nickel.

- The mean free length in pure metals at room temperature is of the order of 100 Å.

4.4 Energy Band Theory

Assumption. The assumption that a constant potential field exists in the free electron theory was questioned by many researchers. Potential energy of an electron is a function of its position with respect to the ion-cores. Considering *Heisenberg's uncertainty principle* and Bragg's *diffraction pattern* of electrons, the potential energy of an electron cannot be neglected as compared to its dimension. This is due to the fact that the 'standing wave' of an electron gives rise to a periodic variation in its amplitude. The probability of finding an electron remains maximum at the crest of the waveform. The two possible waveforms are *sine wave* and *cosine wave*, formed due to the superimposition of travelling de Broglie waves. These waveforms satisfy Bragg's law,

$$2d\sin\theta = n\lambda \tag{4.14}$$

and are shown in Fig. 4.5. Thus the difference in potential energy cannot be assumed to be zero, and the energy will be given by

$$E = E_k + E_p \tag{4.15}$$

4.4.1 Critical Conditions

At critical conditions of $k = \pm n\pi/d$ for $n = 1, 2, 3$ ... etc, the electron is described by a standing wave. When the waveforms are either sine or cosine type, the potential energy shows deviations. This deviation results in break of E vs k curve (Fig. 4.6*a*), giving rise to an energy gap. This energy gap Eg is between different orbits K, L, M,... of an electron, and is shown in Fig. 4.6*b*.

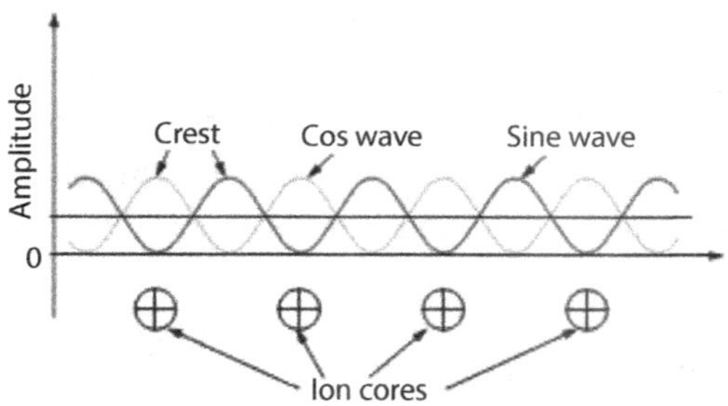

Figure 4.5 Sine and cosine standing waveforms obtained due to super-imposition of travelling waves. The electron is likely to be located at the crest.

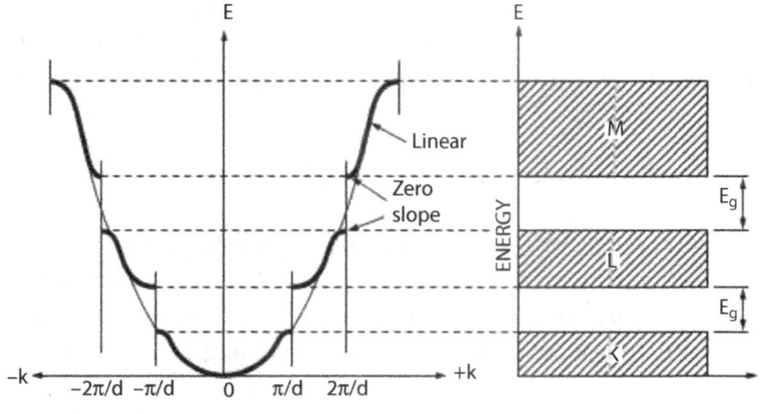

Figure 4.6 (*a*) Consideration of potential energy of electron yields a break in E vs k curve at critical values, and (*b*) energy gap Eg.

Table 4.1 Energy gap in some materials

Material	Energy gap E_g (eV)		Material	Energy gap E_g (eV)	
	at 0 K	at 300 K		at 0 K	at 300 K
Sn(gray)	0.08	—	InSb	0.23	0.17
Ge	0.74	0.66	InAs	0.42	0.36
Si	1.17	1.12	GaAs	1.52	1.42
C (diamond)	5.48	5.47	GaP	2.34	2.26
CdSe	3.42	3.35	SiC	3.03	2.99
ZnO	1.85	1.70	AlP	—	2.45

4.4.2 Magnitude of Energy Gap

Magnitude of this energy gap is an indication of the difference in potential energy for electron locations of two different waveforms. The two closely spaced energy levels are known as energy bands. The slope of E vs k curve is linear except at the critical values of k where this slope becomes zero. Energy gap in some materials is given in Table 4.1.

4.5 Brillouin Zone Theory

The discontinuous plot *of E* vs k in Fig.4.6*a depicts* only unidirectional motion of an electron. *Brillouin zone* theory considers motion of electron in all possible directions such as [100], [110] etc., and all possible reflections in Bragg's equation 4.14, where value of glancing angle θ can vary between 0° to 90°. Critical values of k correspond to $\pm n\pi/(d \sin \theta)$ instead of $\pm n\pi/d$ which was used for unidirectional case. Here d is interplaner spacing.

Effective Zones on Different Planes and Directions

- Now consider a set of parallel planes (100) in a cubic crystal. For first order reflection ($n = 1$), the critical condition will be $k = \pm\pi/(d \sin \theta)$. As direction [100] is perpendicular to plane (100), hence movement of an electron along direction [100] will have critical condition for $k = \pm\pi/(d \sin 90^\circ) = \pm\pi/d$.
- Now if direction of electron motion is changed to [110], then for reflection from planes (100), the critical condition

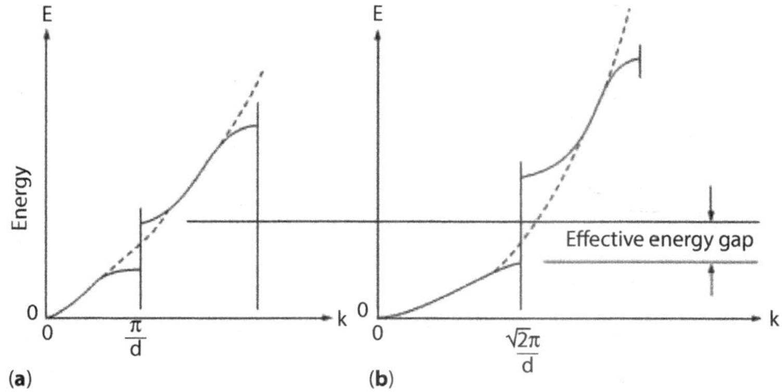

Figure 4.7 Effective energy gap for electron moving along (*a*) direction [100], and (*b*) [110] direction.

will be obtained at $k = \pi/(d \sin 45°) = 2\pi/d$. In this case, the Bragg angle is 45°.

Effective energy gap. Figures 4.7*a-b* show *E-k* curve for electron moving along [100] and [110] directions. The two curves are for first order reflection from set of (100) planes. The gaps occur at $k = \pi / d$ for [100] motion and at $k = \sqrt{2}\pi / d$ for [110] motion. The effective energy gap is depicted in above figures. This effective gap is known as forbidden gap (or zone), and are those energy levels that an electron cannot take.

4.5.1 Meaning of Different Brillouin Zones

This theory demarcates different Brillouin zones as follows.

i. First Brillouin zone,
ii. Second Brillouin zone, etc.

Brillouin zones are the ranges of *k* lying between the forbidden gaps. A Brillouin zone may be obtained by plotting the values of *k* for different sets of planes. First and second Brillouin zones, for a two-dimensional crystal, are shown in Figs. 4.8*a-b* respectively. The first Brillouin zone contains all energy levels upto first discontinuity in *E-k* curve. Range of *k* for this discontinuity is from $-\pi/d$ to $+\pi/d$. Similarly second Brillouin zone contains all energy levels between first and second discontinuities lying

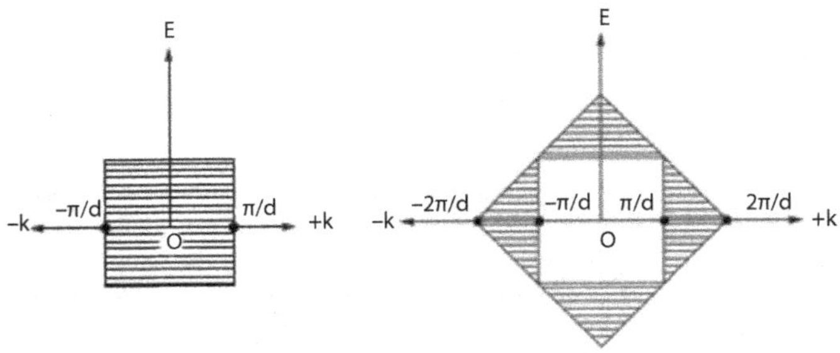

Figure 4.8 (*a*) First Brillouin zone, and (*b*) second Brillouin zone, in a two dimensional crystal.

between $\pm 2\pi/d$. Similar explanation may be given for third Brillouin zone and so-on.

4.5.2 First and Second Brillouin Zones

Brillouin zone is a 3-dimensional representation of the allowable values of k. In a 3-dimensional crystal, the critical value of k depends on the relative direction of the moving electron to the crystal lattice, due to the changing value of θ, and also because different sets of planes may diffract the electron. For 1-dimensional lattice, the critical values of k in Fig. 4.9 would be

$$k_n = \frac{n\pi}{a} \tag{4.16}$$

where $n = 1, 2, 3 ....$ and a is the distance between atoms. The region between k_1 and k_2 is called the *first Brillouin zone.*

The second one is the region from k_2 to k_3. Figure 4.9 shows the two zones. In a 2-dimensional square lattice, the components of k are k_x and k_y. Diffraction will occur whenever k satisfies Eq. 4.16. If we apply Eq. 4.19 and remember that both the vertical and horizontal planes can diffract, we arrive at

$$k_x n_1 + k_y n_2 = (\pi / a)(n_1 + n_2) \tag{4.17a}$$

where n_1 is the integer for diffraction by the vertical planes, and n_2 for the horizontal planes.

For the first zone, one integer is ±1 and the other is zero. This is the minimum value of Eq. 4.17*a*. For the second zone, each integer is ±1. The

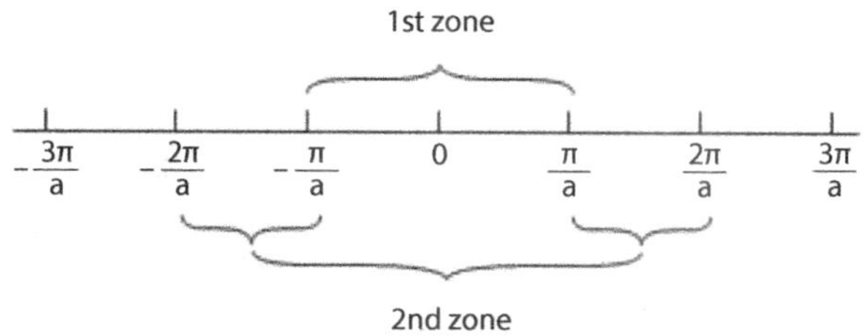

Figure 4.9 The first two Brillouin zones for a 1-dimensional lattice.

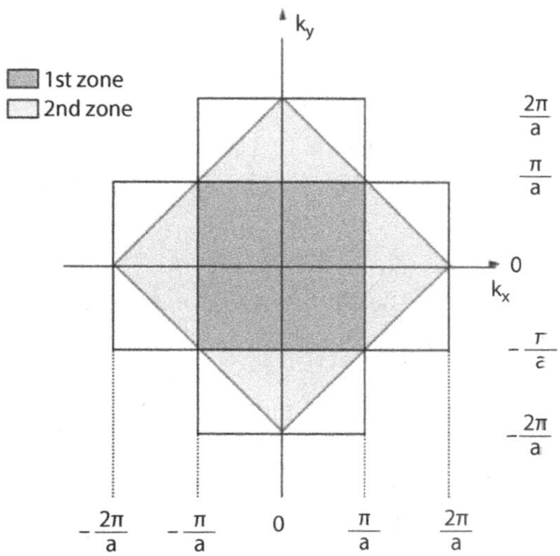

Figure 4.10 The first two Brillouin zone for a square lattice.

first two zones of the square lattice are shown in Fig. 4.10. The higher zones correspond to higher values of n_1 and n_2.

4.5.3 Brillouin Zones for Simple Cubic Lattice

The Brillouin zones of a simple cubic lattice in 3-dimensions can be calculated using an equation of the form as given below.

$$k_x n_1 + k_y n_2 + k_z n_3 = (\pi / a)(n_1 + n_2 + n_3) \tag{4.17b}$$

From this equation, it follows that the first zone for a simple cubic lattice *is a* cube (Fig. 4.11) whose walls intersect the k_x, k_y and k_z axes at the points $+\pi/a$ and $-\pi/a$. The second zone in 3-dimensions is made up of pyramids

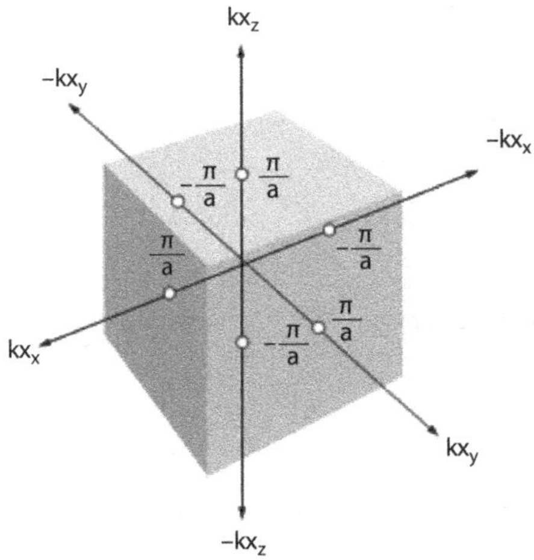

Figure 4.11 The first Brillouin zone for a cubic lattice. The second zone consists of altitude π/a and square bases $2\pi/a$ wide, resting on the cubic faces of the first zone.

which rest on each face of the first zone cube in a manner similar to the triangles of the second zone in Fig. 4.10.

4.5.4 Brillouin Zones for BCC, FCC and HCP Lattices

In a similar manner, the first and second Brillouin zones for the BCC, FCC and HCP lattices can be calculated to obtain the polyhedra shown in Fig. 4.12. The equations which define the boundaries of these zones are all based on the Bragg equation for the reflection of a wave by periodic lattice. Thus, the Brillouin zones in metals are polyhedra whose plane surfaces are parallel to the reflecting planes which are responsible for *X*-ray diffraction.

In both BCC and FCC metals, each Brillouin zone holds as many quantum states (without considering electron spin) as there are atoms or primitive unit cells in a crystal. For a crystal of *N* unit cells, there are *N* states in the first zone. The quantum rule apply to the filling of quantum states in a Brillouin zone. Electrons occupy the states of lowest energy first. Two electrons of opposite spin occupy each state. If there are *N* atoms in the crystal, then for mono-valent metals the *N*/2 quantum states of lowest energy are filled. The first Brillouin zone would therefore be half filled. For multivalent metals, the situation becomes more complicated because the zone may overlap. When the zones overlap, it is impossible to fill up one zone without starting to fill the next one.

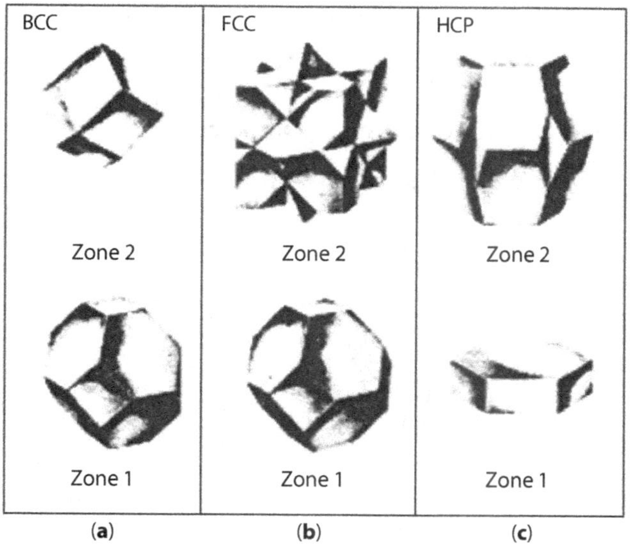

Figure 4.12 The first two Brillouin zones for the (*a*) BCC lattice, (*b*) the FCC, and (*c*) the HCP lattice.

CONDUCTORS

4.6 Conductors

All metals and alloys fall in the category of conductors. They are used in various applications such as

- electricity transmission and distribution lines,
- electrical contacts viz. relays, brushes, switches etc.,
- resistors, and
- heating elements.

Gold is the best conductor of electricity followed by silver, copper and aluminium. Keeping in view the cost factor, copper and aluminium are the natural choices although silver is used for contacts in aircrafts.

- Aluminium conductor reinforced with steel (*ACSR*) is an improved material for transmission lines.
- Oxygen-free-high conductivity (*OFHC*) copper conductor is very suitable for low temperature applications. It is a high

purity copper whose resistivity is 1×10^{-10} ohm m at 4.2 K and zero electric field.

4.6.1 Characteristics of a Good Conductor

A good conductor should possess the following characteristics.

i. high electrical and thermal conductivity,
ii. high melting point,
iii. good oxidation resistance,
iv. low cost,
v. good wear and abrasion resistance, and
vi. better mechanical properties.

Presence of phosphorus, sulphur and iron in copper is undesired as they lower its electrical conductivity.

4.7 Factors Affecting Conductivity (and Resistivity) of Metals

Resistance and resistivity. Flow of current through a conductor means the flow of electrons through it under an applied electric field. The property of a conductor that helps it to resist the flow of electrons is called *resistance*. The electrical resistance R *of* a conductor is related to its length l and area of cross-section a by

$$R = \rho \frac{l}{a} \tag{4.18}$$

where ρ is called *specific resistance* or *resistivity*.

The electrical resistance and the resistivity of materials are influenced by various factors such as given below.

1. Temperature,
2. Impurities, and
3. Plastic deformation.

Example 4.4 Resistance of a 200 metre long copper wire is 21 ohm. Its diameter is 0.44 mm. Determine its specific resistance.

Solution. Given are: $l = 200$ m, $R = 21\ \Omega$, $d = 0.44$ mm $= 0.44 \times 10^{-3}$ m

$\therefore$ $$A = \frac{\pi d^2}{4} = \frac{\pi(0.44\times10^{-3})^{-2}}{4}\text{m}^{-2}$$

$\because$ $$R = \rho\frac{l}{A}\text{, therefore } \rho = \frac{RA}{l}$$

$\therefore$ $$\rho\,\Re\frac{21\times\pi(0.44\times10^{-3})^2}{4\times200}\quad 1.597\times10^{-8}\,\text{ohm}\quad\text{m}$$

4.7.1 Temperature Effect on Conductivity

The electrical resistance of a conductor varies with its temperature. With a rise in temperature, the resistance of pure metals increases whereas it decreases for semiconductors, insulators and electrolytes. The variation of resistance with temperature in a material can be expressed as

$$R_{T2} = R_{T1}(1 \pm a dT) \tag{4.19}$$

where R_{T2} and R_{T1} are the resistances of the material at temperatures T_2 and T_1 respectively, $dT = T_2 - T_1$ is the change in temperature, and α is temperature coefficient of resistance of the material. Here $+\alpha$ stands for pure metals and $-\alpha$ for insulators and electrolytes. The variation of resistivity with temperature is similar to Eq. 4.19, and is given by

$$\rho_{T2} = \rho_{T1}(1 \pm a dT) \tag{4.20}$$

- The resistivity of an alloy is generally greater than the resistivity of its constituent metals.

Example 4.5 Resistance of a conducting wire is 57.2 Ω at 70 °C and 50 Ω at 25 °C. Estimate its temperature coefficient of resistance.

Solution. Given are: R_{70} = Resistance at 70°C = 57.2 Ω, R_{25} = Resistance at 25°C = 50 Ω

As $R_T = R_0(1 + aT)$, so writing above equation for temperatures 25°C and 70°C, we have

$$R_{25} = R_0(1 + 25a) \tag{i}$$

$$R_{70} = R_0(1 + 70a) \tag{ii}$$

On dividing Eq. (*ii*) by Eq. (*i*), we get

$$\frac{R_{70}}{R_{25}} = \frac{1+70a}{1+25a}$$

or

$$\frac{57.2}{50} = \frac{1+70a}{1+25a}$$

or

$$57.2 + 57.2 \times 25a = 50 + 50 \times 70a \qquad (iii)$$

On solving Eq. (*iii*), we get

$$(70 \times 50 - 25 \times 57.2)a = 7.2$$

or

$$a = \frac{7.2}{2070} = 3.478 \times 10^{-3}\,\text{K}^{-1}$$

∴ $$\alpha = 0.003478\ \text{K}^{-1}$$

Electrical properties of conducting materials. Table 4.2 illustrates the resistivity, temperature coefficient and other related properties of some conductors.

Table 4.2 Resistivity and temperature coefficient of some materials

Material	Electrical resistivity (ohm m)	Temperature coefficient ⟨(1/K)
Gold	22×10^{-9}	0.0035
Silver	15×10^{-9}	0.0040
Copper	17×10^{-9}	0.0044
Aluminium	28×10^{-9}	0.0042
Nichrome	1100×10^{-9}	0.0001

4.7.2 Nordheim Equation for Impurity and Alloying Effects on Resistivity

The interstitial and substitutional impurities present in a solid solution cause local disorder. This disorder induces scattering of electrons which results in a change in their mobility and mean free length. Consequently, mean free length decreases and hence the electrical resistivity increases. Effect of alloying or of an impurity on resistivity may be determined by *Nordheim equation* given as

$$\rho_A = \rho_M + C\rho_l \tag{4.21}$$

where ρ_A and ρ_M are resistivities of alloy and base metals respectively, concentration of impurity is C atomic percent and ρ_l is resistivity of alloying or impurity metal.

Illustration. As an illustration, the resistivity of copper increases by 1.3 ohm-cm when 1% nickel is added to it. Thus for Cu-Ni system, Eq. 4.21 may be written as

$$\rho_{\text{Cu-Ni}} = \rho_{\text{Cu}} + (C \times 1.3)\mu\text{ohm - cm} \tag{4.22}$$

4.7.3 Effect of Plastic Deformation and Cold Working

We have already discussed that the materials get strain-hardened and the number of dislocations increase when they are plastically deformed. A similar behaviour is noticed when a metal is cold worked. These effects induce scattering of electrons, decrease in mean free path and hence an increase in the resistivity of the material.

4.7.4 Matthilseen Rule of Total Resistivity

According to *Matthilssen additive rule*, the total resistivity of a material is the sum of thermal based, impurity based, and plastic deformation based resistivities ρ_t, ρ_c and ρ_p respectively. So

$$\rho = \rho_t + \rho_c + \rho_p \tag{4.23}$$

where $\rho_c + \rho_p = \rho_0$ is called *residual resistivity*. The quantity $\rho_t = 0$ at absolute zero temperature, and $\rho_0 = 0$ at high temperatures. Prediction of Eq. 4.23 is less accurate at higher impurity concentrations and temperatures.

Example 4.6 The resistivity of copper is 0.015 μohm m at 300 K. Addition of each atomic percent of nickel and silver to it causes an increase in resistivity by 0.012 μ ohm m and 0.0016 μ ohm m respectively. Compute the resistivity of Cu-Ni-Ag alloy at 300 K when 0.25 atomic percent nickel and 0.4 atomic percent silver is added into copper.

Solution. For this case the Eq. 4.21 may be written as

$$\rho_{\text{Cu-Ni-Ag}} = \rho_{\text{Cu}} + C_1\rho_{\text{Ni}} + C_2\rho_{\text{Ag}}$$

where $\rho_{Cu} = 0.015 \times 10^{-6}$ ohm m, $\rho_{Ni} = 0.012 \times 10^{-6}$ ohm m
$\rho_{Ag} = 0.016 \times 10^{-6}$ ohm m, $C_1 = 0.25$ atomic %, and
$C_2 = 0.40$ atomic %
Therefore

$$\begin{aligned}\rho_{\text{Cu-Ni-Ag}} &= (0.015 \times 10^{-6}) + (0.25 \times 0.012 \times 10^{-6}) \\ &\quad + (0.40 \times 0.016 \times 10^{-6}) \\ &= (0.015 + 0.003 + 0.0064) \times 10^{-6} \\ &= 0.0244 \times 10^{-6} \\ &= 2.44 \times 10^{-8} \text{ ohm m}\end{aligned}$$

4.8 Thermal Conductivity

The transport of heat flux Q through a solid of cross-sectional area A is proportional to the thermal gradient dT/dx, and is given by

$$Q \propto -A\frac{dT}{dx}$$

Therefore

$$Q = -K_t A\frac{dT}{dx} \tag{4.24}$$

where K_t is a proportionality constant which is known as *thermal conductivity* of the solid. The negative sign indicates a drop in thermal gradient for increasing length x. The conduction of heat in solids is due to the following activities.

1. Thermally excited lattice vibrations, and
2. Thermal motion of the free electrons.

Hence the total thermal conductivity of a solid is the sum of lattice conductivity K_l and electronic conductivity K_e. Thus

$$K_t = K_l + K_e \tag{4.25}$$

- In metals, $K_e >> K_l$ due to the motion of free electrons, therefore $K_t \approx K_e$.
- In dielectrics, $K_l >> K_e$ due to the atomic or molecular vibrations of the lattice, therefore $K_t \approx K_l$.
- In semiconductors, both K_l and K_e have their contributions, therefore for them Eq. 4.25 holds true.

4.8.1 Salient Features of Different Materials Regarding Thermal Conductivity

Important conclusions regarding thermal conductivity of different classes of materials can be enumerated as below.

i. The thermal conductivity of polycrystalline metals is much lower than that of single crystals due to scattering of electrons along grain boundaries.
ii. Thermal conductivity of metals decreases due to strong magnetic field and high temperatures.
iii. The thermal conductivity of metals is lower than their alloys due to scattering of electrons and phonons in the presence of alloying (foreign) atoms.
iv. In semiconductors, the thermal conductivity increases with increasing temperature due to flow of more number of excited electrons.
v. Thermal conductivity of amorphous solids such as plastics and glasses increases with a rise in temperature.
vi. Thermal conductivity of metals, alloys, semiconductors and dielectrics are in the decreasing order.

Table 4.3 shows thermal conductivity of some solids.

Table 4.3 Thermal Properties of Some Materials

Material	Linear coefficient of expansion ⟨ (10^{-6}/K)	Thermal conductivity K_t (W/m-K) at 25°C	Melting point T_m (K)
• Aluminium (pure)	23.7	300	930
• Al alloy with 5% Cu	23.2	170	780
• Copper (pure)	17.7	425	1355
• Brass	18.8	200	1280
• Mild steel	15.2	65	1785
• 18-8 stainless steel	10.5	25	1765
• Teflon	140.5	0.20	600
• Nylon	107.0	0.40	500
• Alumina	6.9	65.00	2320

4.9 Heating Effect of Current

When an electric current flows through a metal (i.e. a conductor), the metal becomes hot after some time. It is due to the collision of free electrons moving through the lattice of metals in a random manner. During their movement (drift), they collide with the atoms in the lattice of conducting metals. As a result of such collisions, the electrons lose some of their kinetic energy and manifests as vibrational energy of lattice. This results in generation of heat, and hence the phenomenon is termed as 'heating effect of electric current'. The heat produced is measured in the unit of calorie (cal) or kilocalorie (kcal), and may be expressed by Joule's law. Joule's law is explained in the next article.

4.9.1 Joule's Law of Electrical Heating

According to Joule's law, if a current of I ampere flows through a conductor of resistance R ohm for t second, then the amount of work required (W.D.) to maintain the current flow will be given by

$$\text{W.D.} = I^2Rt \text{ joule} \tag{4.26}$$

$$= VIt \text{ joule as } R = V/I$$

$$= Pt \text{ joule as power } P = V \times I$$

$$= \frac{V^2t}{R} \text{ joule, as } I = \frac{V}{R}$$

Heat equivalent of this work may be expressed as

$$H = \frac{W.D.}{J} \tag{4.27a}$$

where J is mechanical equivalent of heat whose value is 4186 joule/kcal ≈ 4200 joule/kcal.

Hence, using Eq. 4.26, the Eq. 4.27a may be written as

$$H = \frac{I^2Rt}{4200} \text{ kcal} \tag{4.27b}$$

4.9.2 Applications of Heating Effect

Heating effect of current is desired for several electrical applications but undesirable for many others. It is undesirable in the sense that heating causes reduction in mechanical strength of conducting material and induces variation in its resistance. However, it is desired for the following important applications.

- Electric furnace heating in metallurgical industry for melting of metals.
- Heating of electric kettle, heater, boiler, immersion heater etc.
- Heating of filament of incandescent lamp, arc lamp etc.
- Heating of d.c. locomotive for industrial hauling uses.
- In calorimetry.
- In fuses, which is used as protection device against flow of excessive current.
- In hot wire ammeter for measuring alternating current (a.c.).

4.10 Thermoelectric Effect (or Thermoelectricity)

It is a well-known fact that the electrical energy can be converted into heat energy and the heating effect of electric current is *irreversible*. Similarly, the heat energy can also be converted into electrical energy but this is a *reversible* effect, and is known as thermoelectric effect.

Thermoelectric effects are the phenomenon occurring in metals in which the flow of electrons (i.e. current flow) begins on application of thermal potential difference. This flow of current is different from the flow of current in metals when an electrical potential difference (gradient) is applied. The term thermoelectric implies that the flow of electricity is caused due to thermal effect, which can be explained in a general way as follows.

The explanation of thermoelectric effect is entirely different from the explanation of electrical heating effect. The thermoelectric effect (or thermoelectricity) can be studied through the following phenomena.

1. Seebeck effect
2. Peltier effect
3. Thomson effect

These phenomena are interrelated. Whereas the Seebeck effect is related to existence of a thermoelectric electromotive force (e.m.f.), the Peltier effect answers as to what is the source of energy of this Seebeck emf.

The Seebeck and Peltier effects explain the phenomena in a two metal junction (also called thermocouple), but the Thomson effect presents thermoelectric phenomena in a single metal, provided it is unequally heated.

4.11 Seebeck Effect

Seebeck discovered that if the junctions of a closed circuit of two different metals is maintained at different temperatures, then an electric current flows in the circuit. The two-metal circuit system is referred to as 'thermocouple'. The flow of current implies that there acts an e.m.f. (electromotive force) in the circuit. This e.m.f. is known as *thermoelectric emf.*

For understanding the mechanism of Seebeck effect, let us consider two different metals A and B in contact such that they form a junction as shown in Fig. 4.13a. Now let the electron density (i.e. the number of free electrons per unit volume of metal) in A is greater than that B. Then according to kinetic theory (of thermodynamics), the electrons will diffuse from positive A to negative B and a potential difference will be created at the junctions H (hot) and C (cold). This potential difference will allow the flow of electrons from A to B until a state of equilibrium is reached and the migration of electrons ceases completely.

4.11.1 Seebeck Series

Depending upon the behaviour of pairs of metals, the metals may be arranged in a series called Seebeck series. This series is as given below.

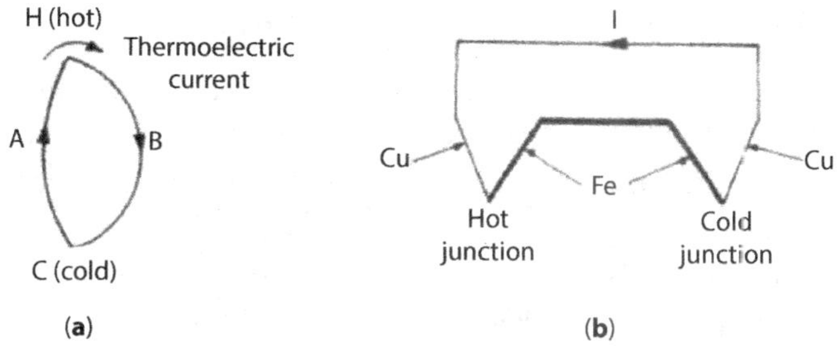

Figure 4.13 Explanation of Seeback effect (*a*) a thermocouple of two metals A and B showing flow of thermoelectric current across the hot junction, and (*b*) an iron-copper thermocouple showing flow of current in accordance with arrangement of metals in Seeback series.

Bi, Ni, Co, Pd, Pt, Cu, Mn, Ti, Hg, Pb, Sn, Cr, Mo, Rh, Ir, Au, W, Cd, Fe, As, Sb, Te

The Seebeck series enables us to get the idea for the magnitude of e.m.f produced and the direction of current in thermocouple. The current flows from the metal occurring earlier in series to the metal located later in series across the hot junction. For example in copper and iron junction, the current flows from copper to iron across the hot junction. Similarly in *Bi-Sb* thermocouple, the current will flow from *Bi* to *Sb* across the hot junction.

4.11.2 Seebeck e.m.f.

The thermoelectric e.m.f. of a thermocouple varies with a change in temperature of the hot junction, or with temperature difference in the hot and cold junctions. For majority of thermocouples, the variation is of parabolic nature. In general, the Seebeck e.m.f. is expressed by

$$E_s = aT + bT^2 \tag{4.28a}$$

where T is the temperature difference of two junctions in kelvin, a and b are characteristic constants of the given thermocouple.

Seebeck Voltage. It is the difference in voltages of hot and cold junctions. Thus if V_A and V_B are the voltages at hot and cold junctions, then

$$\mathrm{V}_{\mathrm{Seebeck}} = V_A - V_B \tag{4.28b}$$

4.11.3 Applications of Thermoelectric Effect

Some important applications of thermoelectric effect are the following.

1. Temperature measurement of ovens and furnaces by means of a thermocouple.
2. Temperature (radiation) measurement by thermopiles. Thermopiles consist of a number of thermocouples in series.
3. Detection of radiations by means of *radiomicrometer.* A radiomicrometer is a combination of a thermocouple and a sensitive galvanometer.
4. Measuring a.c. and d.c. by thermo-milliammeter.

4.12 Peltier Effect

It explains the mechanism of flow of current (electrons) from a metal *A* to another metal *B*, when an external thermal potential difference is applied on their junction. If the electron density in metal is greater than that in metal *B*, the movement of electrons begins from *B* to *A*. Since it involves work to be done against difference in electronic pressure, hence *energy is absorbed at the junction;* and the junction gets cooled. The reverse is also true, *i.e.* when electrons move from *A* to *B*, energy is available at the junction in the form of heat (i.e. heat is generated), and hence the junction gets heated.

4.12.1 Peltier Coefficient

The energy absorbed or liberated when a unit charge passes through the junction is termed as Peltier coefficient P_c, and is expressed as

$$Pc = \int_{V_1}^{V_2} e_p dV \tag{4.29}$$

where V_1 and V_2 are the volumes corresponding to unit charge and e_p is electronic pressure. According to kinetic theory, the electronic pressure is expressed as

$$e_p = \frac{1}{3} nmv^2 \tag{4.30a}$$

in which n is number of electrons per unit volume of a conductor, m the mass of an electron, and v^2 is its mean square velocity. Knowing that the kinetic energy $\frac{1}{2}mv^2 = \alpha T$, the Eq. 4.30a may be written as

$$e_p = \frac{2}{3}n\alpha T \tag{4.30b}$$

where α is a constant and T is the absolute temperature.

Since the volume V corresponding to unit charge is given by $V = 1/ne$, hence in Eq. 4.29 for Peltier coefficient, the values given in Eq. 4.31 can be substituted.

$$V_1 = \frac{1}{n_1 e}, \quad V_2 = \frac{1}{n_2 e}, \quad \text{and} \quad dV = V_2 - V_1 \tag{4.31}$$

On substituting the Eqs. 4.30b and 4.31 in Eq. 4.29, the value of Peltier coefficient is finally found as

$$Pc = \int_{V_1}^{V_2} \left(\frac{2}{3} \cdot \frac{1}{eV} . \alpha T \right) dV = \frac{2}{3} \frac{\alpha T}{e} \int_{V_1}^{V_2} \frac{dV}{V}$$

$$= \frac{2}{3} \frac{\alpha T}{e} . \ln \left(\frac{V_2}{V_1} \right) \tag{4.32a}$$

$$= \frac{2}{3} \frac{\alpha T}{e} . \ln \left(\frac{n_1}{n_2} \right) \tag{4.32b}$$

This shows that the Peltier coefficient is directly proportional to the absolute temperature.

4.13 Thomson Effect

This phenomenon refers to the absorption or evolution of heat due to flow of current (electrons) in a *single unequally heated conductor.* Depending upon the absorption or evolution of heat in a conductor whose one end is at higher temperature than the other, the hotter end will get *heating* effect and the colder end will get *cooling* effect. The involved phenomenon can be explained as follows.

i. Let the current is passed through a conductor whose one end is at a higher temperature than the other, then the transfer of electrons occur from colder to hotter end. Since the energy of an electron is proportional to the absolute temperature ($1/2\ mv^2 = \alpha T$) the energy of electrons moving towards hotter parts is increased and causes *heating effect.*

ii. Since the electronic pressure is higher at the hotter end due to $e_p \propto T$, the electrons move towards colder parts where energy is less, thereby causing *cooling effect.*

4.13.1 Types of Materials on the Basis of Thomson Effect

Both the above processes (effects) are observed in most of the materials. But in some materials, it is not observed at all. Accordingly, the materials are grouped into following categories, viz, those having

1. **Positive Thomson Effect.** Where heat is evolved when a current flows from hot to cold side such as in Ag, Cu, Zn, Cd, Sb etc.
2. **Negative Thomson Effect.** Where heat is absorbed when a current flows from cold to hot side such as in Fe, Ni, Co, Pt, Bi etc.
3. **Zero Thomson Effect.** Where heat is neither evolved not absorbed such as in Pb.

The Thomson coefficient C_T is determined from

$$C_T = -T\frac{d^2E_s}{dT^2} \tag{4.33}$$

where E_s is Seeback e.m.f.

4.13.2 Materials for Thermocouples and Thermopiles

The choice of metals for the thermocouples depends upon the accuracy and range of temperature measurement. Suitable materials in this respect are given in Table 4.4.

4.14 Wiedemann-Franz Law and Lorentz Relation

Statement. According to Wiedemann-Franz law, the electrical conductivity of solids can be related to their thermal conductivity. It is because all

Table 4.4 Materials for Thermocouples and Thermopiles

Range of temperature measurement	Suitable pair of metals	Junction to be heated is enclosed in a
Thermocouple:		
upto 1000°C	Fe-Constantan	Steel tube
upto 1400°C	Pt and Pt-Rh	Quartz tube
upto 1700°C	Mo-W	Reducing atmosphere of N_2 and H_2
Thermopile:		
for very small temperature differences	Sb-Bi	—

the solids conduct heat and electricity. Whereas the thermal conductivity in ionic, covalent and molecular solids is primarily through lattice vibrations; the transport of thermal energy in metals and alloys is mainly by free electrons.

Proof. To prove the above statement, we need to derive the expressions for electrical and thermal conductivities. We have already derived the electrical conductivity in art 4.3.5 in the form of Eqn. 4.11 and is given as $\sigma = ne^2t/m$. We shall now derive the expression for thermal conductivity.

4.14.1 Determining the Thermal Conductivity

We have studied earlier that the transport of heat flux Q through a solid of cross-sectional area A is proportional to the thermal gradient dT/dx, and is given by

$$Q = -K_t A \frac{dT}{dx} \tag{4.33}$$

where K_t is a proportionality constant and is known as thermal conductivity of the solid.

As the temperature gradient is dT/dx along x-direction, the average energy of electrons is a function of x. Now let the average energy per electron be $E(0)$ in the plane $x = 0$, therefore at a distance Δx from this plane, the energy of electrons will be

$$E(\Delta x) = E(0) + \frac{dE}{dx}\Delta x$$
$$= E(0) + \frac{dE}{dT}.\frac{dT}{dx}\Delta x$$

Now let v_{xi} is the velocity of an electron in x-direction and the location of plane of its immediate preceding collision is $-\Delta x_i$, then the number of electrons passing the plane $x = 0$ will be $n_i v_{xi}$, where n_i is the number of electrons per m^3. Thus the density of total heat current through $x = 0$ plane will be the summation extending over all the electrons per m^3. Hence

$$Q = \sum_{i=1,2,...} \left[E(0) - \frac{dE}{dT}\frac{dT}{dx}\Delta x_i \right] v_{xi} \tag{4.34}$$

Knowing that the average velocity of electrons/unit volume is zero (i.e. $v_{avg} = 0$), the quantity $E(0)$ in Eqn. 4.34 vanishes. We are thus left with

$$Q = -\sum_i \frac{dE}{dT}\frac{dT}{dx}\Delta x_i v_{xi} \tag{4.35}$$

or
$$Q = -n\frac{dE}{dT}\frac{dT}{dx}(v_x \Delta x)$$

where n represents the total number of conduction electrons/m^3 and $(v_x\Delta_x)$ is the average value of $\Sigma v_{xi}\Delta x_i$.

4.14.2 Consideration of Electron Collision

Considering now an electron moving with a velocity v in the direction making θ angle with x-axis (Fig. 4.14), its velocity component along x- axis will be $v_x = v\cos\theta$. Similarly the distance Δ_x between the plane $x = 0$ and the plane of previous collision of electron, actually covers a length Δr (Fig. 4.14). Since $\Delta r = \Delta x/\cos\theta$, the Eqn. 4.35 can be written as

$$Q = -n\frac{dE}{dT}\frac{dT}{dx}(v\cos\theta.\Delta r\cos\theta)$$
$$= -\frac{n}{3}\frac{dE}{dT}\frac{dT}{dx}(v\Delta r) \tag{4.36a}$$

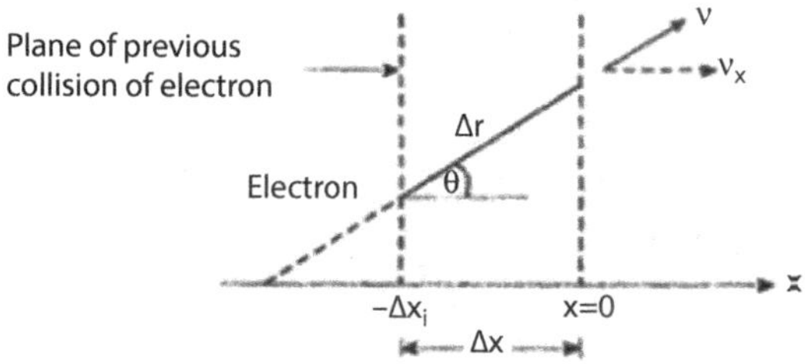

Figure 4.14 The geometry of electron movement and collision.

$$\because \qquad \overline{\cos^2\theta} = \frac{\int_0^{\pi} \cos^2\theta \sin\theta d\theta}{\int_0^{\pi} \sin\theta d\theta} = \frac{1}{3}$$

4.14.3 Consideration of Fermi Energy

In such analysis, the consideration of mean free path or relaxation time of electrons with Fermi energy is of greater importance. Therefore v in Eqn. 4.36a may be replaced by Fermi- velocity v_F, and Δr may be replaced by λ which is mean free path of scattering corresponding to electrons with v_F. Thus Eqn. 4.36a modifies to

$$Q = -\frac{n}{3}\frac{dE}{dT}\frac{dT}{dx}(v_F\lambda) \tag{4.36b}$$

In above equation the quantity n(dE/dT) represents the specific heat of electron gas (i.e. free electrons cloud) at constant volume *i.e.* C_v and $\lambda = v_F\tau_r$ in which τ_r is relaxation time. Therefore, Eqn. 4.36b modifies again as

$$Q = -\frac{1}{3}C_v v_F^2 \tau \frac{dT}{dx} \tag{4.37}$$

Since from the consideration of Fermi-Dirac probability distribution, one may find that

$$C_p = \frac{n\pi^2 k^2 T}{m v_F^2}$$

therefore $$Q = -\frac{1}{3}\frac{n\pi^2k^2T\tau}{m}\frac{dT}{dx} \tag{4.38}$$

On comparing this equation with Eqn. 4.33 for unit area A, we find

$$-K_t\frac{dT}{dx} = -\frac{1}{3}\frac{n\pi^2k^2T\tau}{m}\frac{dT}{dx}$$

or $$K_t = \frac{1}{3}\frac{n\pi^2k^2T\tau}{m} \tag{4.39}$$

4.14.4 Lorentz Number

If we compare Eqn. 4.39 for thermal conductivity and Eq. 4.11 for electrical conductivity, we find that

$$\frac{K_t}{\sigma} = \frac{\frac{1}{3}\frac{n\pi^2k^2T\tau}{m}}{\frac{ne^2\tau}{m}}$$

or $$\frac{K_t}{\sigma T} = \frac{\pi^2}{3}\frac{k^2}{e^2} = (\text{say } L) \tag{4.40}$$

Since each quantity on right side of this equation is a constant, it implies that $K_t / \sigma T$ for all metals is a universal constant. This constant $(\pi^2k^2/3e^2) = L$ is called **Lorentz number.** Its value is 2.45×10^{-8} WΩ/ (Kelvin)^2.

Example 4.7 Find the thermal conductivity of a metal containing 2.5×10^{22} electrons/cm^3 at 27°C. The average collision time between electrons and the obstacles is 3×10^{-14} s. Take Lorentz number = 2.44×10^{-8} ohm W/K^2. The mass of electron is 9.1×10^{-31} kg. Assume any data else, if required.

Solution : Given are

$n = 2.5 \times 10^{22}/\text{cm}^3 = 2.5 \times 10^{28}/\text{m}^3$, $t = 3 \times 10^{-14}$ s,
$m = 9.1 \times 10^{-31}$ kg.
$L = 2.44 \times 10^{-8}$ ohm W/K^2, and
$T = (27 + 273) = 300$ K. We take $e = 1.6 \times 10^{-19}$C.

We first determine the electrical conductivity given as

$$\sigma = \frac{ne^2t}{m}$$

On substituting the appropriate values, we get

$$\sigma = \frac{(2.5\times10^{28})\times(1.6\times10^{-19})^2\times(3\times10^{-14})}{9.1\times10^{-31}}$$

$$= 21.09 \times 10^6 \text{ (ohm m)}^{-1}$$

Now using Eqn. 4.40 for Lorentz number, we find

$$\frac{K_t}{\sigma T} = 2.44\times10^{-8} \text{ ohm W / K}^2$$

or $K_t = \sigma_T \times (2.44\times10^{-8})$
$= (21.09 \times 10^6) \times (300) \times (2.44 \times 10^{-8})$
$= 154.378$ W/mK

Example 4.8 A cold rolled silver sheet containing small amounts of Mg or Al is suitable for electrical contact applications- Justify.

Solution: The addition of Mg or Al in solid solution of cold rolled silver sheet enhances its strength and surface hardness by about three times. Fine precipitate of MgO or Al_2O_3 forms at about 700°C to 800°C when oxygen dissolves in silver, and diffuses in the sheet. This reduces the electrical resistivity i.e. the electrical conductivity increases. Hence the suitability for use in electrical contacts is established.

4.15 Solved Examples

Example 4.9 Explain the relaxation process and relaxation time.

Solution. Relaxation is a time dependent phenomenon that occurs due to collision of electrons with the obstacles present within the crystal (material). Jumping of atoms by diffusion, atomic vibration, flow of grain boundaries, oscillation and damping are some relaxation processes. Time interval of applied forces/fields and the time taken for completion of these relaxation processes may be comparable in some cases and quite different in other cases. A relaxation process may be too slow or too fast as compared to force/field (application) time interval. Relaxation processes are specified by relaxation time t_r, which is expressed as

$$\text{p} = 1 - \text{e}^{(-\text{t}/\text{tr})} \qquad (4.41)$$

where p is the fraction of relaxation process and t is any arbitrary time. When $t = t_r$, then $p = 0.63$ i.e. the relaxation time is the time taken to complete 63% of a process. For different processes, t_r varies from 10^{-13} second to 10^{7} second.

Example 4.10 What is electron scattering? How does it influence the behaviour of electron and the conducting crystal?

Solution. Collision of electrons with obstacles within a conducting material is an inherent phenomenon. Depending upon various factors, they scatter over certain angle θ, after collision. This is called *electron scattering.*

The phenomenon of electron scattering causes loss in its velocity after collision. It also causes transfer of heat energy from electron to crystal lattice. Probability of scattering remains a constant in case the scattering is spherically symmetric.

Example 4.11 What are the sources of presence of obstacles in conducting materials? Why do we consider an average value of collision time in electrical designing?

Solution. The obstacles in materials remain present in different forms. They may be present in natural form such as cementite in steel and iron, oxides in copper and aluminium; or may be present in manmade (unnatural form) such as grain boundaries. These obstacles are located randomly in different orientations within the (conducting) materials. That is why the collision of electrons with them occurs at different intervals of time. Therefore, the collision time is determined as an average value, and is defined as the average time elapsed between two successive collisions.

Example 4.12 Discuss the differences between Joule effect and Peltier effect.

Solution. The above two effects are different in the following respects.

S. No.	Joule effect	Peltier effect
1.	It is irreversible effect.	It is reversible effect.
2.	Heat is always evolved.	Heat is evolved at one junction and absorbed at the other.
3.	Heat is produced all along the conductor.	Heat is produced at one junction only.
4.	Heat generated is proportional to the square of current.	Heat generated is proportional to the current.

Example 4.13 Establish a relation between temperature coefficient of resistance α_2 at t_2 temperature and α_1 at t_1 temperature.

Ans. With the help of Eq. 4.19, the relation among resistances R_1, R_2 and R_3 at temperatures t_1, t_2 and t_3 respectively may be written as

$$R_2 = R_1[1+\alpha(t_2 - t_1)],$$
$$R_3 = R_1[1+\alpha(t_3 - t_1)],$$
and
$$R_3 = R_2[1+\alpha(t_3 - t_2)],$$

A simple manipulation of above equations will yield a relation given as

$$\alpha_2 = \frac{1}{\frac{1}{\alpha_1} + (t_2 - t_1)} \tag{4.42}$$

Example 4.14 Prove that $\alpha_2 = \frac{1}{\frac{1}{\alpha_1} + (T_2 - T_1)}$ where each term has its usual meaning.

Alternatively $\alpha_1 = \frac{\alpha_0}{1+\alpha_0(T_1 - T_0)}$

Solution. The value of R_0 in terms of R_T during cooling is given by

$$\begin{aligned} R_0 &= R_T[1+\alpha_T(0-T)] \\ &= R_T(1-\alpha_T.T) \end{aligned} \tag{i}$$

where α_T is temperature coefficient at T°C.
We know that

$$R_T = R_0[1+\alpha_0(T-0)] \tag{ii}$$

where α_0 is temperature coefficient at 0°C.
From Eq (i), we find that

$$\alpha_T = \frac{R_T - R_0}{R_T(T-0)} \tag{iii}$$

On substituting the value of R_T from Eq. (ii) in Eq. (iii), we get

$$a_T = \frac{R_0[1+a_0(T-0)]-R_0}{R_0[1+a_0(T-0)](T-0)}$$

$$= \frac{a_0}{1+a_0 T} \quad (iv)$$

On generalization, Eq. (*iv*) may be written as

$$a_1 = \frac{a_0}{1+a_0 T_1}$$

or $$\frac{1}{a_1} = \frac{1+a_0 T_1}{a_0} \quad (v)$$

Similarly $$\frac{1}{a_2} = \frac{1+a_0 T_2}{a_0} \quad (vi)$$

On subtracting Eq. (*v*) from Eq. (*vi*), we get

$$\frac{1}{a_2} - \frac{1}{a_1} = (T_2 - T_1)$$

or $$\frac{1}{a_2} = \frac{1}{a_1} + (T_2 - T_1)$$

$\therefore$ $$a_2 = \frac{1}{\frac{1}{a_1}+(T_2-T_1)} = \frac{a_1}{1+a_1(T_2-T_1)}$$

Quick Revision Summary

Drift velocity v_d. It is the additional velocity acquired by an electron above its normal velocity v_n when accelerated by an applied electric field.

Mobility μ is defined as the proportionality constant between the drift velocity and applied field.

Relaxation time is defined as the time taken to complete 63% of a process.

Residual resistivity ρ_r is defined as the temperature independent part of the resistivity of a conductor. Impurities and lattice defects in materials are responsible for it.

Resistivity ratio. It is defined as the ratio of resistivity at room temperature to that at any lower temperature. The lower temperature is generally taken as 4.2 K. Thus

$$\rho_{\text{ratio}} = \rho \text{ at } 300\text{K}/\rho \text{ at } 4.2\text{K}$$

Wave number k. It is the wave function of an electron which is assumed to be moving in plane waveform along x-direction. It is expressed as $k = 2\pi/\lambda$ where λ is de-Broglie's wavelength.

Review Questions

1. What assumptions have been made in the free electron theory? Derive the relation for kinetic energy of free electron of M-shell as a function of its wave number.
2. Explain briefly the following.
   a. Drift velocity and its relation with current density.
   b. Mean free path and its effect on conductivity.
   c. Effect of temperature on resistivity.
   d. Collision time.
3. Explain the energy band theory stating the assumptions made therein. Discuss the conclusions of this theory.
4. Why is the energy versus wave number curve discontinuous in energy band theory? Plot this curve and show the forbidden gap on it.
5. Explain the Brillouin zone theory. Why is it called zone theory? How is first Brillouin zone different from second Brillouin zone?
6. Discuss the Brillouin zone for a simple cubic lattice, BCC, FCC and HCP lattices.
7. Classify solids on the basis of energy gaps. Name various conducting materials and compare their conductivity ranges, properties, and characteristics.
8. Enumerate the characteristics of a good conductor. Discuss the effects of different factors on resistivity of a conductor.

9. Discuss the following.
   a. Nordheim equation and Matthilssen rule of electrical resistivity.
   b. Lattice and electronic thermal conductivity.
   c. Electrical heating of current and Joule's law.
   d. Effect of plastic deformation on conductivity of metals.
10. In conducting materials, prove that

$$\frac{1}{\mu_e} = \frac{1}{\mu_I} + \frac{1}{\mu_L}$$

where μ_e = overall drift mobility, μ_I = ionized impurity scattering limited mobility, μ_L = lattice vibration scattering limited mobility.

11. If α_1 is the temperature coefficient of resistance (TCR) at temperature T_1 and α_0 is the TCR at T_0, show that

$$a_1 = \frac{a_0}{1 + a_0(T_1 - T_0)}$$

12. Discuss the various thermoelectric effects and write their applications. Explain Seebeck effect. Peltier and Thomson effects.
13. State the following thermoelectric effects.
    a. Seeback Effect
    b. Peltier Effect and
    c. Thomson Effect
14. Derive an expression for heat development in a current carrying conductor. Also explain the factors responsible for it.
15. Derive the Widemann-Franz law and Lorentz relation. Write all approximations and the assumptions made therin.
16. Write the statement of Wiedemann-Franz law. What are the effects of considering electron collision and Fermi energy?
17. Write notes on the following.
    a. Thermocouples and thermopiles.
    b. Relaxation process and relaxation time of electrons.
    c. Electron scattering phenomenon.

Numerical Problems

1. An electric field of 100 V/m is applied in copper. Calculate (*a*) the average thermal velocity, (*b*) the mobility, and (*c*) the drift velocity for electrons; when the particles have three-dimensional movement. The resistivity of copper at 298 K is 1.72×10^{-8} ohm m.
2. An aluminium alloy of 2.8×10^{-8} ohm m resistivity is of 1.12 mm diameter. What length of it will have a resistance of 0.028 ohm?
3. A 200 μm diameter wire is made of a metal whose conductivity is 2.42×10^{7}/ohm m. For what length will it keep the resistance of 0.526 ohm? Calculate its resistivity also.
4. A conductor has to carry a current of 25 A in which the power dissipation must not exceed 6 watt per metre length. Calculate the diameter of conductor whose resistivity is 4.71×10^{-8} ohm m.
5. A 1.27 m long magnesium flat is 2.03 mm wide and 0.15 mm thick. Its conductivity is 2.31×10^{9} per ohm m. Determine its resistance at 20°C. If this temperature is raised twenty times, how many times the resistance increases ?
6. The resistivity of pure copper is 1.56 ohm cm, of copper-nickel alloy (1 atomic percent Ni) is 2.81 ohm cm, and copper-silver alloy (3 atomic percent Ag) is 1.93 ohm cm. Determine the resistivity of (Cu + Ni + Ag) alloy containing 2% each of the nickel and the silver elements.
7. The number of conduction electrons in silver is 6×10^{28} per m^3. Determine the mobility of conduction electrons and the drift velocity in silver under an electric field of 1 V/m. Take standard values of electron charge and its mass.
8. The conductivity of platinum at room temperature is 9.9×10^{6}/ohm m, its atomic mass 195 amu, and the specific gravity is 21.45. Calculate (*a*) the number of free electrons, and (*b*) the collision time for electron scattering.
9. The resistivity of gold is 2.1×10^{-8} ohm m at room temperature. Its density is 19320 kg/m^3 and the atomic mass 197 amu. The free electrons acquire an incremental velocity of 0.79 m/s in certain field gradient. Calculate this field gradient.

10. Determine the mean free path of free electrons in pure silver when the collision time for phonon scattering is 10 ns at –269°C. The Fermi energy for silver is 5.5 eV.
11. The drift velocity of an electron moving in a silicon crystal is 420 m/s under an electric field gradient of 3 kV/m. Find the mobility of electron at room temperature.
12. An alloy containing 90 wt % Au and 10 wt % Cu is sometimes used in low voltage dc electrical contacts. Predict the resistivity of the alloy and compare it with experimental value of 108 nΩm. Atomic masses of Cu and Au are 63.55 and 197.0 respectively. Room temperature resistivity of gold is 22.8 nΩm and Nordheim's coefficient = 450 nΩm.
13. A uniform silver wire has a resistivity of 1.54×10^{-8} ohm m at room temperature. For an electric field along the wire of 1 volt cm^{-1}, compute the (*a*) average drift velocity of the electrons, assuming there are 5.8×10^{28} conduction electrons per m^3. Also calculate the (*b*) mobility, and (*c*) relaxation time of the electrons.
14. A copper wire has a resistivity of 1.8×10^{-8} ohm m at room temperature (300°K). Assming the copper is very pure, estimate the (*a*) resistivity at 700°C, and the (*b*) percentage change in the resistivity from room temperature to 700°C.
15. A copper wire is 1 m long and has a uniform cross-section of 0.1 mm^2. The resistance of the wire at room temperature is found to be 0.172 ohm. What is the resistivity of the material?
16. The temperature difference between the inside and outside of a glass window is 72° Fahrenheit. The glass has a thermal conductivity of 0.0025 calories $sec^{-1}cm^{-1}$ $degree^{-1}$ and is 1 mm thick. Find the energy loss in joules through the window per m^2 per hour.
17. The electrical resistivities of copper and kanthal (an alloy of iron, chromium and aluminium) at room temperature are respectively 1.7×10^{-8} and 1.4×10^{-6} ohm m. Assuming the Wiedemann- Franz law holds for these materials, find the electronic contributions to the thermal conductivities of these materials. (In alloys such as kanthal, the contribution to thermal conductivity from the lattice vibrations is comparable to that from the electrons)

18. The velocity of electrons in a solid at Fermi level is 9×10^{15} m/s. Obtain the Fermi level of the solid. What this solid is likely to be? Assume data suitably, if required.

Objective Questions

1. In an imperfection free crystal, the resistivity will be

   a. infinite b. zero
   c. negative d. unity

2. An example of a metal having zero Thomson effect is

   a. zinc b. cobalt
   c. lead d. rhodium

3. Which of the following pairs are correctly matched?

1. Nordheim equation	Effect of alloying on resistivity
2. Lorentz number	$\mu_h = T^{2.4}$
3. Wiedemana-Franz law	$n_e >> n_h$
4. Matthilssen rule	Total resistivity is the sum of different effects

   Select the correct answer using the codes given below.

   Codes :

   a. 2 and 3 b. 1 and 4
   c. 1 and 2 d. 1, 2 and 4

True and False Type Questions

4. Behaviour of electrons is different in conductors, semiconductors, and insulators. *(T/F)*
5. The potential energy is assumed constant in free electron theory. *(T/F)*
6. Drift velocity of an electron is equal to its normal velocity. *(T/F)*
7. Conductivity of pure copper will be more than the conductivity of brass (copper alloy) *(T/F)*
8. Mean free length of an electron in gold is longer than that in nylon. *(T/F)*

Fill in the Blank Type Questions

9. Valence electrons are .... bound in conductors.
10. Value of energy gap in silver will be ....
11. Phenomenon of heating effect of current can be explained by ....
12. The resistivity based on impurity and plastic deformation is approximately zero at .... temperature.
13. .... voltage is the difference in voltages at the hot and cold junctions of the thermocouple.

Multiple Choice Type Questions

14. The value of wave number may be
    a. zero b. positive
    c. negative d. all of the above
15. With increase in temperature, the mean free path
    a. decreases b. inreases
    c. remains unchanged d. vanishes
16. Highest electrical resistivity exists in
    a. platinum wire b. nichrome wire
    c. silver wire d. kanthal wire

5

Conductive Materials: Types and Applications

5.1 Mechanically Processed Forms of Electrical Materials

Several electrical materials are processed mechanically before use to impart special qualities in them. By doing so they perform better in their purpose. A brief explanation of such materials is given below.

5.1.1 Cladded Metals

These are metal to metal laminates of two different metals in which both are meant to serve different specific purposes. Stainless steel cladding on mild steel is one such example in which mild steel being cheaper is used to make structural part while the stainless steel being chemical resistant remains in contact with the chemicals.

5.1.2 Bimetals

These are combinations of two dissimilar materials having different coefficients of expansion. Due to unequal expansion and contraction, thermal stresses are induced in them on thermal or electrical heating. Use of bimetallic strips is made in *thermostatic controls*. A bimetallic strip is normally composed of copper and steel.

5.1.3 Sintered Materials

These materials are made by 'sintering' method of 'powder metallurgy' technique. Sintering involves fabrication of products by application of temperature and pressure on metal powders. The metal powder is prepared in the particles form of 1 μm to 200 μm sizes. Sintering method of manufacturing is very much suitable for making intricate parts. It imparts strengthening effect and uniform distribution of particles.

Copper and silver particles are mixed with molybdenum, tungsten and their carbides by this technique to make *electrical contacts*. That is why they are termed as sintered contacts. Hard magnets made of powdered manganese bismuthide (MnBi) are also the example of this kind.

5.1.4 Hot Rolled and Cold Rolled Metals

Rolling is a manufacturing process to make the circular and square bars, thin strips; channel, angle, and I-sections etc. Hot and cold working are the temperature dependent processing of metals. When the rolling is done above Recrystallization temperature of a metal, this is called *hot rolling* while it will be known as cold rolling if processed below Recrystallization temperature. Recrystallization temperature of metals lies in the range of 0.3 to 0.5 T_m and for alloys in the range of 0.5 to 0.7 T_m where T_m is melting point of metals and alloys in kelvin.

Steel and aluminium made products are generally made by rolling process. Depending upon the temperature of rolling, they are called as either hot rolled steel or cold rolled steel. In electrical applications they are used for the following purposes.

- Hot rolled steel for sheet, strip and wire making
- Cold rolled grain-oriented sheet steel (a magnetic material) for making transformer cores
- Cold rolled aluminium for wire, flat, and foil making.

5.1.5 Hard Drawn and Soft Drawn Metals

'Drawing' is another manufacturing process to make rods, bars, tubes, wires etc. If the metals are hot worked, they remain soft and are known as *soft drawn metals.* When the drawing process is done below Recrystallization temperature *i.e.* the metals are cold worked, they exhibit hard nature. Therefore, they are called *hard drawn metals.*

Aluminium, copper, and their alloys are generally made by drawing process. Some important applications of hard and cold drawn electrical materials are as follows.

- Hard drawn copper and aluminium for overhead transmission line, hollow tubes for jacketing, etc.
- Cold drawn lead and lead alloys for sheathing of cables

5.1.6 Annealed Metals

Annealing is a heat treatment process meant to (i) soften the metals, (ii) improve the ductility of metals, (iii) enhance the machinability of metals, (iv) refine the grain structure, and (v) relieve the metals from residual (internal) stresses. This process is performed on various metals and alloys of copper, aluminium, tin, steel etc. Some salient applications of annealed metals for electrical uses are as follows.

- Annealed copper such as in transformer winding, coils
- Annealed aluminium such as in insulated conductors, low voltage cables
- Annealed tin such as in fuses, tinning of copper for high temperature uses.

5.2 Types of Conducting Materials

Depending upon the desired quality of conductivity (reciprocal of resistivity), vividity of metals and types of applications, the conducting materials may be classified as follows.

I. On the Basis of Quality of Resistivity

1. Low resistivity (or high conductivity) materials such as Au, Ag, Cu, Al, brass, bronze etc.
2. High resistivity (or low conductivity) materials such as W, Pt, C, Ni, Ta, nichrome, manganin etc.

II. On the Basis of Nature of Materials

1. Metals such as Cu, steel, Al, Pt, Ir, Pd, Rh etc.
2. Non-metals such as conducting polymer

3. Alloys such as brass, bronze, rose metal, constantan etc.
4. Reinforced composites such as Aluminium Conductor Steel Reinforced (ACSR).

III. On the Basis of Applications

1. Cable purpose materials such as Cu, Al, ACSR, OFHC (Oxygen Free High Conductivity) Cu
2. Filament purpose materials such as W, carbon (graphite)
3. Contact purpose materials such as Au, Pt, Ag, Mo, Ir, W etc.
4. Soldering purpose materials such as Cu, Zn, Sn, Ag, Cd
5. Fusible (or fuse) purpose materials such as Pb, Sn, Ag etc.
6. Sheathing purpose materials such as Pb, p.v.c., Pb-Cd alloy
7. Sealing purpose materials such as Fe with Si, Mn, Mo.

IV. On the Basis of Temperature Requirements

1. High temperature conducting materials such as ACSR
2. Low temperature conducting materials such as OFHC copper
3. Very low (or cryogenic) temperature conducting materials such as Ni-based steel alloy.

V. On the Basis of State of Material

1. Solid conducting materials such as metals, alloys
2. Liquid conducting materials such as Hg
3. Gaseous conducting materials such as N_2 in gas-filled pressure cables.

5.3 Low Resistivity Materials

Low resistivity materials possess lower value of resistivity *i.e.* higher conductivity. Hence, they are suitable for those applications where the voltage drop and power loss are to be kept to a minimum value. Consequently, they find use in the windings of alternators, motors, transformers; in domestic wiring, and in electricity transmission from power stations to other places. Gold, silver, copper and aluminium are the examples of low resistivity materials. Amongst these materials, copper is commercially most acceptable.

5.3.1 Characteristics of Low Resistivity Materials

Besides having a lower value of resistivity, the low resistivity materials are also required to possess the following characteristics.

Low coefficient of temperature resistance α. The value of α should be low to avoid (or minimize) the variation in voltage drop and power loss, with change in temperature. If the value of α is high, the voltage drop and power loss will increase. A low value of α helps in reducing the heat in windings of electrical machines and in increasing the resistance of transmission lines during hot season.

High resistance to corrosion. This is a desired requirement to prevent corrosion due to environmental effects. The conducting material should not easily corrode, particularly when used in atmospheric exposure without insulation.

Good solderability. Conductors are often required to be jointed. The joints should offer a minimum contact resistance. Joining is normally done by soldering because it offers minimum contact resistance. All materials do not possess good solderability. Hence, the solderability of conducting materials should be such as to develop minimum resistance.

High mechanical strength. Overhead transmission and distribution lines are subjected to stresses and strains under their self weight and also due to winds. The conductors used in the windings of electrical machines and transformers develop mechanical and thermal stresses also when loaded. Therefore, the low resistivity materials should be mechanically strong enough to resist these stresses.

Good ductility. The conducting materials should be ductile enough to enable themselves to be drawn into different shapes and sizes.

5.3.2 Copper and its Types

Due to its high conductivity and reasonable cost, copper is most widely used metal for electrical purposes. It is a crystalline, non-ferrous, nonmagnetic (diamagnetic), reddish coloured metal. It possesses several advantageous properties such as given below.

- It is a ductile metal having a ductility of more than 15%. By virtue of this property, it can be easily drawn into thin bars and wires. Hence, it is very useful for making cables, strands, and conductors.
- Its ultimate tensile strength is high enough (300–350 MPa) which makes it substantially strong to sustain mechanical loads.

- Its melting point is sufficiently high (1083°C) that makes it suitable for use at high temperatures also.
- When exposed to atmospheric environment, it forms a protective layer of copper oxide (CuO). Thus, the copper is highly resistant to corrosion which is a desired property for bare/open overhead conductors.
- It can be easily brazed (a kind of welding) which is a necessary requirement in electrical wiring and other connections.

Types of copper. Copper is available in two distinct forms. These are

1. Annealed copper, and
2. Hard drawn copper.

Annealed copper is more ductile than the hard drawn copper. It can withstand severe bending and forging stresses without failure. It is used as power cables, winding wires for electrical machines and transformers, and in making coils. *Hard drawn* copper possesses high mechanical strength. It is suitable for overhead transmission wires and bus-bars etc.

The properties of above copper forms and their suitable uses are compared as follows.

Description	Annealed copper	Hard drawn copper
• Conductivity	Higher	Lower
• Tensile strength	Less	More
• Hardness	Less	More
• Nature	Flexible	Springy
• Tenacity	Higher	Lower
• Specific gravity	8.89	8.93
• Resistivity (at 20°C in ohm-m)	1.72×10^{-8}	1.77×10^{-8}
• Applications	Low voltage power cables, insulated conductors, coils, flexible wires; windings of transformers, motors and alternators	Overhead conductors, bus-bars, high-voltage cables, under-ground cables etc.

5.3.3 Types of Aluminium and their Applications

This is also a widely used conducting material. Its conductivity is lower to that of copper, but it is cheaper than copper. It is a crystalline, nonferrous, weak magnetic (paramagnetic) metal of white colour. It possesses several advantageous properties as stated below.

- It is a lightweight metal having a specific gravity of 2.7 only.
- It is a ductile metal by virtue of which it can be easily drawn into thin bars and wires. Therefore, aluminium is very useful for making cables, strands, and conductors.
- Its ultimate tensile strength is sufficiently high (about 50-70 MPa) which is lower than copper. Therefore, it does not find favour in making windings for transformers, motors and alternators.
- Its melting point is 660°C due to which it is suitable for moderate temperature uses only.
- Under atmospheric exposure, it forms a protective layer over its surface. This layer is of aluminium oxide (Al_2O_3) that makes aluminium as corrosion resistant. However, the protective layer acts as insulation layer due to higher resistivity, and hence lowers the conductivity of aluminium.
- It can be easily soldered which is an essential requirement in wiring and other connections. However, due to insulating nature of aluminium oxide layer, the soldered joint may not be as strong as desired.
- Its electrical contacts suffer from the possibility of loose connections due to its softness.
- Pure aluminium is a very soft metal. But when alloyed with alloying elements like Si, Mg, Ni, Co etc., it becomes hard and tough, and gains higher mechanical strength.

Types of Aluminium. Aluminium may be used in the following forms.

1. Rolled aluminium
   i. hot rolled
   ii. cold rolled
2. Hard drawn aluminium.

Rolled aluminium is manufactured by rolling process. When rolling is accomplished at above recrystallization temperature (about 195°C), it produces hot rolled aluminium, and when the process is carried below recrystallization temperature the produced product is cold rolled aluminium.

Applications. Salient applications of aluminium as electrical material are the following.

- Overhead transmission line
- Domestic wiring
- Bus-bars
- Rotor bars of squirrel-cage induction motor
- Electrolytic capacitors for filter and by-passing purposes from a superimposed d.c. voltage.

5.3.4 Comparison among Different Low Resistivity Conducting Materials

A comparison of electrical, mechanical, thermal and physical properties for important low-resistivity conducting materials is given in Table 5.1 for a ready reference.

5.3.5 Copper Alloys (Brass and Bronze)

Several alloys of copper also fall in the category of low-resistivity conducting materials. Main among them are brass and bronze. Brass is basically an alloy of copper and zinc, while bronze is an alloy of copper and tin. But they are prepared with the addition of other elements also. Brass and bronze are available in different compositions and so possess varying properties. Consequently, they are of different types viz. admiralty brass, muntz metal, delta metal, phosphor bronze, bell metal, gun metal etc. But amongst these, each type is not suitable for electrical applications. Various kinds of brass and bronze, and other non-ferrous alloys that find use in electrical industry are summarized below and given in Table 5.2.

Brass. It is an alloy of copper containing 40% Zn. Its conductivity is lower than that of copper. It has a high tensile strength and is fairly resistant to corrosion. It can be easily pressed into a desired shape and size, can be drawn into wires, and can be easily brazed. Brasses are widely used in the following applications.

- plug-points
- socket-outlets
- lamp holders
- fuse holders
- switches
- knife switches
- sliding contacts for rheostats and starters, etc.

Table 5.1 Comparison of properties of low resistivity metals

Description	**Copper**	**Aluminium**	**Steel**
• Resistivity (ohm-m)	17×10^{-9}	28×10^{-9}	—
• Coefficient of linear expansion (/K)	17.7×10^{-6}	23.7×10^{-6}	15.2×10^{-6}
• Coefficient of temperature resistance at 20°C	0.0039	0.0035	0.004 - 0.005
• Resistance to corrosion	More	More	Less
• Tensile strength	More	Less	Maximum
• Ductility (about)	55%	50%	18-25%
• Density (kg/m^3)	8960	2700	7800
• Melting point (°C)	1083	660	1539
• Thermal conductivity at 25°C (W/m-K)	425	300	65
• Salient feature	Can be drawn into thin wires	Can be drawn into wires upto 0.1 mm diameter	Can be drawn into thick wires
• Comparative cost	12	8	1
• Working efficiency	Best	Better	Good
• Life	Highest	Higher	High

Bronze. It has a composition of 10% Sn in 90% Cu. Its conductivity is lower than that of pure copper. Bronze components are generally made by forging process. It is corrosion resistant and possesses high strength. Different types of bronze are generally used in the following applications.

- Beryllium bronze for making current carrying springs, sliding contacts, knife-switch blades etc.
- Phosphor bronze for making springs, bushings etc.
- Cadmium bronze for making commutator segments.

5.4 High Resistivity Materials

High resistivity materials possess a high value of resistivity *i.e.* lower conductivity. When used as conductors, they require a higher potential difference to set-up the movement of electrons. More heat is also generated in them due to

Table 5.2 Compositions and applications of some non-ferrous alloys

Alloy	Composition	Application
Mg-alloy		
Elecktron	Al 3-12%, Mn 0.03%, Zn 2%, rest Mg	Parts of aircrafts and automobiles, collapsible tubes, coating on steel utensils, solder, coating of petroleum containers and mild steel sheets
Al-alloy		
Duralumin	Cu 4%, Mg 0.5%, Si0.7%, rest Al	Electrical cables, components of aeroplanes and automobiles
Y-alloy	Cu 4%, Mg 1.5%, Ni 2%, Fe 0.6%, rest Al	Casting of engine parts
Cu-alloy: Brass		
Muntz metal	Cu 60%, Zn 40%	Brass castings, stampings and extruded parts
Yellow brass	Cu 90%, Zn 10%	Screw, wires, hardwares
Admiralty brass	Cu 70%, Zn 29%, Sn 1%	Marine uses, pump parts, ship parts
Catridge brass	Cu 70%, Zn 30%	Bullet shots, cistern, storage batteries, military ammunition, foils
Arsenial copper	As 0.3%, rest Cu	Heat exchangers, condensers
Cu-alloy : Bronze		
Ounce metal	Cu 85%, Sn 5%, Zn 5%, Pb 5%	Ornamental fixtures, pipe fittings, low pressure valves
Beryllium bronze	Be 1.5-2%, Sn 8%, rest Cu	Bellows, diaphragms, springs
Aluminium bronze	Al 10%, small quantities of Ni, Mn, Fe, rest bronze	Parts containing corrosive liquids
Alpha bronze	Cu 95%, Sn 5%	Coin making, springs
Admiralty gunmetal	Cu 88%, Sn 10%, Zn 2%	Bearings, steam pipe fittings
Phosphor bronze	Cu 84-89%, Sn 10-15%, P 0.1 to 0.3% + Ni + Pb	Parts exposed to sea water, boiler fittings
Bell metal	Cu 70%, Sn 30%	Casting of bells

heating effect of current. Such materials are used to make resistors, filament of lamps, heating elements, and in energy conversion (electrical energy into heat energy) appliances *e.g.* heaters, hot plates, ovens and furnaces etc.

5.4.1 Characteristics of High Resistivity Materials

High resistivity materials possess a high value of resistance. Following characteristics are desired in them.

Lower coefficient of temperature resistance α It is an essential requirement for precision applications where a high degree of accuracy in measurements is desired. A lower value of α minimizes the heating effect of current and increases the accuracy of measurement. Invar (36% Ni-steel alloy) is one such example whose $\alpha \approx 0$. It is used to make devices and instruments for very accurate measurements.

Ability to withstand higher temperature without oxidation. Since a high resistivity material has to sustain high temperatures for longer duration in use, the oxide layer should not form on the heating element. In case the oxide layer is formed, efficiency of heat radiation will be affected and the amount of radiation will reduce.

High melting point. It is required for the purpose of minimizing heating effect of current. This is an essential requirement in resistance elements that are subjected to high temperatures for long durations, and they are required to withstand these temperatures without melting.

High mechanical strength. High resistivity materials are used as thin wires also in several applications such as precision wire-wound resistors etc. Therefore, they are required to have a high tensile strength to avoid 'breaking' during manufacturing, assembly, and uses.

Adequate ductility. High resistivity materials are also used in the form of wires and coils as elements of heaters, ovens, and starters etc. Since the size of wires may vary from very thin to thick, hence the materials should be capable of being drawn into wires of different sizes. This necessitates an adequate ductility in materials.

5.4.2 Nickel

It is a crystalline, non-ferrous, ferromagnetic metal of silvery-white colour. Its hardness matches with the hardness of soft steel but ductility is less than that. Other salient features of nickel are listed as below.

- It is capable of high quality polishing, thereby provides luster to the products on which it is polished.
- It is reasonably malleable and can also be rolled provided the carbon content is in small amount (upto 0.05% or less).
- It is resistant to acidic attacks, but dissolves readily in nitric acid.
- Its electrical resistivity at 20°C is 1.05×10^{-7} ohm-m and thermal conductivity is 54 W/m-K.

Nickel is extensively used for nickel-plating of metals to provide protective coating against corrosion. *Carbonized nickel* is used to make anodes of power tubes for rapid conduction of heat.

5.4.3 Tantalum

It is a crystalline, non-ferrous metal of specific gravity 16.6 and melting point 2996°C. Its hardness is about 500 BHN (Brinell hardness number). It is used to make *electrolytic capacitors.* The electrolytic capacitor is made by placing tantalum and aluminium in a suitable solution and making them positive electrode of a galvanic cell. Consequently a thin insulating surface film is formed, that can withstand a considerable voltage and is also capable of a high electrostatic capacity.

5.4.4 High Resistivity Alloys

Besides pure metals discussed above, several alloys are also used for high resistivity electrical applications. Main among them are the following.

1. *Nichrome,* a nickel-chromium alloy, having a composition of about 79-80% Ni + 19-20% Cr + 1-1.5% Mn + some Fe.
2. *Constantan,* a copper-nickel alloy, having a composition of about 60% Cu + 40% Ni.
3. *Manganin.,* a copper-manganese alloy, having a composition of about 86% Cu + 12% Mn + 2% Ni.

A comparison of salient properties of some high resistivity alloys is illustrated in Table 5.3.

5.4.5 Salient Applications of High Resistivity Materials

High resistivity (conducting) materials are used in numerous applications, important among them are given below.

Table 5.3 Comparison of properties of some high resistivity alloys

Description	Nichrome	Constantan	Manganin
• Resistivity (ohm-m at 20°C)	110×10^{-8}	52×10^{-8}	48×10^{-8}
• Melting point	1540°C	1300°C	1020°C
• Density (kg/m^3)	8250	8900	8190
• Permissible working temperature	1100°C	850°C	700°C
• Mechanical workability	Can be drawn into thin wires	Can be drawn into thin wires	Can be drawn into wires
• Temperature coefficient (/°C at 20°C)	0.0001×10^{-4}	0.00003×10^{-4}	0.00002×10^{-4}
• Important applications	Heating elements for electric furnaces and ovens, room heaters, electric irons and presses	Resistance elements for rheostats, starters of electric motors	Wire-wound shunts and precision resistances, coils for precision measuring instruments

- Shunts in electrical measuring instruments
- Wire-wound precision resistances
- Resistance boxes
- Starters for electric motors
- Loading rheostats
- Heating elements for heaters, ovens, starters etc.
- Filaments for incandescent lamps.

5.5 Contact Materials

Contact materials are used as 'contact points' and 'contact surfaces' in electrical equipments, appliances, devices and instruments. They operate under severe conditions, aggressive constituents of surrounding medium, and under frequent reversal of mechanical rubbing while 'making and breaking' the electrical circuits. They are also subjected to *arcing* and *sparking* due to ionization of surrounding medium between the contacts, when the contacts are separated.

Contact materials work under contrasting conditions. On one hand they have to allow easy flow of current, but on the other they have to operate through oxide layers. These layers are formed due to oxidation and chemical reactions of metals. Oxide layer increases the resistivity and reduces effectiveness of contact. Therefore, they have to work under mechanical pressure also to break the oxide layer.

They are also subjected to the problem of erosion due to fusion and evaporation on burning. Erosion causes deposition of eroded metal on one electrode (contact) and *crater* in the other. Hence a proper selection of contact material is very essential.

5.5.1 Requirements of a Good Contact Material

A good contact material should satisfy the following requirements.

1. It should have a high electrical conductivity.
2. It should have a high resistance to corrosion and oxidation.
3. It should have a high wear resistance.
4. It should have a high melting point.
5. It should be sufficiently hard so as not to allow pitting of the surfaces.
6. It should be able to sustain mechanical pressure (applied at the contact surfaces) to avoid wearing away of the surfaces.
7. It should be able to provide smooth contact by offering low contact resistance.
8. It should be able to resist chemical attack and the effect of oil, grease etc.
9. It should be easily attachable (by cementing, welding, or other techniques) on the surface of support.
10. It should be able to withstand arcing and sparking.
11. It should be mechanically strong.
12. It should have a stable resistance.

Contact surfaces between spark plug gap, and in contact breaker points between C.B. gap are well known examples of contact materials in automobiles (scooters, motorcycles, cars etc.).

5.5.2 Types of Contact Materials

Depending upon the amount of power flowing in a circuit, nature of the materials used and their manufacturing techniques, the contact materials may be classified into following types.

I. On the Basis of Power Flowing in a Circuit

1. Lightly loaded contacts for very low amperage and very low voltage services.
2. Moderately loaded contacts for low amperage and low voltage services.
3. Heavily loaded contacts for high amperage and high voltage services.

II. On the Basis of Nature of Materials Used

1. Metallic contacts such as Cu, Pt, Au, Ag
2. Alloy contacts such as Pt-Ir, Pd-Os, Cu-Ag
3. Bimetal contacts such as in relays and switches
4. Metallic oxide contacts such as CdO with Ag

III. On the Basis of Manufacturing Technique of Materials

1. Forged metal/alloy contacts such as of heterogeneous mixtures like Ag-refractories
2. Sintered metal/alloy contacts such as Cu-W, Ag-Mo, WC-Mo
3. Electrodeposited metal contacts such as in instruments
4. Sheet form contacts such as in heavy duty cut rollers

IV. On the Basis of Mounting Method on Useable Surface

1. Block type contacts such as in vibrators
2. Weldable contacts such as in automotive horns
3. Cemented contacts such as in telephone relays

V. On Operational Basis

1. Make and Break type contacts such as in ignition systems.
2. Fixed type contacts such as in electric razors.

A lightly loaded contact material is used when the circuit current and voltage are limited to about 5 ampere and 100 volt, but a moderately loaded contact material is suitable for use upto about 25 ampere and 500 volt. For higher power level uses, the contacts are classed as heavily loaded. All these levels of contacts are made of different metals and alloys, as given below.

• Lightly loaded contact materials	Noble metals such as gold, silver, platinum, palladium; rhodium, tungsten, molybdenum
• Moderately loaded contact materials	Gold-silver alloy, copper-silver alloy, cadmium-copper-silver alloy, platinum- iridium alloy
• Heavily loaded contact materials	Silver-palladium alloy, copper-cadmium alloy, molybdenum-platinum alloy, iridium-rhodium-platinum alloy

Contact materials are manufactured by different techniques. Therefore, they are called as forged metal contacts such as tungsten, sintered metal contacts e.g. *cermets* (metal-ceramic composite), and electrodeposited metal contacts such as lightly-loaded non-arcing contact surfaces having electrodeposits of Au, Ag, Pt, Pd, Rh etc.

5.5.3 Common Contact Metals

Contact materials in pure metal form are less wear-resistant than the alloy form, but possess superior conductivity and corrosion resistance. Hence, they are preferred for those applications where wear-resistance is not of greater importance. Properties of some common contact metals are briefly explained below.

Fine silver of 96.9% purity. It is used where a high resistance against wear is not required. It

- possesses a superior electrical conductivity (6.6×10^7/ohm m)
- possesses a superior corrosion resistance (anodic to Pt and Au only)
- has a tendency to arc
- oxidizes in presence of arc
- Its oxide (AgO) breaks at about 20°C, therefore affects the stability of contact resistance
- combines with sulphur which is undesired
- is used to form alloys with copper, tungsten etc.

Silver contacts are suitable for low contact pressure applications such as in light and precise devices.

Platinum. It is milky white coloured metal. It is next to gold at cathodic end of electromotive-force series. Therefore, it corrodes the least. Its other features are

- It does not oxidize in air
- It has no tendency to arc
- It may form 'needles' and 'bridges' with low currents.
- It is malleable and hence can be transformed to sheet form, which is desirable for use as contact material.
- It is corrosion and chemical resistant.

Palladium. It is moderately hard metal having a specific gravity of 12.02. Its melting point is 1552°C. Its other features are

- It is cheap and is used to make alloy with silver
- It may oxidize in air
- It has no tendency to arc
- It has less resistance to erosion.

Tungsten. It is a very hard metal having a high melting point of 3410°C. Its specific gravity is 19.3 and hardness above 150 BHN. It possesses the following salient properties.

- It has no tendency to arc
- It has a least tendency of erosion
- It does not weld during service
- It reacts with air and forms the oxide film
- It enhances the red hardness *i.e.* hardness at high temperatures.

Molybdenum. It is a hard metal having a melting point of 2610°C. Its specific gravity is 10.22 and hardness above 200 BHN. Its other features are

- It suffers from erosion problem
- It forms loosely packed oxide in the air which obstructs normal working by closing the contacts.

Rhodium. It is a hard metal and possesses the following features.

- Melting point = 1966°C
- Young's modulus = 372 GPa
- Specific gravity = 12.44
- Structure is FCC

5.5.4 Salient Applications of Contact Materials

Some important applications of contact materials are the following.

- Push button contacts in telephones
- Commutator segments in small d.c. motors
- Circuit breaker contacts
- Contacts at brushes and collector rings

- Magnet contacts in magneto system of automobiles and aeroplanes
- Contacts for oil-immersed motor starters
- Auxiliary contacts of switches, etc.

5.6 Fusible (or Fuse) Materials

A fuse (or fusible cutout) is a wire conductor connected to an electrical circuit or electrical device for the purpose of its protection against excessive current. It is designed to carry a pre-determined amount of current, and melts if more than the stipulated current passes through it. When it melts, the circuit breaks and damage to the connected device by excessive current is avoided.

The fuse wire is attached to a fuse holder as shown in Fig. 5.1. The fuse holder is made of an insulating material such as porcelain. The insulating material has to be inflammable also to avoid burning due to excessive current. Fuse wires are generally made of pure metals and alloys having low-melting point. In metal form, they are made of tin, lead, bismuth, cadmium, copper, and silver. However, the alloy form is more common in use. Tin-lead alloy is one such example.

5.6.1 Requirements of Fuse Materials

A good fuse material should possess the following characteristics.

1. Low resistivity (i.e. high conductivity).
2. Low melting point.
3. Less generation of metal vapours on melting.

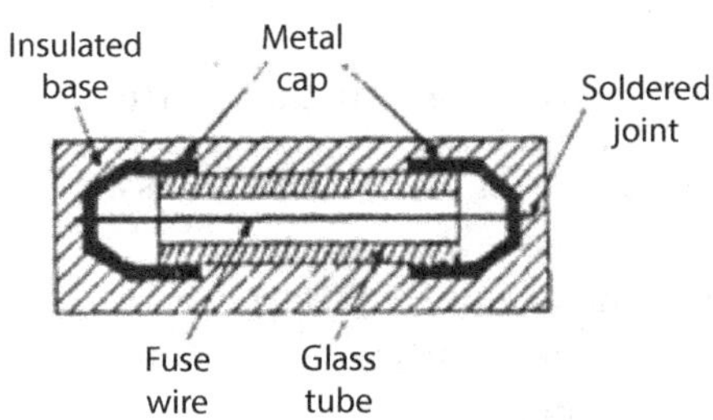

Figure 5.1 A fuse wire attached in a cartridge type fuse

4. Ability to quench the arc produced on burning of fusing element.
5. High mechanical strength.
6. Oxidation resistance at moderate to high temperatures.
7. Ability to carry a large proportion of fusing current without damage or deterioration.

Generally thin wires release less metal vapours on melting. This is advantageous because the metal vapours lower the conductivity. Thus, they become helpful in quenching the arc produced at the time of burning (blowing) of fusing element.

5.6.2 Fusible Metals and Alloys

The following metals and alloys are generally used as fuses (fusing element or fusible material or fuse cutout).

i. *Lead.* It can carry about 67% of its fusing current continuously without damage. Its melting point is 327°C.
ii. *Tin.* Its melting point is 232°C.
iii. *Lead-tin alloys* such as Tinman solder (66% Sn + 34% Pb).
iv. *Tinned copper.* The copper wire is tinned (coated with Sn), reduces the tendency of oxidation at high temperatures. It can carry about 53% of its fusing current without deterioration. Its melting point is less than the melting point of copper which is 1083°C.
v. *Silver.* It can carry more than 90% of its fusing current. It is least affected by oxidation. Its melting point is 961°C.
vi. *Bismuth based alloys* such as Wood's metal (50% Bi + 25% Pb + 12% Sn + 13% Cd), Rose metal (50% Bi + 28% Pb + 22% Sn), Newton metal etc. Their melting point varies from 60°C to 95°C.

Amongst these the lead, tin, and their alloys are used for general applications while the tinned copper is more suitable for high temperature applications. Silver being a costlier metal, *is* generally not used for general applications, but finds use in protecting the delicate instruments and also on high voltage lines.

An important application of fusible alloys is in *safety systems against explosion and fire*, and for *sprinkler system in L.P.G. refueling plant.*

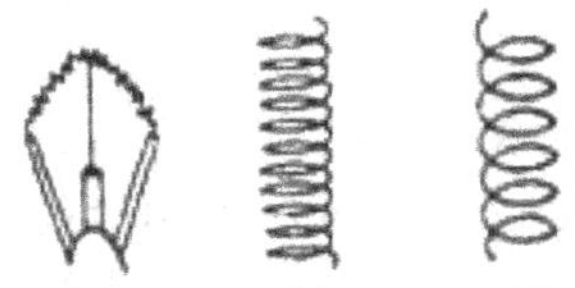

Figure 5.2 Different forms of filaments (a) straight, and (b) (c) coiled

5.7 Filament Materials

Tungsten and carbon are the commonly used filament materials. Amongst these, tungsten finds greater use due to its high melting point (3410°C) and high strength. Although the carbon also possesses a high melting point of 3550°C but its strength is lower than that of tungsten. The filaments are used in Fig. 5.2. Each filament of these forms has its own advantages and disadvantages in respect of operating life and scattering of light etc. For example, the coiled filament concentrates the light, whereas the straight filament has a tendency to sag if the lamp is installed in horizontal position.

5.7.1 Requirements of a Good Filament Material

A good filament material is required to possess the following properties.

1. High melting point.
2. High tensile strength.
3. High ductility so as to be drawn into very thin wires.
4. Ability to be converted into filamentary shape.
5. Should not be brittle at high temperatures.
6. Good oxidation resistance at elevated temperatures.
7. High fatigue resistance against thermally induced fluctuating stresses.

5.7.2 Tungsten Filament

This is a non-ferrous, very hard metal having a melting point of 3410°C. It is used to make filaments of incandescent lamps and of heater in electron tubes etc. It possesses several favorable properties that make it suitable for filament constructions. These properties are

- Its resistivity is 5.5×10^{-8} ohm m at 25°C which is about twice of the value of resistivity of aluminium.

- It can be drawn into very thin wires of a few micron sizes which is essential for making the filaments.
- This is a strong material having tensile strength equal to 1.1 to 4.1 GPa which on transformation into thin wire further increases to about 2 times. Thus, it is stronger in its filament form than its bulk form.
- It is capable of working satisfactory above 2000°C in vacuum and also in the atmosphere of inert gases (neon, argon, helium etc.). As this is a realistic situation of incandescent bulbs working, hence tungsten is most appropriate metal for filament purpose.
- It oxidizes easily in presence of oxygen, so is not fit for applications in oxygen atmosphere.

5.8 Carbon As Filamentary and Brush Material

Electrical carbon materials are basically available in raw carbon forms, which are processed to obtain carbon products. The processing involves the following operations.

i. Grinding of raw carbon materials into powder form.
ii. Mixing the binding agent (binder) to the powder.
iii. Moulding the powder-binder paste to a desired shape.
iv. Baking the moulded shape.

The raw carbon generally exists in coal and graphite forms. The binders used are coal-tar etc. Moulding is done in a smooth surfaced mould and baking is performed at about 900°C. Commonly used electrical carbon materials are the following.

1. Raw graphite
2. Carbon graphite
3. Copper graphite
4. Electrographite

Various characteristics of electrical carbon materials are the following.

- Their resistivity lies in the range of 5 to 60 ohm-meter.
- They have negative coefficient of temperature resistance.

- Their resistances are influenced by the contact pressure. It decreases with increase in pressure.
- They have low value of coefficient of friction ($\mu \approx 0.2$ to 0.3).
- Their conductivity can be enhanced by mixing additives like copper, brass and bronze powders.

5.8.1 Carbon Graphite

It is mainly used in making carbon brushes. Depending upon the functional pressure (force), the current density passing through them and their circumferential velocity, the carbon brushes are made of different hardness, resistivity, and friction qualities. Table 5.4 presents specific details in this regard.

It can be observed that the carbon graphite has maximum resistivity while the copper graphite has least resistivity. The carbon graphite is hardest among all types. Its high coefficient of friction and low wear rate makes it most suitable brush material.

5.8.2 Main Applications of Carbon Materials

Main applications of carbon materials in electrical machines, electrical and electronic devices are the following.

- Brushes of generator and alternator.
- Arc lamps, arc welding.
- Electrodes of electric arc furnaces.
- Contacts.
- Anodes and grids in electron valves (tubes).

Table 5.4 Characteristics of different kinds of carbon as brush material

Kind of carbon	Hardness (BHN)	Resistivity (ohm-m)	While operating at 8 kN pressure, 25 m/s peripheral velocity, and 20 A/cm^2 current density		
			Coefficient of friction μ	Wear (mm)	Contact drop (voltage)
Raw graphite	7–21	8–30	–	0.4–0.5	0.6–1.8
Carbon graphite	20–45	40–60	0.28–0.30	0.10	1.5–2.5
Copper graphite	5–25	0.03–15	0.20–0.25	0.15–0.8	0.11–2.2
Electrographite	3–50	5–25	0.23–0.30	0.4–0.6	1.1–2.5

- Carbon pile resistances (resistors).
- Non-wire resistors, film type resistors, solid type resistors.
- Telecommunication equipments, membranes.
- Lamp filaments.
- Battery cell elements.

5.9 Conductors, Cables, and Wires: Types and Materials

Conductors are used to carry current. They may have to carry a small current of a few milliampere only as in instruments, or a large current of tens of amperes as in transmission cables. Accordingly, the conductors may be small in size (diameter) or of large size. Depending upon the constructional features, shape of cross-section, size and applications; the conductors may be classified as follows.

I. **On the Basis of Constructional Feature**
    1. Straight wire conductor
    2. Stranded conductor
        i. circular
        ii. compact circular
    3. Core cable conductor

II. **On the Basis of Cross-section**
    1. Solid conductor
    2. Hollow (liquid or gas filled) conductor

III. **On the Basis of Material's Specialty**
    1. Single metal conductor
    2. Reinforced conductor such as ACSR
    3. Cladded metal conductor

IV. **On the Basis of Shape**
    1. Circular conductor
    2. Elliptical conductor

Amongst these a single metal, straight wire conductor is a common type for carrying small current. However, it is not suitable to carry a large current because it will then have a large cross-section due to which, the chances of kinking* and breaking are more. To obviate this problem, the

---

* Kinking means permanent bending

large cross-section conductors are made of a number of thin wires of small cross-section and bunched together. These are called *stranded conductors* and are most commonly used types, particularly as transmission conductors. They are flexible, can be easily coiled; and chances of breaking through the insulation are the least.

5.9.1 Stranded Conductors

Stranded conductors are made by twisting the thin wires (known as *strands)* of smaller cross-section. These twisted wires are bunched together by placing them in layers of a particular style. Wires of each layer are then laid helically around the wires of other layer. This process is called *stranding.* Stranding is generally done in opposite directions for successive layers *i.e.* if the wires of one layer are twisted in clockwise (right hand) direction then the wires of next layer are twisted in anti-clockwise (left hand) direction. In this way the stranding is carried-on for different layers.

5.9.2 Types of Stranded Conductors

Stranded conductors can be of two different designs, viz.

1. Circular stranded conductor (Fig. 5.3a), and
2. Compact circular stranded conductor (Fig. 5.3b).

In the former type the strands are circular in cross-section, while in the latter type it is elliptical. The overall dimension of circular stranded type is more than the overall dimension of compact circular type. Therefore, more quantity of insulating and protecting material is required in circular stranded type. They are generally used for *single-phase* applications. The compact circular stranded conductors are used for *three-phase* applications and in making core cables. The 'compact circular stranded conductors' are made by compacting the 'circular strand' by means of pressing them through a set of dies and rollers.

Figure 5.3 Stranded conductors (a) circular type is generally used for 1-phase applications, and (b) compact circular type is used for 3-phase applications.

5.9.3 Specifications of Stranded Conductors

Stranded conductors are specified as 19/2.50, 37/2.06 etc. First numbers 19, 37 etc. indicate the total number of wires in a stranded conductor, and the second numbers 2.50, 2.06 etc. indicate the diameter of each wire in millimeter. Thus a 19/2.50 stranded conductor has 1 wire in the centre, 6 wires in the first layer, and 12 wires in the second layer. Its overall diameter is

$$[1 + 2 \text{ layers} \times 2] \times 2.50 = 12.50 \text{ mm}$$

The cross-section of one such conductor is similar to as shown in Fig. 5.5.

5.9.4 Core Cable

Now-a-days the electricity transmission and distribution is mainly accomplished by 3-phase system. This system requires three stranded conductors, each insulated individually and put together to make a three-core cable. The stranded conductors in three-core cable may have either a circular strand, sector shaped strand, or a solid shaped strand. These constructions are shown in Figs. 5.4a-b-c. Amongst these, the solid shaped conductors provide economy in material and space but the core cables formed by them are less flexible.

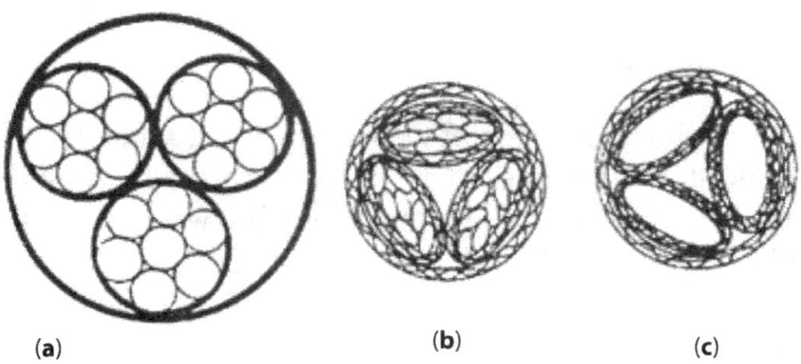

Figure 5.4 A 3-core cable having (a) circular strand, (b) sector shaped strand, and (c) solid shaped strand.

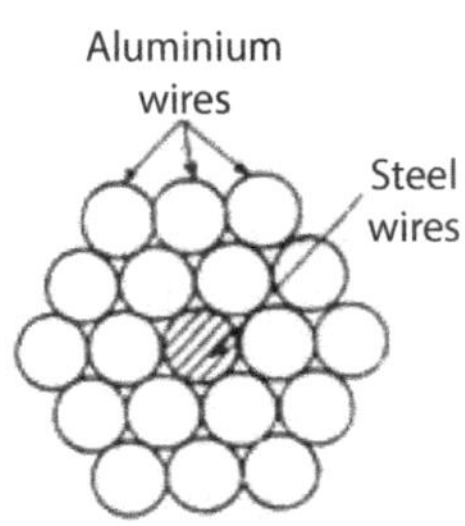

Figure 5.5 Construction of a ACSR conductor.

5.9.5 Reinforced Conductor

To impart strength to the aluminium conductors, they are reinforced with steel as shown in Fig. 5.5. These are called ACSR conductors and are used for overhead transmission lines.

5.10 Solder Materials for Joining Wires and Joints in Power Apparatuses

Solder materials are fusible alloys that are used to join metal surfaces. Soldering may be required to join electrical wiring, joints in electrical and electronic devices, and power apparatuses etc. One such solder joint is shown in Fig. 5.1. Depending upon the composition, properties, and metals to be joined; the solder materials may be classified into following two types.

1. Soft solder for joining Al, Mg etc.
2. Hard solder for joining Cu, brass, bronze, Ni, Ag, Au etc.

A soft solder is usually a Pb-Sn alloy. Hard solders are usually Cu-Zn alloys, Cu-Zn-Sn alloys, and Cu-Zn-Ag alloys etc.

5.10.1 Soft Solder

It is made of varying proportion of Pb and Sn. A soft solder having 72.5% Pb is mechanically most strong but its fusibility is poor. The alloys containing 55 to 60% Pb have melting points ranging between 215°C to 230°C, and are sufficiently fusible. They have good flowability also during soldering of joints. However, a 'half-half' solder containing 50% Pb + 50% Sn is most favourable solder. It melts rapidly, flows freely, and makes a technically sound joint.

5.10.2 Hard Solder

Whereas the process of joining aluminium etc. is called 'soldering', the process of joining copper and brass etc. is termed as *brazing*. Hard solders are used for brazing. They can be of the following kinds.

1. Spelters which are mainly Cu + Zn or Cu + Zn + Sn alloys.
2. Silver solders which are mainly Cu + Zn + Ag alloys.
3. Other copper solders which contain P, Ni, Mn etc. in Cu.

Table 5.5 Different types of solders and their properties

S. No.	Type	Composition (weight %)	Melting point (°C)
1.	Soft solder	50 Pb + 50 Sn	≈ 200
2.	Spelter	45 Cu + 50 Zn + 5 Sn	855
3.	Silver solder	45 Ag + 30 Cu + 25 Zn	675
4.	Other hard solder	57.25 Cu + 40 Zn + 1.5 Sn + 0.75 Mn	540
		75 Cu + 20 Ag + 5 Cd	790

Spelters are used to join strong connections of steel, iron, nickel and copper alloys. *Silver solders* are used for connections in electronic devices. However, they are costlier. Therefore, Cu + P solders are used instead of them. A list given in Table 5.5 presents some solders and their related properties.

5.11 Sheathing Materials

Sheathing materials are used to protect cables against moisture, corrosion and chemical actions. Sheathing of cables is necessary because the insulation of cables will break on moisture penetration through the cracks. Following materials are generally used for this purpose.

1. Lead
2. Lead alloys with Sb, Sn, Cu, Cd
3. P.V.C.

Amongst these, the latter two materials are more common as they do not suffer from the deficiencies that a pure lead sheathing suffers from. The pure lead suffers from several deficiencies such as follows.

- Due to higher specific gravity (11.36), it adds to the weight of the cable.
- It is mechanically weak (strength is about 50 MPa only).
- It cannot withstand vibrations at high temperatures.
- Cracks are developed in them when lead-sheathed cables are laid near the railway tracks and on the bridges.
- It is corroded by various acids, lime, and many chemical substances.

- It is also corroded by fresh concrete, tanners and chalk in the presence of air and water.
- To overcome these deficiencies, an *additional protective sheathing* over the lead is generally required in above situations, particularly when the cables are laid in newly constructed concrete trenches.
- As an alternative to pure lead, the lead alloys and PVC coverings are preferred for use now. They possess much improved properties. PVC is lighter in weight also (specific gravity = 1.35 to 1.45 only).

5.12 Sealing Materials

In manufacturing radio valves, lamps, and X-ray tubes etc., it is essential to produce a perfectly gas tight joints between glass and metal. Sealing materials are used for the same. Since the coefficients of thermal and electrical expansion of glass and metal are different, this factor is also considered in selecting the sealing material. Following materials are generally used for this purpose.

- Ferrous alloys containing 35 to 45% Ni with slight addition of Si, Mo or Mn.

The addition of Si improves oxidation resistance (i.e. corrosion resistance at high temperatures), Mo improves red hardness (i.e. hardness at elevated temperatures), and Mn improves wear resistance.

To provide stronger chemical adhesion between metal and glass; the wires in valves, lamps etc. are coated with a layer of copper before encasing them in the tube.

5.13 Solved Examples

Example 5.1 What are carbon pile resistors?

Solution. These resistors are made up of a number of carbon plates, by piling one over the another. The plates are maintained under pressure. The carbon composition consists of finely ground carbon, resin binder and insulating filler. Such resistors are used in electronic circuits for setting biases, controlling gain, loading the circuits etc. in oscillators, precision voltage dividers and communication systems.

Example 5.2 How can the conductivity of carbon be increased?

Solution. To increase the conductivity of carbon, different kinds of additives are used. The additives are mainly copper or bronze powder carbon moulding compound.

Example 5.3 Permissible working temperature of *manganin* is low (about 60°C to 70°C) but that of the *nichrome* is high (about 350°C to 650°C)-why?

Solution. It is because the resistance coefficient of temperature of manganin is high at higher temperature (about 0.00002/K) than that of nichrome, which is about 0.00001/K. Moreover, the electrical resistivity of manganin is 42×10^{-8} ohm m as compared to 108×10^{-8} ohm m for nichrome. Their other properties, which are also the deciding factor in this regard, are as follows.

Material	Density (kg/m^3)	Tensile strength (MPa)
Manganin	8200	420
Nichrome	8410	1000

Example 5.4 Nichrome is often designated as nichrome I, III, V; and Hastelloy by hastelloy A, B, C, D. What meaning do they convey?

Solution. These designations specify different commercial grades of nichrome and hastelloy. Each grade differs in composition, properties and characteristics; and hence is suitable for different specific applications. The details given below are self-illustrious in this regard.

Alloy	Composition (%)	Application
Nichrome I	Ni 80%, Cr 20%	Thermocouple, heating element of furnaces and ovens, and strain gauges
Hastelloy A	Ni 60%, Mo 30%, Fe 5%	Parts of chemical plants

Review Questions

1. Enumerate various mechanically processed forms of electrical materials. Discuss the utilities of cladded metals, bimetals, sintered and annealed metals. Write applications of hot and cold rolled, and hard and soft drawn metals.

2. Classify different types of conducting materials from various aspects. How do the low resistivity materials differ from the high resistivity materials?
3. Compare the characteristics of low resistivity and high resistivity materials. Name some metals and alloys falling under each category, and describe briefly their properties.
4. Compare different properties of copper and aluminium as conductive materials of electricity in transmission and distribution systems. Mention those applications where aluminium can completely replace the copper, and also where some reinforcement of aluminium is considered essential.
5. Compare the properties and applications of annealed and hard drawn copper as a conductive material.
6. Compare the merits and demerits of hot rolled, cold rolled and hard drawn aluminium as a conductive material, and discuss their suitabilities of applications.
7. Discuss the utilities and applications of brass and bronze as a low resistivity materials.
8. Discuss the advantageous features of Ni and tantalum as conductive materials. What are their limitations?
9. Enlist some high resistivity alloys suitable as conductive materials. Compare the nichrome, constantan and manganin from different viewpoints.
10. Discuss the requirements of a good contact material. Name different types of contact materials and quote their examples. How do the lightly loaded contacts differ from heavily loaded contacts?
11. What is the need of fuse (fusible) material? What are their requirements? Discuss different fusible materials. In what ways the fusible materials are helpful in providing safety against fire and explosion?
12. What are the requirements of a good filament material? Describe tungsten as a filamentary material for use in lamps and compare its properties with carbon filament.
13. Describe different types of carbon which are suitable as filamentary and brush materials. Explain the characteristics and properties of electrical carbon which is most useful for the purpose of alternator brush.
14. Describe various types of conductors and cables. Explain the stranded cables, core cable, and reinforced conductor with the help of suitable diagrams.

15. Differentiate between the following stating their purpose, types and examples.
    a. Soft and hard solder materials.
    b. Sheathing and sealing materials.
    c. Annealed copper and tinned copper.
    d. Nichrome and platinum.
16. Suggest with reasons, the materials generally used for the following applications.
    a. Brushes on commutator for a d.c. generator.
    b. Filament of an incandescent lamp.
    c. Electric contact in an iron-clad switch.
    d. Busbar in a sub-station.
    e. Precision shunts and resistors.
17. Discuss the advantages and limitations of the following uses.
    a. Use of aluminium conductor for 132 kV overhead transmission lines.
    b. Use of tungsten as push button contacts in telephones.
    c. Use of copper graphite as carbon-pile resistors.
    d. Use of 19/2.50 stranded conductor for domestic wiring.
    e. Use of lead as soldering, sheathing and sealing material.
18. What are the advantages and drawbacks of fusible cutouts? Explain 'fusing current' and the factors on which it depends. Derive a relation between fusing current and the wire diameter stating clearly the assumptions made therein.
19. Comment on the difficulties in joining an aluminium conductor with a copper conductor, and state as to how this is accomplished.

Objective Questions

1. For which of the following applications, the annealed copper is a suitable material?

   a. Busbar (b) Domestic wire
   c. Commutator (d) Overhead line

2. Match the items of list I with those of list II, and choose the correct answer from the codes given below the lists.

List I	*List II*
A. Aluminium	1. Overhead conductor
B. Lead	2. Brushes for electrical machines
C. Tungsten	3. Fuse material
D. Carbon	4. Insulating material
	5. Filaments of incandescent lamps

Codes:

	A B C D		A B C D
a.	1 3 2 5	b.	5 2 3 4
c.	1 3 5 2	d.	2 5 3 4

3. Consider the following statements.
   A. Lead is used to make the fuse links.
   B. Copper solders are soft solders.
   C. Light load contacts are made from pure noble metals.
   D. An e.m.f. exists across the junction of two different metals.

   Of these, the correct statements are
   a. A and C b. B and D
   c. B, C and D d. A, C and D

4. In a coaxial cable, braided copper is used as a
   a. conductor b. shield A
   c. dielectric d. jacket

5. An example of a liquid conducting material is
   a. mercury b. iridium
   c. silicon resin d. boiling sodium

6. Which of the following pairs does *not* match?
   a. Low resistivity material Brass
   b. Commutator segments Cadmium bronze
   c. High resistivity material Tantalum
   d. Sealing of glass tubes Palladium

6

Semiconducting Materials: Properties and Behaviour

6.1 Introduction to Semiconductors

These are the solids having energy gap E_g lying in between those of the conductors and insulators. Their conductivity is more than that of the dielectrics but less than that of conductors. They are basically electronic materials. Semiconductors are available in following forms.

1. Elements such as silicon (Si) and germanium (Ge).
2. Compounds such as GaAs, InP, AlSb, CdTe, ZnSe etc.
3. Alloys such as $GaAs_xP_{1-x}$, $HgCd_xTe_{1-x}$ etc.

Semiconductors are also classified as follows.

a. Intrinsic, and
b. Extrinsic.
   i. *n*-type
   ii. *p*-type

The elemental forms of pure Si and pure Ge are intrinsic. In intrinsic form they are not useful. They are, therefore, doped by dopants to make extrinsic semiconductors. Extrinsic forms are directly useful and are widely employed in manufacturing of the solid state devices. They belong to the category of alloys and compounds. Electronics industry requires a purity better than 1: 10^9 in pure Si and Ge.

Extrinsic semiconductors are primarily of *n*-type and *p*-type. They may be in compound form such as GaAs, CdTe etc. or in alloy form such as $HgCd_xTe_{1-x}$. A number of semiconducuting compounds are available in oxide, halide and sulphide forms also. The *n* and *p*-types are used to make *n-p* junction for diode, *n-p-n* and *p-n-p* for transistor, etc.

6.1.1 Properties of Semiconductors

Semiconductors are generally hard and brittle, and possess negative coefficient of temperature resistance. In contrast with conductors whose conductivity increases with purification, the conductivity of semiconductors decreases with purification. Semiconductors are widely used as rectifiers, amplifiers, photocells etc. Their properties are of greater importance in telecommunication, power electronics, computer hardware etc. Complete details about them are given in this and later chapters.

6.2 Different Types of Semiconducting Materials

Although the Si and Ge are most commonly used semiconductors, yet there are several other semiconducting materials used in vivid devices. A detailed classification of them is given below in Table 6.1. In elemental form they belong to various columns of the periodic table; in compound form they are binary, ternary and quarternary; and in alloy form they may be homogeneous or heterogeneous types.

6.2.1 Merits of Semiconducting Materials

Semiconductors possess several favourable properties as listed below.

1. They are much smaller in size. For example, the gauge length of a semiconductor strain gauge may be as small as 0.25 mm or less, and thickness of wafer as less as 0.1 mm or less.
2. They are very light in weight. For example, the weight of a transistor is as low as fraction of a gram. And also a calculator, consisting a large number of transistors on its ICs, weighs 146 gram only including the cells.
3. They operate on low voltage e.g. 9 volt, 3 volt, 1.5 volt etc. For example, the calculator operates on manganese-dry battery of 3 volt only.

Table 6.1 Classification of different kinds of semiconducting materials

Types of semiconductors	Examples
Elemental form	
Column	
II	Zn, Cd
III	B, Al, Ga, In
IV	C, Si, Ge
V	N, P, As, Sb
VI	S, Se, Te
Binary compounds	
II-VI compounds	ZnS, ZnSe, ZnTe, CdS, CdSe, CdTe
III-V compounds	AlP, AlAs, AlSb, GaN, GaP, GaAs, GaSb, InP, InAs, InSb
IV-IV compounds	SiC, SiGe
Ternary compounds	GaAsP, HgCdTe, AlGaAs
Quarternary compounds	InGaAsP
Ternary alloys	$Al_{0.3}Ga_{0.7}As$ (30% is Al and 70% is Ga); $In_{0.53}Ga_{0.53}As$, $In_{0.11}Ga_{0.8}As$, $GaAs_{0.88}Sb_{0.12}$
Quarternary alloys	$In_xGa_{1-x}As_yP_{1-y}$

4. They consume negligible power. For example, the calculator works at an expense of 0.0004 watt only for years together.
5. They have a long life. It may be many hundred hours or a few years.
6. They do not develop any creep effect and hardly show any ageing effect.
7. They are least affected by shock loads and are almost shockproof.
8. They operate instantly and do not require any pre-heating.
9. They operate over a practically fruitful wide range of temperatures.

6.2.2 Characteristics of Semiconducting Materials

Main characteristic of semiconducting materials is that their conductivity increases with increase in temperature and also with increase in voltage. This increase is 10 to 20 times on adding 1 ppm impurity. Some more salient characteristics of semiconductors are given as follows.

1. They have negative temperature coefficient of resistance. It means that

$$\alpha_{\text{semiconductor}} \propto -\frac{dR}{dT}, \text{ and also} \propto -\frac{d\rho}{dT}$$

where ρ is resistivity. The value of α lies in the range of $-4 \times 10^{-6}/°C$ to $6 \times 10^{-6}/°C$.

6.3 Determining the Percentage Ionic Character of Compound Semiconductor

Compound semiconductors are generally composed of mixed bonding. Normally the mixed bounding in them is of ionic-covalent type. It results due to the difference in electronegativities of the elements forming the mixed bond. The *ionic* character thus imparted may be determined from 'Pauling equation' given by

$$I_c = \left[1 - e^{-0.25(E_{NP} - E_{NQ})^2}\right] 100 \tag{6.1}$$

where I_c is percentage of ionic character, E_{NP} and E_{NQ} are the electronegativities of atoms P and Q forming the mixed bond. Difference in electronegativity of about 1.7 to 1.8 corresponds to 50% ionic character.

Some examples of ionic-covalent bonded compound semiconductors are the following.

- GaP, GaAs, GaSb
- InP, InAs, InSb
- ZnS, ZnSe, ZnTe
- CdS, CdSe, etc.

Example 6.1 Estimate the percentage ionic character of the compound semiconductors InAs and GaAs using Pauling's equation. The electronegativities are 1.5 for In, 2.2 for As, and 1.8 for Ga.

Solution: Using Pauling's equation given by Eqn. 6.1, we have

For InAs:

$$\% \text{ of ionic character} = \left[1 - {}^{-0.25(2.2-1.5)}\right] \times 100$$

$$= (1 - 0.884) \times 100 = 11.53\%$$

For GaAs:

$$\% \text{ of ionic character} = \left[1 - e^{-0.25(2.2-1.8)^2}\right] \times 100$$
$$= (1 - 0.96) \times 100 = 4\%$$

6.4 Fermi Energy Level

The Fermi energy level E_F, corresponds to the highest filled energy level at 0 K. The free electrons occupy energies upto E_F at 0 K. When the temperature goes up above 0 K, some lower energy electrons at energy $E(E < E_F)$ move up to higher energy levels $E > E_F$. This motion of electrons under an applied electric field occurs only when the electrons are located in partially filled energy bands. Fermi energy level of a metal can be determined by

$$E_F = \left(\frac{3n}{2\rho_E}\right)^{2/3} \tag{6.2}$$

where n is number of free electrons per unit volume of a metal, and ρ_E is an energy density constant whose value is $6.82 \times 10^{27}/\text{m}^3(\text{eV})^{3/2}$. Number of free electrons in some metals are given in Table 6.2.

6.4.1 Fermi-Dirac Probability Function and Temperature Effect

The thermal behaviour of electrons in an atom can be explained by *Fermi-Dirac probability function* $p(E)$ given by

$$p(E) = \frac{1}{1 + e^{(E-E_F)/kT}} \tag{6.3}$$

Table 6.2 Number of free electrons in some metals

Metal	Number of free electrons (per m^3)	Metal	Number of free electrons (per m^3)
Al	1.67×10^{29}	Au	5.65×10^{28}
Cu	6.30×10^{28}	Zn	3.0×10^{28}
Ag	6.10×10^{28}	Si	2.0×10^{29}

where k is Boltzmann constant, and T the absolute temperature. Fermi-Dirac distribution is shown in Fig. 6.1.

- It illustrates that $p(E) = 0$ or 1 at $T = 0$ K. Variation in $p(E)$ with increasing temperature is also shown.
- At higher temperatures, more and more electrons occupy energy greater than Fermi energy.
- The kinetic energy of electrons in a metal having maximum energy is equal to its Fermi level.
- Only some electrons may attain this level.

The Fermi energy level of some elements is given in Table 6.3. The value of $p(E)$ is 0.5 at cross-over point. This point is invariant at all temperatures.

Thus the Fermi energy level can also be defined as the level at which the probability of occupation by an electron is 50%. For $E > E_F + kT$, in Eq. 6.3, the exponential term in denominator becomes too large than unity. Therefore, Eq. 6.3 may be rewritten as

$$p(E) = e^{-(E-E_F)/kT} \tag{6.4}$$

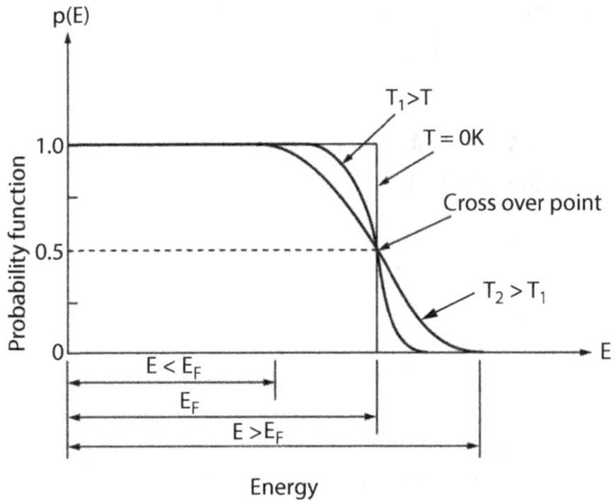

Figure 6.1 The Fermi-Dirac distribution of free electrons as a function of temperatures.

Table 6.3 Fermi energy level of some elements

Element	Be	Al	Zn	Cu	Au	Ag	Cs	K	Na	Li
Fermi energy E_F (eV)	12	11.8	11	7.04	5.51	5.51	1.6	2.1	3.2	4.7

Example 6.2 There is 5% probability for an electron to occupy an energy state which is 0.4 eV above the Fermi energy. Estimate the temperature at which this can happen.

Solution. Using Eqn. 6.3 for Fermi-Dirac probability distribution

$$p(E) = \frac{1}{1 + e^{(E - E_F)/kT}} \qquad (i)$$

in which $p(E)$ = 5% = 0.05, $E = E_F + 0.4$, and T is to be determined. On putting the value of Boltzmann constant $k = 8.614 \times 10^{-5}$ eV/K in Eqn. (*i*), we find

$$0.05 = \frac{1}{1 + e^{0.4/8.614 \times 10^{-5}} T}$$

$$\therefore \qquad 1 + e^{0.0464/(T \times 10^{-5})} = 20$$

$$\text{or} \qquad e^{0.0464/(T \times 10^{-5})} = 19$$

$$\text{or} \qquad \frac{0.0464}{T} \times 10^5 = \ln 19 = 2.944$$

$$\therefore \qquad T = \frac{0.0464 \times 10^5}{2.9444}$$

$$= 1575.8 \text{ K}$$

6.5 Intrinsic Semiconductors

Reason of unsuitability of intrinsic semiconductors. We noticed that the conduction of electrons from valence band to conduction band needs crossing-over the forbidden gap. The applied electric field required for this conduction *is* extremely high in intrinsic semiconductors. For example, the forbidden gap in germanium is 0.7 eV and in silicon 1.1 eV. The distance between the locations of electrons near an ion core and away from it is about 1 Å. Thus semiconduction of an electron from valence band to conduction band will take place when a field gradient of about 0.7 V/1 Å in Ge and 1.1 V/1 Å in Si is applied. This comes out to be an impracticable value of about 0.7×10^{10} V/m and 1.1×10^{10} V/m respectively.

On the other hand, thermal energy at room temperature can excite limited number of electrons across the energy gap. The number of electrons crossing-over the forbidden gap may be calculated by Fermi-Dirac probability distribution,

$$p(E) = \frac{1}{1+e^{(E-E_F)/kT}} \tag{6.5}$$

6.5.1 Energy Diagram of Intrinsic Semiconductor

The energy diagram for an intrinsic semiconductor is shown in Fig. 6.2. It shows that the Fermi level E_F lies in the middle of forbidden energy gap E_g. This Fermi level is also referred to as intrinsic energy level E_i. The energy gap E_g is equal to the difference in the energies of valence band E_v and conduction band E_c. The term $(E - E_F)$ in Eq. 6.4 is therefore equal to $E_g/2$. For pure germanium, $(E - E_F)$ is $E_g/2 = 0.35$ eV, which is 13.5 times higher than kT at room temperature (= 0.026 eV). Therefore, as compared to the exponential term in Eq. 6.4, the value unity is neglected, and Eq. 6.4 can be rewritten as

$$p(E) = e^{(-E_g/2kT)} \tag{6.6}$$

Number of electrons crossing-over to conduction band. If the total number of electrons available in valence band for semiconduction is N, and out of this n numbers cross-over to conduction band, then $n/N = p(E)$. Thus Eq. 6.5 modifies to

$$n = Ne^{(-E_g/2kT)} \tag{6.7}$$

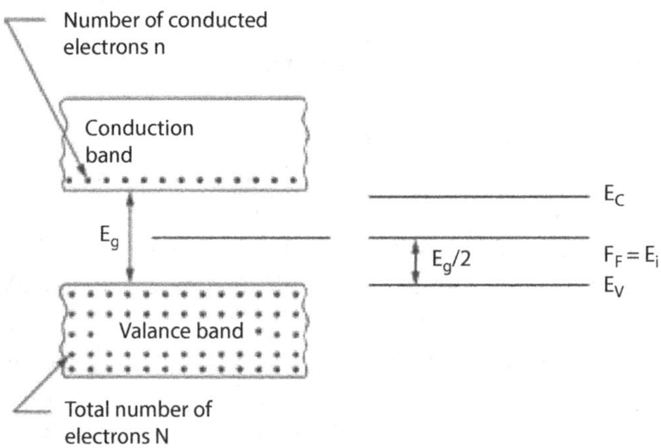

Figure 6.2 Energy diagram of an intrinsic semiconductor showing energy levels of conduction band, valence band, energy gap and intrinsic energy.

6.5.2 Holes, Mobility and Conductivity

Definitions. If n number of electrons cross the gap, n sites become vacant in the valence band. These vacant sites are called *holes*. Thus the number of electrons n_e and number of holes n_h are equal ($n_e = n_h$). Both: electrons and the holes take part in semiconduction. Electrons conduct in the conduction band and the holes in the valence band. They move in opposite directions with certain drift velocity v_d under an applied field gradient ζ. This movement of electrons and holes is known as *mobility*. Mobility of an electron and of a hole is designated by μ_e and μ_h respectively, and is defined as

$$\mu = \frac{v_d}{\zeta} \tag{6.8}$$

Effect of temperatures. Mobility is a temperature sensitive property, and decreases with increasing temperature. They vary with temperature as $\mu_e \propto T^{-2.4}$ and $\mu_h \propto T^{-2.2}$. Its value is more for electrons than for holes ($\mu_e > \mu_h$). Since both these charge carriers (electrons and holes) contribute towards conduction, hence conductivity σ of an intrinsic semiconductor is obtained from

$$\sigma = n_e e_e \mu_e + n_h e_h \mu_h \tag{6.9}$$

where $e_e = e_h = e$ is electronic charge, n_e and n_h are carrier concentrations per unit volume of electrons and holes respectively. Intrinsic carrier concentration of an electron or a hole in germanium at 300 K is $2.5 \times 10^{19}/\text{m}^3$.

Example 6.3 Determine the intrinsic carrier density of pure silicon whose resistivity at room temperature is 3000 ohm m. The mobilities of electrons and holes in silicon at room temperature are 0.14 and 0.05 m²/V s. Electron charge $e = 1.602 \times 10^{-19}$ C.

Solution. We know that the resistivity $\rho = 1/\sigma$, and $n_e = n_h = n$ in an intrinsic semiconductor. Using Eq. 6.9, in which

$$\mu_e = 0.14, \mu_h = 0.05, \text{ and } e_e = e_h = e$$

we get

$$\frac{1}{3000} = n \times 1.602 \times 10^{-19} (0.14 + 0.05)$$

$\therefore$

$$n = 1.095 \times 10^{16}/\text{m}^3$$

Example 6.4 The mobilities of silicon are $\mu_e = 0.17$ m^2/V-s and $\mu_h = 0.035$ m^2/V-s at room temperature. If the carrier density in the material is known to be 1.1×10^{16}/m^3, calculate the resistivity of silicon.

Solution. As $$\sigma = n_e e_e \mu_e + n_h e_h \mu_h$$

where $e_e = e_h$ = electronic charge = 1.6×10^{-19} C,

$n_e = n_h = 1.1 \times 10^{16}$/m^3 (for intrinsic semiconductor)

μ_e and μ_h = 0.17 and 0.035 m^2/V-s respectively.

$$\therefore \quad \sigma = n_e \times e \times (\mu_e + \mu_h)$$

$$= 1.1 \times 10^{16} \times 1.6 \times 10^{-19} (0.17 + 0.035)$$

$$= 3.608 \times 10^{-4}/\text{ohm-m}$$

$\therefore$ Resistivity $$\rho = \frac{1}{\text{Conductivity}} = \frac{1}{3.608 \times 10^{-4}}$$

$$= 2771.618 \text{ ohm-m}$$

6.6 Extrinsic Semiconductors

In the last article we discussed that the conduction in intrinsic semiconductor is due to inherent crystal properties. Contrary to this, the conduction in extrinsic semiconductors occurs due to the presence of *foreign impurities*. Deliberate impuring of intrinsic semiconductors is done by adding impurities to obtain extrinsic semiconductor. The impuring agents are called *dopants*, and the process is known as *doping*. The dopants are elements either of the third column or the fifth column of the periodic table. They increase the conductivity of semiconductors considerably. Based on the type of dopants used, extrinsic semiconductors are classified as

i. *n*-type, and ii. *p*-type.

- The *n*-type semiconductor is also called negative or electron carrier type, and *p*-type as positive or hole carrier type.

6.6.1 *n*-Type Semiconductors and their Energy Diagram

In this case, the fifth column elements such as As, Sb or P are used as dopants. Figure 6.3(*a*) shows a germanium crystal doped with phosphorus.

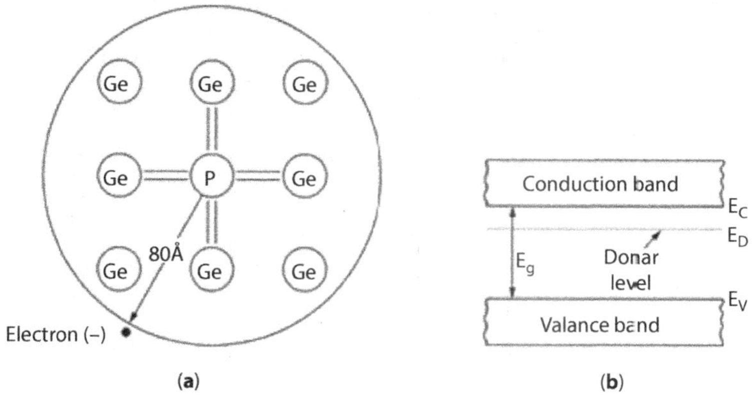

Figure 6.3 *n*-type semiconductor (*a*) germanium doped with fifth column phosphorus, and (*b*) energy diagram showing donar energy level.

Four of the five electrons in the outermost orbit of phosphorus atom forms a tetrahedral bond with four germanium neighbours. The fifth electron remains loosely bound to its parent atom, and moves in the electric field of germanium crystal with an electron orbit of about 80A° radius. The energy of fifth electron is close to the conduction band, and is shown in Fig. 6.3(*b*). This energy level is known as *donor energy level* E_D, as the dopant has donated one of its electron to the semiconductor.

6.6.2 Law of Mass Action

Now the cross-over of fifth electron into the conduction band is much easier than it would have been from the valence band. The energy needed to excite the fifth electron into the conduction band is known as *ionization energy* E_I, and is given in Table 6.4. The number of donar level electrons excited into the conduction band at room temperature is large, as $E_I << E_g$. The number of electrons in the conduction band are more than the number of holes n_h in the valence band (*i.e.* $n_e > n_h$). The electrons remain the majority charge carriers. However, the law of mass action ($n_e \times n_h$ = constant) holds good.

6.6.3 *p*-Type Semiconductors and their Energy Diagram

An intrinsic crystal can also be doped by elements of third column such as Ga, In or Al. Doping of a silicon crystal by an aluminium atom is illustrated in Fig. 6.4(*a*). Aluminium requires a fourth electron in addition to three electrons in its outer orbit to form a tetrahedral bond. The fourth electron

Table 6.4 Ionization Energies of Dopants in Intrinsic Semiconductors

Type	Dopant	Ionization Energy (eV)	
		Silicon	Germanium
n	Sb	0.039	0.010
	P	0.044	0.012
	As	0.049	0.013
p	B	0.045	0.010
	Al	0.057	0.010
	Ga	0.065	0.011
	In	0.160	0.011

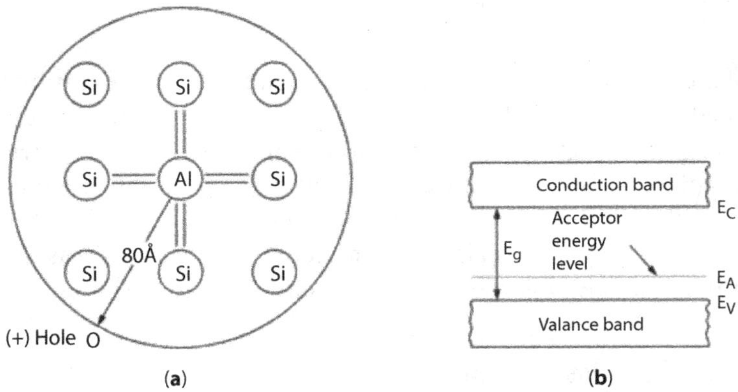

Figure 6.4 *p*-type semiconductor (*a*) silicon doped by third column aluminium, and (*b*) energy diagram showing acceptor energy level.

comes from the neighbouring silicon atom. Migration of electron from silicon atom creates a hole in it. The hole thus created (in silicon) moves around aluminium atom with a positive charge.

Acceptor energy level. Calculations reveal that the holes orbit at a radius of about 80 Å, and are close to the valence band. The energy level possessed by them is known as *acceptor energy level* E_A, and is shown in Fig. 6.4(*b*). The ionization energy needed to excite the holes into the valence band are given in Table 6.4. The holes are majority charge carriers in this case ($n_h > n_e$), and the law of mass action holds goods here also.

- The conductivity of extrinsic semiconductors, both *n*-type and *p*-type, may be found from Eq. 6.8 knowing that $n_e \neq n_h$.

Example 6.5 The resistivity of Ga-Sb at 300 K is found to be 2×10^{-3} ohm-m. The electron and hole mobilities are 0.3 and 0.1 m²/volt-sec respectively. Calculate the carrier density. Given $e = 1.6 \times 10^{-19}$ coulomb.

Solution. $\text{Conductivity}\,\sigma = \dfrac{1}{\text{Resistivity }\rho} = \dfrac{1}{2\times10^{-3}} = 500\,/\,\text{ohm - m}$

As $$\sigma = n_e e_e \mu_e + n_h e_h \mu_h$$

where $e_e = e_h = e$ = electronic charge = 1.6×10^{-19} C,

Assuming $n_e = n_h = n$, and μ_e and μ_h = 0.3 and 0.1 m²/V-sec respectively.

∴ $$500 = n_e \times e\,(\mu_e + \mu_h)$$

or $$n_e \times 1.6 \times 10^{-19}\,(0.3 + 0.1) = 500$$

∴ $$n = 7.8125 \times 10^{21}/\text{m}^3$$

Example 6.6 A current of density 1000 A/m² flows through a n-type germanium crystal. This crystal has a resistivity of 0.05 ohm m and the electron mobility of 0.4 m²/V s. Calculate the drift velocity and time taken by the electrons to travel 100 μm in the crystal.

Solution. The given data are,

$$I_d = 1000\text{ A/m}^2,\ \rho = 0.05\text{ ohm m},$$

travel length $$l = 100\ \mu\text{m} = 100 \times 10^{-6}\text{m},$$

and $$\mu_e = 0.4\text{ m}^2/\text{Vs}.$$

Let t be the time taken by electrons to travel a distance of l.

For a n-type semiconductor we know that $n_e >> n_h$, therefore neglecting the term $n_h\, e_h\, \mu_h$, Eq. 6.9 may be written as

$$\sigma = n_e \cdot e \cdot \mu_e$$

∴ $$\frac{1}{0.05} = n_e \times 1.602 \times 10^{-19} \times 0.4$$

or $$n_e = \frac{1}{0.05 \times 1.602 \times 10^{-19} \times 0.4} = 3.13 \times 10^{20}\,/\,\text{m}^3$$

We also know that

$$I_d = nev_d$$

$$\therefore \qquad v_d = \frac{I_d}{ne} = \frac{1000}{3.13\times10^{20}\times1.602\times10^{-19}} = 20\,\text{m/s}$$

The distance l travelled in time t at drift velocity v_d, by an electron may be obtained from

$$l = t\,v_d$$

$$\therefore \qquad t = \frac{l}{v_d} = \frac{100\times10^{-6}}{20} = 5\times10^{-6}\,\text{sec} = 5\text{ms}$$

Example 6.7 A p-type germanium rod, 10 mm long and 1 mm in diameter, has a resistance of 100 ohm. Determine the impurity concentration in the rod. The mobility of holes in germanium = 0.19 m²/V s.

Solution. The area of cross-section of rod is found as

$$a = \frac{\pi d^2}{4} = \frac{\pi(1\times10^{-3})^2}{4} = 7.85\times10^{-7}\,\text{m}^2$$

The resistivity of the rod is obtained as

$$\therefore \qquad \rho = \frac{aR}{l} = \frac{7.85\times10^{-7}}{10\times10^{-3}}\times100 \qquad (i)$$

$$= 7.85\times10^{-3}\text{ ohm m}$$

In a p-type semiconductor, the holes are majority carriers as $n_h >> n_e$. Therefore the term $n_e \cdot e \cdot \mu_e$ can be neglected from Eq. 6.9.

Knowing that the resistivity $\rho = \frac{1}{\sigma}$, the Eq. (i) can be written as

$$\rho = \frac{1}{\sigma} = \frac{1}{n_h.e.\mu_h}$$

$$\therefore \qquad 7.85\times10^{-3} = \frac{1}{n_h\times1.602\times10^{-19}\times0.19}$$

$$\text{or} \qquad n_h = \frac{1}{7.85\times10^{-3}\times1.602\times10^{-19}\times0.19}$$

$$= 4.18\times10^{21}/\text{m}^3$$

Intrinsic concentration is neglected, and assuming that each impurity atom produces one electron in the rod, the impurity concentration N_h may be approximated as

$$N_h = n_h = 4.18 \times 10^{21}/\text{m}^3$$

6.7 Effective Mass

Rest mass. The mass of an electron and a hole is one of the basic properties in design of semiconductor devices. The mass of these carriers in vacuum is called the rest mass, Fig. 6.5(*a*). In real applications, the electrons or hole carriers move within a solid and not in the vacuum. While moving, they collide with the atoms of semiconductor, which causes their periodic deceleration. These carriers are also subjected to a complex crystalline field in addition to the applied field. Hence, their mass in movement is different from their mass at rest, and is known as *effective mass m*. The effective mass of an electron and a hole is denoted by* m_e^* *and* m_h^* respectively, Fig. 6.5(*b*). This mass is taken into consideration during design calculations for semiconductor systems.

The density of state, effective mass and the rest mass for intrinsic semiconductors is shown in Table 6.5.

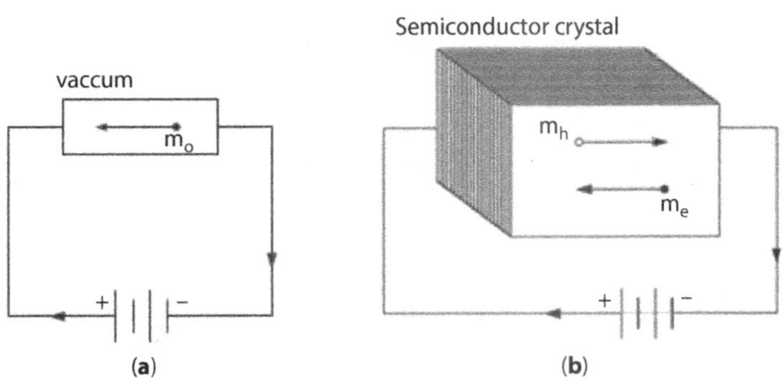

Figure 6.5 (*a*) Rest mass of electron carrier in vacuum, and (*b*) effective mass of electron and hole carriers in a semiconductor crystal.

Table 6.5 Rest mass and effective mass m*

Intrinsic semiconductor	Effective mass of electron m_e^*	Effective mass of hole m_h^*
• Silicon	1.1 m_0	0.59 m_0
• Germanium	0.55 m_0	0.37 m_0

6.8 Density of State

Density of state means the population density of electrons in a metal. It has relevance to Fermi- Dirac distribution. In the last article we discussed that the Fermi probability function $p(E)$ determines the probability of energy level E occupied by an electron. It tells us about the energy level but not about the number of electrons in those levels. The density of state $N(E)$ indicates the number of electrons n_e across the energy band. This number is not uniform across the energy band, rather it is greatest at the centre of the band. The product of $p(E)$, $N(E)$ and the number of electrons for metals are related by

$$n_e = \int_0^{E_F} p(E)N(E)dE \tag{6.10}$$

This relation is illustrated in Fig. 6.6 over a range of band energy at 0 K and temperature above 0 K. It illustrates that only a small fraction of electrons within the energy range of kT can be excited above Fermi level. Here k is Boltzmann constant and T is absolute temperature. The effective density of energy states can be found by employing the quantum mechanics. If the effective density of states at the conduction and at the valence bands are N_C and N_V respectively, then

$$N_C = 2\left(\frac{2\pi^{\vee}\,{}^{*}_{e}}{h^2}\right) \tag{6.11a}$$

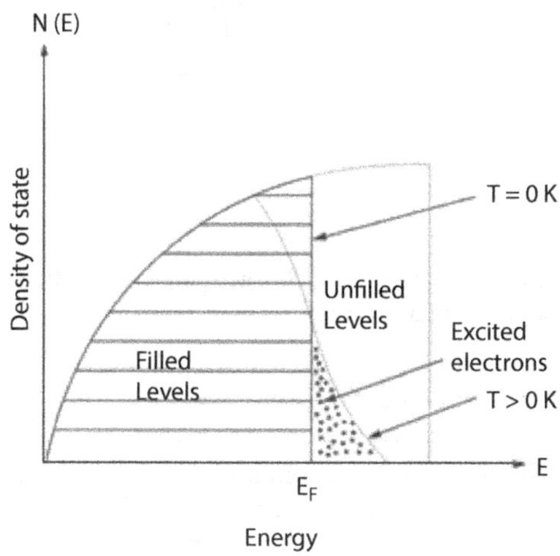

Figure 6.6 Illustration of density of state showing a small fraction of excited electrons possessing $E > E_F$ at $T > 0$ K.

and
$$N_V = 2\left(\frac{2\pi m_h^* kT}{h^2}\right)^{3/2} \tag{6.11b}$$

where m_e^* and m_h^* are the effective mass of an electron and a hole respectively; and h is the Planck's constant. The number of negative and positive charge carriers n_e and n_h in their respective bands may be found from

$$n_e = N_C e^{-(E_V - E_F)/kT} \tag{6.12a}$$

and
$$n_h = N_V e^{-(E_F - E_V)/kT} \tag{6.12b}$$

where E_F is the Fermi energy. Now the product of positive and negative charge carriers is

$$n_e \times n_h = n_i^2 \tag{6.13}$$

and the term $\sqrt{N_C N_V}$ depends on the band structure of the semiconductor. For a specific material, the product

$$\sqrt{N_C N_V} = \text{constant} \tag{6.14}$$

Example 6.8 Calculate the conduction electron and hole density at room temperature in Si-doped with 10^{19} donors/m^3, if the intrinsic carrier concentration is 1.5×10^{16}/m^3.

Solution. Using product of mass relation between the number of conduction electrons n and number of holes p as

$$np = n_i^2$$

we write
$$p = \frac{n_i^2}{n} = \frac{(1.5\times10^{16})^2}{10^{19}} = 2.25\times10^{13}\ /\ \text{m}^3$$

Example 6.9 A silicon sample is doped with 10^{17} arsenic atoms per cubic centimeter. Calculate the hole concentration at 300 K. What will be the location of Fermi energy level relative to intrinsic energy level? Take intrinsic carrier concentration as 1.5×10^{10}/cm^3 at room temperature.

Solution. Given data are: total number of donor atoms

$$N_D = 10^{17}/\text{cm}^3 \text{ and } n_i = 1.5 \times 10^{10}/\text{cm}^3$$

Since $N_D >> n_i$, we can approximate

$$n_e = N_D$$

Using Eq. 6.13, the hole concentration n_h is obtained as

$$n_h = \frac{n_i^2}{n_e} = \frac{(1.5\times10^{10})^2}{10^{17}}$$

$$= \frac{2.25\times10^{20}}{10^{17}} = 2.25\times10^3\ /\ \text{cm}^3$$

The relative locations of E_F and E_i are found from

$$E_F - E_i = kT\ln(n_e / n_i)$$

Since the value of kT at room temperature (300 K) = 0.0259 eV

$$\therefore \quad E_F - E_i = 0.0259\ln\left(\frac{10^{17}}{1.5\times10^{10}}\right)$$

$$= 0.407\ \text{eV}$$

Thus E_F is located at 0.407 eV away from E_i, towards E_c side as given below.

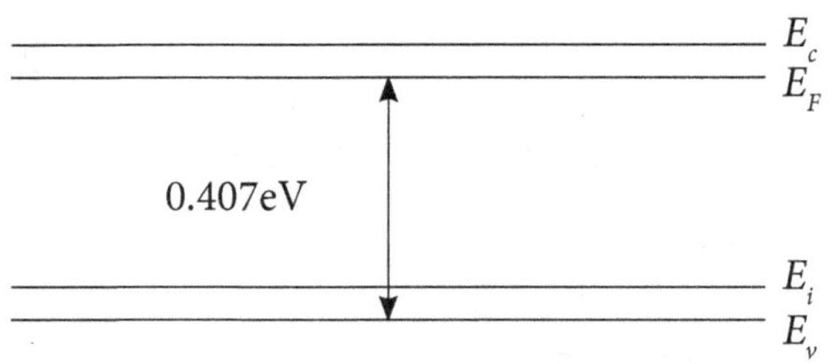

6.9 Temperature Dependency of Carrier Concentrations

The carrier concentration is a function of temperature. It is evident from the following relations.

$$n_o = n_i e^{(E_F - E_i)/kT} \tag{6.15a}$$

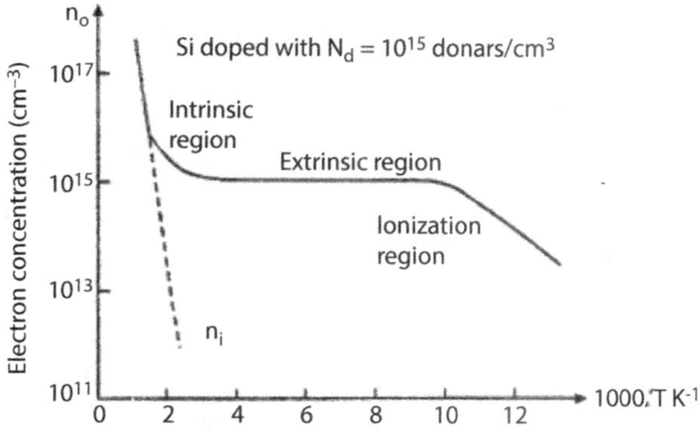

Figure 6.7 Depiction of dependency of electron concentration as a function of reciprocal of temperature.

$$p_o = n_o e^{(Ei-EF)/kT} \tag{6.15b}$$

Here n_o and p_o *are* electron and hole concentration respectively in conduction band. Figure 6.7 depicts the variation of electron concentration as a function of reciprocal of temperature for a n-type extrinsic Si. It has a donor concentration of $N_d = 10^{15}/cm^3$. The curve shows three distinct regions.

1. *Ionization region.* It occurs at large values of 1000/T *i.e.* at low temperatures where donor electrons are bound to donor atoms. At about 1000/T= 10 *i.e.* 100 K, all donor atoms are ionized and $n_o \approx N = 10^{15}/cm^3$.
2. *Extrinsic region.* It occurs when every available extrinsic electron is transferred to the conduction band, hence $n_i \approx N_d$ and n_o becomes almost constant.
3. *Intrinsic region.* It occurs at smaller values of 1000/T *i.e.* at higher temperatures when $n_i >> N_d$.

Amongst the above three regions, the extrinsic region is desirable for the operation of semiconductor devices. It is due to its constant characteristic. The extrinsic range can also be extended beyond the highest temperature at which the device is to operate. This is accomplished by either

- generating the thermal electron-hole pair, or
- doping

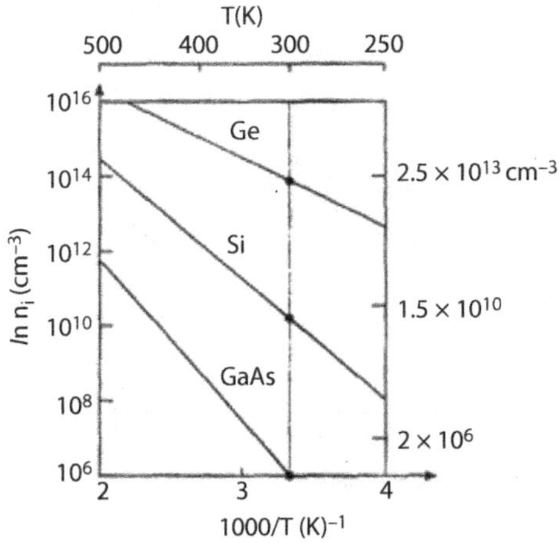

Figure 6.8 Depiction of variation in intrinsic carrier concentration on a semi-log curve, as a function of reciprocal of temperature.

In most semiconductor devices, the control of the carrier concentration by doping is preferred.

6.9.1 Temperature Dependency of n_i

Intrinsic carrier concentration n_i has a strong temperature dependence. It can be visualized from the following relations.

$$n_i = \sqrt{N_C N_V}.e^{-E_g/2kT} \tag{6.16}$$

$$N_C = 2\left(\frac{2\pi m_e^* kT}{h^2}\right)^{3/2} \quad \text{and} \quad N_V = 2\left(\frac{2\pi m_h^* kT}{h^2}\right)^{3/2} \tag{6.17}$$

On combining above equations, we obtain

$$n_i = 2\left(\frac{2\pi kT}{h^2}\right)^{3/2}.(m_e^* m_h^*)^{3/4} e^{-E_g/2kT} \tag{6.18}$$

It shows an exponential dependence of n_i on T. Figure 6.8 depicts this variation on a semi-log curve. The ln n_i vs 1000/T plot. It appears to be linear on the pretext that the effects of temperature dependency and E_g are neglected.

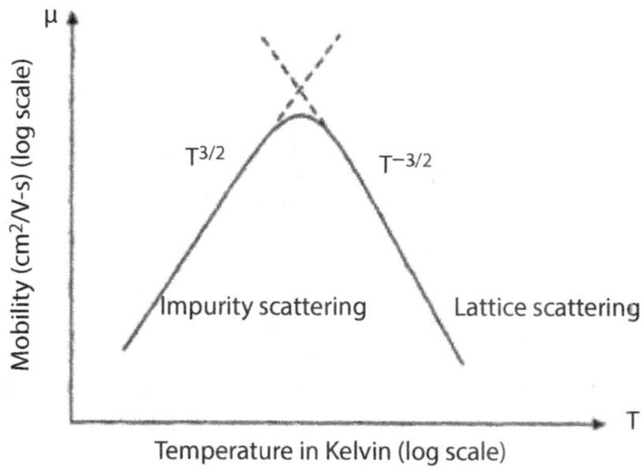

Figure 6.9 Effect of temperature on mobility of carriers.

6.10 Effects of Temperature on Mobility of Carriers

The mobility of electrons or holes is influenced by scattering. Main sources of scattering in a semiconductor are phonons and ionized impurity atoms. The mobility of electron and hole carriers are governed by following scattering mechanisms, which themselves result from temperature.

1. *Lattice scattering.* In this mechanism, a carrier moving through a crystal is scattered by inherent vibration of the lattice, caused by temperature. The frequency of such scattering increases with increase in temperature, therefore the thermal agitation of the lattice also becomes greater. Hence the mobility decreases, as shown in Fig. 6.9.
2. *Impurity scattering.* In this mechanism, the scattering dominates at low temperatures. At lower temperatures, as the atoms are less agitated, therefore the thermal motion of the carriers is slow. Hence, there is an increase in mobility with increasing temperature, Fig. 6.9.

The approximate temperature dependencies for both the above mechanisms are as follows.

$$\text{(i) for lattice scattering } \mu_L = \text{a T}^{-1.5} \tag{6.19a}$$

$$\text{and (ii) for impurity scattering } \mu_I = \text{b T}^{+1.5} \tag{6.19b}$$

where a and *b* are material constants.

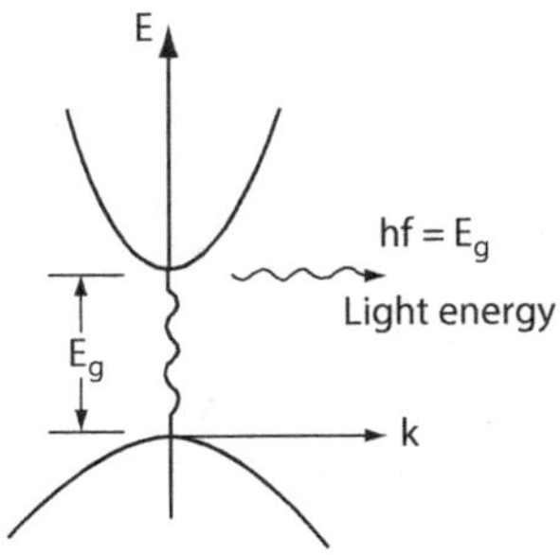

Figure 6.10a Direct electron transition accompanied with photon emission

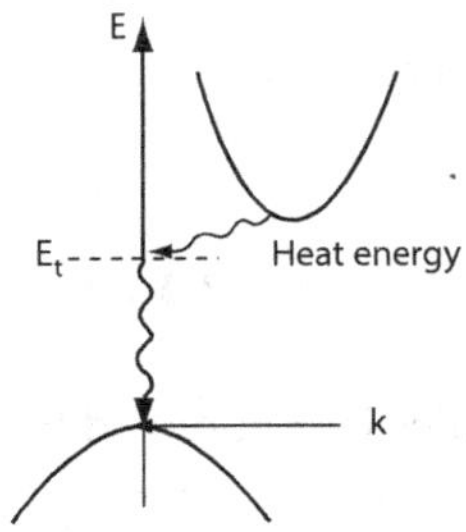

Figure 6.10b Indirect electron transition via a defect level with heat emission and a change in the value of k.

6.10.1 Effects of Doping on Mobility

The effect of doping (i.e. impurity concentration) on mobility of carriers is shown in Fig. 6.10. It shows a decrease in mobility of electrons and holes for Si, Ge, and GaAs with increasing doping concentration. The reason of decrease in mobility is attributed to a rise in impurity scattering due to the presence of larger dopants.

6.11 Direct and Indirect Energy Band Semiconductors

The *E* vs *k* curve shown in Fig. 4.6a (see art. 4.4.1) represents the wave function of the electron in x-direction only. It is typically true for a single electron travelling through a perfectly periodic lattice. However, since the periodicity of most lattices is different in different directions, the *E* vs *k* curve is also different for various crystal directions. For illustration, we cite the examples of GaAs and Si according to which

- the band structure of GaAs for k = 0 has a maximum in valence band and a minimum in conduction band. Therefore, an electron making a transition from conduction band to valence band can do so without any change in the value of *k*.

However,

- the band structure for Si has a maximum in valence band at a different value of *k* than a minimum in conduction band. Therefore, a transition from conduction band to valence band requires some change in the value of *k*.

Accordingly, the energy bands in semiconductors have two different classes viz.

i. direct energy band, and
ii. indirect energy band

Thus the semiconductors falling in above categories are also known as

1. Direct semiconductors, and
2. Indirect semiconductors

Different semiconducting materials falling in each of these categories are listed below.

Direct energy band semiconductors	
GaAs	ZnS
GaN	ZnSe
GaSb	ZnTe
InP	CdS
InAs	CdSe
InSb	CdTe

Indirect energy band semiconductors	
Si	GaP
Ge	PbS
SiC(α)	PbSe
AlP	PbTe
AlAs	
AlSb	

6.11.1 Differences between Direct and Indirect Semiconductors

These are given as below.

S. No.	Description	Direct semiconductor	Indirect semiconductor
1	When an electron in conduction band falls to an empty state in valence band	The energy difference E_g is given-off as a photon of light as shown in Fig. 6.10a.	The electron cannot fall directly to valence band, however, it undergoes a change in its • momentum, and • energy
2	During transition	There is no change in the value of k ($k = 0$) as shown in Fig. 6.10a.	There is a change in the value of k as shown in Fig. 6.10b. Also the energy is given up as heat to the lattice.
3	Suitability of applications.	Semiconductor light emitters and lasers for materials capable of *direct* band-to-band transitions.	Semiconductor light emitters and lasers for materials capable of *vertical* transitions between defect states.

6.12 Variation of E_g with Alloy Composition

Often we require a change in the value of E_g for certain semiconductor, to make it suitable for a particular application. For that the composition of semiconductor has to be varied. It is done by alloying the semiconductor in varying proportion. In the same way the band structure of ternary and quarternary alloys is also altered by varying their composition. Its mechanism is explained through Figs. 6.11a-b-c, which illustrate the band structures of binary compounds GaAs and AlAs, and effect of changing the fraction x in ternary compound $Al_xGa_{1-x}As$.

Figure 6.11a shows that for GaAs the direct minimum is lower than the indirect minima, but this is just opposite (see Fig. 6.11b) for AlAs. Figure 6.11c depicts the variation of Al fraction x in conduction bands of AlGaAs. The range of x for compositions GaAs ($x = 0$) to AlAs($x = 1$) shows the smallest band gap E_g at $x = 0.38$. Thus the ternary AlGaAs is

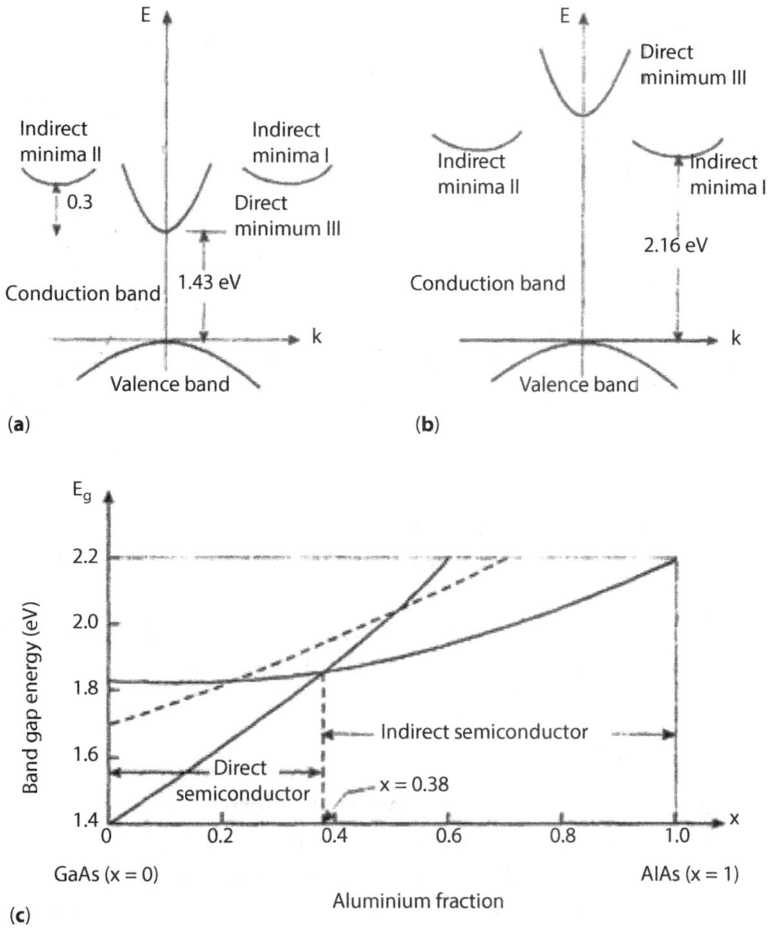

Figure 6.11 Illustration of variation in E_g by changing the alloy composition (*a*) GaAs shows direct minimum is lower than indirect minima, (*b*) AlAs band diagram, and (*c*) effect of variation of Al fraction *x* in conduction band of ternary alloy $Al_xGa_{1-x}As$.

- direct semiconductor upto $x = 0.38$, and
- indirect semiconductor for $x > 0.38$

6.12.1 Effect of Alloying on $GaAs_{1-x}P_x$

In another illustration, the variation in energy band for a ternary alloy $GaAs_{1-x}P_x$, as a function of alloy composition is shown in Fig. 6.12. It is illustrated that the GaAsP behaves as

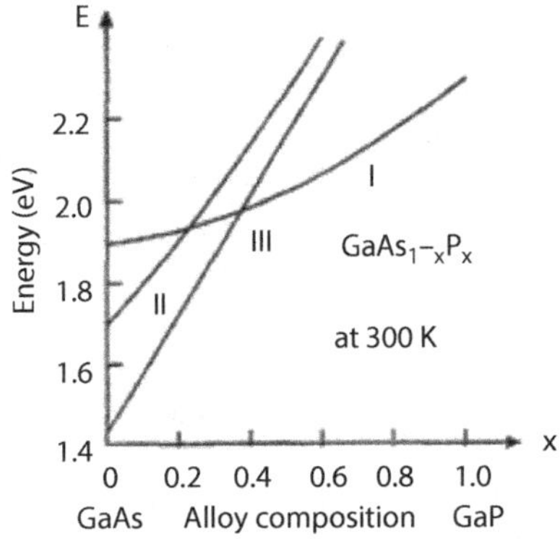

Figure 6.12 Conduction band energy of ternary alloy $GaAs_{1-x}P_x$ a function of variation in fraction x of alloy composition.

- direct semiconductor upto $x = 0.45$, and
- indirect semiconductor for $x > 0.45$.

The ternary alloy is a direct semiconductor due to GaAs and is indirect semiconductor due to composition of GaP. It is about $GaAs_{0.55}P_{0.45}$.

6.12.2 Applications

As the alloy composition decides whether the ternary compound will be a direct semiconductor or indirect, the percentage composition becomes the criteria for selection of materials. For example, since the light emission is most effective for direct materials in which electrons drop from conduction band to valence band without changing *k* and momentum, hence

- LEDs are made of $GaAs_{1-x}P_x$ having $x < 0.45$, and
- Red LEDs are made for $x \approx 0.4$.

6.13 Degenerate Semiconductors

We know that the amount of impurities doped in semiconductor is in a very small fraction. It is generally of the order of 1: 10^9 to 10^{12}. Although the ppm (1ppm = 0.0001%) addition of dopants has a tremendous effect on conductivity and other properties of semiconductors, but it has almost no effect on the following.

i. On total atomic density of the material.
ii. On charge transport within the donor or acceptor levels because of a very few impurity atoms, and that too very widely spaced throughout the material.
iii. On interactions between the impurities.

In fact, the donor or acceptor energy levels appear to be composed of discrete and non-interacting energy states.

6.13.1 Effect of Heavy Doping

If doping in semiconductor is continued uninterrupted, a limit will reach at which the impurities will be closely packed within the material lattice and will interact with each other. In this condition when the concentration is high (say, about 10^{20} donors/cm^3), following occurrences may be noticed within the material.

- Some effect on total atomic density of material.
- Charge transport within the donor or acceptor levels.
- Considerable interactions between the impurities.
- Non-discrete (i.e. continuous) donor or acceptor energy levels.

In such condition, the energy levels form a band which may overlap the bottom of conduction or valence band, as the case may be.

6.13.2 Degenerate Types

Depending upon the nature of heavy doping, the semiconductors may be called as either

1. degenerate n-type, or
2. degenerate p-type.

It is known as *degenerate n-type* when the electron concentration n_e in conduction band exceeds the effective density of states $N(E)_{eff}$ i.e. $n_e > N(E)_{eff}$. In this case the Fermi level does not lie within the energy band gap, rather lies within the conduction band as shown in Fig. 6.13a. The semiconductor is known as *degenerate p-type* when the acceptor concentration is higher than the effective density of states *i.e.* $n_p > N(E)_{eff}$. In this case the Fermi energy level lies in the valence band as shown in Fig. 6.13b.

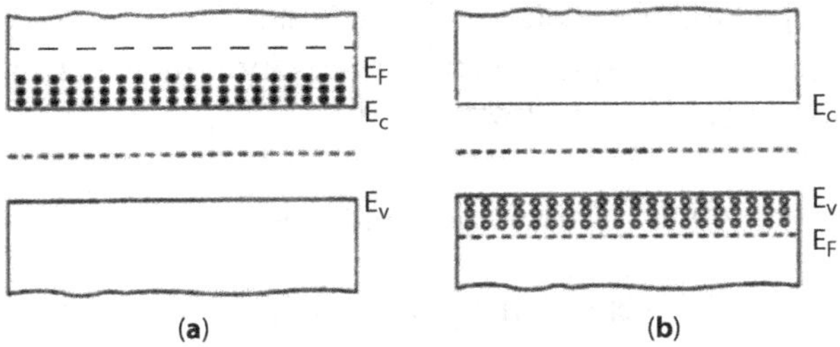

Figure 6.13 The Fermi energy level lies (a) within the conduction band in degenerate n-type semiconductor, and *(b)* within the valence band in degenerate p-type semiconductor.

The filled or empty energy states in conventional and degenerate semiconductors are also different. These are mentioned below.

Energy states	**In conventional semiconductors**	**In degenerate *n*-type semiconductor**	**In degenerate *p*-type semiconductor**
• Below E_F	Mostly filled	—	—
• Above E_F	Mostly empty except for a small distribution	—	—
• Region between E, and E_F	—	For most part, filled with electrons.	—
• Region between E_v and E_F	—	—	Almost completely filled with holes

6.14 Hall Effect

Meaning. The electrons in a n-type semiconductor and the holes in a p-type semiconductor usually drift in the direction x of an applied electric field. When a magnetic field is applied in z- direction which is perpendicular to the direction of drift of holes or electrons, they tend to deflect. This is an undesirable effect and needs to be prevented. Establishment of an electric field along y-direction to prevent deflection of holes or electrons is known as Hall effect.

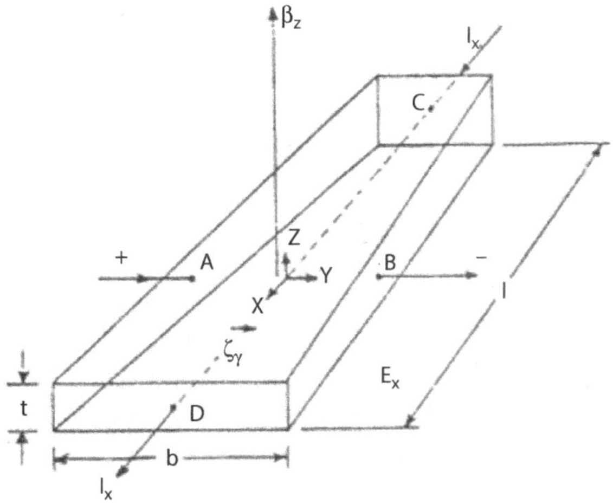

Figure 6.14 Establishment of electric field along y-direction to prevent deflection of holes along x-direction is called Hall effect.

6.14.1 Explanation of the Phenomenon

To illustrate this phenomenon, let us take a p-type semiconductor bar of length l, width b and thickness t lying in the field of an electric current I_x and a magnetic flux density β_z as shown in Fig. 6.14. A single hole in the bar experiences a force F_y along y-direction whose magnitude is given by

$$F_y = e_h(\zeta_y - \upsilon_x \beta_z) \tag{6.20}$$

where e_h is the charge on hole, ζ_y is the electric field strength along y-direction, and v_x is the velocity of drifting holes.

6.14.2 Hall Voltage

To maintain steady state flow of holes down the length of bar, it is desired that the holes do not deflect. Therefore, F_y in Eq. 6.20 must be zero and hence

$$\zeta_y = \upsilon_x \beta_z \tag{6.21}$$

Establishment of ζ_y along the width of the bar is known as Hall effect. The resulting voltage V_{AB} along y-direction is called Hall voltage and is given by

$$V_{AB} = \zeta_y b \tag{6.22}$$

6.14.3 Significance of Hall Effect, Hall Coefficient etc.

The idea of Hall effect is utilized to determine

1. nature of an unknown semiconductor; whether n-type or p-type,
2. mobility of the semiconductor,
3. conductivity, and
4. resistivity of the semiconductor.

Hall coefficient. An unknown piece of semiconductor whether n-type or p-type can be investigated by measuring the Hall coefficient H_C. It is found from

$$H_c = \frac{1}{e_h h_0} \tag{6.23}$$

where h_o is hole concentration. A negative value of H_C indicates n-type and a positive value confirms p-type semiconductor.

Relation between Hall voltage and drift current. The electric field gradient ζ_y can also be expressed as

$$\zeta_y = I_{ds}\beta_z H_c = \frac{I_{ds}\beta_z}{e_h h_0}$$

therefore

$$h_0 = \frac{I_{ds}\beta_z}{e_h \zeta_y} \tag{6.24}$$

where I_{ds} is the drift current density along x-direction and is equal to I_x/bt. We can also write Eq. 6.24 as

$$h_0 = \frac{(I_x / bt)\beta_z}{e_h(V_{AB} / b)} = \frac{I_x\beta_z}{e_h t V_{AB}} \tag{6.25}$$

Resistivity of semiconductor. All the quantities of R.H.S. of Eq. 6.25 can be measured experimentally. Then the resistivity ρ may be obtained from

$$\rho = \frac{Rbt}{l} \tag{6.26}$$

where R is the resistance of semiconductor.

Determining the mobility. As $R = (V_{CD}/I_x)$ and conductivity $\sigma = 1/\rho$, hence mobility of hole may be determined from

$$\mu_h = \frac{\sigma}{H_c} = \frac{1}{\rho H_c}$$
$$= \frac{l}{RbtH_c} = \frac{I_x l}{V_{CD} btH_c} \quad (6.27)$$

Hall Angle. It is the non-vanishing angle made by the total electric field in a semiconducting sample, with the x-axis along which the carriers drift. **Hall Generator** is a Hall effect based measuring device which is used to measure high amplitude a.c. and impulse currents, and very high direct currents. Such high currents are used in testing of cables lightening arresters, circuit breakers etc. Hall generators employ semiconducting materials in their construction. It is because the Hall coefficient of semiconductors is high as compared to metals whose Hall coefficient is very small.

Example 6.10 An aluminium piece 15 mm thick, 60 mm wide and 180 mm long is placed under a magnetic field of 0.6 tesla. It carries a current of 25 ampere. If the hole mobility and electrical conductivity are 0.0012 m²/Vs and 3.8×10^7 per ohm m respectively, determine the *(a)* Hall coefficient, *(b)* Hall voltage, and *(c)* resistance of metal.

Solution: Given are

$l = 180$ mm $= 180/1000\ m = 0.18$ m, $\mu_h = 0.0012$ m²/Vs

$b = 60$ mm $= 0.06$ m, $t = 15$ mm $= 0.015$ m

$\beta_z = 0.6\text{T} = 0.6$ Wb/m², $I_x = 25$ A, $\sigma = 3.8 \times 10^7$/ohm m

a. The Hall coefficient is found as

$$H_c = \mu_h / \sigma = 0.0012 / 3.8 \times 10^7$$
$$= 3.166 \times 10^{-11}\ \text{Vm/AT}$$

b. The Hall voltage is found from

$$V_{AB} = \frac{H_C I_x \beta_z}{t}$$
$$= \frac{3.166 \times 10^{-11} \times 25 \times 0.6}{0.015}$$
$$= 3166 \times 10^{-11} = 3.166 \times 10^{-8}\text{V}$$

c. The resistance is found as

$$R = \frac{\rho l}{bt} \text{ i.e. } R = \frac{l}{\sigma bt}$$

$$\therefore \quad R = \frac{0.18}{(3.8\times10^{7}\times0.06\times0.015)} = 5.263\times10^{-6} \text{ ohm}$$

6.15 Analysis of Drift and Diffusion Currents

We have seen earlier that the conduction mechanism in semiconductors is associated with the 'drift of electrons' and 'diffusion of carriers'. Consequently, following two kinds of currents flow in a semiconductor.

1. Drift current due to electric field, and
2. Diffusion current due to gradient of carrier concentration

Flow of these currents apply to both: to the electrons as well as to the holes. Since the conductivity associated with conduction electrons is $\sigma_e = n_e \cdot e \cdot \mu_e$ and with conduction holes is $\sigma_h = n_h e\mu_h$, therefore the current density associated with drift of electrons and holes, due to the applied field E may be written as

$$J_e = n_e e\mu_e E \tag{6.28a}$$

and

$$J_h = n_h e\mu_h E \tag{6.28b}$$

Diffusivity. According to diffusion theory given by Fick's 1st law for steady state condition, the net flow of carriers along x-axis (in one- dimensional case) is given by

$$\frac{d\eta}{dt} = -D_x A\frac{d\eta}{dx} \tag{6.29}$$

where η is density of carriers, $d\eta/dt$ is the number of carriers diffusing per unit time into a cross-sectional area A normal to the direction of diffusion x, $d\eta/dx$ is concentration gradient, and D_x is diffusion coefficient along x-direction. The term D_x is popularly known as *diffusivity.* It is a constant and is the system characteristic. The minus sign in above equation indicates that when concentration increases, the flow of particles due to diffusion occurs in negative x-direction. Equation 6.29 *may* also be written as

$$J_x = \frac{1}{A}\frac{d\eta}{dt} = -D_x \frac{d\eta}{dx} \tag{6.30}$$

where J_x is diffusion flux per unit cross-sectional area per unit time.

Thus the hole current may be written as the sum of contributions from electric field and from diffusion process, as follows.

$$J_h = n_h e\mu_h E - eD_{xh} \frac{d\eta_h}{dx} \tag{6.31a}$$

where D_{xh} is diffusion coefficient of the holes. The term e for charge appears in diffusion contribution because of the fact that the electric current is equal to the particle current times the charge e per particle. A similar expression may be written for electron also, as

$$J_e = n_e e\mu_e E + eD_{xe} \frac{d\eta_e}{dx} \tag{6.31b}$$

In this case the diffusion contribution term is positive because the charge for electron is -e (not +e as in the case of holes).

Total current. The sum of Eqs. 6.31a and 6.31b yields the total (or resultant) current. It is

$$J_h + J_e = (n_h e\mu_h E + n_e e\mu_e E) - \left(eD_{xh} \frac{d\eta_h}{dx} - eD_{xe} \frac{d\eta_e}{dx} \right) \tag{6.32}$$

$$= \text{Drift current} - \text{Diffusion current}$$

- The diffusion current plays an important role in rectifier and transistor actions.

6.15.1 Einstein Relation

If a semiconductor is subjected to an electric field E and a concentration gradient $d\eta/dx$ such that the resultant current given by Eqs. 6.28a-b is zero; an important relation can be established between the diffusion coefficient D_x and mobility μ of the carriers. This relationship is known as Einstein relation. Under the condition of no current, the system remains in thermal equilibrium, and the gradient of hole density is given by

$$\frac{d\eta_h}{dx} = -\frac{e}{KT}\eta_h \frac{dV}{dx} \tag{6.33}$$

where K is Boltzmann constant and $dV/dx = -E(\text{x})$. Here V is potential and E is electric field as a function of direction x.

Since the hole current vanishes in thermal equilibrium, hence using Eqs. 6.31a and 6.33, we obtain

$$D_{xh} = \left(\frac{KT}{e}\right)\mu_h \tag{6.34a}$$

Similarly for electrons, the relationship exists as

$$D_{xe} = \left(\frac{KT}{e}\right)\mu_e \tag{6.34b}$$

Equations 6.33a-b are Einstein relations and can be used to calculate diffusivity at different temperatures, if mobilities are known. These values for germanium at room temperature are

- $D_e = 0.0093\ \text{m}^2\text{s}^{-2}$, and
- $D_h = 0.0044\ \text{m}^2\text{s}^{-2}$

6.16 Continuity Equation

Continuity equation describes the behaviour of minority carriers in rectifying junctions and transistors. The minority carriers (electrons) in *p*-type semiconductor and holes in *n*-type semiconductor play an important role in understanding the theory of above devices and their design. In *p*-type semiconductor, the basis of derivation for continuity equation is

$$\begin{pmatrix}\text{Time rate of}\\ \text{increase of}\\ \text{electrons}\end{pmatrix} = \begin{pmatrix}\text{Rate of thermal}\\ \text{generation of}\\ \text{electrons}\end{pmatrix} + \begin{pmatrix}\text{Rate of increase}\\ \text{due to recombination}\end{pmatrix}$$

if there is no electron current or hole current. And if there is an electron current flowing through it, the formulation of basic equation will involve

$$\begin{pmatrix}\text{Total rate of change}\\ \text{of electron density } n_e\end{pmatrix} = \begin{pmatrix}\text{Rate of change}\\ \text{of } n_e \text{ for } J_e = 0\end{pmatrix} + \begin{pmatrix}\text{Rate of change of}\\ n_e \text{ due to } J_e\end{pmatrix}$$

Here J_e is current density. A rigorous analysis ultimately leads to following continuity equations which must be satisfied by the minority carriers in their operations.

i. In *p*-type semiconductor where electrons are the minority carriers,

$$\frac{d\eta_e}{dt} = \frac{\eta_{eo} - \eta_e}{\tau_e} + \frac{1}{e}\frac{dJ_e}{dx} \tag{6.35a}$$

ii. In n-type semiconductor where holes are the minority carriers,

$$\frac{d\eta_h}{dt} = \frac{\eta_{ho} - \eta_h}{\tau_h} + \frac{1}{e}\frac{dJ_h}{dx} \tag{6.35b}$$

Here J_e and J_h are the current densities as given by Eqs. 6.28a and 6.28b; η_{eo} and η_{ho} are thermal equilibrium densities; and τ_e and τ_h are the *lifetime* (time constant) of electrons and holes respectively.

6.17 Solved Examples

Example 6.11 Semiconductor materials are the natural choices for electronic devices-why?

Solution. Semiconductor materials are the natural choices because their conductivity and other electrical properties can be easily varied to suit the requirements of desired electronic devices. For example, their conductivity can be varied by following means.

- by controlling the impurity content
- by optical excitation
- by change in temperature

Example 6.12 Why the impurity concentration in extrinsic semiconductor is kept very small?

Solution. The electrical conductivity of pure semiconductors is primarily of electron carrier type. The hole conductivity is comparatively weaker. This makes the semiconductors extremely sensitive to the presence of any impurity. Hence impurity concentration is kept very small, about 1: 10^6 to 1: 10^9.

Example 6.13 What is electron scattering? How does it influence the behaviour of electron and the semiconducting crystal?

Solution. Collision of electrons with obstacles within a conducting material is an inherent phenomenon. Depending upon various factors, they scatter over certain angle θ, after collision. This is called *electron scattering.*

The phenomenon of electron scattering causes loss in its velocity after collision. It also causes transfer of heat energy from electron to crystal lattice. Probability of scattering remains a constant in case the scattering is spherically symmetric.

Example 6.14 Calculate (a) the Hall coefficient and (b) conductivity of an Al sample whose electron mobility is 1.065×10^{-3} m²/Vs, number of free electrons is 10^{19}/m³, and relaxation time is 6×10^{-15} s. Take standard values of electronic charge and mass.

Solution: Given are $\mu_e = 1.065 \times 10^{-3}$ m²/Vs, $\tau = 6 \times 10^{-15}$ s

and $$\mu_e = \sigma H_c, \text{ and } \sigma = \frac{ne^2\tau}{m_e}$$

$\because$ n = Number of free electrons = 10^{19}/m³

$$\therefore \quad \mu_e = \frac{ne^2\tau H_c}{m_e}$$

i.e. $$1.065\times10^{-3} = \frac{10^{19}\times(1.6\times10^{-19})^2\times(6\times10^{-15})H_c}{9.1\times10^{-31}}$$

or $$H_c = \frac{(1.065\times10^{-3})\times(9.1\times10^{-31})}{10^{19}(1.6\times10^{-19})^2\times(6\times10^{-15})}$$

$$= 6.309 \times 10^{-1} \text{ Vm}^3/\text{A Wb}$$

$$\therefore \quad \sigma = \frac{\mu_e}{H_c} = \frac{(1.065\times10^{-3})}{(6.309\times10^{-1})}$$

$$= 5.924 \times 10^{-2} \text{ (ohm m)}^{-1}$$

Example 6.16 The magnetic flux density in a Hall model is 0.5 tesla. The Hall coefficient has been observed to be 3.66×10^{-4} m³/C. If the resistivity of semiconductor at room temperature is 0.00893 ohm-m, determine the Hall angle.

Solution: Given are β_z = 0.5 tesla = 0.5 Wb/m², Hall coefficient H_c = 3.66×10^{-4} m³/C, and ρ = 0.00893 ohm m. We have to determine the Hall angle θ_H.

As $$\tan\theta_H = \mu\beta_z$$
$$= (\sigma H_C)\beta_z$$
$$= \left(\frac{H_C}{\rho}\right)\beta_z$$

$$\therefore \qquad \tan\theta_H = \frac{3.66 \times 10^{-4}\,\text{m}^3/C \times 0.5\,\text{Wb}}{0.00893\ \text{ohm m} \qquad \text{m}^2}$$

$$= 2.049 \times 10^{-2}$$

$$\text{or} \qquad \theta_H = 1.17°$$

Example 6.17 Explain as to why the ionization energy of impurity carrier is higher in silicon than in germanium.

Solution. The ionization energy is the amount of energy needed to detach the valence electron from the impurity atom. The total energy of an orbiting electron is the sum of its kinetic energy and potential energy given as

$$E_{\text{total}} = E_{KE} + E_{PE} = \frac{1}{2}mv^2 - \frac{q^2}{4p\varepsilon_0\varepsilon_r r}$$

Above equation shows that for other factors being constant, the ionization energy is inversely proportional to dielectric constant ε_r. Since ε_r = 16 for germanium and 12 for silicon, hence ionization energy is higher in silicon than in germanium.

Example 6.18 What are the sources of obstacles present in conducting materials? Why do we consider an average value of collision time in electrical designing?

Solution. The obstacles in materials remain present in different forms. They may be present in natural form such as cementite in steel and iron, oxides in copper and aluminium; or may be present in manmade (unnatural) form such as grain boundaries. These obstacles are located randomly in different orientations within the (conducting) materials. That is why the collision of electrons with them occurs at different intervals of time. Therefore, the collision time is determined as an average value, and is defined as the average time elapsed between two successive collisions.

Example 6.19 Fermi velocity of the electron of a certain metal from among the list given below is 0.85×10^6 m/s. Identify this metal. Take rest mass of electron = 9.109×10^{-31} kg.

Metal	Al	Cu	Ag	Na	K	Cs
E_F (eV)	11.8	7.04	5.51	3.2	2.1	1.6

Consider (a) rest mass, and *(b)* variable mass of electron.

Solution: (a) Since the $KE = \frac{1}{2}mv^2$, the Fermi energy may be calcualted as below.

$$\begin{aligned} E_F &= \frac{1}{2} m_O V_F^2 \\ &= \frac{1}{2} \times \left(9.109 \times 10^{-31}\right) \times \left(0.85 \times 10^4\right)^2 \\ &= 3.290 \times 10^{-19}\,\text{J} \\ &= \frac{3.290 \times 10^{-19}}{1.620 \times 10^{-19}} = 2.05\,\text{eV} \end{aligned}$$

As this value is close to 2.1, hence the metal in question is **potassium (K).**

(b) Considering the mass varying with velocity, it is found from Einstein equation given as

$$\begin{aligned} m &= m_O \sqrt{1 - \left(\frac{v}{c}\right)^2} \\ &= \left(9.109 \times 10^{-31}\right) \sqrt{1 - \left(\frac{0.85 \times 10^4}{3 \times 10^5}\right)} \\ &= \left(9.109 \times 10^{-31}\right)\left(0.99999\right) = 9.108 \times 10^{-31}\ \text{kg} \end{aligned}$$

$$\begin{aligned} E_F &= \frac{1}{2} m v_F^2 \\ &= \frac{1}{2} \times \left(9.109 \times 10^{-31}\right)\left(0.85 \times 10^4\right)^2 \\ &= 3.290 \times 10^{-19}\,\text{J} \\ &= \frac{3.290 \times 10^{-19}}{1.602 \times 10^{-19}} = 2.05\,\text{ev} \end{aligned}$$

Hence, the metal in question is **potassium**.

Example 6.20 What is the utility of phosphor in fabrication of a colour T.V. screen?

Solution. The colour of the light emitted by phosphor depends mainly on the impurities present in it. It is so because many radiative transitions involve impurity levels within the band gap of certain materials such as ZnS. Hence, selection of proper phosphors is useful in fabrication of colour T.V. screen.

Quick Revision Summary

Amorphous materials. They are non-crystalline solids having no periodic structure at all. SiO_2 glass used as raw material to prepare a single-crystal Si is its example.

Binary, ternary and quarternary compounds. The binary compounds are made of two-elements such as GaAs, the ternary compounds are made of three-elements such as GaAsP, and quarternary compounds are made of four-elements such as InGaAsP.

Degenerate *n*-type semiconductors are those in which the electron concentration in conduction band exceeds the effective density of states *i.e.* $n_\epsilon > N(E)_{eff}$.

Density of state means the population density of electrons in a semiconductor over a certain energy range. It may be expressed as effective density of states N_c located at conduction band edge, or N_v located at valence band edge.

Direct band gap semiconductors are those in which the most probable recombination process occurs in such a way that the electrons and holes have the same momentum, and the wave number $k = 0$.

Doping. The intentional addition of donor or acceptor impurities to a semiconductor.

Drift velocity v_d. It is the additional velocity acquired by an electron above its normal velocity v_n, when accelerated by an applied electric field.

Effective mass. It is the mass of the carriers when they move in an actual crystal on application of external electric field. Effective mass is different from the mass of carriers at rest.

Electron hole pair (EHP). These are electron and hole of conduction band, created by the excitation of a valence band electron into the conduction band.

Hall angle. It is the non-vanishing angle made by the electric field in a semiconducting sample, with the axis along which the carrier drifts.

Intrinsic semiconductor is a perfect crystal having no impurities or lattice defects. Ultra-high pure Si and Ge crystals are intrinsic.

Wave number k. It is the wave function of an electron which is assumed to be moving in plane waveform along x-direction. It is expressed as $k = 2\pi/\lambda$ where λ is de-Broglie's wavelength.

Review Questions

1. What are different types of semiconductor materials? Enumerate their uses.
2. Describe the merits and characteristics of semiconducting materials.
3. Differentiate between 'Fermi energy level' and 'Fermi-Dirac probability function'. What are their importance in the study of electronic behaviour of solids?
4. Define Fermi energy. What is its importance for semiconductor materials and devices? Explain the Fermi-Dirac probability distribution. How does the temperature influence upon this distribution?
5. Enumerate different types of semiconductors. Intrinsic semiconductors are not suitable for applications in electronic devices—why? Show that the Fermi level for a pure germanium lies in the middle of its forbidden gap.
6. Explain as to how the electrons and holes both conduct in a pure silicon crystal. How is the conductivity influenced by mobility of electrons and holes?
7. Differentiate between n-type and p-type semiconductors. Name various dopants used to prepare them from intrinsic elements.
8. Draw the energy diagrams for both the above types, show salient levels on them, and explain the mechanism of conduction in them.
9. Draw energy band diagrams for Arsenic-doped-Silicon, and Indium-doped-Germanium. Explain the important differences between them.
10. Describe 'effective mass' and 'density of states', and discuss their importance in designing of the solid state devices.
11. Describe the effects of temperature on (a) carrier concentration, and (b) mobility of carriers. Also discuss the effects of doping on mobility.

12. Describe the various element form of semiconducting materials. Compare the properties and characteristics of Se and Sb.
13. Discuss various compound and alloy forms of semiconductor materials. Write the properties and applications of oxides, sulphides, halides, tellurides and sellurides.
14. Enlist various semiconductor devices and state the characteristic property and principle of semiconductor which is utilized in their working.
15. Enumerate various semiconductor devices and give a brief account of all of them.
16. Explain the phenomenon of Hall effect. Discuss the method of identifying an unknown piece of semiconductor as *n*-type or *p*-type semiconductor.
17. Define the Hall voltage and Hall coefficient.

Numerical Problems

1. Determine the resistivity of an intrinsic germanium sample at 300 K when intrinsic concentration is 25×10^{19} per cubic metre. Take electron and hole mobilities at room temperature as 0.39 m^2/V s and 0.19 m^2/V s respectively.
2. The intrinsic carrier density of silicon at 30°C is 11×10^{16} per m^3. The energy gap is 1.17 eV, and the electron and hole mobilities are 0.14 m^2/V and 0.05 m^2/V s respectively at the same temperature. Calculate the (*a*) resistivity of silicon, and (*b*) number of electrons available in valence band for thermal excitation.
3. A silicon sample is doped with 10^{23} arsenic atoms per cubic meter. Calculate the hole concentration at room temperature. Determine the location of Fermi energy level with respect to intrinsic energy level. The number of intrinsic concentration is 15×10^{16} per cubic meter.
4. 7 gm of *p*-type germanium doped with aluminium is melted with 7 gm *n*-type germanium doped with arsenic. In both the types the conductivity is 10/ohm m. Calculate the conductivity of the resulting material. Assume the required data suitably.

5. In the above problem, if *p*-type germanium is 7 gm and *n*-type is 6 gm, calculate the conductivity of the resulting material. What type of semiconductor will it be?
   [**Hint:** Obtain concentration of donor and acceptor level. As the acceptor impurity is greater, hence the resulting material is of *p*-type]
6. In a silicon doped with antimony, 54.35% donor electrons are promoted to the conduction band at 320 K. Calculate the donor ionization energy in the semiconductor.
7. The wavelength of the light from a GaP laser is 5490Å. Compute the energy gap of it.
8. A silicon sample is doped with 10^{17} phosphorus atoms per cubic centimeter. A current of 1 mA passes through 100 nm thick sample which is placed in a magnetic field of 10^5 Wb/cm^2. Taking mobility as 700 cm^2/V s, calculate (*a*) resistivity, (*b*) Hall coefficient, and (*c*) Hall voltage.
9. In the above problem, predict the type of silicon sample.
10. The hole and electron mobilities in a silicon single crystal at 300 K are 0.025 m^2/Vs and 0.17 m^2/Vs respectively. Determine the diffusion coefficient of holes and electrons at this temperature.

Objective Type Questions

1. Which of the following pairs are correctly matched?

1. Dopant in n-type semiconductor	A. Arsenic
2. Ratio of m_h/m_o	B. More than one
3. Thin film microelectronics	C. Interconnections of electronic circuits
4. MOSFET	D. A type of thermistor

   Select the correct answer using the codes given below.

   Codes:

   a. 1 and *B*, 3 and *D* b. 2 and *D*, 2 and *B*
   c. 1 and *A*, 3 and *C* d. 3 and *B*, 4 and *A*

2. Hall effect is utilized to determine
   a. type of semiconductor b. energy gap
   c. Fermi energy level d. endoscopic investigations

3. Density of state indicates the
   a. number of electrons across the energy band
   b. donar energy level
   c. mobility of compound semiconductors
   d. effective mass

4. Consider the following statements.
   A. Semiconductors are bipolar solids and have homopolar bond.
   B. Compound semiconductors generally have ionic-covalent bond.
   C. During transition, there is a change in the value of k for direct energy band semiconductors.
   D. Carrier concentrations are generally independent of temperature in extrinsic regions.

   Of these, the correct statements are
   a. B, C and D
   b. C and D
   c. A, B and C
   d. A, B and D

5. Probability of occupation by an electron at Fermi energy level is
   a. 100%
   b. 50%
   c. 0%
   d. between 1 to 99%

6. An electron with velocity *u* is placed in an electric field *E* and magnetic field *B*. The force experienced by the electron is given by
   a. $-eE$
   b. $-eu \times B$
   c. $-e(u \times E + B)$
   d. $-e(E + u \times B)$

7. Given below are some statements.
   A. The phosphorus is a V column elemental form of semiconductor.
   B. Unit cell of Si is an agglomerated structure.
   C. Zinc blende is the structure of many compound semiconductors.
   D. The number of effective atoms per unit cell in Ge is 8.

   Of these, the correct statements are
   a. *A* and *B*
   b. *B* and *C*
   c. *A*, *C* and *D*
   d. *A*, *B* and *D*

8. Consider the following statements.
   Extrinsic semiconductors show high electrical conductivity because the impurities are
   1. of high conductivity
   2. highly mobile
   3. highly charged

   Which of these statements are *not* correct?
   a. 1 and 2 b. 2 and 3
   c. 1 and 3 d. 1, 2 and 3

9. Consider the following statements.
   A. A semiconductor is an amorphous solid.
   B. Semiconductors are usually metallic in appearance.
   C. Semiconductors behave like dielectrics at extremely low temperatures.
   D. Temperature coefficient in semiconductors may be negative, positive, or zero.

   Of these, the correct statements are
   a. *A* and *B* b. *B* and C
   c. *C* and *D* d. *B*, *C* and *D*

7

Semiconducting Materials: Types and Applications

7.1 Element Form Semiconducting Materials

7.1.1 Silicon

It is the element of IVth column in periodic table. Its structure is diamond cubic (DC) and the atoms are covalent bonded. The bond is directional in nature with a bond angle of 109°. It is most important and widely used material among all known semiconducting elements. This is dull whitish coloured monocrystal possessing the following salient properties.

- Atomic mass = 28.09 amu
- Atomic radius = 1.18 Å
- Young's modulus = 103 GPa
- Lattice constant = 5.43 Å
- Electronic configuration $1s^2 2s^2 2p^6 3s^2 3p^2$
  i. Properties of silicon are sensitive to the presence of interstitial oxygen, which acts as donor and tends to destabilize the material at high temperatures. It is because of the creation of neutral complexes due to the combination of oxygen with other impurity atoms.

ii. Its properties are influenced by nuclear radiations which induce such defects that readily combine with oxygen to produce active centres.
iii. It allows diffusion of metallic impurities such as Fe, Ni, Co, Mn, Cu etc. at moderate to higher temperatures.
iv. Only 1 atom in about 10^{12} contributes a free electron (and also a hole) to the crystal because of broken covalent bonds.
v. It is suitable for services over a wide temperature range due to its higher energy gap (more than that of germanium).

7.1.2 Germanium

Similar to Si, it is also an element of IVth column in periodic table. Its structure is diamond cubic and atoms are covalent bonded. The bonds are directional in nature with a bond angle of 109°. It is a grey coloured brittle material of metallic look. Its mechanical properties are similar to glass. Germanium possesses the following salient properties.

- Atomic radius = 1.22 Å
- Lattice constant = 5.66 Å
- Atomic mass = 72.59 amu
- Young's modulus = 99 GPa
- Electronic configuration = $1s^2 2s^2 2p^6 3s^2 3p^6 3d^{10} 4s^2 4p^2$

i. It is capable of creating deep impurity energy levels.
ii. It has a great sensitivity with elements like Au, Ni, Cu, Li etc., which can move within its crystal with high speeds at low temperatures.
iii. The doping elements like In, As, Ga, Sb diffuse slowly in it.
iv. It is prepared by horizontal or vertical pulling from the melt for high purity/quality.
v. Since it contains the order of 10^{22} atoms/cm^3, but at room temperature (300 K) the $n_i \approx 10^{13}$/cm^3; therefore only 1 atom in about 10^9 contributes a free electron (and also a hole) to the crystal due to broken covalent bonds.
vi. Its carrier mobility is higher than silicon, so it is useful for high frequency services.

7.1.3 Selenium (Se)

It is an element of group VI in periodic table. It exists in both crystalline and non-crystalline forms. Its structure is hexagonal. It is covalent bonded with a bond angle of 104°. Its salient properties are as follows.

- Atomic number = 34
- Melting point = 217°C
- Specific gravity = 4.79
- Young's modulus = 58 GPa

It finds use in the following main applications.

i. Xerographic printing
ii. Photoelectric light meters
iii. TV cameras
iv. Photovoltaic cells
v. Photocells in cinematography

7.1.4 Antimony (Sb)

It is an element of group V in periodic table. Its structure is rhombic and the atoms are covalent bonded. Due to directional nature of bond, the bond angle is 96°. Its other properties are as follows.

- Atomic number = 51
- Melting point = 630°C
- Young's modulus = 55 GPa
- Structure is rhombus

It finds use in the following applications.

i. Photon detection
ii. Laser

7.1.5 Other Elements

These are used to make various compounds and alloy semiconductors whose details are given in art. 7.2. A summary detail of these elements is given in Table 7.1.

Table 7.1 Properties of various semiconducting elements

Element	Symbol	Atomic number	Specific gravity	Melting point (°C)	Young's modulus (GPa)
Arsenic	As	33	5.72	817	39
Bismuth	Bi	83	9.80	271	34
Boron	B	5	2.34	2030	440
Indium	In	49	7.31	156	10.5
Phosphorus	P	15	1.83	44	4.6
Sulphur	S	16	2.07	119	19.5
Tellurium	Te	52	6.24	450	41
Tin	Sn	50	7.30	232	52

Their structure and other geometrical details are displayed in periodic table given on front and back cover pages of this book.

7.1.6 Comparison between Silicon and Germanium

We have described earlier that the silicon and germanium are covalent bonded. Each atom has four bonds. Now, a comparison of these two extensively used semiconductor elements is given in Table 7.2.

7.2 Formulated (Compound and Alloyed) Semiconducting Materials

These are made by doping the element semiconductors. Various dopants used for this purpose are P, As, Sb, In, Ga, Al etc. A brief description of some important formulated semiconducting materials is given below.

7.2.1 Gallium Arsenide (GaAs)

This is a compound of gallium (a third column element) and arsenic (a fifth column element) in periodic table. It is made by zone refining technique. Its melting point is about 1250°C. Production of GaAs is a difficult task due to generation of high vapour pressure in arsenic at about 1200°C. It has high mobility and large energy gap. Its other properties are as given below.

Table 7.2 Comparison between Silicon and Germanium

Description	Silicon	Germanium
• Atomic number	14	32
• Type of structure	DC	DC
• Valency	4	4
• Electrical resistivity at 20°C (ohm m)	3000	0.47
• Energy gap E_g : at 0 K : at 300 K	1.17 eV 1.12 eV	0.74 eV 0.66 eV
• Instrinsic carrier concentration (at 300 K)	$1.1 \times 10^{16}/m^3$	$2.5 \times 10^{19}/m^3$
• Mobility μ : of electron (at 300 K): of hole (at 300 K)	$0.14\ m^2/Vs$ $0.05\ m^2/Vs$	$0.38\ m^2/Vs$ $0.18\ m^2/Vs$
• Dielectric constant ε_r	12	16
• Number of atoms	$5.0 \times 10^{22}/cm^3$	$4.4 \times 10^{22}/cm^3$
• Number of intrinsic carrier (at 300 K)	$1.5 \times 10^{10}/cm^3$	$2.5 \times 10^{13}/cm^3$
• Melting point	1410°C	937°C
• Specific gravity	2.33	5.32
• Coefficient of thermal expansion α	$4.2 \times 10^{-6}/°C$	$6 \times 10^{-6}/°C$
• Thermal conductivity (W/cm°C)	0.8	0.55
• Heat capacity (cal/gm°C)	0.17	0.08
• Limiting operational temperature	≈ 200°C	80–85°C
• Cost factor	Costlier	Cheaper

i. E_g at 300 K = 1.42 eV
ii. $\varepsilon_r = 13.2$
iii. Specific gravity = 5.31
iv. Hole mobility = $0.025\ m^2/Vs$
v. Electron mobility = $0.80\ m^2/Vs$

However, this compound is not competitive with silicon and germanium in respect of purity and structural perfection. It is useful materials for several devices, important amongst them are given as follows.

- Switching and parametric diodes
- Tunnel diodes

- Semiconductor lasers
- Hot electron diodes.

7.2.2 Indium Antimonide (InSb)

This is a compound of indium (In) an element of IIIrd column, and antimony (Sb) an element of Vth column in periodic table. Electron mobility of InSb at room temperature is highest ($\approx$10 m^2/Vs) among all known semiconductors. It has a low melting point (525°C) and is easier to produce in single crystal form. Its electrical resistivity at 20°C is $\approx$ 20000 ohm m. Indium antimonide finds use in the following main applications.

- Infrared detectors
- Hall effect devices
- Laser diodes
- Tunnel diodes
- Infrared filter material
- Transistors

7.2.3 Oxides, Sulphides, Halides, Tellurides and Sellurides

These are extrinsic semiconductors of compound form. Some of them are *n*-type and others are *p*-type. Zinc oxide (ZnO) is a *n*-type and Cuprous oxide (Cu_2 O)is a *p*-type semiconductor. Sulphides such as PbS (Gelina), CdS, Cu_2S, ZnS; and halides such as CuBr, CuI, CuCl etc. have deviations from stoichiometric compositions. Tellurides such as Bi_2Te_3, PbTe and sellurides such as PbSe, CdSe and other elements are also the semiconductor compounds. These are used in the following main applications.

- BaO: in oxide coated cathodes
- CdS, CdSe, CdTe: in photoconductivity based automatic door opener, street light switching, burglar alarm
- $PbSO_4$, CdS, PbS: in photoconductive devices such as photocells of TV camera and cinematography
- GaP, GaAs, GaSb: in semiconductor lasers
- CsSb: in photomultipliers.

7.2.4 Cadmium Sulphide (CdS)

It is used to make photoconductors of high dissipation capability and excellent sensitivity in visible spectrum, and to prepare cadmium sulphide cell by depositing a layer of CdS. This layer generally contains Ag, Sb, In etc. as impurity. Its energy gap is 2.4 eV. CdS finds use in the following main applications.

- As constituent of cathode-ray phosphorus
- To measure a fixed amount of illumination as with lightmeter
- To record modulating light intensity on sound track
- As ON-OFF light relay in digital and control circuits.

7.2.5 Silicon Carbide (SiC)

It is hard and refractory in nature. Its energy gap is large, about 3 eV. Its melting point is also very high (about 2400°C). Due to above properties, the SiC is useful as

- High-temperature rectifiers
- High-temperature transistors.

7.2.6 Lead Sulphide (PbS)

It is used to make commercial photoconductive cells and finds use in the following main applications.

- Infrared detection
- Infrared absorption measurements.

7.3 Lattice Structures of Some Compound Semiconductors

The basic lattice structure of many semiconductors is diamond cubic (DC), which is a characteristic of Si and Ge. In many compound semiconductors; although the atoms are arranged in DC structure, but they are different on alternating sites. Consequently, they differ from DC and are of following kinds.

1. Zinc blende: which is typical of III-V compounds, and
2. Wurtzite: which is typical of II-VI compounds.

The zinc blende and wurtzite lattice structures are shown in Figs. 7.1*a*-*b*. In zinc blende structure, for example, if one FCC sub-lattice is composed of

Table 7.3 Electronic properties of compound semiconductors at room temperatures

Compound System of III-V Columns	Symbol	Energy Gap E_g (eV)	Effective mass		Mobility (m^2/Vs)		Application
			Electron (m_e^*/m_o)	Hole (m_h^*/m)	Electron (μ_e)	Hole (μ_h)	
Indium phosphide	InP	1.29	0.07	0.40	0.46	0.015	Photon detection
Indium arsenide	InAs	0.33	0.02	0.41	3.30	0.040	—
Indium antimonide	InSb	0.16	0.013	0.60	7.80	0.075	Photon detection
Gallium phosphide	GaP	2.24	0.82*	0.60*	0.05	0.002	Laser
Gallium arsenide	GaAs	1.40	0.072	0.54	0.80	0.025	Diode, laser
Gallium antimonide	GaSb	0.77	0.047	0.50	0.40	0.140	Laser
Cadmium sulphide	CdS	2.42	—	—	—	—	Photoconductor in green light
Cadmium selluride	CdSe	1.74	0.13*	0.45*	0.08	—	Photoconductor in red light
Cadmium telluride	CdTi	1.45	—	—	—	—	Photoconductor in infrared region
Lead sulphide	PbS	0.38	—	—	—	—	Photoconductors
Lead selluride	PbSe	0.32	—	—	—	—	Photoconductors
Lead telluride	PbTe	0.33	—	—	—	—	Photoconductors
Silicon carbide	SiC	2.70	—	—	0.04	0.005	—

* At 4 K

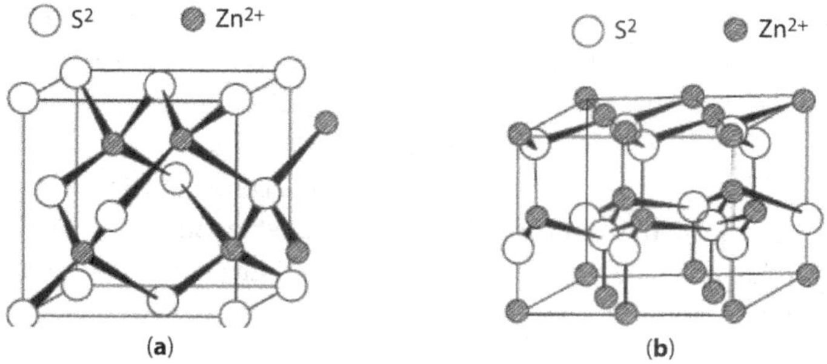

Figure 7.1 Structures of (*a*) zinc blende, and (*b*) wurtzite.

Ga atoms and the intermediatory sub-lattice is of As, it will form a GaAs zinc blende. Most of the compound semiconductors have this type of lattice. Main examples of semiconductors falling under each kind is given below.

- Zinc blende: AlP, AlAs, AlSb, GaP, GaAs, GaSb, InP, InAs, InSb, ZnSe, ZenTe, Cd Te
- Wurtzite: SiO_2, CdSe

Some semiconductors possess the structure of both kinds, zinc blende and wurtzite. These are

- ZnS
- GaN
- CdS

7.3.1 Structure of Zinc Sulphide

In zinc sulphide (ZnS), each Zn^{2+} ion is tetrahedrally surrounded by four S^{2-} ions and each S^{2-} ion is tetrahedrally surrounded by four Zn^{2+} ions. Coordination number of both ions is 4, therefore it is called a 4 : 4 arrangement. Zinc sulphide exists in two different forms *viz.* (*i*) zinc blende, and (*ii*) wurtzite. Their structures are shown in Figs. 7.1*a-b*. Zinc blende is related to FCC structure while wurtzite relates to HCP structure.

7.4 Solar Cells

A solar cell is used to convert solar energy into electrical energy. It is a direct energy conversion system, based on photovoltaic action. A solar cell is a self-generating arrangement which employs semiconductor contacts against metals. The current generated depends on the illuminated area. As an illustration, the current generated is about 10-100 mA from a junction of about 1 cm^2 area. Usually the amount of current generated from a single solar cell is too less for a certain application, therefore many solar cells are used together to form a 'solar battery'. The power produced by a solar battery is significant. The arrays of *p-n* junction solar cells are widely used to supply electrical power to space satellites, as shown in Fig. 7.2.

7.4.1 Working Principle

Working of a solar cell is based on photovoltaic action. As the power can be delivered to an external circuit by illuminated junction, therefore the solar energy can also be converted into electrical energy. Accordingly, when the light is incident upon the semiconductor-metal combination (see Fig. 7.3), an internal voltage is generated that causes current to flow through the internal circuit even though not a single battery is connected externally. The e.m.f. thus generated depends on the radiant energy received by the solar cell.

7.4.2 Construction and Working

The most commonly used photovoltaic cells are of the barrier-layer type like iron-selenium cells or Cu-CuO_2 cells. In the iron-selenium cell as shown in

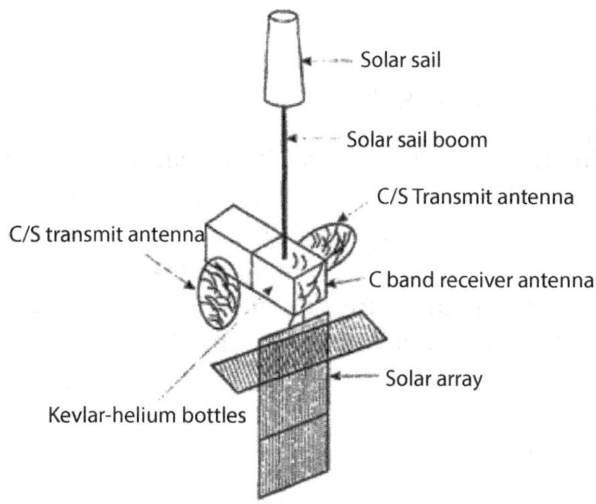

Figure 7.2 A solar cell showing the arrays of *p-n* junctions in a satellite.

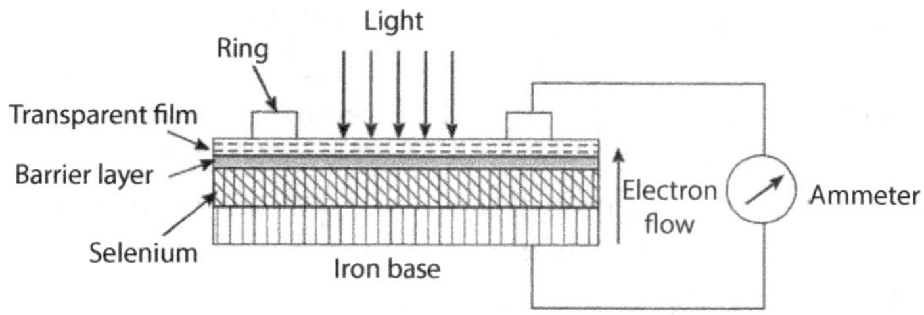

Figure 7.3 A solar cell showing (*a*) the arrays of *p-n* junctions in a satellite, and (*b*) a typical construction.

Fig. 7.3, the selenium layer is placed on an iron disc and then an extremely thin transparent layer of gold or silver is formed on the selenium to act as a front electrode. The barrier layer is formed by cathode-sputtering the semi-transparent film on the selenium. A contact ring on the silver layer acts as one electrode and the iron base as the other. When luminous flux or radiant flux falls on the semiconductor *i.e.* selenium, it ejects electrons which travel from selenium to the front silver electrode through the barrier layer as shown in Fig. 7.3. The flow in the opposite direction is not permitted by the (boundary)barrier layer because it acts as a rectifier. The e.m.f. generated is almost directly proportional to the incident flux.

7.4.3 Factors Affecting the Efficiency of Solar Cells

There are three main factors which influence the efficiency of solar cells. These are

1. the energy gap E_g of the emitter/absorber material.
2. the absorption coefficient α of the absorber, and
3. the diffusion length of minority carrier.

Effect of energy gaps. E_g determines the threshold frequency f_0 and the maximum current which can be generated. The energy gap also determines the maximum emf generated by the solar cell. It thus follows that the energy gap would largely control the maximum efficiency that can be obtained using a given absorbing material. The maximum conversion efficiency is found to occur around the energy gap of 1.4 eV for GaAs material.

Effect of absorber. The degree of absorption of solar flux by the absorber depends on the thickness of specimen and the absorption coefficient α of the material. The minimum thickness required to absorb almost all the useful photon flux having energy greater than E_g depends on the nature of the material. The absorption is higher for direct band materials than for the indirect band materials because of reduced transition. Silicon being an indirect band semiconductor requires a minimum thickness of 300 mm from the consideration of optical absorption. It has been observed that the absorbing material for high efficiency must have high values of α over most of the region of high solar spectral density.

It is also observed that the direct band GaAs has higher values of a as compared to Si. Amorphous Si has a higher value of α as compared to the crystalline state. This occurs on account of the presence of large density of states in the energy gap in amorphous state which is of the order of 10^{24} to 10^{26} m^{-3} eV^{-1}. Recently, hydrogenated amorphous silicon has been found to possess high absorption coefficient ($\approx 5 \times 10^6$ m^{-1}) with low density of states in the gap ($\approx 10^{21} m^{-3} eV^{-1}$). This material can also be doped and made *p*- or *n*-type amorphous semiconductor. Anti-reflection coatings are also used to reduce reflection loss at the top surface of the cell.

Effect of diffusion length. Third factor which controls the efficiency of solar cells, is the diffusion length of minority carriers. After the electron-hole pairs are produced by absorption of light quanta from the solar flux, the carriers have to be diffused to the junction to get separated. Therefore, the diffusion length has to be greater than the thickness of absorption layer. Since this thickness is already a fixed value from the consideration of maximum absorption, the diffusion length has to be optimised by controlling the physical parameters of the material. The diffusion length L_{diff} is proportional to $\sqrt{\mu\tau}$ where μ is mobility and τ is lifetime of the charge carrier. The value of τ depends on the cross-section for recombination and the density of defects; μ depends on temperature, density and types of defects. All these factors have to be taken into account while choosing a suitable material for the solar cell.

7.4.4 Solar Cell Fabrication and Materials

A *p-n* junction solar cell is shown in Fig. 7.4. This has three main parts *viz.* the emitter/ absorber, the junction and the collector. The emitter/absorber region is that in which the light is absorbed and electron-hole pairs are created.

The fabrication of a solar cell necessarily produces surface states at the top layer of the cell. These surface states reduce the lifetime of the electron-hole pairs. Consequently in this region, the recombination takes place even

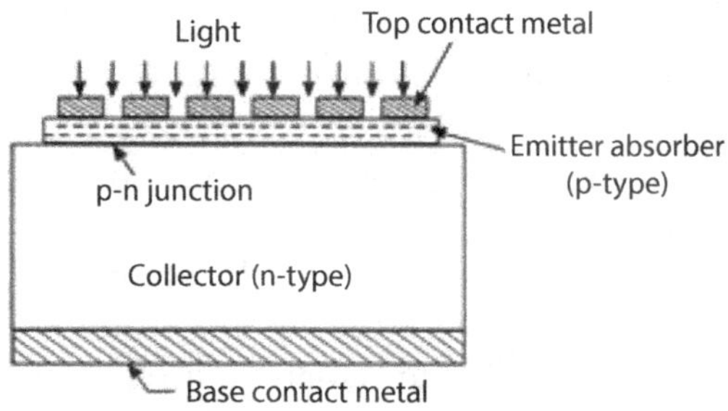

Figure 7.4 Basic construction of a solar cell.

before collection can occur. This is called a dead layer. In a material like GaAs, the presence of a dead layer of 1 mm thickness can reduce the available power by more than 90%. From this consideration, Si is the best choice.

Some of the best solar cell materials with efficiencies lying between 7 to 10% are the following.

Absorber material	Collector material
Si	Si
Cu_2S	CdS
CdTe	CdS
Cu_2S	ZnCdS
$CuInSe_2$	CdS
GaAs	AlGaAs

Whereas the Si cells lose efficiency at high temperatures, the GaAs compound can be used at 100°C or more. The GaAs-AlGaAs heterojunction cell can operate at elevated temperatures, which is a common requirement of solar concentration systems.

7.4.5 Advantages and Limitations of Solar Cells

Advantages. A solar cell possesses several advantages which are given below.

1. It is a self-generating device, therefore it does not require any external power source such as a battery or electric motor etc.
2. The internal e.m.f. and the current generated by it are large enough to be measured on a galvanometer.

3. These cells can supply power to a space satellite for a long period of time. This is a distinct advantage over the acid or alkaline batteries.
4. It is pollution-free energy conversion system.

Limitations. However, the solar cells have certain limitations also. These are

i. It does not convert all solar radiation into electric energy.
ii. Efficiency of a well-made Si cells is about 10% only under full illumination.
iii. The efficiency of Si cells lowers down at high temperatures.
iv. Its use in terrestrial applications is less effective than outer space applications because a considerable amount of solar radiation is absorbed by the atmosphere.

7.4.6 Applications of Solar Cells

Solar cells are used in a wide range of outer space and terrestrial applications. Besides them, they are used in a variety of applications. The main among them are the following.

- In direct-reading illumination meters.
- In portable exposure meters.
- In low-resistance relays for 'ON-OFF' operations.
- In various industrial monitoring operations.
- In space satellites and deep space vehicles.
- For supplying electrical power to various electronic equipments aboard a satellite.

7.5 Semiconductor Lasers

The word LASER is an abbreviation of "*light amplification by stimulated emission of radiation*". It is a quantum electronic device that generates intense electromagnetic radiation. It is a source of highly directional, monochromatic, coherent light. This light can be a continuous beam of low or medium power, or a very short burst of intense light. The short burst may deliver a power worth millions of watt. The laser is of great interest in optics and optoelectronics, particularly in fiber-optic communications.

7.5.1 Merits of Semiconductor Lasers

Although the ruby laser (chromium ions in alumina), CO_2 laser, He-Ne laser etc. are still used in many applications; but the semiconductor lasers are more efficient than these lasers. GaP, GaAs and GaSb are most significant compounds for this purpose. The reasons for superiority of semiconductors (or p-n junction) lasers over non-semiconductor lasers are the following.

1. They are remarkably miniature in size (typically of the order of 0.1 mm × 0.1 mm × 0.3 mm).
2. Their efficiency is high.
3. Their output can be easily modulated by controlling the junction current.
4. They operate at low power as compared to ruby or CO_2 lasers.
5. Their power output is comparable with He-Ne lasers.
6. They can provide a portable source of radiation.
7. They provide an easily controllable source of low-power coherent radiation.

The semiconductor lasers are particularly suitable for fiber-optic communication systems.

7.5.2 Characteristics and Working

In semiconductors lasers, light emission is obtained through a *pn* junction as a result of recombination of electrons and holes. An extremely narrow and perfectly straight junction line provides beneficial laser action. Miniature semiconductor lasers employ various methods of excitation such as injection of electrons and holes through the *pn* junction, optical pumping, and a beam of fast electrons. A higher energy gap of semiconductor material results in emission of shorter wavelengths as shown in Fig.7.5.

7.5.3 Laser Applications

Lasers are being put to a number of uses in different fields. A few examples of these applications are listed below.

1. **Communication.** Modulated laser beams are being used for transmitting messages. Due to high degree of coherence, the loss of transmitted energy is comparatively much less.

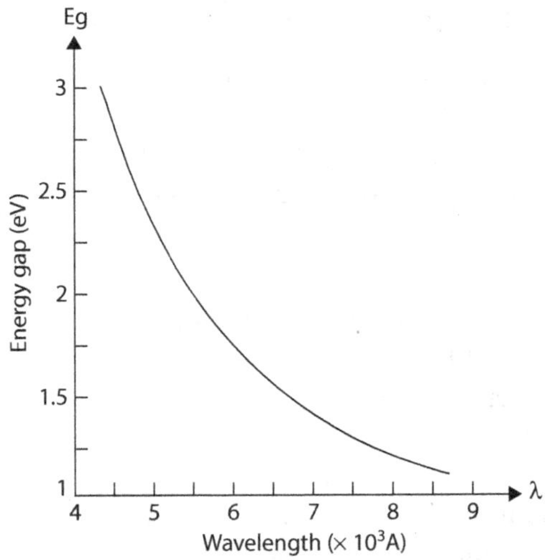

Figure 7.5 Effect of forbidden gap on the emission of wavelength in a semiconductor laser.

2. **Surgery.** Laser beam is being used extensively for bloodless surgery. *e.g.* lasers can be used
    - to weld the detached retinas
    - for drilling the teeth
    - for removal of tumors
    - for removal of infected cell
    - for preventing the tooth decay by depositing hard material on the surface of the tooth.
3. **Measurements of long distances.** It enables us to determine the distance of the moon from the earth with an error within 0.6 m.
4. **Nuclear fusion.** The laser beams can be used to induce the nuclear fusion. By concentrating the laser beam on to a very narrow spot, the temperature may rise to about 10^8 K and nuclear fusion can occur at this temperature.
5. **Scientific research.** A modified version of the Michelson Morley experiment was conducted to test for the ether drift. Here the beams of two infrared lasers of slightly different frequencies were combined by means of a beam splitter and the beam frequency was determined. No change in beat frequency was observed on rotating the apparatus through 90°. Thus the null result of Michelson Morely experiment could be confirmed.

6. **Industry.** Lasers can be focussed onto a very fine beam, which result in raising the temperature to about 1000 K. It can be used for drilling the holes and fusing or melting of metals.
7. **Weather forecasting.** Pictures of the clouds, wind movements etc. can be obtained with the help of laser beam, and the data so obtained can be used in weather forecasting.
8. **National defence.** Lasers can be used for guiding the missiles and satellites. They can also be used for detecting the enemy aeroplanes. The laser beams can also be used as weapons to destroy the enemy tanks, aeroplanes etc.
9. **Other uses.** The other uses of laser are
   - in holography for the production of three-dimensional images.
   - in cutting of microelectronic circuits.
   - for cutting and sealing the nerves.
   - to raise the temperature of localized area of tissue *i.e.*, physiotherapy.

Laser can also be used for the following purposes.

- Unblocking the heart valves.
- Breaking the kidney stones, etc.

7.5.4 Materials for Semiconductor Lasers

A good material for the purpose of semiconductor lasers must fulfil certain requirements. Main among these are the following.

1. It must be efficient light emitter.
2. It should be amenable to the formation of *p-n* junctions.
3. It should be amenable to the formation of heterojunction barriers.

Based on various desirable properties, the following materials are suitable for semiconductor lasers applications.

II-VI compounds. These are very efficient for the purpose of light emission, but formation of their junction is difficult. However using nitrogen (N) as acceptor, the junctions can be grown easily in the following materials by crystal growth techniques like MBE (molecular beam epitaxy) and MOVPE (metal-organic vapour-phase epitaxy).

- ZnS
- ZnSe

- ZnTe
- Alloys of these

Lasers made of these materials emit in the region of green and blue-green light spectrum.

Large bandgap semiconductors. Semiconductor compounds having a large band gaps (about 2 to 5 eV) offer good candidature as semiconductor lasers materials. Main among these are the following.

- InN ($E_g = 2$ eV)
- GaN ($E_g = 3.4$ eV)
- AlN ($E_g = 5$ eV)

This covers a wide range of light emission from blue to UV spectrum.

Ternary alloy system. These are well suited for the lasers used in fibre-optic communication systems. Main among these are the following.

- AlGaAs
- PbSnTe
- InGaSb

The PbSnTe is able to provide laser output in the wavelength range from 7 μm to above 30 μm at low temperatures. The InGaSb is suitable for use at intermediate wavelengths.

Quaternary alloy system. These are specifically versatile in fabricating the laser diodes by allowing flexibility in lattice matching. They also provide a choice of wide range of wavelength. Main examples of these materials are the following.

- InGaAsP

Short wavelength emitters. These blue/UV semiconductor lasers are very suitable for storage purposes such as of compact discs (CDs), digital versatile discs (DVDs) etc. By reducing the laser wavelength *i.e.* by using short wavelength emitters, the storage density of these devices can be increased many times. An example of a recent semiconductor laser has used the undermentioned material.

- InGaN

This laser has multi-quantum-well heterostructures.

7.6 Optical Materials in Light Emitting Diodes

Light-emitting diodes are the devices that act as the generator of light. They are the light source for fibre optic systems. Their working is based on the effect called as 'injection electroluminescence'. According to this effect, the semiconductors characterized by direct recombination give-off light from their junctions under forward-bias. It is in contrast with the semiconductors having indirect band gaps (*e.g.* Si or Ge), in which only heat is released on recombination and not the light.

7.6.1 Construction and Working of LED

It is a *p-n* junction which when forward biased, emits light. Charge carrier recombinations occur at the junction as the electrons cross from *n*-side and recombine with holes on *p*-side. When this recombination takes place, charge carriers give up energy in the form of light and heat. If the semiconductor material used is translucent, the light is emitted and the junction acts as source of light.

Cross-sectional view of a typical LED is shown in Fig. 7.6. In it, the recombination takes place in *p*-type material. Thus *p*-region becomes surface of the device. For maximum light emission, a metal film anode is deposited around the edge of *p*-type material. The cathode connection is usually a gold film at the bottom of *n*-type region. This helps in reflecting the light on surface. Most common *p*-type semiconductor materials used for manufacture of LED are gallium arsenide and phosphide, which emit red or yellow light, and gallium arsenide which emits green or red light emission.

7.6.2 Advantages, Applications and Specifications of LEDs

The merits of LEDs over conventional incandescent and other types of lamps are the following.

1. Small size and weight.
2. No warm-up time.
3. Low working current and voltage
4. Less power consumption
5. Very fast switching
6. Immune to mechanical vibration
7. Long life
8. Less fragile

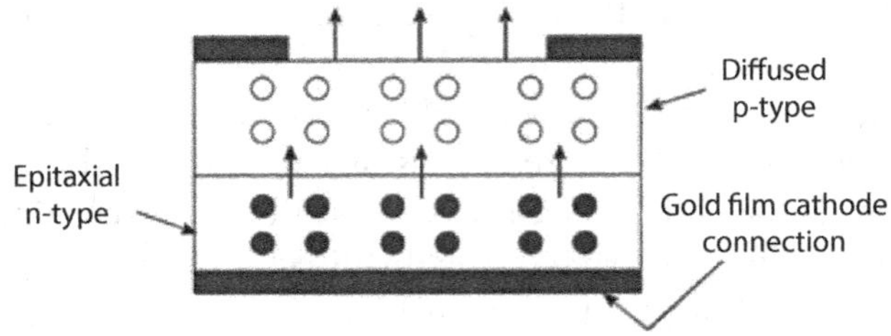

Figure 7.6 Cross-section of a typical LED.

7.6.3 Applications and Specifications of LEDs

LEDs are used in following main applications.

- As display devices
- In optical communication systems
- Several LEDs are added in series to form segmental and dot matrix displays.

Specifications of LED. The main specifications of a LED are as follows.

i. Colour
ii. Wavelength at peak emission
iii. Forward voltage
iv. Forward current
v. Power dissipation
vi. Luminous intensity
vii. Maximum reverse current

7.6.4 Light Emitting Materials

A large variety of light emitting semiconductors are known to exist. They include *n*-type, *p*-type, III-V and II-VI group materials, ternary and quaternary compounds and alloys. They are available in a wide range of photon energy emission extending from ultraviolet to infrared and visible range of spectrum. A detailed list of them is given below.

n-type binary compounds

- ZnSe
- ZnS
- CdSe
- CdS

p-type binary compounds

- ZnSe
- ZnTe

Infrared emitter compounds and alloys

- Ga As
- InP
- GaN (E_g = 3.4 eV) for blue wavelength
- InSb (E_g = 0.18 eV)
- Mixed alloys of GaAs and InP

Ternary compounds

- AlGaAs
- AlGaP system for yellow and green wavelengths
- Gallium arsenide-phosphide ($GaAs_{1-x}P_x$) alloy

Quaternary compounds

- InGaAsP system
- AlGaAsSb system

Multilayer heterojunction systems

- (Zn, Cd) system
- (S, Se) system

For a good light-emitting material its band gap energy should be high. This band gap in compound/ alloy semiconductors can be increased sufficiently by using ternary and quaternary compounds. For example, the $In_xGa_{1-x}As_yP_{1-y}$ quaternary alloy when grown on InP substrate will result in band gaps ranging from 0.75 eV to 1.35 eV.

7.7 Materials for Optical Fibres

Optical fibres are extremely fine and flexible rods of glass. They are optical transmission medium used for the purpose of communicating the light. These fibres function as an optical waveguide with a transmitter and a detector. Light propagates through it by total internal reflection and refraction. During operation, they are placed between a light source and detector for transmission of signals. The operation of optical fibres is based on the principle of 'total internal reflection and refraction' at the interface between high refractive index core and lower refractive index cladding of pure glass (see Fig. 7.7).

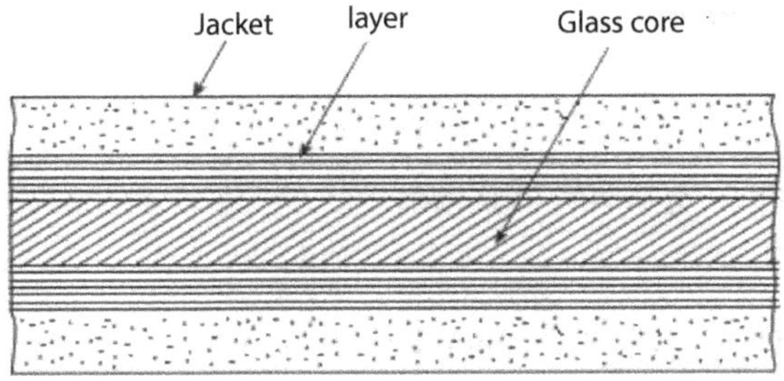

Figure 7.7 Constructional details of an optical fibre

7.7.1 Construction

Optical fibres are prepared from doped high-silica glass, plastic and silicate glasses. It consists of a glass core surrounded by a cladded (or sheathed) layer encased in a jacket, as shown Fig. 7.7. A single optical fibre has a diameter of about 3 μm with plastic, glass or rubber jacket of about 65 to 145 μm diameter. An optical fibre is highly pure glass of about 2 μm diameter having high refractive index. It is coated with a layer of glass having different chemical combination and lower refractive index. These fibres are bundled together which can bend easily without breaking. These bundles are also called *light pipes*. Transparent plastics are also used instead of glass fibres.

7.7.2 Types of Optical Fibres

On the basis of construction and operation, the optical fibres can be categorized into following two types.

1. *Monomode fibres* in which the diameter of inner core is comparable with the wavelength of light. Hence, *only one mode* of light propagates through it.
2. *Multimode fibres* in which the diameter of core is sufficiently larger than the wavelength of light. Hence, the light propagation takes place in a *large number of modes*. The multimode fibres, on the basis of refractive index μ of the core, may be further divided into following two types.
   i. step-index type, and
   ii. graded-index type

Both these types are shown in Figs. 7.8*a*-*b*-*c* and 7.9*a*-*b*-*c*.

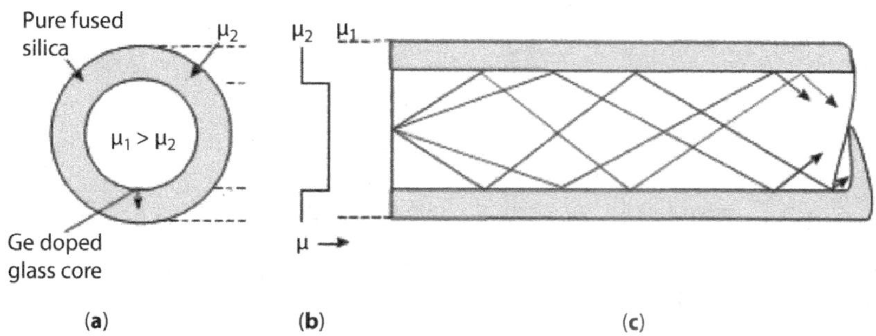

Figure 7.8 Step-index type multimode optical fibre showing (*a*) cross-section of a core having larger refractive index μ_1, (*b*) its refraction profile, and (*c*) mode pattern.

Step-index type fibre. The step-index fibre shown in Fig. 7.8*a* comprises of an outer layer of fused SiO_2 having refractive index μ_1 and Ge-doped glass core with slightly greater refractive index μ_2 (*i.e.* $\mu_1 > \mu_2$). Its index of refraction profile is shown in Fig. 7.8*b*. Such fibre maintains the light beam in central core with little loss at the surface. The light is transmitted along the length of the fibre by internal reflection at the step of the refractive index. Its mode pattern is shown in Fig. 7.8*c*.

The losses in the fibre at a given wavelength is determined by considering the intensity of signal at a distance x along the fibre. It is given by

$$I(x) = I_o e^{-\alpha x} \tag{7.1}$$

where α is the *attenuation coefficient* which is similar to absorption coefficient. Its value is different for different wavelengths. It is expressed in the unit of decibel per kilometer (dB/km). For a fused silica optical fibre, its value lies approximately between 0.6 to 11 dB/km.

Chromatic dispersion. The refractive index is influenced by the wavelength. Therefore, different optical frequencies travel at slightly different velocities down the fibre. This behaviour is known as the 'chromatic dispersion' of the data pulse. Further, the different modes propagate at different path lengths. Such dispersion can be reduced by grading the refractive index of the core. Example of a parabolically graded core is shown in Figs.7.9*a-b-c*. In this case the various modes are continually refocused, which result in reduced difference in the path lengths.

7.7.3 Suitable Materials and their Requirements

Construction of optical fibres requires a high technology application of glass making. They require following important qualities in them.

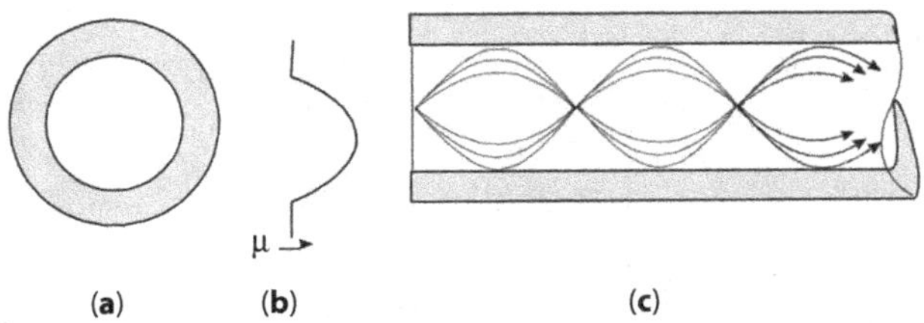

Figure 7.9 Graded-index type multimode optical fibre showing *(a)* cross- section of the core, *(b)* its refraction profile, and *(c)* mode pattern.

i. High purity, having impurities to the level of less than 1 ppm.
ii. Low loss.
iii. Low-dispersion.
iv. High refractive index for core materials.
v. Low refractive index for cladding material.

Following materials are generally used to construct the optical fibre.

- Fibre of highly pure glass.
- Core made of vitreous high-silica glass doped with germanium oxide (Ge_2O) *i.e.* germania.
- Jacket of plastic, glass or rubber.

7.7.4 Advantages and Applications

The optical fibres possess several advantages over conventional cables and microwave links. These are given below.

1. They are of miniature size.
2. They are immune to cross-talks.
3. They are least subjected to electromagnetic interference.
4. They are capable of carrying much more information for a given cross-section of fibre, than can be transmitted via the conventional coaxial cable.
5. They can guide the optical signals over long distances (many kilometres) without the necessity of perfect alignment between source and detector.

7.7.5 Applications of Optical Fibres

Optical fibres find major use in following fields.

i. In telecommunication of high-frequency pulses such as in
   - telephone
   - data transmission
ii. In medical applications such as *endoscopy*, through which the condition inside the human stomach can be detected. It can also be used for
   - detection of ulcer
   - detection of cancer
   - detection of wounds

To accomplish above functions, an endoscope is used with a TV monitor.

7.8 Choicest Materials for Different Semiconductor Devices

Various semiconducting materials mentioned above possess varying properties, and hence are suitable for application in vivid devices. Main among these materials and their suitability for different devices are given below.

Devices/products/services	Suitable semiconducting materials
• Diodes	Si, Ge, GaAs
• CPU, microprocessor	AlGaAs chips
• Transistors	Si, Ge
• Photocells	Se, CdS, PbS (i.e. Galena or galenite)
• Rectifiers	Si, Se, CuO
• ICs	Si chip, GaAs and Si hybrid chips
• Light detectors	InSb, CdSe, PbTe, HgCdTe
• High-frequency devices	Ge
• Flourescent screens such as T.V. screens	ZnS, Phosphor coating of oxides and sulphides of Zn, Cd, Be
• Infrared detectors	Si, Ge
• Nuclear radiation detectors	Si, Ge
• Gunn diode (a microwave device)	GaAs, InP
• Semiconductor lasers	GaAs, AlGaAs, GaP, GaSb
• Solar cells/batteries	Photovoltaic action materials such as Se
• Cinematography	Photocell effect based materials such as Se, CdS, $PbSO_4$

Devices/products/services	Suitable semiconducting materials
• Xerox-type photocopier	Se
• Hyperhigh-speed computers	New generation AlGaAs chips
• Sensor elements for guided missile	Hg-Cd-Te crystals
• Automatic door opener	Photoconductivity based materials such as CdSe, CdS, CdTe
• Stroboscope disk (flashing light called stobotron)	Opto-electronic polymers
• Red phospher for TV tubes	Yttrium (Y)
• Diamond transistors	Phosphorus-doped diamond film *n*-type semiconductor
• Semiconductor lenses and mirrors for high power lasers	Synthetic diamond
• Light absorption and optics	Single crystal diamond
• Light-emitting diode (LED)	
• for visible green colour light	GaP, CdS
• for visible red colour light	GaAsP, CdSe
• for visible yellow light	GaP
• for visible blue light	SiC
• for invisible infrared light	GaAs, InSb
• for ultraviolet region of light	ZnS
• Avalanche photodiode	InAlAs, InGaAs, GaAsSb
• Photon detector	InP, InAs, InSb
• Photoconductors	
• in green light	CdS
• in red light	CdSe
• in infrared region	CdTi

7.9 Solved Examples

Example 7.1 The energy band gap of binary compounds GaAs and GaP are about 1.43 eV and 2.3 eV respectively. Discuss their suitability as electronic devices.

Solution. The 1.43 eV band gap corresponds to near infrared region of the light spectrum, while 2.3 eV band gap corresponds to wavelengths in green portion of the light spectrum. Hence GaAs is suitable to make microwave devices like the Gunn diodes, while GaP is suitable for light emitting diodes (LEDs) purposes.

Example 7.2 High-frequency transistors are generally made of germanium and not of silicon, but silicon is used in electronic devices operating over a wider temperature range-why?

> **Solution.** The carrier mobility of silicon is lower than that of germanium. Therefore, it is inferior to germanium for high frequency services. But the energy gap in silicon is higher than that of germanium, so it offers a greater temperature range service for electronic devices.

Example 7.3 How does a metallurgical grade Si (MGS) differ from an electronic grade Si (EGS)? Which grade is more suitable for making semiconductor devices?

> **Solution.** A MGS is an output product obtained from the reaction of SiO_2. The MGS contains impurities such as Al, Fe and other heavy metals of the order of 100s to 1000s ppm. The MGS is an input product to obtain the EGS. EGS is highly pure semiconductor containing impurities of the order of a few ppb (1 ppb* = 10^{-3} ppm). Due to its high level purity, the EGS grade is more suitable for making semiconductor devices.

Example 7.4 Which material is most suitable for producing blue LEDs? Explain why? Write its applications also.

> **Solution.** Blue LEDs are short wavelength emitters. GaN is most suitable material for it, as it gives a very high efficiency. Blue LEDs are used in outdoor displays and TV screens. They are combined with LEDs of other colours also to form very intense white light sources. Luminous efficiency of such light source exceeds over the conventional light bulbs.

Example 7.5 Describe stepwise the method of fabrication of a *p-n* junction laser, briefly.

> **Solution.** A *p-n* junction laser is fabricated as follows.

i. The junctions are formed with a highly doped, direct semiconductor. The GaAs compound is an appropriate choice for it.
ii. The junctions are isolated by etching or cutting.
iii. Individual junction is then cleaved (or cut) into devices to make the front and back faces flat and parallel.
iv. Each individual junction is now mounted over the header, and lead is drawn out from it.

Finally, the adequate heat sinking is provided to the device for large forward current levels.

Example 7.6 What is meant by optoelectronic pair? What is its importance?

Solution. An optoelectronic pair means a pair of a *light emitter* and a *photodiode.* It provides electrical isolation between input and output. In an optoelectronic isolator, both these devices are mounted on a ceramic substrate and packed together to form a compact unit. This unit, while maintaining isolation, passes the information.

Example 7.7 Briefly explain the process of photolithography.

Solution. Photolithography is a process of planar technology to fabricate an IC in which a desired pattern of doping is transferred onto the silicon wafer. The entire process involves (i) coating of SiO_2 layer with a film of photosensitive emulsion called *photoresist,* (ii) blackening the portions to be doped on a transparent sheet, placing this sheet on photoresist layer and then exposing to ultraviolet light (iii) removing the mask and dipping the wafer in trichloroethylene, and (iv) *etching* the SiO_2 with hydrofluoric (HF) solution.

Example 7.8 What is an 'integrated circuit'?

Solution. An integrated circuit (IC) is an assembly of interconnected electrical components which is fabricated as a single unit on a substrate of semiconducting material. These are fabricated together with the necessary elecrtrical connections. Various circuit elements such as transistors, diodes etc. are arranged in such a way that the whole integrated circuit performs an electrical circuit function.

Modern ICs are typically of 5 mm square or less and a large number of identical ICs are fabricated simultaneously on silicon wafers 100 μm in diameter, which are subsequently separated into individual chips.

Example 7.9 In performing diffusion process during fabricating a *p-n* junction, the Si wafers are placed inside a diffusion furnace and the impurity atoms into the gas. Which impurity source materials are used to introduce B, P etc?

Solution. Common source materials for introducing B and P during diffusion in Si are the following.

- *For boron* B : B_2O_3, BBr_3, $BC1_3$
- *For phosphorus* P : PH_3, P_2O_5, $POC1_3$

Example 7.10 Write the specifications of a p-n junction diode.

Solution. A *p-n* junction diode is specified by the following quantities/parameters.

- *Construction material e.g.* Si-diode, Ge-diode etc.
- *Power rating i.e.* the maximum power that can be dissipated safely, *e.g.* 5 mA diode, 10 mA diode etc.
- *Maximum forward current* $I_{F\,max}$ that the diode can carry, *e.g.* 1 mA, 2 mA etc.
- *Maximum reverse current* $I_{r\,max}$ *e.g.* 1 μA, 2 μA etc. It should be a minimum possible value.
- *Peak inverse voltage* that the diode can withstand under reverse- biasing.
- *Junction capacitance e.g.* 1 pF (10^{-12} F), 2 pF etc.
- *Packaging style*

Quick Revision Summary

Bipolar solids. These are the solids which conduct current by means of two charge carrying particles of opposite sign (+ and –). For example, the semiconductors conduct current by means of negative charge carrier *i.e.* electrons and positive charge carriers *i.e.* holes both, so they are bipolar solids.

Compensation. The neutralization of donor impurities by acceptor impurities and vice-versa.

Efficiency of solar cell. It is defined as the percentage of incident solar energy that can be converted into useful electrical energy. This efficiency is generally very less, about 10% only.

Forward bias. Bias applied to a rectifying junction in the conducting direction.

Homopolar bonds are those in which the atoms share electrons with each other. Such bonds are also known as covalent bond. Si and Ge are examples of such bonds.

Microelectronics. It refers to the extreme miniaturization of circuits. In its most extreme form this technology involves the integration of circuits, junctions, etc. into a single small chip of solid material.

Photocell. A junction whose sensitivity to light is due to the extra minority carriers excited by the radiation.

Photoconductor. A semiconductor whose conductivity is sensitive to light due to the creation of conduction electron-hole pairs by absorption of photons.

Thermistor. A semiconductor thermometer, which uses the temperature dependence of the carrier densities.

Zinc blende. It is a type of lattice generally found in III-V semiconductor compounds in which the basic structure is diamond cubic (DC), but the arrangement of atoms is different on alternating sites.

Review Questions

1. Describe various element form of semiconducting materials. Compare the properties and characteristics of Si and Ge.
2. Discuss various compound and alloy form of semiconductor materials. Write the properties and applications of oxides, sulphides, halides, tellurides and sellurides.
3. Enumerate various semiconductor devices and give a brief account of any five of them.
4. Describe the lattice structures of some compound semiconductors. Discuss the structure of Zinc Sulphide.
5. Sketch the Zinc blende and Wurtzite structures and explain them.
6. What is solar cell? What is its working principle? Describe the construction and working of a solar cell.
7. Discuss the factors that affect the efficiency of a solar cell. Suggest suitable materials for them.
8. Explain the advantages and limitations of a solar cell. Write their various applications.
9. What is light-emitting diode? What are its applications? Enlist the various materials used in their construction.
10. Describe the construction and working of any LED. Discuss the advantages and specification of LED.
11. Explain the principle and working of a semiconductor laser.

12. What is optical fibre? What are its different types? Describe the construction and working of an optical fibre.
13. Discuss the requirements of a good optical fibre material. Also suggest suitable materials for them.
14. What are the advantages and limitations of optical fibres? Write their applications.
15. What are the merits of semiconductor lasers on conventional lasers? Elaborate the properties and applications of lasers.
16. Suggest suitable materials for semiconductor lasers.
17. Write notes on the following.
    a. Metallurgical grade Si and electronic grade Si.
    b. Recrystallization methods of producing pure single-crystal Si.
18. Enlist at least 10 different semiconducting devices/products mentioning the most suitable materials for them.
19. Suggest the choicest semiconductors for producing the following devices.
    a. ICs
    b. Automatic door opener
    c. Infrared detectors
    d. LEDs
    e. Photoconductors
    f. Xerox-type photocopier
20. What are the reasons of the following.
    a. High-frequency transistors are made of Ge, and not of Si.
    b. Solar cells are made of Se.
    c. Hyper high-speed computers are made of new generation AlGaAs chips.

Objective Questions

1. Match list I with list II, and choose the correct answer from the codes given below the lists.

List I	*List II*
A. LEDs	1. Watches
B. LCDs	2. Fibre-optic communication systems
C. Semiconductor lasers	3. Optical switches
	4. InAlAs
D. Photodiodes	5. Numeric display in calculators

Codes:

	A B C D		A B C D
a.	1 2 3 5	b.	5 1 2 3
c.	2 3 4 1	d.	3 5 4 2

2. Which of the following pairs of devices and materials does *not* match?
   a. Solar cell : $CuInSe_2$ as absorber and CdS as collector
   b. LED: GaAs-AlGaAs
   c. Optical fibre: Vitreous high-silica glass doped with germania
   d. Semiconductor laser: Nd YAG

3. The losses in optical fibre at a given wavelength is determined by
   a. $I(x) = I_o e^{-\alpha x}$ b. $I(x)= I_o e^{\alpha x}$
   c. $I(x) = I_o e^{-2\pi x}$ d. $I(x) = I_o e^{2\pi x}$

4. Which of the following pairs does *not* match?

a.	These utilize the voltage dependency	Varistors effect of semiconductors.
b.	Use of selenium	TV cameras
c.	Use of GaAs	Tunnel diodes
d.	Semiconductor used for LED in infrared region of the light spectrum	SiC (α)

5. Match list I with list II, and choose the correct answer from the codes given below the lists.

List I	*List II*
A. InSb	1. 10^{-2} to 10^{-3} ohm m
B. Resistivity of Si	2. 10^{-8} to 10^{-9} ohm m
C. Photoengraving	3. Simple orthorhombic
D. Structure of As	4. For cutting windows in oxide layer
	5. III-V compound
	6. II-VI compound

Codes:

	A B C D		A B C D
a.	5 1 4 3	b.	6 2 4 3
c.	5 2 3 4	d.	6 1 3 4

6. Consider the following statements.
   A. Semiconductors are bipolar solids and have homopolar bond.
   B. Compound semiconductors generally have ionic-covalent bond.
   C. During transition, there is a change in the value of k for direct energy band semiconductors.
   D. Carrier concentrations are generally independent of temperature in extrinsic regions.

   Of these, the correct statements are

   a. *B*, *C* and D b. C and D
   c. A, B and *C* d. *A*,*B*and*D*

8

Semiconducting Materials: Processing and Devices

Processing

8.1 Production of Element Form Of Silicon (Si)

The raw feedstock for Si crystal is silicon dioxide (SiO_2). SiO_2 is reacted with carbon *C* in the form of coke in an arc furnace at very high temperatures (≈1800°C) to reduce SiO_2 according to the following reaction:

$$SiO_2 + 2C \rightarrow Si + 2CO \tag{8.1}$$

This forms the metallurgical grade Si (MGS) which has impurities such as Fe, Al and heavy metals at levels of several hundred to several thousand parts per million (ppm). It is worth mentioning that 1 ppm of Si corresponds to an impurity level of $5 \times 10^{16}/cm^{+3}$. While the MGS is clean enough for metallurgical applications such as using Si to make stainless steel, it is not pure enough for electronic applications. It is also not a single crystal. The MGS is refined further to yield semiconductor-grade or electronic-grade Si (EGS), in which the levels of impurities are reduced to parts per billion or ppb (1 ppb = 10^{-3}ppm). This involves reacting the MGS with dry HCl according to the following reaction to form trichlorosilane, $SiHCl_3$, which is a liquid having boiling point of 32°C.

$$Si + 3HCl \rightarrow SiHCl_3 + H_2 \quad (8.2)$$

Along with $SiHCl_3$, chlorides of impurities such as $FeCl_3$ are also formed which fortunately have boiling points that are different from that of $SiHCl_3$. This allows a technique called fractional distillation to be used, in which the mixture of $SiHCl_3$ and the impurity chlorides is heated, and the vapours are condensed in different distillation towers held at appropriate temperatures. Thus the pure $SiHCl_3$ is separated from the impurities. $SiHCl_3$ is then converted to highly pure EGS by reaction with H_2,

$$2SiHCl_3 + 2H_2 \rightarrow 2Si + 6HCl \quad (8.3)$$

8.2 Semiconductor Crystal Growth

Element form of Si and Ge are obtained by chemical decomposition of compounds such as $SiC1_4$, $SiHC1_3$ and GeO_2 etc. After isolation from these compounds, the Si and Ge are purified. They are melted (melting points of Si is 1410°C and that of Ge is 937°C) and then casted into ingots. When the casted ingots are cooled down, the Si and Ge are obtained in polycrystalline form. It is now the turn of crystal growth. For growth of a single crystal, a very careful control has to be maintained at the boundary of 'solid form of material during cooling' and 'molten material'. Different methods are adopted for growth of crystals from molten material. These are listed below.

1. Bridgman method (or the method of growth from the melt)
2. Czochralski method
3. Zone refining and floating zone growth
4. Liquid phase epitaxy
5. Molecular beam epitaxy

To meet the requirement of level of this book, the first two methods will be described now in subsequent sections.

8.2.1 Bridgman Method

Common technique for growing single crystal from the melt involves selective cooling of molten material so that the solidification occurs along

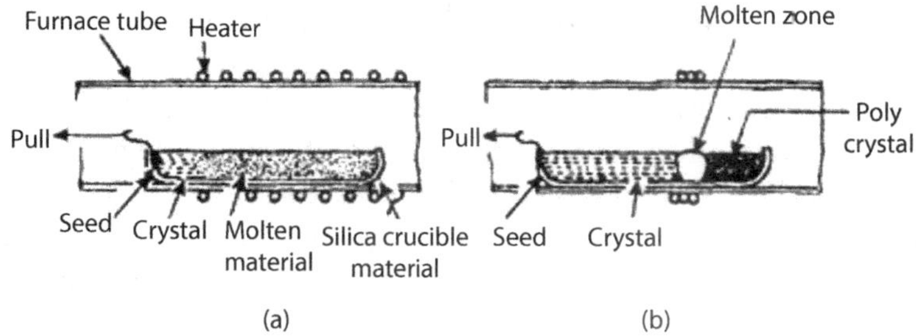

Figure 8.1 Bridgman method of crystal growth from the melt in a crucible.

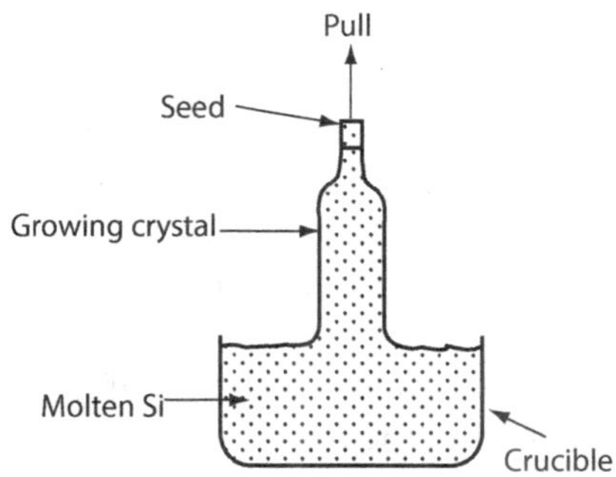

Figure 8.2 Czochralski method of crystal growth.

a particular crystal direction as shown in Fig. 8.1a. Here the silica crucible containing molten Si or Ge is pulled slowly through the furnace. Consequently, the solidification begins at one end and slowly proceeds down the length of the bar. To increase crystal growth, a small seed crystal is placed at the end which is cooled first. Ge, GaAs and other semiconductor crystals are grown using this method.

As shown in Fig. 8.1b, a small region of polycrystalline material is melted and the molten zone is moved down the crucible at such a rate that a crystal is formed behind the zone, as it moves. This method is not suitable for growth of Si, as it has high melting point and sticks to the crucible.

8.2.2 Czochralski Method

In this method, a seed crystal is lowered into the molten material and is raised slowly, allowing the crystal to grow on to the seed. The crystal is rotated slowly as it grows to provide slight stirring of the melt and to correct any temperature variation. This arrangement is shown in Fig. 8.2.

8.3 Processing of Semiconducting Materials

We have discussed earlier that a ppm (parts per million) addition of impurity (dopant) alters the electrical conductivity of semiconductors tremendously, hence this addition has to be minutely controlled to get a refined quality of extrinsic semiconductor. This necessitates an effective application of different refining techniques. The semiconductors can be refined by following methods.

1. Zone refining method
2. Recrystallization method
   i. Bridgman's method
   ii. Czochralski's method.

Zone refining method is based on phase separation. Ultrapure single crystal silicon can be produced by this method.

In *Bridgman's recrystallization method,* the material is melted in a vertical cylinder that tapers conically to a point at the bottom. The vessel is gradually lowered in the cold zone where crystallization begins at the tip and the growth continues from the nucleus. In *Czochralski's method,* a single crystal is introduced on the surface of the melt and is drawn slowly upward, in the cold zone. The grown crystal usually acquires the form of container through which it has been drawn.

8.4 Zone Refining

Although it is necessary to have acceptor or donor impurities, yet a practical semiconductor must be as for as possible chemically pure and crystallographically perfect. Thus, the *p-n* junction should occur in a single crystal because a grain boundary at the junction might contribute extra levels, thereby can complicate the matters. Recombination centres traps, and local distortion of the band structure may exist around a grain boundary. If a *p-n* junction occurs at a grain boundary, the operation of rectification is hindered or prevented.

Dislocations also have similar effects. In transistor materials, the lifetime of excess minority carriers should be as large as possible because, the operation of transistors depends on minority carrier injection across the base from the emitter base junction, to the base collector junction. If the excess minority carriers are arrested in their passage through the base by traps or recombination centres, the transistor operation is impossible. Impurities and imperfections are, therefore, highly undesirable. For practical purposes, it is also advantageous to control the resistivity of the material.

The level of undesirable impurities must, therefore, be of the order of parts per billion (ppb) before the donor is added, if real control of resistivity is so achieved. To attain the highest purity possible, zone refining is employed after chemical processing. Without zone refining, very little transistor grade material would be available. The zone refining process is based on the segregation of dissolved impurities during *non-equilibrium solidification.* Figure 8.3 describes the zone- refining process in nearly pure material where only one impurity is present.

Since the concentration of the impurity in solid and liquid are very small, the liquidus and solidus curves can be approximated by two straight lines. The ratio between the impurity concentration in the solid and that in the liquid is then a constant; $C_s/C_l = C_{seg}$. The constant C_{seg} is called the *distribution or segregation coefficient.* If the solidifying liquid is of composition C_l, the first solid to emerge from the liquid has the composition C_{seg} C_l. If C_{seg} is less than one, as in Fig. 8.3, the solid is purer than the liquid. The material can be purified by partially solidifying and then pouring off the liquid. The first material to solidify is again the purest. A rod solidified from one end by moving the solid liquid interface continuously, leaves the purest and most useful material at the initial end.

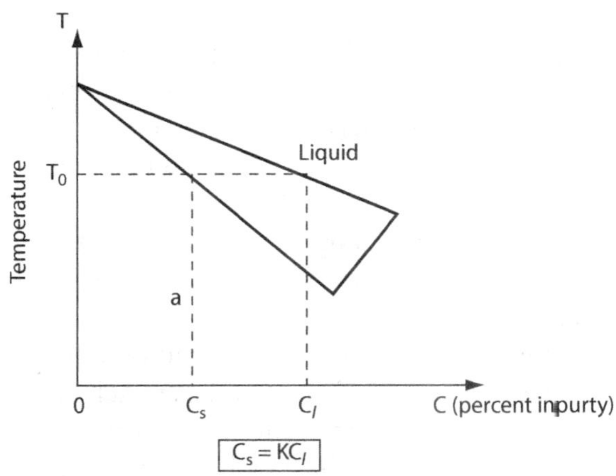

Figure 8.3 The straight liquidus and solidus imply that the ratio of impurity concentration in the solid to that of the liquid is a constant C_g, called segregation.

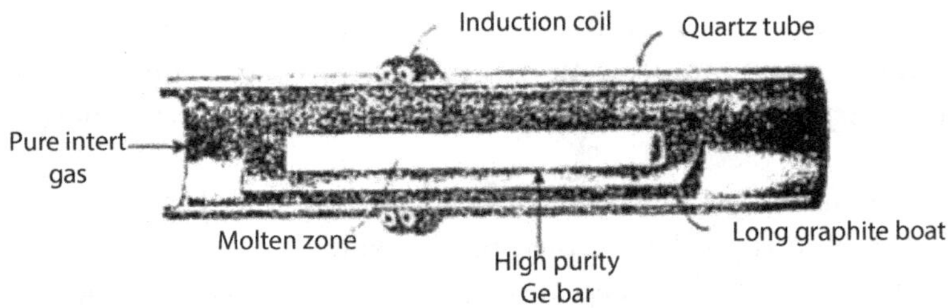

Figure 8.4 Schematic set-up of a zone-refining apparatus. The molten zone moves down the bar.

8.4.1 Zone Refining Apparatus

A schematic representation of a zone-refining apparatus is shown in Fig. 8.4. Only a section of the bar of semiconductor material is melted at any one time. The source of heat is usually a high frequency induction coil. As the coil is moved along the bar, the molten zone moves with it. The liquid which no longer lies in the hot region, solidifies. The length of the zone remains constant if the power input of the induction coil and the environment of the bar remain constant.

8.5 Manufacturing of Wafers

The single-crystal ingot is mechanically processed to manufacture wafers. It involves several steps, as listed below.

1. The first step involves mechanically grinding of the more-or-less cylindrical ingot into a perfect cylinder. This is important because in a modern integrated circuit fabrication, many processing tools and wafer handling robots require tight tolerances on the size of the wafers.
2. Then using X-ray crystallography, the crystal planes in the ingot are identified. Most Si ingots are grown along the <100> direction.
3. For such ingots, a small notch is ground on one side of the cylinder to delineate a {110} face of the crystal. This is useful because for <100> Si wafers, the {110} cleavage planes are orthogonal to each other. This notch then allows the individual integrated circuit chips to be made oriented along {110} planes. Due to this; the chips are sawed apart, there is less chance of spurious cleavage of the crystal.

4. Next, the Si cylinder is sawed into individual wafers about 775 μm thick, by using a diamond-tipped inner-hole blade saw, or a wire saw.
5. The resulting wafers are mechanically lapped and ground on both sides to achieve a flat surface, and to remove the mechanical damage due to sawing. The flatness of the wafer is critical from the point of view of "depth of focus" or how sharp an image can be focussed on the wafer surface during photolithography.
6. The Si wafers are then rounded or chamfered along the edges to minimize the likelihood of chipping the wafers during processing.
7. Finally, the wafers undergo chemical-mechanical polishing using a slurry of very fine SiO_2 particles in a basic NaOH solution to give the front surface of the wafer a mirrorlike finish.

The wafers are now ready for integrated circuit fabrication. The economic value added in this process is impressive as from sand (SiO_2), we obtain Si wafers on which we can make hundreds of microprocessors. Thus the very cheap sand is converted into very costly products costing several hundred dollars.

8.5.1 Photolithography

Photolithography is a process of planar technology to fabricate an IC in which a desired pattern of doping is transferred onto the silicon wafer. The entire process involves:

i. coating of SiO_2 layer with a film of photosensitive emulsion called *photoresist*,
ii. blackening the portions to be doped on a transparent sheet, placing this sheet on photoresist layer and then exposing to ultraviolet light
iii. removing the mask and dipping the wafer in trichloroethylene.

8.6 Semiconductors Fabrication Technology

The development of large-scale integrated circuits (LSICs) and decreasing size of electronic devices are remarkable achievements of semiconductor technology. A single piece of semiconductor can connect more than

1000 transistors on an area of about 1 square centimeter. This has become possible due to the advances in fabrication and processing techniques for semiconductors. In this regard, the processes of microelectronics may be grouped as follows.

1. Semiconductor microelectronics, and
2. Thin film microelectronics.

Diodes, *pnpn* switches, transistors and resistors belong to semiconductors of microelectronics group; while the interconnections of various electronic circuits and capacitors are the product of thin *film microelectronics.*

8.6.1 Microelectronic Circuit Construction

A semiconductor microelectronic circuit requires following sequential processes in its construction.

i. Production of highly purified *n* or *p* type silicon.
ii. Single crystal growth from the melt.
iii. Manufacturing of 0.25 to 0.40 mm thick *wafer.*
iv. Polishing of wafers to a thickness of 0.15 to 0.20 mm.
v. Oxidation to isolate pockets in a wafer.
vi. *Photoengraving* for cutting windows in the oxide layer.
vii. Diffusion by doping via a vapour phase of B_2O_3, B_2H_6 or PH_3 in a stream of hydrogen, results in pn and *pnp* junctions.
viii. *Epitaxy* for growing a new layer of same crystal orientation using hydrogen as carrier gas, silane (SiH_4), and dichlorosilane (SiH_2C1_2) etc. as vapour sources.
ix. Chemical processing such as etching to remove mask and oxide layers.
x. *Masking i.e.* repeating the above operations a number of times on the surface of the wafer. The surface is then separated into chips. The circuit components per chip exceeds 1 million.
xi. Interconnections by metallizing with aluminium which is deposited by vacuum evaporation.

8.6.2 Thin Film Circuit Fabrication

In fabricating thin film microelectronic circuits, glazed ceramics and glass substrates are used. The *substrate* is a polished surface on which many hundred I.Cs. may be located. Deposition of I.Cs. on thin films can be done by one or more of the following processes.

i. Vacuum deposition,
ii. Silk screening,
iii. Vapour plating,
iv. Electron beam decomposition,
v. Plasma decomposition,
vi. Sputtering, and
vii. Anodization.

Thin film resistors employ metallic film while the thin film conductors employ film of gold or aluminium, and capacitors use a dielectric film. *Ion implantation* is a most modern method of doping microelectronic devices such as in MOSFET.

8.7 Fabrication of a Semiconductor P-N Junction

The main steps involved in the fabrication of a p-n junction are given below in Fig. 8.5. The diagrams show a typical case in which only four diodes per wafer are shown for simplicity. Also the relative thicknesses of the oxide, photoresist, and Al-layers are exaggerated.

8.8 Transistor Manufacturing Processes

Various techniques adopted to manufacture a transistor are displayed step-wise in Fig. 8.6. The details are self-explanatory.

DEVICES

8.9 Semiconducting Devices and Their Operating Principle

Semiconductors are solid materials of crystalline nature. Their electrical conductivity is superior to those of insulators but inferior to conductors. Thus, these are intermediatory between insulators and conductors. Electrically the semiconductors are neither good conductors like copper

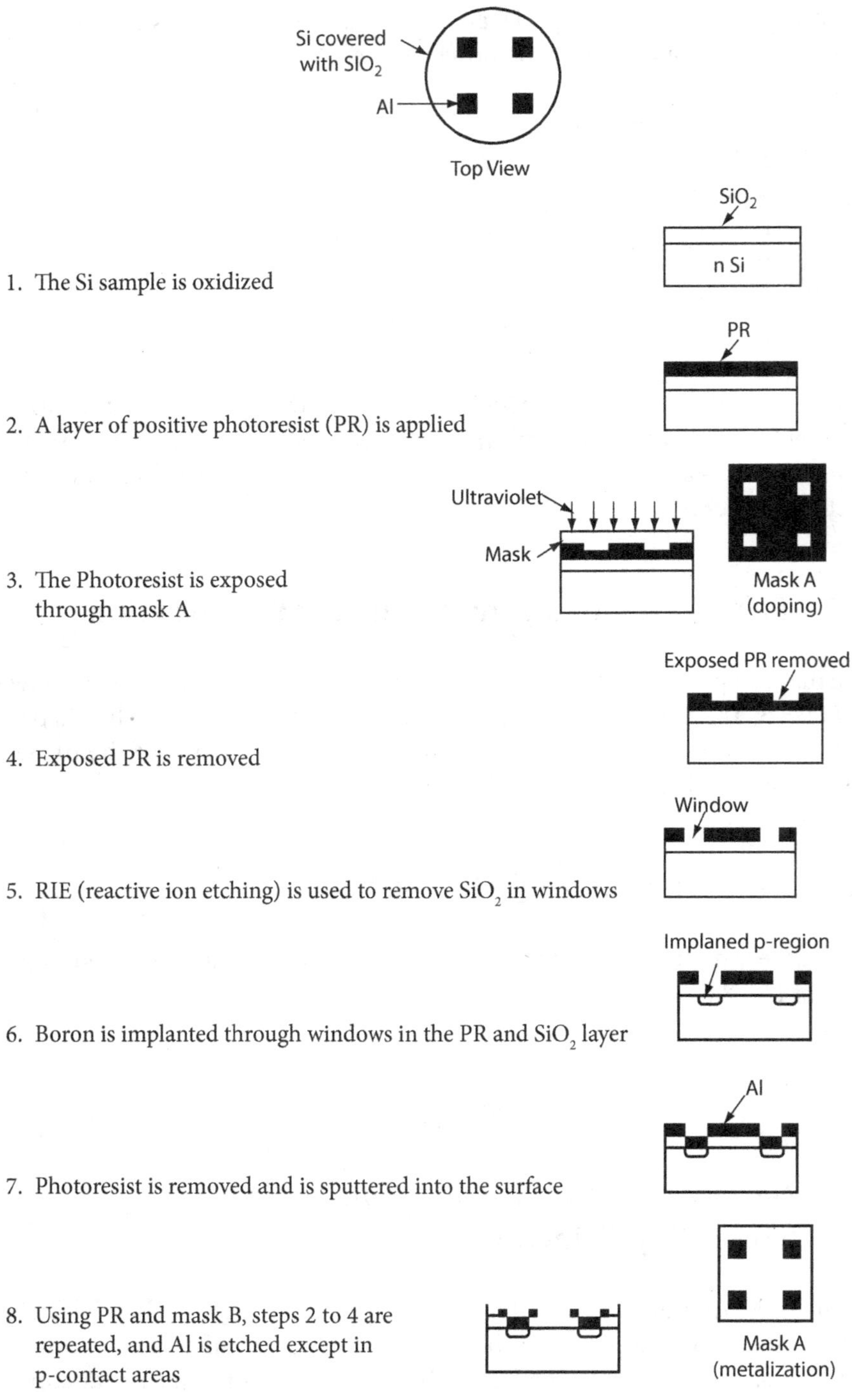

Figure 8.5 Simplified description of steps in the fabrication of p-n junctions.

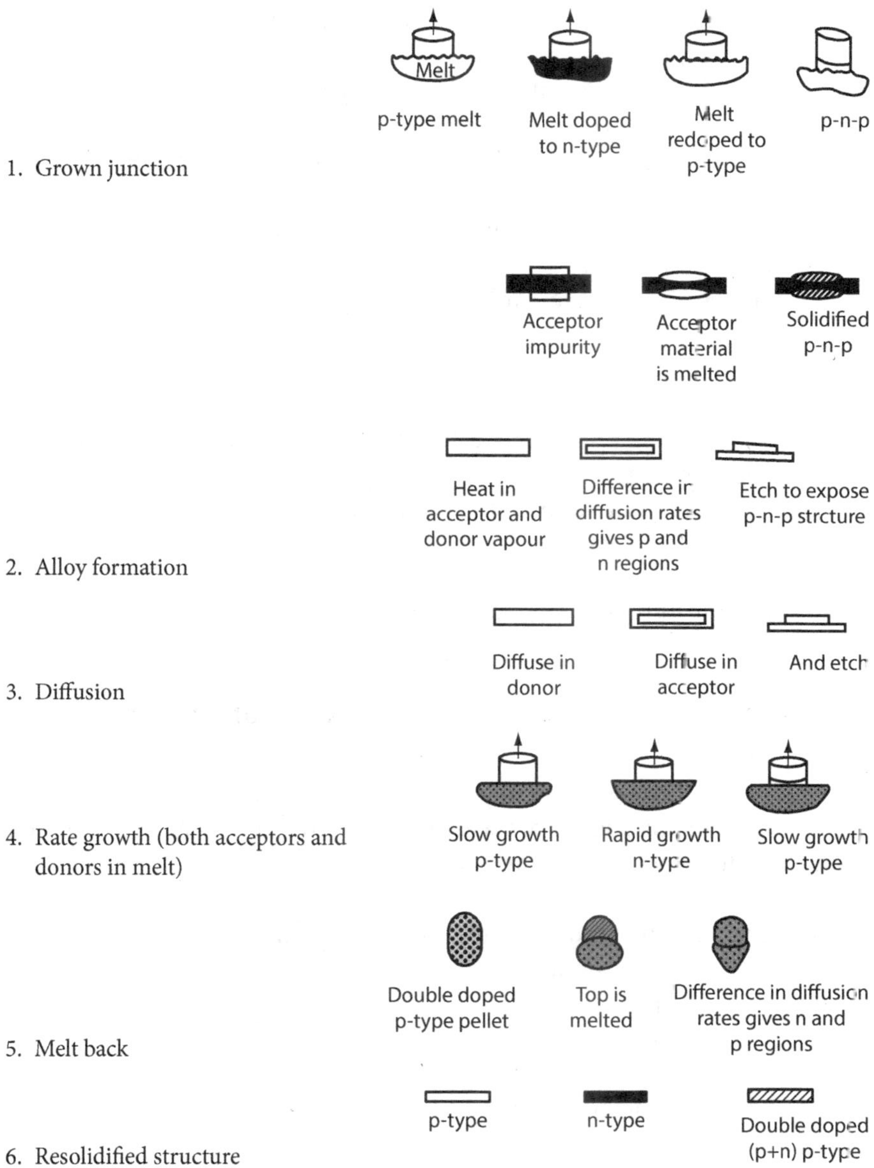

Figure 8.6 Schematic arrangement of transistor manufacturing technique.

and aluminium, nor good insulators like mica and porcelain. Compound semiconductors find wide use in/as following important devices.

1. Thermistors: These utilize the temperature dependency effect of semiconductors.

2. Varistors: These utilize the voltage dependency effect of semiconductors.
3. Rectifiers: These utilize the impurity dependency effect of semiconductors.
4. Strain gauges: These utilize the change in resistance effect of semiconductors.
5. Zener diodes: These utilize the electric field effect of semiconductors.
6. Transistors: These utilize the amplification effects of semiconductors.
7. Photoconductive cells: These utilize the light illumination effect of semiconductors.
8. Photovoltaic cells: These utilize the optical characteristics of semiconductors.
9. Hall effect generators: These utilize the carrier drift effect in semiconductors.

Details on some of these topics are given later in appropriate sections.

8.10 Important Applications of Semiconductor Devices

Compound semiconductors are widely employed to fabricate a number of semiconductor devices. Main among them are listed below.

1. Equilibrium condition junction diodes e.g.
   - Rectifying diodes as
     i. half-wave rectifier,
     ii. full-wave rectifier
   - Zener diode
     i. for meter protection
     ii. as peak clipper
2. Transient condition diodes e.g.
   - Switching diodes
   - Varactor diodes
   - Metal-semiconductor junction Schottky diode
3. Bipolar junction transistors (BJT) e.g.
   - Field effect transistor (FET)
   - Metal-semiconductor field-effect transistor (MESFET)

- Metal-insulator-semiconductor field effect transistor (MISFET)

4. Optoelectronic devices e.g.
   - Photodiodes
   - Photodetectors
   - Photocells
   - Solar cells
   - Light emitting diodes (LEDs)
   - Semiconductor lasers
   - Optical fibres
5. Negative conductance microwave devices e.g.
   - Tunnel diodes
   - Impact diode
   - The Gunn diode
6. Power devices e.g.
   - p-n-p-n diode
   - Silicon controlled rectifiers (SCRs)
   - Silicon controlled switch (SCS)
   - Bialateral devices such as
     i. Diac, ii. Triac
7. Integrated circuits (ICs)
8. Transducers
9. Microprocessors

8.11 Brief Description of Some Semiconductor Devices

Brief description of some devices is given below.

Junction. The boundary between *p* and *n* semiconductors is known as *junction.* The region of small thickness at the junction is called *depletion layer.* The *rectifying action* takes place in this region.

Transistor is a two junction *(n-p* and *p-n)* system exhibiting amplifying action of electron signals. Revolution in the electronics industry may be attributed to miniature, cheap, reliable, instant acting and less power consuing solid stae transistors. Metal oxide semiconductor field- effect transistor (MOSFET) is a modern transistor used in street lights and as burglar alarm. CdS, CdSe and CdTe are commonly used photoconducting semiconductors.

Photodiodes. A photodiode is a two terminal device, which responds to photon absorption.

Photoconductors (or photo detectors) are used to detect and measure the quanta of light such as in automatic door opener, in switching the street lights, and as burglar alarm. CdS, CdSe and CdTe are commonly used photoconducting semiconductors.

Photocells convert the light energy into electrical energy. They are fabricated from CdS, Se and $PbSO_4$. Photocells are used in cinematography, fire-alarms and television cameras etc.

Solar cells made from semiconducting materials are of immense utility in satellites and space-going vehicles. They are also used in calculators, solar power generation, and solar autovehicles.

Light-emitting diode is an incoherent light source that is used as a light source in fibre optic systems and other devices.

Laser is the source of a highly directional, monochromatic coherent light which is used as a light source for various optical and electronic devices.

Optical fibre is a means of transmitting optical signals from a source to a detector.

Thermistors are used to determine the temperature of various systems in process industries, ovens and furnaces etc. They are made from sintered manganese oxide that contains dissolved germanium or lithium.

Integrated circuits are manufactured on a single chip that contains diodes, transistors, resistors and capacitors etc. They are generally monolithically constructed using either unipolar or bipolar techniques. If we open our calculators, we will see I.Cs in it. Fabrication of I.Cs is discussed under the section on microelectronics.

8.12 P-N Junction Diode

P-N diode is a two-terminal electronic device consisting of *P-N* junction, formed by Si or Ge crystals. The P-type and N-type regions are referred to as anode and cathode respectively. A P-N junction diode is a one- way device as it conducts current in one direction only. In other (reverse) direction, it offers a very high resistance.

By themselves, P-type and N-type semiconductors are of very limited use separately, but when they are joined together the crystal structure maintains continuity at the boundary and offers appreciable properties. Special fabrication techniques are used to form a *P-N* junction. *P-N* diodes are commercially available in a variety of physical structures with different kinds of terminals and having identification marks. Practically all semiconductor devices contain at least one *P-N* junction.

8.12.1 Applications of P-N Diode

P-N junction diodes are used in following main applications.

1. As rectifier for converting a.c. into d.c.
2. In d.c. power supplies such as for transistor radio.
3. As zener diode for operating in breakdown region (2 to 200 V) without damage, such as in
   i. voltage regulator
   ii. meter protection
   iii. peak clipper
   iv. reshaping a waveform
4. As tunnel diode for conducting large current even when the forward bias (see Fig. 8.7b) is very low (0.05 V or so), such as in
   i. ultrahigh speed switches
   ii. logic storage memory devices
   iii. microwave oscillators
   iv. relaxation oscillator circuits
5. As light-emitting diode (LED) to emit light when in forward-bias, such as in
   i. burglar alarm circuit
   ii. video displays
   iii. optical communication
   iv. numeric displays in calculators

8.12.2 Biasing

Behaviour of a *P-N* junction diode is different in its inherent form, Fig. 8.7a (i.e. when not connected to a battery) from its behaviour under 'biasing' (i.e. when connected to a battery). It can be connected to a battery in following two ways.

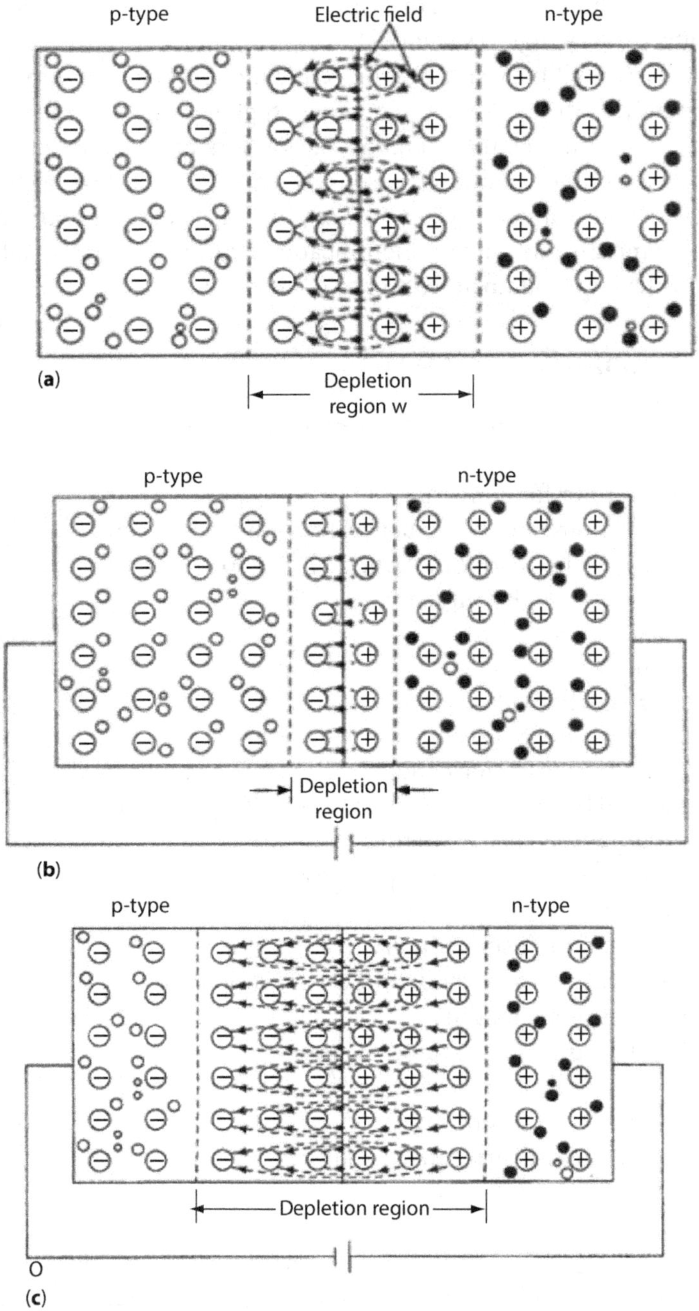

Figure 8.7 A P-N diode (a) when not connected to a battery, (b) under forward-bias, and (c) under reverse-bias.

1. Under forward biasing, and
2. Under reverse (or backward) biasing.

Under forward biasing, the positive terminal of battery is connected to P-side and negative terminal to N-side, Fig. 8.7b. But if the connection is just opposite of it, *i.e.* negative terminal connected to P-side and positive terminal connected to *N-side,* it is known to be in reverse biasing as shown in Fig. 8.7c.

8.13 Working of P-N Diode When not Connected to a Battery

Figure 8.7a shows a single crystal of P-N junction without any external voltage. In it the holes and electrons are the mobile charges, but the ions are immobile. The crystal is electrically neutral, and the P-region and N-region are separate. As soon as the *P-N* junction is formed, the following processes are initiated.

8.13.1 Diffusion of Holes and Electrons in P-N Diode

Holes from P-region diffuse into N-region and combine with free electrons available there. Similarly, the free electrons from N-region diffuse into P-region and combine with holes. Diffusion occurs for a short duration due to thermal energy and also due to difference in concentration of holes and electrons in their regions.

8.13.2 Set-up of Barrier in P-N Diode

After a few recombination of holes and electrons in close vicinity of the junction, a restraining force called Barrier is set up automatically. This barrier stops further diffusion of holes and electrons from one side to the other.

8.13.3 Formation of Depletion (or Space Charge) Region in P-N Diode

Recombination of holes and electrons eliminates a hole and a free electron from their respective regions. Consequently, the negative acceptor ions in P-region and positive donor ions in *N*- region are left uncompensated. Additional holes trying to diffuse into N-region and additional electrons trying to diffuse into P-region are repelled by uncompensated positive and negative charges, respectively. As a result of which further recombination of holes and electrons cannot occur.

The region containing the uncompensated -acceptor and donor ions is called depletion region. It is named so because there is a depletion of mobile charges (holes and electrons) in this region. The electric field between the acceptor and donor ions is called a barrier. The physical distance between two sides of the barrier is called width of the barrier and the difference in potential between them is referred to as height of the barrier.

The width of the barrier is of the order of fraction of millimeter and the barrier height is of the order of tenths of a volt. For silicon *P-N* junction, the barrier potential is 0.7 V and for germanium it is about 0.3 V.

8.13.4 Flow of Drift and Diffusion Current in P-N Diode

The barrier discourages the diffusion of majority carriers across the junction but helps the minority carriers (few free electrons in P-region and a few holes in N-region) to drift across the junction. The minority carriers are constantly generated due to thermal energy. The drift of minority carriers is counterbalanced by diffusion of the same number of majority carriers across the junction. Since the majority carriers have sufficiently high kinetic energy, therefore they overcome the barrier and cross the junction.

8.14 Different Types of P-N Junction Diodes

There are several types of p-n junction diodes. They either emphasize different physical aspects of diode such as geometric scaling, doping level, or just an application of diode in a special circuit, or different devices like the Gunn diode, laser diode and the MOSFET etc.

Normal (p-n) diodes are usually made of doped silicon or rarely of germanium. Before the development of modern silicon power rectifier diodes, earlier the cuprous oxide and later selenium were used. They had low efficiency, much higher forward voltage drop, and required a large heat sink. They were much larger than a silicon diode of the same current rating. The vast majority of all diodes are the p-n diodes found in CMOS integrated circuits, which include two diodes per pin and many other internal diodes. These are enumerated below.

- Zener Diodes
- Avalanche Diodes
- Thermal Diodes
- PIN Diodes
- Schottky Diodes
- Super Barrier Diodes

- Gunn Diodes
- Light-emitting Diodes (LEDs)
- Transient Voltage Suppression Diode (TVS)

Details of these diodes is beyond the scope of this text, and can be referred to in other related books.

8.14.1 The Gunn Diode Materials and Fabrication

These are similar to tunnel diodes in that they are made of materials such as GaAs or InP that exhibit a region of negative differential resistance. With appropriate biasing, the dipole domains are formed and travel across the diode, allowing high frequency microwave oscillators to be built.

The Gunn diodes are made of a number of such materials which have an appropriate band structures for electron transfer. They are also made from such materials whose band structure can be altered to exhibit properties which are favourable for electron transfer. Various materials suitable for making the Gunn diodes are the following.

- GaAs
- InP
- GaAsP
- CdTe
- ZnSe
- InAs

Although a Gunn diode can be made of bulk samples, but the use of ingot material as a substrate for epitaxial layer is desired. It is because the material properties of epitaxial layers are superior to those of bulk samples. External contacts of Gunn diodes are made by evaporating a thin layer of following materials on each surface.

- Au-Sn
- Au-Ge

8.15 Junction Transistors

A transistor is a semiconductor device. It is a junction device. Unlike a diode which has only one junction, a transistor has *two* junctions. It is one of the most important micro-electronic devices which has completely revolutionalized the scenario of modern electronic industry. The transistor was invented in 1948 by John Bardeen, Walter Brattain and William Shockley at Bell Laboratory in USA. Since then, there has been a rapidly expanding efforts to utilize and develop many types of semiconductor devices such as FET, MOSFET, UJT, SCR, etc.

Transistors offer several advantages over earlier used gas or vacuum tubes. A few of them are the following.

- They are much smaller in size, and are light in weight.
- They consume little power, resulting in greater circuit efficiency.
- They use very low operating voltages.
- No heater or filament is required. Hence, there are no heating delays and no heating power needed.
- They have long life with no ageing effect.
- They are essentially shock-proof.

8.15.1 Different Categories of Transistors

Different types of transistors are used in vivid applications. They serve different functions in numerous applications such as in amplifiers, oscillators etc. Broadly, they are classified into two main categories, given as below.

1. Field effect transistors (*i.e.* FET)
2. Bipolar junction transistors (*i.e.* BJT)

A detailed classification is shown in Fig. 8.8, and their description follows in subsequent sections.

8.16 Bipolar Junction Transistor (BJT)

A bipolar junction transistor is a sandwiched form of construction in which, one type of semiconductor (say n-type) is placed between two layers of other type of semiconductor (i.e. *p*- types). Thus, the formation will be a *p-n-p* type construction. Similarly when the p-type semiconductor is sandwiched between the two n-types, it forms a *n-p-n* type transistor.

8.16.1 Construction of BJT

A transistor is basically formed of a Si or Ge crystal containing three separate regions These regions are separated by two junctions. These regions and junctions are the following.

1. Emitter *E*
2. Base *B*, and
3. Collector C.

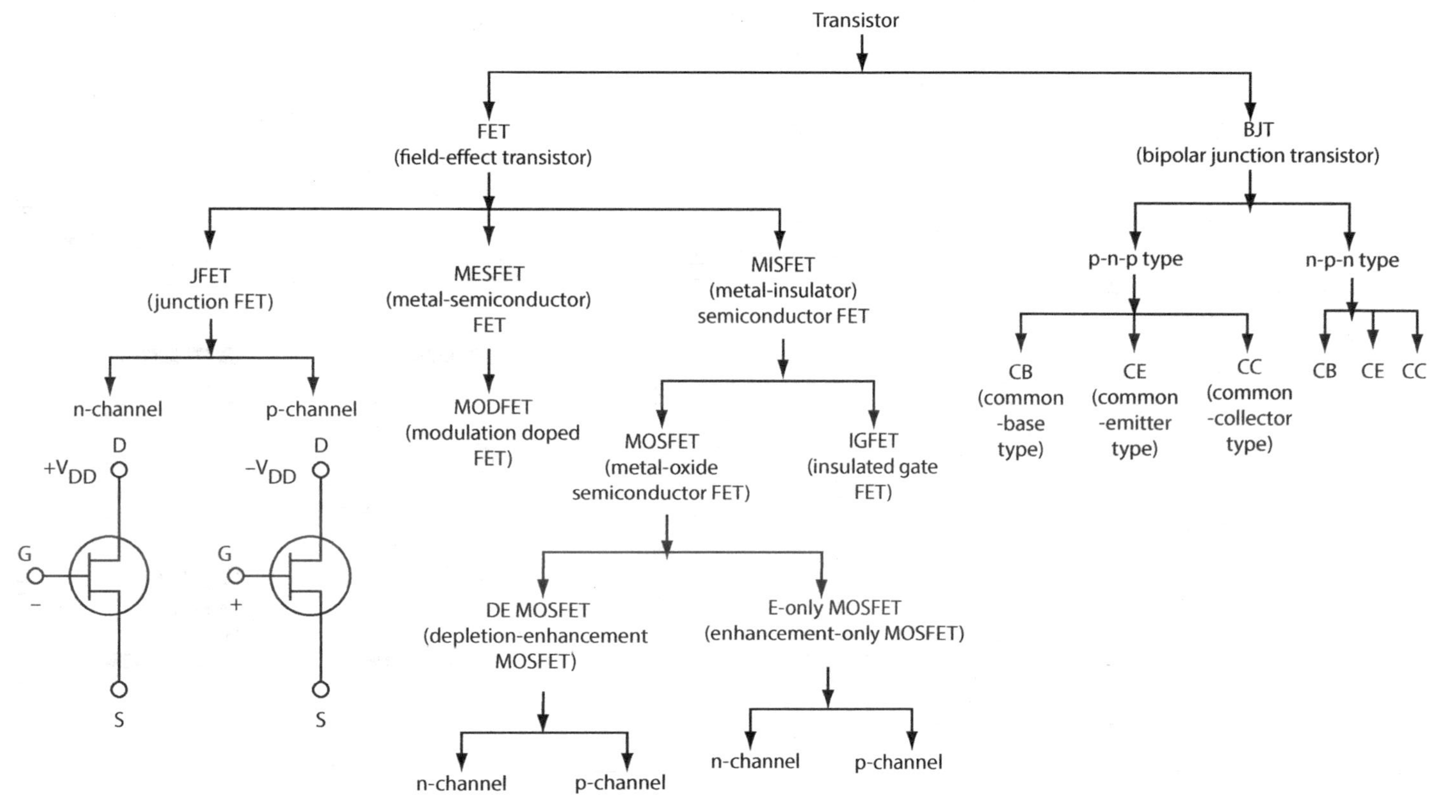

Figure 8.8 Classification of transistors

The junctions are of following types.

1. Emitter-base (or EB) junction, and
2. Collector-base (or CB) junction.

These are shown in Figs. 8.9a-d. Figures 8.9a-b show a *n-p-n* transistor and Figs. 8.9c-d depict a *p-n-p* transistor.

Constructional features. Salient features of their construction are the following.

- The middle region is called the *base* and the two outer regions are called the *emitter* and the *collector.* Although the two outer regions are of the same type (n-type), their function cannot be interchanged. They have different physical and electrical properties.
- In most transistors, the collector region is made physically larger than the emitter region since it is required to dissipate more heat.
- The base *is* very *lightly doped,* and is very thin. The emitter is *heavily doped.* Doping of the collector is intermediate between the heavy doping of emitter and light doping of the base.

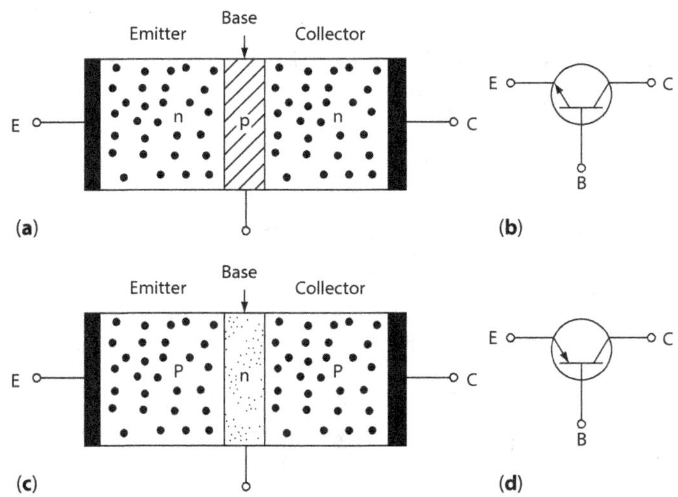

Figure 8.9 Construction of a BJT (a) *(b) n-p-n* type and its symbol, and *(c) (d) p-n-p* type and its symbol.

- The function of the emitter is to emit or inject electrons (holes in case of a *p-n-p* transistor) into the base. The base passes most of these electrons (holes in case of *p-n-p)* on to the collector. The collector serves the function of collecting these electrons (holes in case of a *p-n-p)* from the base.

8.17 Field-Effect Transistor (FET)

FET is an abbreviation of field-effect transistor. It is a unipolar, 3- terminal solid-state device in which the current is controlled by an electric field. The FET was developed in early 1960s. There are several kinds of FETs *viz.* JFET, MOSFET, MISFET, ICIFET etc.

8.17.1 Advantages of FETs over BJTs

The FETs possess many advantages over BJTs. These are listed below.

1. FETs have high input impedance.
2. FETs have high power gain.
3. FETs can be fabricated easily for different devices, particularly the ICs.
4. They are of very small size.

8.17.2 Differences between FETs and BJTs

A FET differs from a junction transistor. Unlike BJT whose operation depends on both; the electrons and holes, the operation of FET depends on the flow of majority carriers only. It is, therefore, also known as *unipolar transistor* (UPT).

8.17.3 Applications of FETs

FETs are used in almost all those applications where bipolar transistors are also used. Besides them, the FETs find exclusive use in following applications.

1. In logic circuits where it is kept 'OFF' when the input is zero. It can be turned 'ON' with very little power input.
2. As voltage-variable resistor (VVR) in amplifiers and tone controls etc.
3. In large-scale integration (LSI) and computer memories.

4. As input amplifiers in the measuring and testing equipments, and also in
   - oscilloscopes
   - electronic voltmeters, etc.
5. For mixer operation of FM and TV receivers.

8.18 Metal-Semiconductor Field-Effect Transistors (MESFET)

A MESFET is a metal-semiconductor junction. It uses a reverse-biased Schottky barrier instead of a *p-n* junction. The MESFET devices are generally made of III—V group compounds such as GaAs, InP etc. instead of Si. It is because the metal-semiconductor FET device possesses several meritorious features such as given below.

1. They have high operational speed.
2. They have high power levels.
3. They can be fabricated to close geometrical tolerances.
4. They can be made very small. Gate lengths of less than 0.25 μm are very common
5. They have higher mobilities.
6. They have higher carrier drift velocities than Si.
7. They can be operated at higher temperatures.
8. They have better high frequency operations since the drift time and capacitance may be kept a minimum in them.

These devices are useful in following applications.

- High-speed digital circuits
- High-speed microwave circuits
- High-frequency services
- GaAs integrated circuits
- Very small size gates having lengths ≤ 0.25 μm.

8.18.1 Basic Construction of MESFETs

JFETs have low noise and relative insensitivity to cosmic radiation. Therefore, they are ideally suitable for amplification in GHz frequency range such as for the satellite communication systems. However, their main disadvantages are their low gain bandwidth product. Since these devices

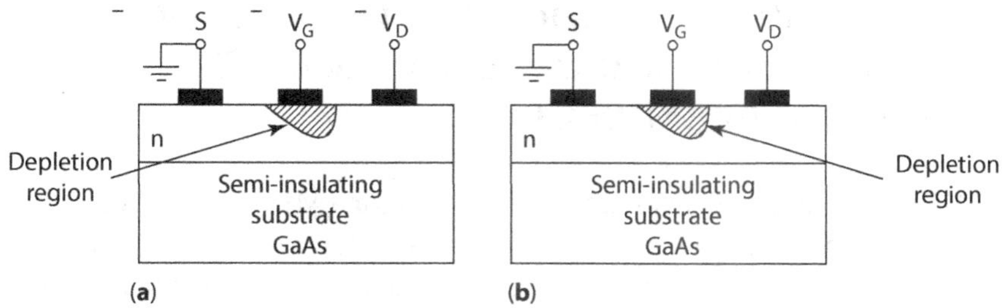

Figure 8.10 *(a-b)* Cross-section of two basic types of GaAs MESFET. (a) depletion mode device and (b) enhancement mode device.

employ a Schottky barrier metal gate instead of the *p-n* junction, they are known as metal-semiconductor field-effect transistors (MESFETs).

Figures 8.10*a-b* show the cross-section of a GaAs MESFET. The device is fabricated by growing an *n*-type epitaxial layer on a high-resistivity GaAs substrate. The substrate may be either of *n*-type or of *p*-type. The source and drain ohmic contacts are made by depositing a suitable metal by vacuum evaporation and by alloying. The metal gate offers a number of advantages viz. the formation of a Schottky barrier contact can be achieved at much lower temperatures than those required for *p-n* junction formation. And also, the gate length can be reduced to submicron dimensions by controlling the length of the metal electrodes.

8.18.2 Basic Types of MESFETs

Depending on the thickness of the epitaxial layer, a MESFET may be designed to operate either as a

1. depletion mode (normally on) device, or
2. an enhancement mode (normally off) device.

In a *depletion mode device*, the thickness of the epitaxial layer is more than the zero-bias depletion region width of the Schottky barrier gate, and the transistor has a conducting channel at $V_G = 0$. Thus, the gate is biased negative to deplete the channel, as shown in Fig. 8.10*a*.

In the *enhancement-type device* (Fig. 8.10*b*), the epitaxial layer is kept thin, and the built- in voltage of the metal gate Schottky barrier junction is sufficient to deplete the channel completely at $V_G = 0$. Conduction in the channel occurs only for small positive values of V_G.

Enhancement-type devices are useful in high-speed, low-power applications; but the majority of MESFETs are the depletion type.

8.19 Insulated Gate Field Effect Transistor (IGFET) or Metal-Insulator-Semiconductor Field-Effect Transistor (MISFET)

A MISFET is a metal-insulator-semiconductor (MIS) transistor. This is one of the most used electronic device. In this device the channel current is controlled by a voltage applied at gate electrode and an insulator is used to isolate this gate from the channel. Due to this reason, this device is also known as 'insulated-gate field-effect transistor (IGFET). The MISFET is used in following main applications.

- Digital ICs, in which it is switched from no conducting channel 'OFF' state to conducting channel ('ON' state).

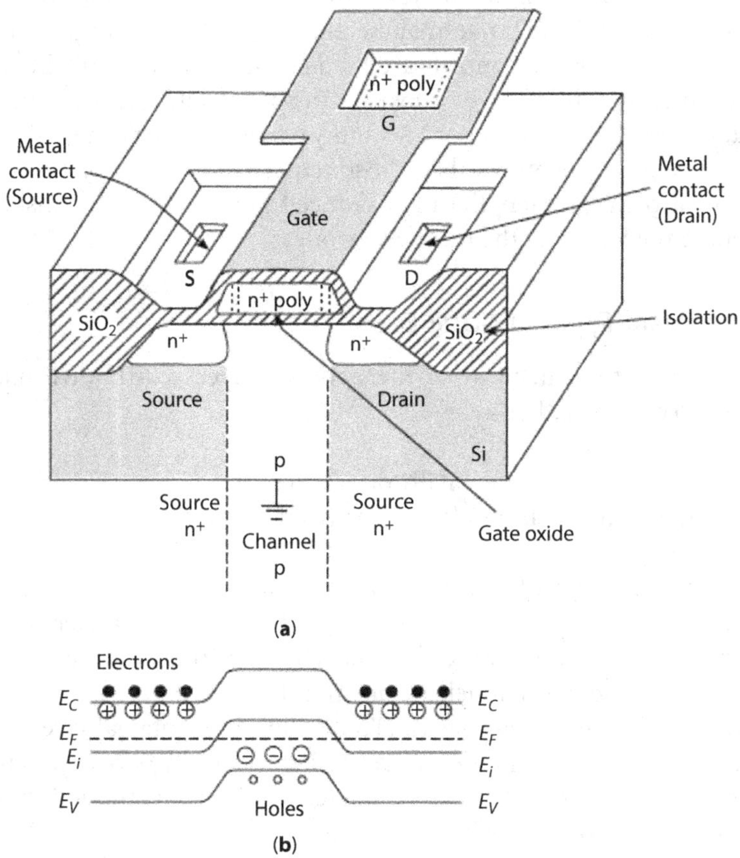

Figure 8.11 An enhancement type n-channel MOSFET showing its (a) fabricational details, and *(b)* equilibrium band diagram along channel.

8.19.1 Construction of IGFET

Since the materials used are Si, SiO_2 and metal; this transistor is commonly known as MOSFET. Fabrication details of a typical MISFET is shown in Fig. 8.11a. It is a *n*-channel construction comprising of the following components.

- *Source S* made of a metal contact.
- *Drain D* made of a metal contact.
- *Isolation* made of SiO_2 .
- *Gate electrode* made of a heavily doped polysilicon.
- *Substrate* made of p-type Si.

The n^+ source and drain regions are implanted (or diffused) into the substrate. Any current cannot flow from drain to source until a conducting n-channel is placed between them.

The MOSFET is surrounded on all sides by a thick SiO_2 layer. It is for the purpose of providing critical electrical isolation between the adjacent transistors on an *IC*. The equilibrium band diagram is shown in Fig. 8.11b.

8.20 Charge Coupled Devices

A charge coupled device (CCD) is an integrated circuit etched onto a silicon surface forming the light sensitive elements, called *pixels*. Photons incident on this surface generate charge that can be read and turned into a digital copy of the light patterns falling on the device. CCDs come in a wide variety of sizes and types. They are used in many applications such as

- cell phone cameras
- high-end scientific applications

A CCD stores and displays the data for an image in such a way that each pixel (picture element) in the image is converted into an electrical signal, the intensity of which is related to a colour in the colour spectrum such as for conversion into a digital value. This is achieved by '*shifting*' the signals between the stages, within the device, one at a time. CCDs move signals between capacitive *bins* in the device, with the shift allowing for the transfer of signals between bins. Figure 8.12 shows various CCDs of different sizes. The largest is mounted on a 15 cm wafer.

Figure 8.12 Showing various types of CCDs

8.20.1 Salient Uses

CCDs are now-a-days commonly included in 'digital still' and 'video cameras'. They are also used in

- astronomical telescopes
- scanners
- bar code readers

The devices are also used in

- machine vision for robots
- in optical character recognition (OCR)
- in processing of satellite photographs
- in enhancement of radar images, especially in meteorology.

8.21 Solved Examples

Example 8.1 Energy band gap in a Ge crystal is 0.75 eV. Determine the wavelength at which Ge starts to absorb light.

Solution. Given are $E_g = 0.75 \text{ eV} = 0.75 \times 1.6 \times 10^{-19}$ J

$$\because \qquad E_g = hf = \frac{hc}{\lambda}$$

$$\therefore \qquad \lambda = \frac{hc}{E_g}$$

$$= \frac{(6.62 \times 10^{-34}) \times (3 \times 10^8)}{0.75 \times 1.6 \times 10^{-19}} = 16560 \text{ Å}$$

Example 8.2 Find the cutoff wavelength for InP having bandgap energy 1.35 eV. Take appropriate values suitably. Energy band gap for indium phosphide is 1.35.

Solution. Given is: E_g = 1.35 eV = 1.35 × 1.6 × 10^{-19} eV

⊠

$$\lambda = \frac{hc}{E_g}$$

$$= \frac{(6.63\times10^{-34})\times(3\times10^{8})}{1.35\times1.6\times10^{-19}} = 0.902\ \mu m$$

Example 8.3 Find the maximum energy of the emitted photoelectrons when the light of frequency 1.5 × 10^9 MHz falls on the surface of a material whose threshold frequency is 1.2 × 10^9 MHz.

Solution. As $E = hf$

$$\therefore \quad E_{max} = h(f - f_o) \text{ joule}$$

$$= (6.62\times10^{-34})\times\left[(1.5\times10^{15}) - (1.2\times10^{15})\right]$$

$$= 6.62\times10^{-34}\,(1.5 - 1.2)\times10^{15} \text{ joule}$$

$$= \frac{6.62\times10^{-34}\times0.3\times10^{15}}{1.602\times10^{-19}} \text{eV}$$

$$= 1.24 \text{ eV}$$

Example 8.4 What is meant by trapping centre?

Solution. An impurity or a defect centre in semiconductor is referred to as trapping centre (or trap).

Example 8.5 What is lattice absorption edge?

Solution. In pure semiconductors, the absorption coefficient drops rapidly at a wavelength near to or in the intermediate infrared region, and the material becomes transparent at larger wavelengths. This marked drop in absorption coefficient is called *lattice absorption edge* or only 'absorption edge'. The absorption edge for some semiconductors are given as follows.

- InSi, at λ = 1.1 μm
- InGe, at λ = 1.65 μm

Example 8.6 Write the two major factors which govern the operation of semiconductor devices.

Solution. The operation of semiconductor devices mainly depends upon the following two factors.

- Injection of minority carriers, and
- Extraction of minority carriers.

Example 8.7 Explain briefly the following terms.

(*i*) Depletion region (*ii*) Space-charge region
(*iii*) Barrier (*iv*) Width and height of barriers.

Solution. The terms are defined as follows.

1. In a *p-n* junction, the region containing the uncompensated acceptor and donor ions is called *depletion region.* There is a depletion of mobile charges (holes and free electrons) in this region.
2. Since this region has immobile (fixed) ions which are electrically charged, it is also the called *space-charge region.*
3. The electric filed between the acceptor and donor ions is called a *barrier*.
4. The physical distance from one side of the barrier to the other side is referred to as *width* of the barrier. The difference of potential from one side of the barrier to the other side is referred to as *height* of the barrier.

With no *external* batteries connected, the barrier height is of the order of tenths of a volt. For a silicon *p-n* junction, the barrier potential is about 0.7 V, whereas for a germanium *p-n* junction it is approximately 0.3 V.

Example 8.8 What is the effect of doping on breakdown of a *p-n* junction diode?

Solution. The effect of doping on breakdown of a *p-n* junction diode is as follows.

- It suffers Zener breakdown when heavily doped.
- It suffers avalanche breakdown when lightly doped.

Example 8.9 Under which conditions does a *p-n* junction possess the diffusion capacitance and transition capacitance?

Solution. a *p-n* junction posesses above types of capacitance under following conditions.

- Diffusion capacitance when forward-biased.
- Transition capacitance when reverse-biased.

Example 8.10 Display the various physical structures of diodes and their terminal identifications.

Solution. Figure 8.13*a* shows different physical structures and Fig. 8.13*b* shows various terminals identifications for diodes.

Example 8.11 Some *p-n* diodes are designed with a very narrow *n*-region—why?

Solution. By making a diode with narrow n-region, the stored charge problems can be reduced. If the *n*-region is much narrower than the hole diffusion length L_{Dh}, then a very little charge will be stored in the junction. Consequently, the time required to switch the diode 'off and on' will be much less.

Example 8.12 How can the switching of a *p-n* diode be made faster by choosing an appropriate material? Explain its reason also.

Solution. The switching process of a *p-n* diode can be made faster by adding gold (Au) atoms in Si. It is because the gold acts as recombination centre, thereby increases the recombination rate and decreases the carrier lifetime.

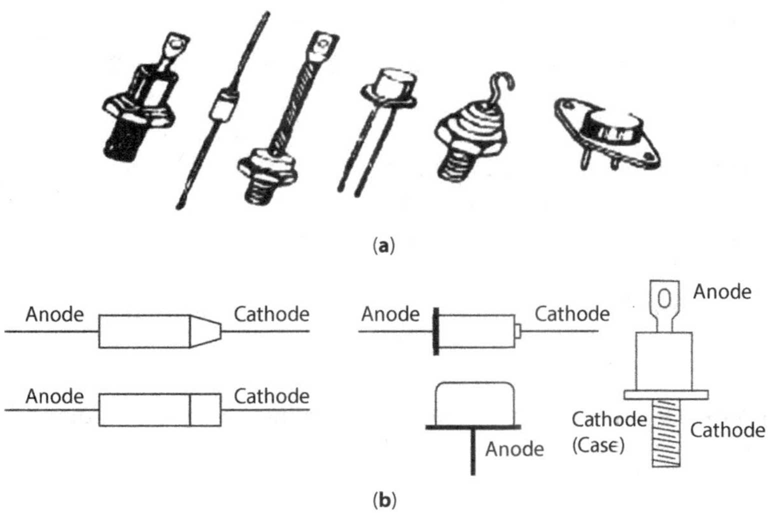

Fig. 8.13

Example 8.13. What is the main difference between FET and BJT?

Solution. The FET is a majority carrier device and is therefore called as *unipolar transistor*; whereas the BJT operates by injection and collection of minority carriers. Since the electrons and holes both play role in this device, it is called as a *bipolar transistor*.

Example 8.14. Suggest the materials used to form (a) the source and drain ohmic contacts for MESFET, and (b) for Schottky barrier gate.

Solution. The following materials are suggested for these purposes.

(a) For source and drain ohmic contacts : Au-Ge
(b) For barrier gate : Al

Example 8.15. Use of *n-p-n* transistors is more popular than p-n-p transistors—why?

Solution. The *n-p-n* transistors are used more than the *p-n-p* transistors because the electrons have higher mobility than holes (i.e. $\mu_e > \mu_h$).

Quick Revision Summary

Bias. The voltage applied to a rectifying junction or any other two electrodes of an electronic device.

Biased junction. Biasing is a method of connecting *p-n* junction with the battery. It can be of two kinds *viz.* **1.** Forward biasing, and **2.** Reverse (or backward) biasing. When the positive terminal of the battery is connected on positive side of junction and negative terminal on negative side of junction, it is called *forward-biasing*; while for the connection in just opposite way, the junction is known to be in reverse- biasing.

Breakdown of *p-n* junction is a situation in which the reverse current rises sharply to a very high value. This value can be limited only by the resistance connected in series with the junction.

Compensation. The neutralization of donor impurities by acceptor impurities and vice-versa.

Czochralski (crystal pulling) technique. The withdrawal of a crystal from the melt as fast as it grows.

Depletion (or transition) region. It is a region of small thickness at the junction in which the rectifying action takes place.

Epitaxy means an 'arranged layer'. It is a word from Greek literature. The lattice structure of a newly grown layer is an exact extension of the substrate crystal structure of semiconductor. Epitaxy process is one among the sequential processes used in the construction of a microelectronic circuit.

Equilibrium condition refers to a state in which there is neither any external excitation nor flow of any net current across the *p-n* junction.

Field-effect transistor (FET). A FET is a three-terminal unipolar solid-state device in which the current is controlled by an electric field.

Forward bias. Bias applied to a rectifying junction in the conducting direction.

Junction transistor. It is a two junction device having their separate regions *viz.* base, emitter and collector. These regions are separated by base-emitter and base-collector junctions.

Junction. The boundary between *p* and *n* semiconductors is known as junction.

Liquidus. It is the boundary between a liquid phase and a (solid + liquid) phase. Knowing this boundary in terms of temperature and composition of two alloying elements, is important for zone refining of semiconductors.

Masking. It is the process of repeating oxidation-photolithography-doping epitaxial growth metallization a number of times on wafer surface during making of an integrated circuit (IC).

***p-n* diode.** It is a semiconductor diode made by joining *n*-type and *p*-type extrinsic crystals. By doing so the favourable properties of each type is contributed in the buffer zone (or depletion region). It is a one-way device because it conducts in one direction only.

Reverse bias. Bias applied to a rectifying junction in the insulating direction.

Seed crystal. A piece of single crystal which is used as a nucleus to grow a larger crystal.

Solidus. It is the boundary between a solid phase and a (solid + liquid) phase. Knowing this boundary in terms of temperature and composition of two alloying elements, is important for zone refining of semiconductors.

Steady state conditions are those in which the responses of *p-n* junctions do not depend on time, and the operation of devices is time- independent.

Unipolar (solids) are those which conduct current by means of mobile charges of one sign only (+ or –).

Zener diode. It is a reverse-biased heavily doped two-terminal *p-n* junction that operates in breakdown region.

Zone refining. It is a method of purifying the material from the impurity contents present in it. Zone refining is based on the principle of phase separation. Highly pure single-crystal Si is produced by this method.

Review Questions

1. How is the element form of silicon produced? Explain the Bridgman method of semiconductor growth.
2. Explain the meaning of crystal growth. What are its different techniques? Explain any one of them.
3. Describe the working of 'Czochralski method' of crystal growth in semiconductor material with the help of suitable diagram.
4. What is meant by zone refining? Why is it necessary for semiconducting materials?
5. Describe the technique of zone refining with the help of a suitable sketch and apparatus. How is this process accomplished?
6. Discuss any one technique of 'zone refining' in semiconductor material.
7. What are different processing techniques of semiconducting materials? Briefly describe them.

8. Describe in brief, the stepwise methodology to manufacture the semiconductor wafer.
9. Write notes on the following.
   a. Epitaxy
   b. Masking
   c. Photolithography
10. What do you mean by microelectronics? How is a micro-electronic circuit constructed? Also explain briefly as to how the thin film microelectronic circuit is fabricated?
11. With the help of suitable diagrams, describe the stepwise method of fabricating a *p-n* junction.
12. With the help of suitable diagrams, describe stepwise the method of fabricating a transistor.
13. Write the most be-fitting semiconducting materials for following applications.
    a. CPU and microprocessors
    b. Integrated circuits
    c. Solar cells/batteries
    d. Automatic door opener
14. Enumerate various semiconducting devices and state their operating principle.
15. Describe the construction and working of a P-N junction diode in its inherent state stating as to how *(i)* diffusion of holes and electrons occur in it, *(ii)* barrier is set-up, and *(iii)* depletion layer is formed. Write its salient applications also.
16. What is P-N junction? What happens when P-N junction is biased in forward direction?
17. What are the advantages of junction transistor in electronic (semiconductor) devices? Bring out a broad classification of transistors.
18. Differentiate between the following kinds of transistors.
    a. BJT and FET
    b. MESFET and MISFET
    c. MODFET and MOSFET
    d. JFET and IGPET
19. Explain as to why an ordinary junction transistor is called 'bipolar'? How is it different from a `unipolar' transistor?
20. Sketch and explain the construction and working of a MESFET and IGFET with the help of suitable diagrams.
21. Explain the fundamentals of a BJT operation. What is meant by transistor biasing? How is forward-biasing different from reverse- biasing?
22. What is a Charge Coupled device? Write its salient uses.

Objective Questions

1. The correct sequence of processes involved in fabrication of a *p-n* junction diode is (out of many, only 4 to 5 processes are written)

   a. thermal oxidization, photolithography, diffusion, etching
   b. chemical vapour deposition, ion implantaton, rapid thermal processing, diffusion
   c. metallization, etching, photolithography, oxidization
   d. oxidization, rapid thermal processing, ion implantation, etching, metallization

2. Consider the following statements.

   A. A *p-n* diode conducts current in one direction only.
   B. A Zener diode may be used for meter protection.
   C. Barrier potential is 0.3 V in silicon *p-n* junction.
   D. Drift current flows due to electric field.

   Of these, the correct statements are

   a. A and B b. C and D
   c. B,C and D d. A, B and D

3. In an unbiased *p-n*-junction, the junction current at equilibrium is

   A. due to diffusion of minority carriers only
   B. due to diffusion of majority carriers only
   C. zero, because equal but opposite carriers are crossing the junction
   D. zero, because no charges are crossing the junction

4. In a *p-n* junction diode, if the junction current is zero, this means that

   A. the potential barriers has disappeared
   B. there are no carriers crossing the junction
   C. the number of majority carriers crossing the junction equals the number of minority carriers crossing the junction.
   D. the number of holes diffusing from the *p*-region equals the number of electrons diffusing from the *n*-region.

5. When a *p-n* junction is reverse-biased,
   a. if offers high resistance
   b. its depletion layer becomes narrow
   c. its barrier potential decreases
   d. it breaks down.
6. The number of minority carriers crossing the junction of a diode depends primarily on the
   a. concentration of doping impurities
   b. magnitude of the potential barrier
   c. magnitude of the forward-bias voltage
   d. rate of thermal generation of electron-hole pairs.
7. Avalanche breakdown in a semiconductor diode occurs when
   a. forward current exceeds a certain value
   b. reverse bias exceeds a certain value
   c. forward bias exceeds a certain value
   d. the potential barrier is reduced to zero
8. The forward bias applied to a *p-n* junction diode is increased from zero to higher values.
   Rapid increase in the current flow for a relatively small increase in voltage occurs
   A. immediately
   B. only after the forward bias exceeds the potential barrier
   C. when the flow of minority carriers is sufficient to cause an avalanche breakdown
   D. when the depletion area becomes larger than the space-charge area
9. A zener diode
   a. has a high forward-voltage rating
   b. has a sharp breakdown at low reverse voltage
   c. is useful as an amplifier
   d. has a negative resistance
10. A field-effect transistor (FET)
   a. uses a high-concentration emitter junction
   b. uses a forward-biased *p-n* junction
   c. has a very high input resistance
   d. depends on minority-carrier flow

9

Dielectric Materials: Properties and Behaviour

9.1 Introduction to Dielectric Materials

We have described in chapter 4 that the solids having an energy gap of 3 eV or more are termed as insulators. In such materials it is almost impossible to excite the electrons from the valence band to the conduction band by means of an applied electric field or normal thermal energy. *Insulators*, also known as *dielectrics,* are therefore very poor conductors of heat and electricity. However, they allow movement of some electrons at abnormally high temperatures causing a small flow of current.

i. Dielectrics are characterized by their
   - high specific resistance,
   - negative temperature coefficient of resistance, and
   - large insulation resistance.
ii. The insulation resistance is affected by
   - moisture,
   - temperature,
   - applied electric field, and
   - age of dielectrics.

iii. Dielectric materials may be
   - solid,
   - liquid, and
   - gas.

In solid form they may be polymeric such as nylon, pvc, bakelite, polyethylene and wood, or may belong to the ceramic family such as glass, silica, mica, porcelain and steatite etc. A detailed classification is given in the next article.

Uses. Dielectric materials are used in

- electrical insulation,
- as capacitors,
- in strain gauges and
- sonar devices etc.

Formvar is a suitable insulating material for low temperature applications. It is the trade name of polyvinyformal.

9.2 Classification of Dielectric (or Insulating) Materials

Depending upon various considerations, the electrical insulating materials may be classified into following types.

I. On the Basis of State of Material

1. Solid insulating materials
   i. Polymeric materials such as bakelite, pvc, rubber, etc.
   ii. Ceramic materials such as mica, porcelain, silicate, glass, etc.
2. Liquid insulating materials
   i. Organic liquids such as petroleum products : mineral oils, asphalt, transformer oil etc.; agro products : wood oil, wax etc; pyranols, silicone oil.
   ii. Inorganic liquids such as purified water, liquid He, SF_6, liquid N_2.
3. Gaseous insulating materials
   i. Electropositive gases such as N_2
   ii. Electronegative gases such as H_2, freon

iii. Mixture gases such as air
iv. Hydrocarbon gases such as CH_4, C_2H_6
v. Oxide gases such as CO_2, SO_2

II. On the Basis of their Availability

1. Natural materials
   i. in solid state such as varnish, resins, silica
   ii. in liquid state such as bitumen, vegetable oil, liquid O_2
   iii. in gaseous state such as air, N_2, CO_2, SF_6
2. Synthetic materials
   i. in solid state such as silicone resins, glass wool, askarel
   ii. in liquid state such as substitutents of hydrocarbons: sowol, aroclors; halogen-free liquids: silicone oil, transformer oil
   iii. in gaseous state such as freon, sodium vapour, CCl_4

III. On the Basis of Maximum Working Temperature

1. Class Y (workable upto 90°C) such as cotton, pvc, valcanized rubber, paper, silk
2. Class A (workable upto 105°C) such as paper : impregnated or immersed in oil, resins
3. Class E (workable upto 120°C) such as glass tape, enamels
4. Class B (workable upto 130°C) such as mica, asbestos, glass
5. Class F (workable upto 155°C) such as mica, asbestos, glass etc. with bonding substance capable of operation at 155°C
6. Class H (workable upto 180°C) such as varnished glass fibre, bonded materials with silicone resins etc.
7. Class C (workable above 180°C) such as porcelain, mica, quartz, glass, teflon, silicone liquids, SF_6.

IV. On the Basis of Form of Material and Special Purposes

1. Fibrous materials such as nylon, glass fibres
2. Impregnating materials such as wax paper-aluminium foil
3. Coating materials such as polyethylene, teflon
4. Filling materials such as glass wool, asbestos wool
5. Bonding materials such as elastomers, glass-reinforced-mica
6. Adhesive materials such as plastic films, resins.

V. On the Basis of Types of Services

1. for high voltage services such as N_2 (in gas-filled cables alongwith oil-treated paper insulation), vacuum

2. for low voltage services such as air, porcelain
3. for high temperature (thermal stability) services such as phlogophite mica, glass tapes
4. for low temperature services such as polyvinylformal (formvar)

Amongst these, the solid dielectrics are most common. Liquid dielectrics perform other functions also such as arc extinguishing, cooling etc. and are used for special purposes. The gaseous dielectrics are cheap and easily available.

9.3 Main Properties

The main properties of dielectric materials are the following.

1. Dielectric constant,
2. Dielectric strength,
3. Dielectric losses, and
4. Surface and volume resistivity

These are discussed in the subsequent sections.

9.4 Dielectric Constant

The dielectric constant ε_r (or relative permittivity) is a measure of polarization of dielectric materials. It is the ratio of absolute permittivity ε and the permittivity of free space ε_o, and is given by

$$\varepsilon_r = \frac{\varepsilon}{\varepsilon_0} \tag{9.1}$$

Here ε_r is dimensionless. The value of $\varepsilon_r = 1$ for air or vacuum, and $\varepsilon_r > 1$ for solids. Its value for diamond is 5.68, for silicon is 12, and for germanium is 16.

9.4.1 Factors Affecting Dielectric Constant

Dielectric constant is influenced by the following main factors.

1. Frequency f of applied field, and
2. Temperature T.

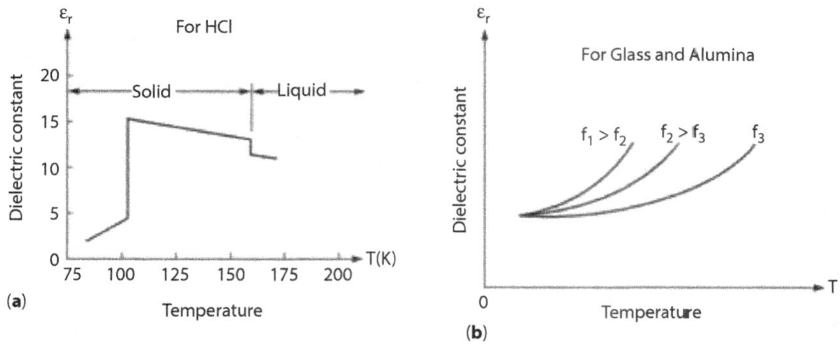

Figure 9.1 Dielectric constant as a function of (*a*) temperature for hydrogen chloride, and (*b*) temperature and frequency for glass and alumina

It may decrease or increase with an increase in frequency and temperature. Variation of ε_r as a function of temperature T for HCl is shown in Fig. 9.1*a*. In solid state, its value suddenly goes-up beyond 100 K, and then lowers-down gradually. Figure 9.1*b* shows variation in ε_r for glass and alumina.

Here the dielectric constant increases with increase in temperature and frequency. Dielectric constant of some materials and the effect of frequency on them is illustrated in Table 9.1. With increase in frequency, it remains constant for some materials and lowers-down for others.

9.5 Dielectric Strength

The voltage per unit thickness that can be sustained by an insulating material before its breakdown is called as *dielectric strength*. A good insulating material possesses high dielectric strength. Dielectric strength of some materials is shown in Table 9.1.

9.5.1 Types of Dielectric Breakdown

Dielectric breakdown occurs on account of the following factors. These are called as

1. Intrinsic breakdown.
2. Thermal breakdown.
3. Electrochemical breakdown.
4. Discharge breakdown, and
5. Defect breakdown.

Table 9.1 Dielectric properties of Some Materials

Dielectric material	Dielectric constant ε_r at frequency		Dielectric strength (MV/m)	Dielectric loss (tan δ)
	60 Hz	10^6 Hz		
• Polyethylene	2.3	2.3	4	0.0004
• Elastomer	4.0	2.7	25	0.003
• Fused silica	4.0	3.8	10	0.0001
• Nylon 6,6	4.0	3.5	15	0.02
• Waxed paper	4.2	—	—	—
• Bakelite	4.4	4.4	15	0.028
• Transformer oil	5.0	2.5	10	0.0001
• Porcelain	6.0	6.0	5	0.02
• Steatite($MgO.SiO_2$)	6.0	6.0	12	0.001
• Soda-lime glass	7.0	7.0	10	0.005
• P.V.C.	7.0	3.4	2	0.05
• Mica ($Al_2(OH)_2$. $AlSi_3O_{10}.K_2$)	8.0	5.0	100	0.0005

Intrinsic breakdown occurs when electrons in the valence band cross the forbidden gap under the influence of applied voltage, and enter into the conduction band. A large conduction current and extremely high local field is created in the process. This failure is also called as Zener breakdown. Intrinsic breakdown can happen at a lower voltage if impurities are present in the dielectric material.

Thermal breakdown occurs at high temperatures due to poor dissipation, and hence accumulation of non-dissipated heat produced by electrical energy. Failure occurs by melting, and can be avoided if heat dissipation is more than heat generation. This breakdown is more severe in d.c. field than in a.c. field.

Electrochemical breakdown takes place when the leakage current increases due to larger mobility of ions at raised temperatures. The dielectrics convert into their oxides, and insulation resistance decreases due to this chemical action. Breakdown of rubber is an example of this kind.

Discharge breakdown occurs due to the presence of gas bubbles in the solid and their bombardment on application of applied field. Gaseous atoms get ionized at lower potential than the solid atoms, and hence causing deterioration.

Defect breakdown occurs on the surface of dielectric materials due to detrimental effects of moisture on the cracks and pores. *Glazing* of the surface may eliminate this breakdown. Fireproof silica, high strength mica, and other chemically inert materials are used to provide good surface finish.

Example 9.1 The dielectric strength of a natural rubber is 40000 volts/mm at 60 Hz. Calculate the thickness of insulation on a wire carrying 33 kV to sustain the breakdown.

Solution. The required thickness of insulation will be

$$= \frac{33\text{kV}}{40000\,\text{volts / mm}} = \frac{33 \times 1000 \text{ volt}}{40000 \text{ volts / mm}}$$

$$= 0.82 \text{ mm}$$

9.6 Dielectric Loss

Duration t of the applied voltage V as compared to relaxation time affects the behaviour of dielectric materials. It may result into loss of electrical energy under certain conditions. There can be three possible conditions.

i. If $t < t_r$, polarization will not occur. Therefore, the question of electrical loss does not arise as the applied voltage V lags behind the charging current I by 90°, Fig. 9.2*a*.
ii. If $t > t_r$, then polarization completes instantaneously and no electrical losses occur. Here also, V lags behind I by 90°, Fig. 9.2*b*.
iii. (*iii*) If $t \approx t_r$, resonance takes place, and V lags behind I by $(90 - \delta)°$, Fig. 9.2*c*. Here the loss of electrical energy occurs whose measure is $\tan \delta$ where δ is the *loss angle*.

$\tan \delta$ is called as the *power factor* or *loss factor* of a dielectric. The component $I \cos\delta$ is known as ideal capacitor part and $I \sin\delta$ as electric loss part of the current.

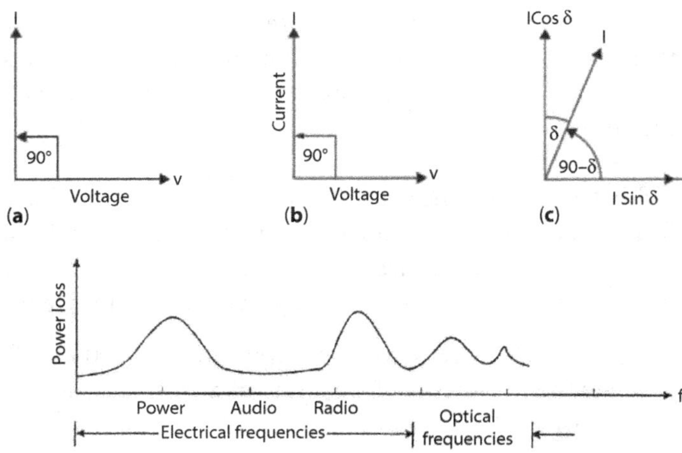

Figure 9.2 Electrical loss (*a*), (*b*) does not occur when period of voltage is much different than the relaxation time, and (*c*) occurs when the period of voltage matches with the relaxation time. δ is loss angle. (*d*) Effect of frequency on power loss

9.6.1 Factors Affecting Dielectric Loss

The dielectric loss is influenced by following main factors.

- Temperature
- Frequency
- Applied voltage
- Humidity

i. A low value of δ is desired for electrical capacitors to be used at high frequency. Values of tan δ for some dielectrics are listed in Table 9.1. The electrical loss is least in transformer oil.
ii. Dielectric losses increase with an increase in temperature, humidity, applied voltage and its frequency.

The effect of frequency on power loss is shown in Fig. 9.2d.

The dielectric loss P in a capacitor can be calculated from

$$P = VI \tan \delta \tag{9.2}$$

in which voltage V across a capacitor is determined from

$$V = IX_c \tag{9.3}$$

Here X_c is capacitive reactance which is given by

$$X_c = \frac{1}{\omega C} = \frac{1}{2\pi fC} \tag{9.4}$$

where f is the frequency of applied voltage and C the capacitance expressed in the unit of farad.

9.6.2 Calculation of Loss Factor

The loss factor for a dielectric is expressed as the ratio given as

$$\text{Loss factor} = \frac{\text{Energy lost per cycle}}{2\pi \times \text{Maximum energy stored}}$$

For an equivalent parallel circuit consisting of a resistance R and a capacitance C, the phaser diagram will be similar to one as shown in Fig. 9.2c. In it the $I\cos\delta$ component represents I_C, and $I\sin\delta$ component represents I_R. Thus

$$\tan\delta = \frac{I_R}{I_C} = \frac{V/R}{V/X_C} = \frac{X_C}{R} \tag{9.5a}$$

Using Eq. 9.4 for X_C, we can write

$$\tan\delta = \frac{1}{\omega RC} \tag{9.5b}$$

Energy loss per cycle. Since the power loss in parallel resistor is V^2/R, therefore the energy loss (ΔE) per cycle during a time period of $2\pi/\omega$ will be

$$\Delta E = \frac{V^2}{R} \times \frac{2\pi}{\omega} \tag{9.5c}$$

Here V is the rms (root mean square) value.

As the maximum energy stored by the capacitor is equal to $\frac{1}{2}CV_{max}^2$, where $V_{max} = \sqrt{2}V$; therefore the loss factor may be written as

$$\text{loss factor} = \frac{2\pi V^2/R\omega}{2\pi\left(\frac{1}{2}C.2V^2\right)} = \frac{1}{\omega RC} = \tan\delta \tag{9.5d}$$

Example 9.2 Capacitance of a capacitor is 0.025 μF and the power factor is 0.0005. Estimate the dielectric loss of the capacitor when it carries a current of 200 ampere at a frequency of 25 kHz.

Solution. The given values are

$$C = 0.025\ \mu\text{F} = 0.025 \times 10^{-6}\ \text{F},\ \tan\delta = 0.0005 = 5 \times 10^{-4}$$

$$f = 25\ \text{kHz} = 25 \times 10^{3}\ \text{Hz}$$

and

$$I = 200\ \text{A}$$

The voltage across a capacitor is obtained from Eq. 9.3 and 9.4, and is given by

$$V = I \times \frac{1}{2\pi fC} = \frac{200}{2\pi \times 25 \times 10^{3} \times 0.025 \times 10^{-6}}$$

$$= 50\ 930\ \text{volt}$$

Now, Eq. 9.2 yields dielectric loss

$$P = V.I.\ \tan\delta = 50930 \times 200 \times 5 \times 10^{-4}$$

$$= 5095\ \text{Watt}$$

9.7 Polarization

Definition. Insulating materials are subjected to different types of electric fields. These may be static field, alternating and direct current fields, impulsive and frequency switching fields. Their properties are different under different fields. Therefore, we first need to study insulating properties of dielectrics in static field to understand polarization. *The polarization is the sum of total dipole moments produced within the solid on application of electric field.*

It occurs because, on application of an electric field to a dielectric, the positive charges are displaced towards negative end of the field while the negative charges are displaced towards the positive end. Hence, local dipoles are produced in the dielectric due to this displacement. These dipoles keep their moments and are called dipole moments. Total dipole moments within the volume of a solid is called *polarization P.*

9.7.1 Expression for Polarization under Static Electric Field

The electric flux density D and the electric field strength E at a point in a dielectric material are related by

$$D \propto E, \text{ or } D = \varepsilon E = \varepsilon_o \varepsilon_r E \tag{9.6a}$$

By adding and subtracting $\varepsilon_o E$ in Eq. 9.6a, we may write

$$\begin{aligned} D &= \varepsilon_o \varepsilon_r E + \varepsilon_o E - \varepsilon_o E \\ &= \varepsilon_o E + \varepsilon_o E(\varepsilon_r - 1) \\ &= \varepsilon_o E + P \end{aligned} \tag{9.6b}$$

Value of D from Eq. 9.6a is substituted in Eq. 9.6b. It yields

$$\varepsilon_o \varepsilon_r E = \varepsilon_o E + P$$

or

$$P = \varepsilon_o \varepsilon_r E - \varepsilon_o E$$

Thus polarization of a dielectric material is expressed by

$$P = \varepsilon_o (\varepsilon_r - 1) E \tag{9.7}$$

Example 9.3 Determine the polarization produced in steatite ($MgO.SiO_2$) by an electric field of 600 V/m, if its dielectric constant is 6.1. Take $\varepsilon_0 = 8.85 \times 10^{-12}$.

Solution. Using Eq. 9.7, we obtain

$$\begin{aligned} P &= E\varepsilon_o(\varepsilon_r - 1) \\ &= 600 \times 8.85 \times 10^{-12} \times (6.1 - 1) \\ &= 2.7 \times 10^{-8} \text{ C/m}^2 \end{aligned}$$

9.7.2 Measurement of Polarization

Polarization can he measured using a parallel plate capacitor. When the surface area A of the parallel plates and the distance of separation d between them is known, the capacitance C under an applied voltage V under vacuum can be determined by

$$C = \frac{\varepsilon_0 A}{d} \tag{9.8}$$

Capacitance. Capacitance, thus obtained, is for a case when vacuum exists between the plates. As C is now known, ε_r of a dielectric material between the plates can be obtained from

$$C = \frac{\varepsilon_r \varepsilon_0 A}{d} \tag{9.9}$$

On putting the value of ε_r in Eq. 9.7, the value of P can be determined.

Example 9.4 A simple parallel plate condenser is to be made to store 10 μC at a potential of 10 kV. The separation between the plates is to be 5×10^{-4} m. Calculate the area that the plates must have if the dielectric material between the plates is of alumina of dielectric constant 10.

Solution. Given are: $Q = 10\ \mu\text{C} = 10 \times 10^{-6}$ coulomb, $V = 10\ \text{kV} = 10 \times 10^3$ V, $\varepsilon_r = 10$, $\varepsilon_o = 8.854 \times 10^{-12}$, and d = distance between plates = 5×10^{-4} m.

As $\quad Q = CV$

$\therefore \quad C = \frac{Q}{V}$

$$= \frac{10 \times 10^{-6}}{10 \times 10^{3}} = 10^{-9}\ \text{farad}$$

$\because \quad C = \frac{\varepsilon_r \varepsilon_0 A}{d}$

$\therefore \quad A = \frac{Cd}{\varepsilon_r \varepsilon_0}$

$$= \frac{10^{-9} \times 5 \times 10^{-4}}{10 \times 8.854 \times 10^{-12}}$$

$$= 5.647 \times 10^{-3}\ \text{m}^2$$

$$= 56.47\ \text{cm}^2$$

Example 9.5 A barium titanate crystal is inserted in a parallel plate condenser of plate dimensions 10 mm × 10 mm. A capacitance of 10^{-9} F is noticed when the plates are separated by 2 mm. Taking $\varepsilon_o = 8.854 \times 10^{-12}$ F/m, determine the dielectric constant of the crystal.

Solution. The values given are

$$A = 10 \times 10 = 100 \text{ mm}^2 = 100 \times 10^{-6} \text{ m}^2$$

and $$d = 2 \text{ mm} = 2 \times 10^{-3} \text{ m}$$

The dielectric constant of barium titanate crystal is determined from Eq. 9.9 as below.

$$10^{-9} = \frac{8.854 \times 10^{-12} \times \varepsilon_r \times 100 \times 10^{-6}}{2 \times 10^{-3}}$$
$$= 4.43 \times 10^{-13} \varepsilon_r$$

or $$\varepsilon_r = \frac{10^{-9}}{4.43 \times 10^{-13}} = 2257$$

Example 9.6 Two capacitors are made of: one using glass plate ($\varepsilon_r = 6.0$) of thickness 0.25 mm and the other using plastic film ($\varepsilon_r = 3.0$) of thickness 0.1 mm between the metal electrodes. Which one holds greater charge?

Solution. The capacitance of a system of two plates separated by a distance d is obtained from Eq. 9.9. For glass plate and plastic films, the capacitances C_g and C_p respectively are

$$C_g = \frac{\varepsilon_g \varepsilon_0 A_g}{d_g} \quad (i)$$

and $$C_p = \frac{\varepsilon_p \varepsilon_0 A_p}{d_p} \quad (ii)$$

The given data are $\varepsilon_g = 6.0, d_g = 0.25$ mm

and $\varepsilon_p = 3.0, d_p = 0.10$ mm

Taking $A_g = A_p = A$, and on dividing Eq. (*i*) by Eq. (*ii*), we get

$$\frac{C_g}{C_p} = \frac{\varepsilon_g}{\varepsilon_p} \times \frac{d_p}{d_g} = \frac{6 \times 0.10}{3 \times 0.25} = 0.8$$

Thus $C_g = 0.8C_p$, which means the plastic **film holds more charge.**

9.8 Mechanism of Polarization

Following atomic processes are responsible for polarization in a dielectric material.

1. Electronic polarization P_e,
2. Ionic polarization P_i,
3. Orientation (or molecular) polarization P_o, and
4. Space charge (or interfacial) polarization P_s.

9.8.1 Electronic Polarization

It occurs due to displacement of nucleus and electrons of an atom in opposite directions when an electric field E is applied on dielectric (Fig. 9.3*a*). The shifting of electron cloud towards the positive end of the field results in a dipole moment M_d given by

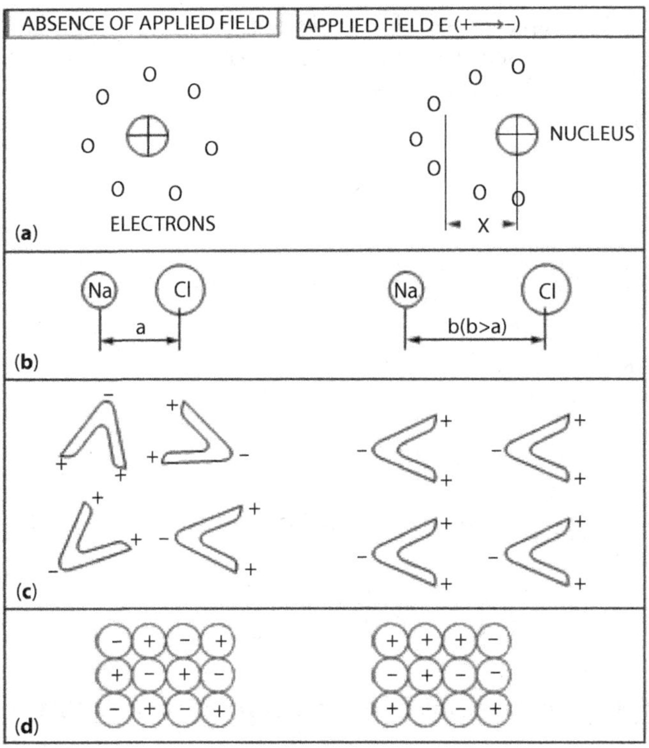

Figure 9.3 Polarization mechanism (*a*) electronic polarization due to shifting of electron cloud, (*b*) ionic polarization due to shifting of ions, (*c*) orientation polarization, and (*d*) space charge polarization

$$M_d = x \times e \tag{9.10}$$

where x *is* the distance between the nucleus and the centre of electron cloud, and e is the electron charge. The dipole moment is proportional to the applied field strength, therefore

$$M_d \propto E$$

or
$$M_d = p_e E \tag{9.11}$$

where constant of proportionality p_e is known as *electronic polarizability* of the atom. It is independent of temperature, and its values for inert gases are of the order of 10^{-40} farad m^2.

9.8.2 Ionic Polarization

It occurs in ionic solids due to displacement of cations and anions in opposite directions. Figure 9.3*b* shows this shifting in sodium chloride (NaCl) in which the distance between the cation Na and the anion Cl increases ($b > a$) after the application of electric field. This polarization is also independent of temperature.

9.8.3 Orientation (Molecular) Polarization

It is temperature dependent polarization process of hydrocarbon molecules. The molecules align themselves in the direction of applied field, Fig. 9.3*c*. However, with increasing temperature, the alignment becomes random. The CH_3Cl molecule, carrying dipole moment even without a field, gets aligned in this way. The positive and negative charges do not coincide in this molecule. Contrary to CH_3Cl, the methane (CH_4) molecule carries no permanent dipole as its positive and negative charges coincide.

9.8.4 Space Charge (or Interfacial) Polarization

It occurs due to ionic diffusion. In real dielectrics, the arrangement of atoms, ions or molecules is never perfect. Impurities remain peresent there and so the array of atoms is not perfect. Some free electrons may also remain present there.

When an electric field is applied upon such dielectrics; the free electrons, impurities and points move through the material to the surfaces. Consequently, the surface of material nearest to the positive end acquires a negative charge, and the surface nearest to the negative end acquires a positive charge. Thus under the influence of applied field, the redistribution of positive and negative charges takes place (Fig. 9.3*d*).

9.9 Comparison of Different Polarization Processes

Total polarization of a dielectric material may be obtained by summing-up the polarization from all the above sources. Hence

$$P_{Total} = P_e + P_i + P_o + P_s \tag{9.12}$$

The behaviour of above processes are summarized below.

Factor	Electronic polarization	Ionic polarization	Orientation polarization	Space charge polarization
1. Temperature dependency	Independent	Independent	Dependent	Dependent
2. Main effect	Shifting of electron clouds with respect to nucleus	Shifting of cations and anions	Alignment of random molecules	Ion diffusion
3. Examples	Inert gases	Ionic crystals	Alcohol methane, CH_3Cl	—
4. Frequency range	All the frequencies upto 10^{15} and above	Upto about 10^{14} Hz	Upto about 10^{12} Hz	Upto about 10^5 Hz
5. Relaxation time	Extremely rapid	Slow	Slower	Slowest
6. Power loss	Low	High	Higher	Highest

9.9.1 Polarizability

When a dielectric is subjected to an applied electric field E, a dipole moment M_d is induced due to it. The induced dipole moment is proportional to the field strength i.e.

$$M_d \propto E$$

$$\therefore \qquad M_d = pE \tag{9.13a}$$

where the proportionality factor p is called polarizability of the atom.

Similar to different kinds of polarization processes; the polarizability is also termed as one of the following types.

- Electronic polarizability p_e
- Ionic polarizability p_i
- Orientation polarizability p_o
- Space charge polarizability p_s

The electronic polarizability is caused due to shift of electron cloud relative to the nucleus, ionic polarizability is caused due to shifting of ions, orientation polarizability is caused due to alignment of molecules in the direction of applied field, and space charge polarizability occurs due to ionic diffusion.

The total polarizability p_{total} is the sum of all polarizabilities. Thus

$$p_{\text{total}} = p_e + p_i + p_o + p_s \tag{9.13b}$$

9.10 Factors Affecting Polarization

The polarization process is affected by following main factors.

1. Time, and
2. Frequency of the applied field

9.10.1 Time Effects and Relaxation Time

Suppose a dielectric material, subjected to a static applied field for a prolonged duration t, develops maximum polarization P_{max}. Then after relaxation time t_r, the polarization decreases to P as given by

$$P = P_{\text{max}}[1 - e^{(-t/t_r)}] \tag{9.14}$$

It is due to the fact that the atoms and molecules get disturbed when an electric field is applied on to dielectric. It takes some time for them to return the equilibrium position from the disturbed position. Time taken in doing

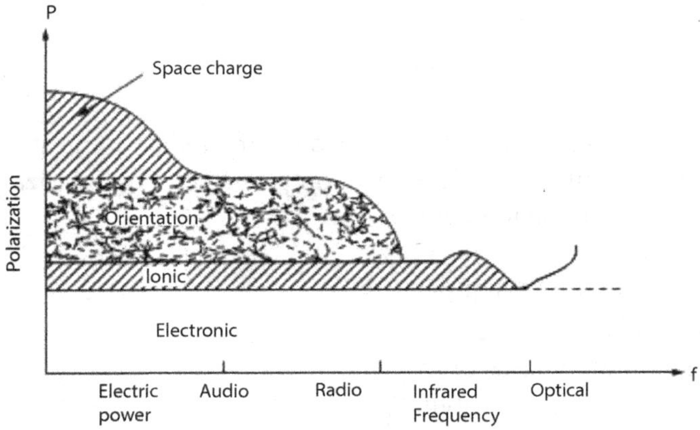

Figure 9.4 Frequency dependency of polarization processes.

so is called *relaxation time*. The relaxation time refers to a polarization process which attains 63% of the maximum possible value of polarization.

Relaxation time in different polarization processes. The relaxation time is negligible for electronic polarization as it is an instantaneous process. The relaxation is slower in ionic polarization as compared to electronic polarization, as the ions involved in this process are heavier than the electron cloud. Orientation polarization is still slower than ionic polarization, and the relaxation completes in *micro* (10^{-6}) to *nano* (10^{-9}) *second*. The slowest process is space charge polarization. Here the ions diffuse through many atomic distance.

9.10.2 Frequency Effects

Polarization processes are largely influenced by the frequency of applied voltage. It is shown in Fig. 9.4 that electronic polarization occurs at all frequencies ranging from that of electric power to optical frequencies. Ionic and orientation processes occur between the infrared and audio frequency ranges. Space charge polarization occurs at ordinary frequencies and is considerable below 100 Hz.

9.11 Spontaneous Polarization

Definition. *Spontaneous polarization is the ability of ferroelectrics and many dielectric materials to become easily and strongly polarized.* The term ferroelectric is similar to ferromagnetic and has analogous characteristics too. Just as ferromagnetic materials which are composed of domains (or regions) with aligned magnetic moments, the ferroelectric materials also comprise of regions with aligned electric field.

Characteristics. Whereas the polarization is a linear function of applied field for dielectric materials, the ferroelectric materials and a number of other dielectric materials do not follow linear law.

- These materials exhibit hysteresis effect similar to as observed in ferromagnetic materials.
- They also exhibit spontaneous polarization which is analogous to spontaneous magnetization in ferromagnetic materials.

To understand the phenomenon of spontaneous polarization, an understanding of polarization curve and loop is essential. Hence, we shall discuss it now.

9.11.1 Polarization Curve

The relation between changing electric field E and its influence on polarization P can be represented by a curve. This is called polarization curve and is shown by *OACD* path in Fig. 9.5. This curve has three regions viz. (*i*) *OA* region of rising curve, (*ii*) *AC* region of almost constant slope, and (*iii*) *CD* region of decreasing slope. The curve becomes almost horizontal at *D* when the dielectric material reaches its saturation polarization P_s. If E is enhanced further, there will be no useful increase in the value of P_s.

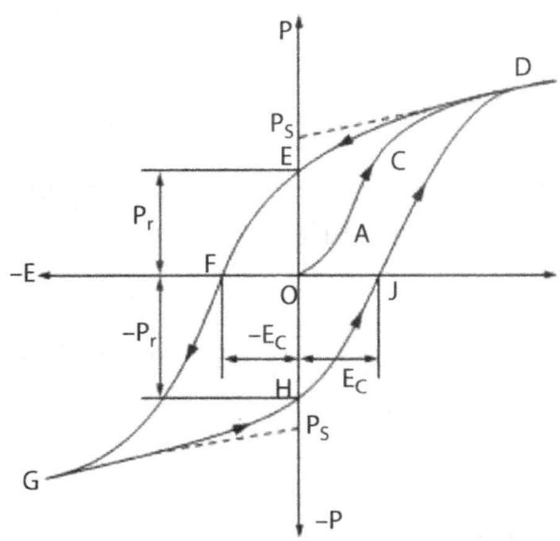

Figure 9.5 Polarization curve and polarization hysteresis loop showing saturation polarization, remanent polarization, and depolarization to explain the spontaneous polarization.

9.11.2 Polarization Hysteresis Loop

The changes in polarization lag behind the variations in electric field applied on it. From Fig. 9.5, it can be seen that when electric field E is decreased and brought back to zero, the curve assumes DE path. When direction of the field is reversed, the curve follows FG path. On further increasing the electric field, the curve follows FG path and reaches a saturation polarization $-P$ in opposite direction. Thus one cycle is complete.

Now when the electric field is applied again on the dielectric, the P-E curve follows HJD path. The curve does not take on to the original $OACD$ path. Hence, the curve $DEFGHJD$ is called *polarization hysteresis loop*. It should be clear that $OP_S = - OP_S$, $OF = OJ$ and $OE = OH = P_r \cdot P_r$ is termed as remanent polarization.

9.11.3 Salient Features of Spontaneous Polarization

Some salient features of the above curve can be highlighted as follows.

- At $E = 0$, the magnitude of polarization is equal to $OE = P_r$. This indicates that the material is able to retain the polarization even in the absence of any applied field. Hence, the material is known to be spontaneously polarized.
- Higher the value of P_r, better is the quality of dielectric material. It means that the dielectric possesses ability of spontaneous polarization.
- F is another important point on this loop. Here OF indicates depolarization field (or coercive field). It means that the dielectric requires an electric field of intensity $E = -E$ to bring down its polarization to zero.
- A strong dielectric will need a higher value of $-E$.

The spontaneous polarization of materials usually vanishes above a certain temperature, called ferroelectric curie temperature.

9.12 Behaviour of Polarization Under Impulse and Frequency Switching

Total polarization P, total polarizability p, and relative permittivity ε_r of a dielectric in an alternating field depend on the ease with which the dipoles can reverse their alignment with each reversal of the field. Such reversal of field occurs on impulse and/or frequency switching. As already discussed earlier in art. 9.9 that some polarization mechanisms permit rapid reversal of dipole alignment while the others do not. In such processes, time required to reach the equilibrium orientation is called *relaxation time* and its reciprocal as *relaxation frequency.*

When the frequency of applied field exceeds the relaxation frequency of a particular polarization process, the dipoles cannot reorient fast enough and hence operation of the process ceases. Since the relaxation frequencies of all four polarization processes differ, their contributions can be separated. The result is already shown in Fig. 9.4 for different frequencies. From the view point of electrical engineering, the frequency dependence of 'orientation polarization' is of greater importance because the dielectric losses in this case vary from zero to many thousand megacycles frequency. Hence, we shall study as to how the polarization is influenced when electric field is suddenly switched-off or switched-on.

9.12.1 Effect on Polarization When Electric Field is Switched-off

We consider a substance subjected to a d.c. field E for a long time and let the orientation polarization in equilibrium with this field is P_o. When the field is *suddenly switched-off* at the instant $t = 0$, the polarization instantaneously does not become zero. It is because some time is elapsed in rotation of dipoles. Now let the polarization as a function of time decays to zero (Fig. 9.6a), then it can be expressed as

$$P_0(t) = P_0 e^{-t/t_r} \tag{9.15}$$

where t_r is relaxation time. Consequently, the rate of change of polarization will be given by

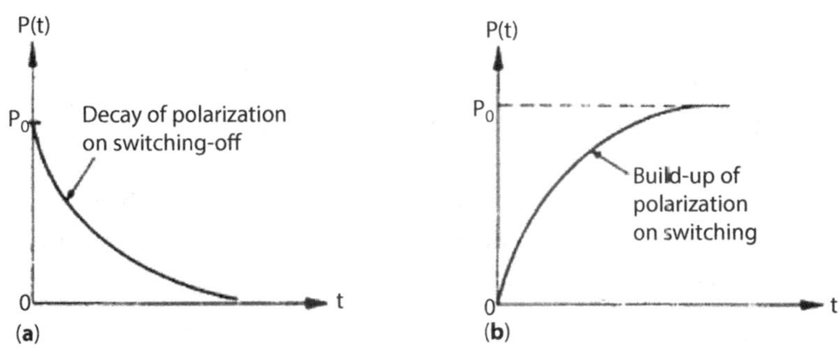

Figure 9.6 Orientation polarization (a) decays on switching- off the field at $t = 0$, and *(b)* builds-up on switching-on the field at $t = 0$.

$$\frac{d}{dt}P_0(t) = \frac{d}{dt}P_0 e^{-t/t_r}$$
$$= -\frac{P_0}{t_r}e^{-t/t_r} = -\frac{P_0(t)}{t_r} \quad (9.16)$$

It shows that, smaller is the instantaneous value $P_0\,(t)$, smaller will be the rate of decay. Since at $t \to \infty$, the ultimate value of $P_0\,(t) = P_0\,(\infty) = 0$, therefore Eq. 9.16 may be written as

$$\frac{d}{dt}P_0(t) = \frac{1}{t_r}[P_0(\infty) - P_0(t)] \quad (9.17)$$

9.12.2 Effect on Polarization When Electric Field is Switched-on

Now, we consider a case in which the external field has been absent for a long time and then an electric field E is switched on at $t = 0$. This results in growth of orientation polarization with time (Fig. 9.6b) which may be expressed as

$$P_0(t) = P_0(1 - e^{-t/t_r}) \quad (9.18a)$$

The rate of increase now becomes

$$\frac{d}{dt}P_0(t) = \frac{P_0}{t_r}e^{-t/t_r} = \frac{1}{t_r}[P_0 - P_0(t)] \quad (9.18b)$$

Since $P_0 = P_0\,(\infty)$ in this case, hence Eq. 9.18 has the same form as that of Eq. 9.16.

9.13 Decay and Build-Up of Polarization Under Alternating Current (A.C.) Field

Let an alternating field E_0 is applied on the dielectric substance. Recalling Eq. 9.7 in which P is polarization under 'static' field E, this is written in a modified way as

$$P_0 = \varepsilon_0(\varepsilon_{r0} - 1)E \quad (9.19)$$

in which ε_{ro} is the part of dielectric constant that measures only orientation polarization. Now at the instant t, the equation for $P_o(t)$may be written as

$$\frac{d}{dt}P_0(t)=\frac{1}{t_r}[\varepsilon_0(\varepsilon_{r0}-1)E_0\cos\omega t-P_0(t)] \tag{9.20}$$

Solution of equation. Solution of this equation is obtained by introducing a relation

$$P_0(t)=\varepsilon_0\,\mathrm{Re}[(\varepsilon_{r0}^{*}-1)E_0e^{j\omega t}] \tag{9.21}$$

where $\varepsilon_{ro}{}^{*}$ is a complex dielectric constant. Here Re refers to real part of the complex quantity, j is complex number $\sqrt{-1}$, and ω is frequency of the applied field.

On substituting Eq. 9.21 in Eq. 9.20, we get

$$\frac{d}{dt}[\varepsilon_0\,\mathrm{Re}[(\varepsilon_{r0}^{*}-1)E_0e^{j\omega t}]=\frac{1}{t_r}[\varepsilon_0(\varepsilon_{r0}-1)E_0\cos\omega t]$$
$$-\varepsilon_0\,\mathrm{Re}[(\varepsilon_{r0}^{*}-1)E_0\cos\omega t] \tag{9.22}$$

After performing the required mathematical operations and on simplification, we obtain the complex quantity as

$$\varepsilon_{r0}^{*}-1=1-\frac{\varepsilon_{r0}-1}{1+j\omega t_r} \tag{9.23}$$

Writing it in real and imaginary parts*, we get

$$\varepsilon_{r0}^{*}-1=(\varepsilon_{r0}-1)\left[\frac{1}{1+\omega^2t_r^2}-j\frac{\omega t_r}{1+\omega^2t_r^2}\right] \tag{9.24}$$

Expression of polarization. Finally from Eqs. 9.21 and 9.24, the expression of polarization is found as

$$P_0(t)=\frac{\varepsilon_0(\varepsilon_{r0}-1)}{1+\omega^2t_r^2}E_0\cos\omega t+\frac{\varepsilon_0(\varepsilon_{r0}-1)\omega t}{1+\omega^2t_r^2}E_0\sin\omega t \tag{9.25}$$

In this relation, the first term is in-phase with the applied field whereas the second term lags by 90^0.

---

* Readers should use this concept to find real and imaginary parts, separately

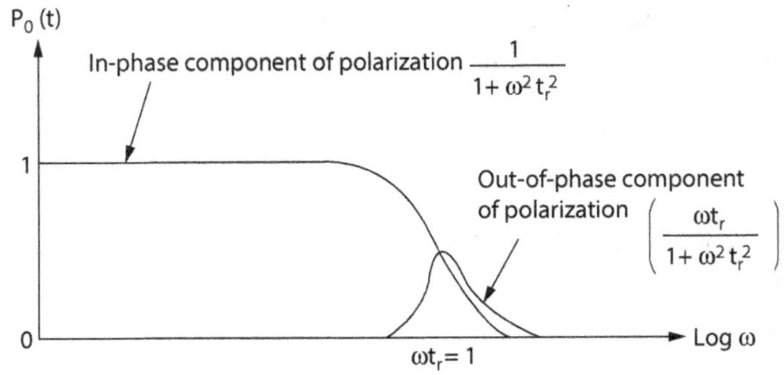

Figure 9.7 Frequency dependence of polarization under impulse and frequency switching shows in-phase and out-of-phase components as a function of frequency of applied field.

9.13.1 Conclusion

Frequency dependence of in-phase and out- of - phase components polarization under impulse and frequency switching are shown in Fig. 9.7. Following conclusions can be drawn from it.

i. In-phase polarization is almost constant upto a large value of ωt_r before unity. It begins to disappear in the range of ωt_r becoming unity. $P_0(t)$ completely vanishes for $\omega t_r >> 1$.
ii. Out-of-phase polarization is of bell shape, having its peak value at $\omega t_r = 1$. It is similar to imaginary part of electronic and ionic polarizability, p_e and p_i respectively. It is a measure of absorption of energy.

9.14 Complex Dielectric Constant

The relationship derived earlier between the polarization *P* and applied electric field *E* given by $P = \varepsilon_0 (\varepsilon_r - 1) E$ is true for static field, but does not hold good for alternating field. Therefore for non-dipolar solids, the polarization equation under alternating field is expressed (see Eq. 9.21) as

$$P_0(t) = \varepsilon_0 \operatorname{Re}\left[\left(\varepsilon_{r0}^* - 1\right) E_0 e^{j\omega t}\right]$$

Here $\varepsilon_{ro}{}^*$ is known as 'complex dielectric constant'.

This is a complex quantity, which after performing the required mathematical operations and simplifications, yields a relation given as

$$\varepsilon_{r0}^* = 1 + \frac{\varepsilon_{r0} - 1}{1 + j\omega t_r} \tag{9.26}$$

in which ε_{ro} is that part of dielectric constant that measures orientation polarization only.

The relation given below for complex dielectric constant is similar to the Clausius - Mosotti expression derived for static case (see art. 9.16).

$$\frac{\varepsilon_r^* - 1}{\varepsilon_r^* + 2} = \frac{1}{3\varepsilon_0} N(p_e^* + p_i^*) \tag{9.27}$$

where N is the number of atoms per m^3 from which the solid may be built-up by means of a 3-dimensional stacking, p_e^* and p_i^* are respectively the complex electronic polarizability and complex ionic polarizability.

Values of real part of complex dielectric constant. The real and imaginary parts of ε_r^* may be defined as

$$\varepsilon_r^* = \varepsilon_r' - j\varepsilon_r'' \tag{9.28}$$

Here ε_r' is a function of frequency. For some dielectrics, it remains constant at all frequencies but for others, it lowers down with increasing frequency. Its value at 100 Hz for porcelain is 8.99, for Bakelite is 4.87, and for polyethylene is 2.25.

9.15 Determining the Internal Field Due to Polarization Inside the Dielectric

The internal (or local) field E_i in a dielectric is defined as the field acting at the location of a given atom. In case of gaseous dielectrics, it is assumed to be equal to the applied field E (i.e $E_i \approx E$ is assumed) but it does not hold good for solids and liquids (i.e. $E_L \neq E$). It is because of the assumption that E_i is approximately equal to E, is quite reasonable for low density molecules. Since the density of gas molecules is low, but for atoms and molecules of solids and liquids it is high; therefore $E_i \approx E$ is true for gases but not for solid and liquid dielectrics i.e. $E_i \neq E$.

9.15.1 Formulation of the Problem for Solid Dielectric

To illustrate, we consider a solid dielectric in which the closely spaced equidistant atoms *A*, *B*, *C*, *D* and *G* are as shown in Fig. 9.8a. Let the atoms

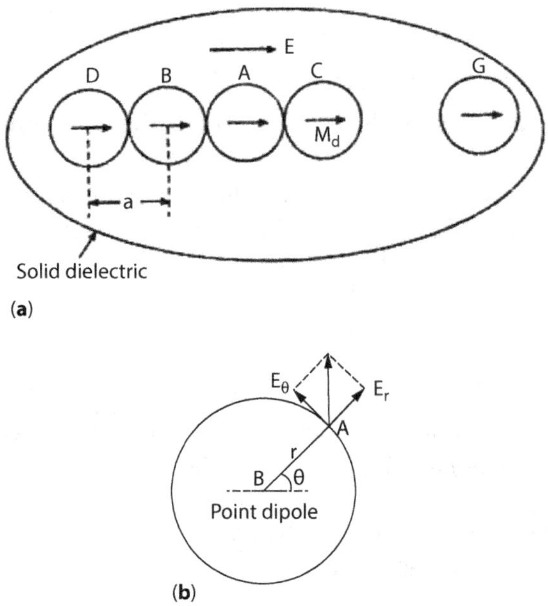

Figure 9.8 (a) A solid dielectric showing atoms in a row, and (b) components of the field around a point dipole.

of polarizability p are subjected to external field E applied in a direction parallel to the row of the atoms. Let the field seen by atom A be E_i which is parallel to E. The field by other atoms will also be the same. Thus the dipole moment induced in each of the atoms of the row will be

$$M_d = pE_i \tag{9.29}$$

Evaluation of E_i. To evaluate E_i, one finds that it is equal to the sum of applied field E and the field produced on atom A by the dipole on all other atoms. The field produced at the centre of A by the point dipole on atom B is obtained in accordance with the 'field theory'. So, the potential $V(r, \theta)$ around a point dipole (Fig. 9.8b) in vacuum is expressed as

$$V(r,\theta) = \frac{1}{4\pi\varepsilon_0}\frac{M_d\cos\theta}{r^2} \tag{9.30}$$

Here V is a function of r and θ, r is distance from the dipole, and θ is angle between M_d and r. Thus the field around a dipole may be expressed by its radial and tangential components given as

$$E_r = -\frac{\partial V}{\partial r} = \frac{1}{4\pi\varepsilon_0}\frac{2M_d \cos\theta}{r^3} \tag{9.31a}$$

$$E_\theta = -\frac{1}{r}\frac{\partial V}{\partial r} = \frac{1}{4\pi\varepsilon_0}\frac{M_d \sin\theta}{r^3} \tag{9.31b}$$

On transforming the situation of Fig. 9.8a into that of Fig. 9.8a, we put r = a and θ = 0 in Eqns 9.31a-b. This yields a contribution of $M_d / 2\pi\varepsilon_0 a^3$ in direction of E.

9.15.2 Contribution of Dipoles on Internal Field

Similarly, it can be shown that the field produced at A by the dipole on atom C will be the same as produced by atom B. Hence on considering the contributions of dipoles of other atoms also and on superimposing them, the internal field at A may be expressed as

$$E_i = E + \sum_{n=1,2,3} \frac{1}{n^3}\frac{M_d}{\pi\varepsilon_0 a^3} \tag{9.32a}$$

On substituting M_d from Eqn. 9.29 into Eqn. 9.32a and noting that the sum of Eqn. 9.32a is approximately equal to 1.2, we can write

$$E_i = \frac{E}{1 - \frac{1.2p}{\pi\varepsilon_0 a^3}} = \frac{E}{1-\lambda} \tag{9.32b}$$

It implies that the actual field seen by an individual atom in the row of atoms is greater than the applied field *E*. Physically it means that the dipoles cooperate with each other i.e. a large dipole moment on a given atom assists in inducing a dipole moment in its neighbouring atom. With increase in polarizability of atoms and decrease in distance between them, the mutual cooperation gets stronger.

9.15.3 Determining E_i for 3-dimensional Case and Lorentz Expression

In three-dimensional case, the expression for internal field is expressed as

$$E_i = E + \frac{\gamma P}{\varepsilon_0} \tag{9.33}$$

where P is dipole moment per unit volume and γ is the proportionately constant for internal field. Although the value of γ is of the order of 1, and for in-row chain of atoms shown in Fig. 9.8a it is 1.2/π, but in the particular case when the atoms in a solid are surrounded cubically by other atoms, the Lorentz internal field comes into picture. In that case the value of γ = 1/3, therefore

$$(E_i)_{\text{Lorentz}} = E + \frac{P}{3\varepsilon_0} \tag{9.34a}$$

This equation is known as **Lorentz expression.** Thus the net internal field is

$$(E_i)_{\text{Lorentz}} - E = \frac{P}{3\varepsilon_0} \tag{9.34b}$$

Example 9.7 Show that the polarizability of an atom of radius 'a' is $4\pi\varepsilon_0 a^3$.

Solution: We know that the internal field on an atom due to polarization is expressed as

$$E_i = \frac{P}{3\varepsilon_0} \text{ (This is Eqn. 9.34b)} \qquad (i)$$

and the dipole moment of polarized atom of radius 'a' is

$$M_d = \frac{4}{3}\pi a^3 \times P \qquad (ii)$$

- We also know that the dipole moment M_d, polarizability p, and internal electric field E_i are related as

$$M_d = pE_i \qquad (iii)$$

Therefore on putting Eqns. *(i)* and *(ii)* in Eqn. *(iii)*, we have

$$\frac{4\pi a^3}{3} \times P = p \times \frac{P}{3\varepsilon_0}$$

or

$$p = \frac{4\pi a^3}{3} \times P \times \frac{3\varepsilon_0}{P}$$

$$= 4\pi\varepsilon_0 a^3$$

9.16 Clausius-Mossotti Relation

Elemental dielectric materials such as diamond, sulphur, phosphorus etc. are made up of only one kind of atoms. Also there are no permanent ions or dipoles in them. Therefore, these materials exhibit only electronic polarization. As the total polarization P of a dielectric is given by

$$P_{\text{total}} = P_e + P_i + P_o + P_s$$

in which $$P_i = P_o = P_s = 0$$

therefore, $$P_{\text{total}} = P_e + 0 + 0 + 0, \text{or say } P = P_e$$

9.16.1 Relation between Electronic Polarizability and Polarization

If p_e is electronic polarizability per atom, N is number of atom/m^3 and E_i is internal field, then the polarization is given by

$$P = N\,p_e\,E_i \tag{9.35a}$$

Using Eqn. 9.33 (as given by $E_i = E + (\gamma/\varepsilon_0 P)$ in Eqn. 9.35a, we write

$$P = Np_e\left[E + \left(\frac{\gamma}{\varepsilon_0}\right)P\right] \tag{9.35b}$$

On simplifying, it can be written as

$$P = \frac{Np_eE}{1 - \left(\dfrac{\gamma Np_e}{\varepsilon_0}\right)} \tag{9.36}$$

The expression for dielectric constant ε_r may be established in terms of atomic quantities by using Eqn. 9.7 given as $P = \varepsilon_0(\varepsilon_r - 1)E$. Hence Eqn. 9.36 is rewritten as

$$\varepsilon_0(\varepsilon_r - 1)E = \frac{Np_eE}{1 - \left(\dfrac{\gamma Np_e}{\varepsilon_0}\right)} \tag{9.37}$$

Consideration of $\gamma = 1/3$. If the value of γ is taken equal to 1/3 as explained in art 9.15.3, Eqn. 9.37 leads to Clausius-Mossotti relation. Thus

$$\varepsilon_0(\varepsilon_r - 1) = \frac{Np_e}{1 - \left(\frac{1}{3}\frac{Np_e}{\varepsilon_0}\right)}$$

or,
$$\varepsilon_0(\varepsilon_r - 1) = \frac{Np_e}{\left(\frac{3\varepsilon_0 - Np_e}{3\varepsilon_0}\right)}$$

or,
$$\varepsilon_0(\varepsilon_r - 1) = \frac{Np_e 3\varepsilon_0}{3\varepsilon_0 - Np_e}$$

or,
$$(\varepsilon_r - 1)3\varepsilon_0 - (\varepsilon_r - 1)Np_e = 3Np_e$$

or,
$$\begin{aligned}(\varepsilon_r - 1)3\varepsilon_0 &= Np_e[3 + (\varepsilon_r - 1)] \\ &= Np_e(\varepsilon_r + 2)\end{aligned}$$

$\therefore$
$$\frac{(\varepsilon_r - 1)}{(\varepsilon_r + 2)} = \frac{Np_e}{3\varepsilon_0} \tag{9.38}$$

Example 9.8 A bakelite insulation is subjected to an alternating current field of 1 MHz frequency. The real part of its dielectric constant is 4.36, while the loss tangent is 0.028. If there are 4×10^{28} atoms/m^3 in it, determine the complex polarizability of bakelite. Assume that the Lorentz field exists in insulator. Take $\varepsilon_0 = 8.854 \times 10^{-12}$ F/m.

Solution: As the Lorentz's field exists in insulator, the relationship between dielectric constant ε_r and polarizability p will be similar to as given by Clausius-Mossotti relation. Hence for A.C. field, Eqn. 9.38 may be expressed as

$$\frac{Np_e^*}{3\varepsilon_0} = \frac{(\varepsilon_r^* - 1)}{(\varepsilon_r^* + 2)} \tag{i}$$

where
$$\varepsilon_r^* = \varepsilon_r' - j\varepsilon_r'' \tag{ii}$$

and the loss tangent is given by

$$\tan\delta = \frac{\varepsilon_r''}{\varepsilon_r'} \qquad (iii)$$

in which δ is loss angle. From Eqn. (iii), $\varepsilon_r'' = \varepsilon_r' \tan\delta$. On substituting it in Eqn. (ii), we get

$$\varepsilon_r^{*} = \varepsilon_r' - j\varepsilon_r' \tan\delta \qquad (iv)$$

i.e. ε_r' is real and $\varepsilon_r' \tan\delta$ is imaginary.

As in the case of insulator under alternating conditions, the real part of complex quantity is a measure of 'dielectric constant' and the imaginary part depicts 'dielectric losses' of the material. Therefore [Re] quantity ε_r' and [Im] quantity $\varepsilon_r' \tan\delta$ are

$$\varepsilon_r' = 4.36 \text{ and } \varepsilon_r' \tan\delta = 0.028$$

i.e. $$\varepsilon_r^{*} = 4.36 - j0.028$$

Now on substituting the appropriate values in Eqn. *(ii)*, we find

$$\frac{4\times10^{28}\times p_e^{*}}{3\times(8.854\times10^{-12})} = \frac{(4.36 - j4.36\times0.028)-1}{(4.36 - j4.36\times0.028)+2}$$

$$= \frac{3.36 - j0.122}{6.36 - j0.122}$$

$$= \frac{(3.36 - j0.122)}{(6.36 - j0.122)}\times\frac{(6.36 + j0.122)}{(6.36 + j0.122)}$$

$$= \frac{3.36(6.36 + j0.122) - j0.122(6.36 + j0.122)}{40.449 + 0.0140}$$

$$= \frac{21.369 + j0.41 - j0.777 + 0.0149}{40.464}$$

$$= \frac{21.384 - j0.366}{40.464}$$

$$= 0.528 - j40.464$$

$$\therefore \qquad p_e^{*} = \frac{(0.528 - j40.464)\times3\times8.854\times10^{-12}}{4\times10^{28}}$$

$$= (3.509\times10^{-40}) - j(0.06\times10^{-40})$$

or $$= (3.509 - j0.06)\times10^{-40}\ \text{Fm}^2$$

9.17 Solved Examples

Example 9.9 A capacitor of capacitance 0.2 μF has a loss factor of 0.004. What will be the power loss when it is connected across a 240 V, 50 Hz supply?

Solution. Given data are

$$V = 240 \text{ volt}, f = 50 \text{ Hz}, C = 0.2\ \mu\text{F}, \tan\delta = 0.004$$

As the loss factor is tan δ, hence representing the losses by a parallel resistor, the power loss may be found as follows.

$$\begin{aligned}\text{Power loss} &= V^2\omega C \tan\delta \\ &= 240^2 \times (2\pi \times 50) \times (0.2 \times 10^{-6}) \times 0.004 \\ &= 14.4 \times 10^{-3} \text{ W}\end{aligned}$$

Example 9.10 The electric polarizability of LiF is 5.11×10^{-41} Fm^2. Determine its dielectric constant at 273 K and 1 atmospheric pressure. Take Avogadro's number as 6.023×10^{23}, gas constant as 8.314 J/gm K, and $\varepsilon_o = 8.85 \times 10^{-12}$.

Solution. We know that the polarization $P = \varepsilon_o (\varepsilon_r - 1) E$ and also $P = N p_e E_i$ by Eqs. 9.7 and 9.35a respectively.

Here p_e = electronic polarizability = 5.11×10^{-41} Fm^2.

$$\therefore \quad N p_e E_i = \varepsilon_o (\varepsilon_r - 1) E$$

or

$$N p_e = \varepsilon_o (\varepsilon_r - 1)$$

i.e. $$(\varepsilon_r - 1) = \frac{Np_e}{\varepsilon_0}$$

As $$pV = nRT \quad \text{(Gas equation where } p = \text{pressure)}$$

and $$\text{No. of mole } n = \frac{\text{No. of atoms}}{\text{Avogadro's no. of atoms}} = \frac{n'}{N_A}$$

$$\therefore \quad pV = \frac{n'}{N_A} RT$$

or

$$\frac{n'}{V} = \frac{pN_A}{RT}$$

where n'/V = N [No. of atoms per unit volume]

$$\therefore \qquad N = \frac{pN_A}{RT} \qquad (ii)$$

Using $p = 1$ bar $= 1.013 \times 10^5$ pascal, $N_A = 6.023 \times 10^{23}$, $R = 8.314$ J/gm K, and $T = 273$ K; Eq. *(ii)* may be written as

$$N = \frac{(1.013\times10^5)\times(6.023\times10^{23})}{8.314\times273}$$
$$= 2.688 \times 10^{25} \text{ atoms/m}^3$$

Putting above value in Eq. (i), as given below, we get

$$(\varepsilon_r - 1) = \frac{Np_e}{\varepsilon_0}$$

$$= \frac{(2.688\times10^{25})\times(5.11\times10^{-41})}{8.85\times10^{-12}}$$

where $\qquad \varepsilon_0 = 8.85 \times 10^{-12}, p_e = 5.11 \times 10^{41}$ (given),

$$\therefore \qquad \varepsilon_r - 1 = 1.552 \times 10^{-4}$$

or $$\varepsilon_r = 1 + 0.0001552$$

$$= 1.0001552$$

Example 9.11 Discuss the polarization of a polyatomic gas and show its dependence on dielectric constant.

Solution. In polyatomic gas, the contribution to total polarization is made by electronic polarization, orientation polarization and ionic polarization. If the gas contains N molecules/m^3 and various polarizability are p_e, p_o and p_i respectively, then

$$P = N\left(p_e + \frac{p_o^2}{3kT} + p_i\right)E \qquad (9.39a)$$

From macroscopic theory we know that $P = \varepsilon_0 (\varepsilon_r - 1)$, therefore

$$\varepsilon_0(\varepsilon_r - 1) = N\left(p_e + \frac{p_o^2}{3kT} + p_i\right) \qquad (9.39b)$$

This equation shows that ε_r is independent of p_e and p_i, but is temperature dependent due to orientation polarizability.

Example 9.12 What is the role of dielectric material in a photocopier machine?

Solution. The Xerox machine is a plate made of a thin layer of amorphous semiconductor on a metal plate. The semiconductor is a compound of *As*, *Se*, and *Te*. It is almost an insulator so that it behaves like a dielectric, but it is also photo-conductive. Therefore in the light, it becomes more conducting. The dielectric plate is highly charged electro-statically by brushing it with wire electrodes charged to about 30 kV. The document to be copied is imaged onto the plate. The regions that are white cause the semiconductor to become conducting, and the surface charge leaks away to the earthed metal backing plate.

However, where the dark print is imaged, the charge persists. The whole plate is dusted with a fine powder consisting of grains of carbon, silica and a thermosetting polymer. Surplus powder is shaken off and it adheres only to the highly charged dark regions. A sheet of paper is then pressed onto the plate by rollers. It picks up the dust particles, and is then treated by passing under an infra-red lamp. This fuses polymeric particles which subsequently set the black carbon and SiO_2 dust to form a permanent image of the printed document. To clear the plate, it is illuminated all over so that it all discharges, the ink is shaken off and it is ready to copy something else.

Example 9.13 How does a Xerox photocopier work?

Solution. Figure 9.9 shows the basic form of the electrostatic photocopier. A high voltage is connected to a series of wires. Because these wires have small radii, the charge leaks from them to a selenium-coated drum, which then becomes charged. The charged drum is then exposed to an optical image of the item being copied. Where light falls on the selenium, it becomes conducting and allows the charge to leak away; but where not illuminated it retains its charge. The result is an electrostatic charge pattern on the drum of the item being copied.

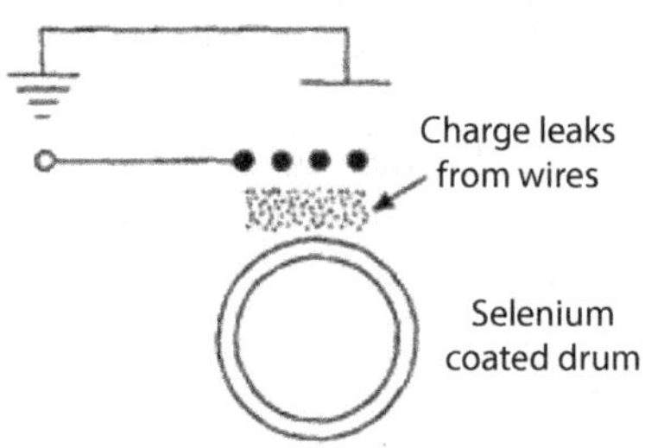

Figure 9.9 Electrostatic photocopier

This is then coated with a powder, the powder sticking as a result of electrostatic polarization to the charged areas. Thus, when the drum presses against a sheet of paper, a copy is produced.

Example 9.14 Is the behaviour of dielectrics same under d.c. field and a.c. field? If not, how does it differ?

Solution. The behaviour of dielectric materials under alternating field is different from the behaviour under static (d.c) field. It is mainly in respect of dielectric constant whose response is frequency dependent. Consequently, the dielectric constant under alternating conditions is a complex quantity whose real part is a measure of dielectric constant and the imaginary part depicts *dielectric losses* of material.

The polarizability p and dielectric constant ε_r under static field,take on to complex values p^* and ε_r^*; under an applied electric field E_0 cos ωt. Here ω is angular frequency of applied field. Thus

$$p^* = p' - jp'' \text{ and } \varepsilon_r^* = \varepsilon_r' - j\varepsilon_r'' \tag{9.40}$$

in which the first terms on right side of equations represent real [Re] parts and second terms represent imaginary [Im] parts. Above quantities are different for different mechanisms of polarization processes.

Example 9.15 What is polarization vector? How is it related to polarization?

Solution. Total polarization P expresses the dipole moment per unit volume of the dielectric material. If there are N number of molecules in this volume, the dipole moment/molecule can be expressed as

$$\vec{P} = NP \tag{9.41}$$

in which the quantity $\vec{P}$ is termed as polarization vector.

Example 9.16 Explain the Lorentz expression for internal field.

Ans. In case of solids and liquids, the atoms or molecules are closely packed, but in gases it is not so. Therefore in solids and liquids the internal field E_i differs from the applied field E, in contrast to gases where E_i equals to E. In three-dimensional case, the expression for internal field is expressed as

$$E_i = E + \frac{\gamma P}{\varepsilon_0} \tag{9.42}$$

where P is dipole moment per unit volume and γ is proportionality constant for internal field. Although the value of $\gamma = 1/2/\pi$ but in the particular case when the atoms in a solid are surrounded cubically by other atoms, the Lorentz internal field comes into picture. In this case the value of $\gamma = 1/3$, therefore

$$(E_i)_{\text{Lorentz}} = E + \frac{P}{3\varepsilon_0} \tag{9.43}$$

This equation is known as Lorentz expression.

Example 9.17 What is Clausius-Mosotti equation?

Ans. Some materials like diamond and sulphur consist of a single type of atoms. These materials exhibit only electronic polarization since they neither contain ions nor permanent dipoles. If p_e is polarizability per atom and N is the number of atoms/m^3, then polarization is given by

$$P = Np_e E_i \tag{9.44}$$

Considering Lorentz's expression for E and $P = \varepsilon_o(\varepsilon_r - 1)$, the above equation may be transformed in terms of ε_r. Then it can be written as

$$\frac{Np_e}{3\varepsilon_0} = \frac{\varepsilon_r - 1}{\varepsilon_r + 1} \tag{9.45}$$

This equation is known as Clausius-Mosotti relation.

Example 9.18 What is electric susceptibility? How is it related to dielectric constant?

Solution. Since the polarization vector $\overline{\text{P}}$ is proportional to total electric field intensity E and is in the same direction as E, hence $\overline{\text{P}}$ can be expressed as

$$\vec{P} = \varepsilon_0 \chi_c \vec{E} \tag{9.46}$$

χ_c in this expression is known as electric susceptibility. It is a characteristic of dielectric materials. It is related to dielectric constant as

$$\chi_c = (\varepsilon_r - 1) \tag{9.47}$$

Example 9.19 What is meant by 'optical absorption' and infrared absorption?

Solution. The dielectric losses in optical frequency region associated with electrons are termed as *optical absorption*. This is the deciding factor for colour of materials ($\approx 10^{10}$ Hz for red colour to $\approx 10^{20}$ Hz for violet colour). The *infrared* absorption refers to dielectric losses on account of ionic vibrations whose frequencies are in infrared region (10^{12} to 10^{14} Hz).

Example 9.20 What are polar and non-polar dielectrics? How do they differ?

Solution. A dielectric that does not possess any permanent dipole is called *non-polar* dielectric, whereas a dielectric having permanent dipole moments is known as *polar* dielectric. The non- polar atoms or molecules have induced dipole moments in the presence of an applied field. These dielectrics are compared below for their characteristics.

Description	Polar dielectric	Non-polar dielectric
• Permanent dipole moments	Present, even in the absence of an electric field	Absent
• Alignment of Dipoles	They align in the direction of applied electric field	No alignment as the dipoles do not exists
• Induced dipole moments	Yes, always present	Yes, in the presence of an applied electric field
• Absorption and emission	Occurs in infrared range	Does not occur
• Polarization	Temperature dependent	Temperature independent
• Centre of symmetry	No, they do not have	Yes, they have
• Examples	HCl, H_2O, CO etc.	O_2, H_2, N_2etc.

Quick Revision Summary

Dielectric (or breakdown) strength. It is the limiting voltage gradient required to cause appreciable current flow or failure of a dielectric.

Electric stress. When experienced by a dielectric or an electrical insulating material, the electric field intensity is often referred to as electric stress.

Induced dipole moment. The induced dipole moment is the difference of orbital dipole moment under an applied magnetic flux and 'orbital magnetic dipole moment in the absence of any magnetic field'. It is related to magnetic dipole moment M per unit volume by $M = \mu_m$ induced $\times N$ where N is the number of atoms per unit volume. The direction of induced dipole moment is opposite to the direction of applied magnetic field.

Interfacial (or space charge) polarization is defined as the polarization that occurs due to diffusion of ions, atoms, molecules and /or free electrons from the material to its surfaces.

Polarization. It is the total dipole moment per unit volume in a dielectric.

Relaxation frequency. It is the reciprocal of the relaxation time.

Relaxation time. It is the time required for a disturbed system to reach *1/e* of the final equilibrium configuration.

Review Questions

1. Classify dielectric materials and quote examples of each type. How do the solid, liquid and gaseous dielectrics compare with each other?
2. What are the characteristics and properties of dielectric materials? What are the effects of high temperature and high frequency on dielectric constant and dielectric strength?
3. Why are the dielectrics glazed? How do the relaxation time and power factor influence the dielectric properties?
4. What are different types of dielectric breakdown and how do the insulators fail in each type?
5. What do you mean by dielectric loss? Explain 'loss angle' and 'loss factor'. How will you calculate the loss factor?
6. Enumerate different favourable qualities a dielectric material should have. How do the surface and volume resistivities influence upon it?
7. Derive an expression to relate polarization, dielectric constant and electrical field strength.

8. Enumerate different kinds of polarization processes. Explain their mechanism, and temperature dependence.
9. Explain the relaxation time. Discuss the effects of frequency of applied field and relaxation time on the polarization of mica, porcelain and soda-lime glass.
10. What is spontaneous polarization? Draw a polarization curve and a hysteresis loop, and explain their salient features.
11. Write notes on the following.
    a. Electric susceptibility
    b. Interfacial polarization
    c. Polar and non-polar dielectrics
    d. Polarizability
12. Derive the expressions to explain the effects of impulse and frequency switching on behavior of polarization . Discuss the phenomena of decay and build - up of polarization under an a.c. field.
13. Prove that the internal field due to polarization inside the dielectric is given by $E'=E+ P/ 3\varepsilon_0$
14. Deduce the Lorentz equation for net internal field for a 3-dimensional case.
15. Derive the Clausius-Mossotti relation and discuss the result.
16. Derive the following expression for Clausius-Mossotti equation

$$\varepsilon_r - 1/ \varepsilon_r + 2 = N\, p_e / 3\, \varepsilon_0$$

17. Discuss the following terms :
    a. Complex dielectric constant
    b. Role of dielectic materials in photocopier

Numerical Problems

1. The electric polarizability of LiF is 5.11×10^{-41} Fm^2. Determine its dielectric constant at 273 K and 1 atmospheric pressure.
2. A ceramic insulator of 3.8 μF capacitance is 60 mm long, 20 mm wide and 10 mm in thickness. If the loss angle is 2.11°, calculate (a) its dielectric constant, (*b*) dissipation factor, (*c*) electric loss, and (*d*) the phase angle. Take $\varepsilon_0 = 8.85 \times 10^{-12}$ F/m.
3. Polyvinylchloride (PVC) insulation of 3 mm thickness is used on a current carrying conductor to withstand the breakdown. The dielectric strength of PVC is 24 000

volt/cm. Taking factor of safety as 2, suggest the voltage in the conductor.

4. In a parallel plate capacitor, the 12 mm × 10 mm plates are placed 4 mm apart in a medium of relative permittivity 7. An electric field of 11 kV at 50 Hz is applied to it. If the resistivity of material of the plates is 3.8×10^{16} ohm-m and the loss angle is 16.7°. calculate the (*a*) d.c. dielectric loss, (*b*) capacitance, and (*c*) a.c. dielectric loss.
5. A capacitor has capacitance of 0.019 farad when uses wax paper (of $\varepsilon_r = 1.85$) between electrodes of aluminium foil. The wax paper is to be replaced by a plastic film (of $\varepsilon_r = 2.15$) of same dimensions. Taking other factors being equal, obtain the change (increase or decrease) in the capacitance.
6. A layer of porcelain is 80 mm long, 20 mm wide and 0.7 μm thick. Calculate its capacitance taking $\varepsilon_r = 6$.
7. An electric field of 10^6 V/m is applied to a krypton atom. Calculate the displacement of electron cloud with respect to the nucleus of the atom. Take required data from the text.
8. What is the total area of dielectric needed for a paper capacitor of capacitance 0.01 μF and its breakdown voltage; if the paper has a thickness of 1 mm, a dielectric strength of 1.6×10^7 V/m and a relative permittivity of 2.5?
9. What is the maximum potential difference that can be applied to a parallel plate capacitor having a dielectric of thickness 2.0 mm, if the dielectric strength is 4×10^7 V/m?
10. What is the charge density on a pair of parallel capacitor plates 4 mm apart with air as the dielectric, when the potential difference between them is 100 V?

Objective Questions

1. Match list I with list II, and select the correct answer using the codes given below the lists in questions 1 and 2?

List I	*List II*
A. Transformer oil	1. Ferroelectrtic
B. Electronic polarizability	2. Glass tape
	3. Transformer core
C. Rochelle salt	4. Temperature independency
D. Class E insulation	5. Organic insulating material

Codes:

	A B C D		A B C D
a.	3 2 4 1	b.	1 2 3 4
c.	5 4 1 2	d.	2 3 5 1

2.

	List I		*List II*
A.	Nitrogen	1.	Class C insulation
B.	Porcelain	2.	Teflon
C.	Coating materials	3.	Ferroelectric materials
D.	Spontaneous polarization	4.	Electropositive gas insulation
		5.	Organic insulating material

Codes:

	A B C D		A B C D
a.	1 3 2 4	b.	4 1 2 3
c.	2 3 1 4	d.	3 4 2 1

3. Which of the following pairs are correctly matched?

1.	Steatite	Magnesia-silica ceramic dielectric
2.	Discharge breakdown	Gas bubbles are responsible for it
3.	Dielectric loss	It decreases with decrease in humidity
4.	Ferroelectric curie temperature	It is related to both ferromagnetic and ferroelectric materials .

Select the correct answer using the codes given below.

a. 2, 3 and 4 b. 1, 3 and 4
c. 1 and 3 d. 1, 2 and 4

4. Displacement current in a dielectric primarily depends upon the

a. Resistivity b. Dipole moment
c. Frequency of operating field d. Mobility

5. Piezoelectricity has been observed in

a. Nickel b. Glass
c. Quartz d. Mica

6. A ferroelectric material exhibits

a. Spontaneous magnetization b. Hysteresis effect with polarization

c. No spontaneous magnetization
d. Superconducting states

7. Insulating materials that can withstand a temperature above 180°C is of
   a. Class A type
   b. Class B type
   c. Class C type
   d. Class H type

8. Polarization in polyatomic gases is the sum of (notations carry their usual meaning)
   a. $P_e + P_o + P_i + P_s$
   b. P_e only
   c. $P_e + P_i$
   d. $P_e + P_o + P_i$

9. Optical absorption and infrared absorption refer to
   a. dielectric losses
   b. anti-ferroelectric behaviour
   c. properties of non-polar molecules
   d. relaxation time when a.c. field is switched-on

10. Dielectric loss is of the order of
    a. 10^4
    b. 10^7
    c. 10^{-4} to 10^{-2}
    d. 10^0

11. Capacitors to be used at high frequencies require a material whose loss angle is
    a. low
    b. high
    c. uniform
    d. one

12. Line insulators are made of
    a. Porcelain
    b. Mica
    c. Marble
    d. PVC

13. Heating in microwave oven is due to
    a. Magnetostriction
    b. Electrostriction
    c. Eddy current
    d. Spontaneous polarization

14. The principle employed in strain gauge working is
    a. Piezoelectric
    b. Capacitive
    c. Resistive
    d. All of the above

10

Dielectric Materials: Types and Applications

Dielectric materials have been classified earlier in chapter 9. Now, we shall study details of their different types and applications.

10.1 Solid Insulating Materials and their Applications

Insulating materials are invariably used in almost all electrical equipments ranging from an ordinary bulb to a complicated power generation system or to a sophisticated electronic system. They are also used to insulate the complex computer system, and instrumentation and control arrangements of an industrial process. Infact, one or the other type of insulating material is an essential need to insulate the conducting components in electrical appliances. Some examples in this regard are given below for an easier understanding.

1. In domestic electric wiring, an insulation covering of plastic, rubber, treated cloth tape or varnish is provided around the copper/aluminium conductor.
2. Flexible micanite is used in slot lining and field coil insulation of electric motors.

3. Cotton and cotton tapes are used to insulate the magnetic coils and armature windings of small machines.
4. Varnishes of different kinds are used to insulate armature coils, field coils and relay coils etc. For that the varnish is heated to 100°C or more and then the coils are dipped into it.
5. Glass and glass tapes are used to insulate windings and their ends for high temperature working.
6. Mica is used as insulation in domestic appliances.
7. High voltage and low voltage porcelains are used in transmission lines, power stations, oil circuit breakers, disconnecting switches, and arrestors etc.

10.1.1 Ceramic Insulating Materials

We have described earlier that the ceramics are non-metallic, inorganic, metallic oxides having non-crystalline (amorphous) structure. They are hard, brittle, and possess poor tensile strength. However, they have high melting point and form very good insulating materials. Several types of ceramics are used for this purpose which can be classified into following main groups.

1. Refractories
2. Silicates
3. Glasses
4. Limes
5. Cements and concretes
6. Clays and clay products
7. Bitumens and asphalts
8. Rocks and stones
9. Asbestos and plaster of Paris

Since the description of all these materials will entail an enormous content, we shall restrict our studies to a few most important and commonly used insulating materials.

10.1.2 Mica

Mica is most widely used insulating material. On the basis of temperature applicability, it is class B type insulating material. It is available in different forms such as given below.

1. Muscovita (Ruby or Potash) mica
2. Phlogopite (Amber or Magnesia) mica.

Muscovita mica is a form of hydrated potassium aluminium silicate, while the *phlogopite* form is magnesia mica containing quartz and iron oxide as impurities. Their properties and applications are given in Table 10.1, showing a comparison between them.

Table 10.1 Properties and applications of different types of mica

S.No.	Properties	Muscovita type mica	Phlogopite type mica
1.	Dielectric constant	6–7	5–6
2.	Dielectric strength	Higher (80 kV/mm at 20°C)	High (60 kV/mm at 20°C)
3.	Power factor (at 25°C)	(1 to 3) × 10^{-4}	(10 to 50) × 10^{-4}
4.	Thermal stability	Good	Greater
5.	Maximum operating temperature	500–600°C	800–900°C
6.	Insulation abilities	Very good	Comparatively poorer
7.	Abrasion resistance	Higher	High
8.	Effect of oil	Adverse	Adverse
9.	Effect of acid	Adverse*	Adverse
10.	Effect of alkalies	No	No
11.	Hardness	Good	Better
12.	Mechanical strength	Strong and tough	Stronger and more tough
13.	Flexibility	Less	More
14.	Structural ability	Good	Greater
15.	Density	≈ 3000 kg/m^3	≈ 2700 kg/m^3
16.	Colour	Colourless, silvery, yellow, green	Amber, grey, yellow, green
17.	Suitability	When electrical requirements are severe	When greater thermal stability is desired
18.	Applications	Capacitors, commutators	Toasters, hot plates, domestic cloth irons

* Particularly the hvdroflouric acid affects more adversely.

10.1.3 Porcelain

This ceramic material is composed of clay, quartz and feldspar. Barium carbonate ($BaCO_3$) is used in high frequency porcelain instead of feldspar. Porcelain is prepared by mixing finely ground (grinded) clay with metallic oxide and water. The paste thus formed is dried and fired at 1200 °C to 1700 °C. Porcelain is used as insulating material in applications like

- Transformer bushing-pins
- Disconnecting switches
- Plugs, sockets, and fuse holders
- Line insulators
- Suspension insulators for transmission and distribution lines.

Line insulators require a proper design to overcome the problems of reduced surface resistivity and flashover due to accumulation of dust, dirt, moisture and rain drops. Porcelains are used for both: the low voltage and high voltage services. Hence, their properties are different from each other. Salient properties of low voltage and high voltage porcelains at 298 K are compared in Table 10.2.

10.1.4 Glass

Glass is an inorganic (ceramic) material made by fusion of different metallic oxides. The fusion product is cooled to a condition in which the

Table 10.2 Properties of low voltage porcelain and high voltage porcelain at 25°C

S. No.	Description	Low voltage porcelain	High voltage porcelain
1.	Power factor (at 50 Hz)	0.007–0.020	0.008–0.025
2.	Dielectric constant (at 50 Hz)	5.5	5.7
3.	Dielectric strength (MV/m)	5	5.5
4.	Volume resistivity (ohm-m)	10^9–10^{12}	10^{10}–10^{12}
5.	Specific gravity	2.4	2.7
6.	Water absorptivity	0.5–2.5%	0.0–0.5%
7.	Softening temperature	1300–1350°C	1400–1450°C

crystallization does not occur. It has a high melting point and is highly resistant to chemical and corrosive environments. Different types of commercial glasses are used for insulation purposes. These are

1. Soda-lime (or window) glass
2. Lead (or flint) glass
3. Borosilicate (or pyrex) glass
4. High silica (or vycor) glass.

These glasses have low dielectric loss and high dielectric strength. They are used as moulded insulating devices such as bushings, laminated boards, capacitors, bulbs, protective shields etc. Various properties of different commercial glasses are illustrated in Table 10.3.

10.1.5 Micanite

It is a sheet form material obtained by glueing the mica with shellac or other binders. To enhance its tensile strength, cloth or paper can also be glued on one or both sides of the sheet. Based on the properties produced-in and

Table 10.3 Electrical/Insulating properties of commercial glasses

S. No.	Description	Soda-lime glass	Lead glass	Borosilicate glass	High silica glass
1.	Dielectric constant	7.0	6.6	4.7	3.4
2.	Power factor (at 1 MHz, 293 K)	0.004	0.0016	0.0046	0.0002
3.	Coefficient of thermal expansion (/°C)	85×10^{-7}	91×10^{-7}	32×10^{-7}	5.5×10^{-7}
4.	Resistivity (ohm m at 298 K)	0.065	0.089	0.081	0.12
5.	Specific heat (cal/gm K at 298 K)	0.20	0.19	0.18	0.17
6.	Density (kg/m^3)	2450	2850	2130	2200
7.	Softening point (°C)	730	630	820	1665

the method of construction etc.; micanites may be of various kinds such as given below.

- Moulded micanite
- Multi-layer micanite
- Hard micanite
- Flexible micanite
- Silicon binder micanite
- Tape and paper micanite, etc.

In their construction, the sheets of about 0.1 mm thickness are normally used. They are generally used for insulating the coils in electric motors and alternators.

10.1.6 Glass Bonded Mica

It is prepared by blending the ground (grinded) mica flakes and powdered glass in 40: 60 to 60: 40 ratio. It possesses high dielectric strength and low dielectric loss. It is suitable for high humidity and high temperature situations.

10.2 Polymeric Insulating Materials

Plastic is an organic material. It is the mixture of resin, catalyst, accelerator, inhibitor and pigment. It can be easily moulded to any form, shape and size to suit the requirements of application. Plastics are available in various usable forms such as sheet, films, rods, tubes, tapes etc. Depending upon the manner in which they are affected by heat, their remouldability and strength; the plastics can be classified into following main categories.

1. Thermosetting (or thermoset) such as
   i. Epoxy
   ii. Polyester
   iii. Melamine
   iv. Phenolics
2. Thermoplastic (or thermoplast) such as
   i. PVC
   ii. Polyethylene
   iii. PTFE (Teflon)
   iv. Nylon
   v. Cellulose acetate
   vi. Polypropylene
   vii. Polystyrene
   viii. Acrylic
   ix. Polysulfone
   x. Acetal, etc.

Thermosets are non-recyclable, but thermoplasts can be recycled again and again. Their characteristics and properties are compared below.

Description	Thermoset	Thermoplast
• Bonding network	3-dimensional, primary in all directions	Chain molecules are secondary bonded
• Effect of heat	Becomes hard and cannot be soften again	Bond breaks and becomes soft
• Consequence of curing	A chemical change occurs during permanent hardening	No chemical change occurs
• Mouldability to any shape	Both: heat and pressure are required	Requires cooling for setting
• Effect of intense heating	They breakdown by degradation	They resoften.
• Melting point	They do not melt	115° C-330° C
• Normal usable temperature	80° C-300° C	60° C-200° C
• Applications *e.g.*	Telephone receivers	Gramophone records

Since the details of plastics are enormous as a subject, therefore we shall restrict our discussion to a few important kinds only which are of greater interest as insulating materials.

10.2.1 Bakelite

It is also known as *phenol formaldehyde* and belongs to the family of epoxy. It becomes hard on heating and cannot be softened again. Its main properties are the following.

- Power factor (at 60 Hz) = 0.008 to 0.032
- Dielectric constant (at 50 Hz) = 4 to 5
- Dielectric strength = 15 to 25 kV/mm
- Heat distortion temperature ≈ 95 to 130° C
- Specific gravity = 1.2
- Tensile strength = 7 MPa
- Elongation = 2% or less
- Water absoptivity = 0.1 to 0.5%.

It possesses a good heat resistance and burns slowly. It has been named after its Belgian-American inventor and chemist L.H. Backeland.

10.2.2 Polyethylene

It belongs to the family of thermoplast kind plastics. It becomes soft on heating and can be remoulded to any other shape. It is available in two types viz. low density polyethylene (LDPE) and high density polyethylene (HDPE). Following are the main properties of polyethylene. The lower values are for LDPE.

- Power factor (at (30 Hz) = 0.0001 to 0.0006
- Dielectric constant (at 50 Hz) = 2.25 to 2.32
- Dielectric strength = 20 to 60 kV/mm
- Heat distortion temperature = 90 to 110° C
- Specific gravity = 0.92 to 0.97
- Tensile strength = 10 to 40 MPa
- Elongation = 50 to 600%
- Water absorptivity = 0.01% or less.

Although its surface and volume resistivities are high, but the heat resistance and flammability are poor.

10.3 Natural and Synthetic Rubber as Insulating Material

Rubber is an organic polymer found in latex (i.e. sap) of certain plants. Latex is treated to obtain *raw* or *natural rubber.* This is then blended with additive compounds to get desired properties. These additives are accelerators (such as lime, litharge), plasticizers (such as stearic acid, vegetable oils), reinforcing agents (such as carbon black, zinc oxide), and sulphur. The processed natural rubber is of the following three kinds.

1. *Chlorinated rubber* which is used for producing adhesives and protective coatings.
2. *Rubber hydrochloride* which is used as packaging insulator.
3. *Cyclised rubber* which is used to manufacture insulating papers with the help of paraffin wax.

10.3.1 Synthetic Rubber

Synthetic rubbers are superior to natural rubbers in many respects such as resistance against temperature, oil, grease etc. and in ageing. These are of several kinds such as given below.

- Neoprenes
- Butyl
- Butadiene
- Hard rubber
- Nitrile (or Buna-N)
- Polyurethane
- Silicone
- Hypalon

The *neoprene* has high resistance to ageing, *butyl* is impermeable to gases, *butadience* is highly abrasion and weather resistant, *nitrile* is excellent to resistance against solvents and greases, *polyurethane* is abrasion resistant, and *silicone* has high thermal stability. Salient properties of some of these synthetic rubbers alongwith the natural rubber are given in Table 10.4.

10.4 Paper as a Fibrous Insulating Material

The structure of paper is fibrous. Hence, it is called fibrous insulating material. It is derived from cellulose which is a constitutent of vegetable plants/wood. Organic contaminations like 'lignin' are removed from wood and some alkaline reagents are added before processing, to obtain desired quality of insulating paper. They are then impregnated with suitable compounds to impart resistivity against water absorption and thermal instability.

Useful properties. The useful properties of insulating papers are their adequate mechanical strength, low dielectric losses, ease of wrapping around the conductor, lightweightness, easier availability-and cheapness. Due to these properties the paper finds use in telephone cables, capacitors etc. Salient properties of cable purpose and capacitor purpose papers are given in Table 10.5.

10.5 Choices of Solid Insulating Materials for Different Applications

For different components of electrical and electronic systems, different types of insulating materials may be useful. But some of them are more suitable than the others. Although it is very difficult to assign a particular insulating material for a specific use, yet the most suitable choices are given below for a practical understanding.

Table 10.4 Salient properties of natural and synthetic rubbers

S. No.	Description	Natural rubber	Neoprene	Butyl	Butadiene	Silicone
1.	Power factor (at 10^6 Hz)	0.002–0.05	0.01–0.06	0.003–0.08	0.003–0.04	0.001–0.01
2.	Dielectric strength (kV/mm)	18–24	4–20	16–32	16–20	12–28
3.	Dielectric constant (at 10^6 Hz)	2.7–5.0	7.5–14.0	2.2–4.0	3.9–10.0	2.8–7.0
4.	Useful temperature	−30°C to 75°C	60–80°C	80–90°C	90–100°C	175–200°C
5.	Specific gravity	0.9–1.5	1.1–1.4	0.9–1.2	1.1–1.6	1.0–1.1
6.	Water absorptivity	1–2%	0.3–1%	0.3–0.5%	0.3–0.4%	0.5–2.5%
7.	Tensile strength (MPa)	20–30	15–25	5–20	20–30	10–13

Table 10.5 Insulating properties of commercial grade papers

Properties	Air resistance (second/ 100 cm³)	Specific gravity	Thickness (micron)	Ash content (%)	Breaking strength (N/mm width)	
					Along fibres	Across fibres
• Cable paper	30–90	0.68–0.70	12–20	1.5	10–15	2.5–5.5
• Capacitor paper	1.5–2.0	0.80–0.93	1–2	0.4	20–30	5–10

Component to be insulated	Insulating material
• Field coils in motors	• Flexible micanite
• Magnet coils	• Cotton, cotton tapes, varnish
• Armature winding	• Cotton, cotton tapes, varnish
• Choke	• Cotton, cotton tapes, varnish
• Small transformers	• Cotton, cotton tapes, varnish
• Relay coils	• Varnish, glass tapes
• High voltage lines	• High voltage porcelain
• Transformer bushes	• Porcelain
• Oil circuit breakers	• Porcelain
• Arresters	• Porcelain
• Disconnecting switches	• Porcelain
• Severe requirements	• Mica (ruby or potash type)
• Capacitors	• Mica (ruby or potash type)
• Commutators	• Mica (ruby or potash type)
• Hot plates	• Mica (magnesia type)
• Toasters	• Mica (magnesia type)
• Domestic irons	• Mica (magnesia type)
• Air cooled transformers	• Class A fibrous material
• Conductors of oil cooled transformers	• High grade manilla paper tape
• Terminals of high voltage machines	• High voltage porcelain
• Coil windings of instruments	• Enamel, cotton
• Indoor installations	• Vulcanized rubber with hemp braiding
• Underground cables	• Paper insulation
• Medical electronics	• Synthetic resin, press board
• Computers	• Synthetic resin, paper
• Industrial electronics	• Glass, ceramics
• Radar	• Glass, synthetic resin
• Overhead power lines	• Porcelain, toughened glass

10.6 Liquid Insulating Materials

Several liquids are known to have insulating properties. They possess high electric strength and widely varying range of permittivity and viscosity. In chapter 9, we have classified the liquid insulating materials in different categories. There, we studied that the liquid insulating materials may be natural or synthetic, organic or inorganic, petroleum or agro products, substituted hydrocarbons or halogen free oils. Purified water; liquid O_2, N_2, He etc.; silicone oil, resins, mineral and transformer oils etc. are common liquid insulators. Besides serving as insulator, they perform a number of other functions also such as stated below.

1. They provide cooling action by convection phenomenon in transformers and oil-filled cables.
2. They assist in arc extinction in oil circuit breakers.
3. They act as insulation between the voltage carrying parts and the grounded container such as in transformers.
4. They provide high capacitance and high permittivity to power capacitors.
5. They fill-up the voids in solids by removing air and gases from them, and thus improve their insulating properties.
6. They provide impregnation to the layered paper insulation in cables, capacitors, and transformers where impregnating compounds or the oils are used.

10.6.1 Requirements of a Good Insulating Liquid

A good insulating liquid has to fulfill certain requirements. Main among them are the following.

1. It should be oxidation resistant.
2. It should not absorb moisture.
3. It should be free from impurities.
4. It should be thermally stable.
5. It should be non-sludging.
6. It should have high permittivity.
7. It should be free from fire hazards.
8. It should not have the tendency of emulsification with water.

10.6.2 Transformer Oil

Transformer oil is a liquid insulator and is mainly used in transformers. It is filled into transformer tank to serve the following purposes.

1. To transfer the heat generated inside the transformer core and windings, to the outer cooling surfaces.
2. To maintain insulation in the primary and secondary windings.

A good transformer oil should possess the characteristics as shown in Table 10.6.

10.6.3 Bubble Theory for Breakdown of Liquid Insulation

Bubble theory is one of the theories of breakdown for liquid insulation (other theory is colloidal theory). It explains that the breakdown occurs due to formation of vapour bubble in liquid. The formation of bubble may be caued due to gas pockets on the surface of electrodes, irregular electrode surfaces, change in temperature and pressure etc. This theory suggests that the *bubble breakdown strength (i.e.* electric field) E_b in a gas bubble which is immersed in a liquid of permittivity ε_L, is given by

$$E_b = \frac{3E_0}{\varepsilon_L + 2} \tag{10.1}$$

where E_0 is field in the liquid in absence of bubble.

10.6.4 Ageing of Mineral Insulating Oils

It refers to degradation of physical as well as electrical properties of oils with time due to ingress of moisture, oxygen and heat. Ageing of oils

Table 10.6 Characteristics of transformer oil

S.No.	Characteristics	Desired limits
1.	Dielectric strength (when applied for 1 minute at 20°C)	12 kV/mm
2.	Dielectric constant (at 50 Hz)	2.2–2.3
3.	Dielectric loss, tan δ: at 50 Hz	10^{-3}
	: at 1 kHz	5×10^{-4}
4.	Resistivity (ohm m)	10^{10}–10^{11}
5.	Specific gravity (at 20°C)	0.89
6.	Permissible water content (ppm)	50
7.	Pour point	40°C
8.	Flash point	160°C
9.	Acid value (mg/gm of KOH)	Nil
10.	Sludge value	1.2%

leads to deterioration of their insulating properties such as loss of electric strength, change in viscosity, formation of sludge etc.

10.7 Gaseous Insulating Materials

The gaseous insulating materials are the gases like air, nitrogen, oxygen, sulphurhexafluoride (SF_6), sodium vapour, freon (CCl_2F_2) etc. Amongst these the air is most important insulator. It is freely available and the cheapest dielectric. Many other gases *e.g.* carbon-dioxide (CO_2), helium (He), neon (Ne), carbon tetrachloride (CCl_4) etc. are also the gaseous dielectrics but they find limited use.

The gaseous dielectrics may be classified into two main groups viz.

1. Electropositive gases such as N_2
2. Electronegative gases such as SF_6, freon, H_2.

In electropositive gases the molecules try to give-up the 'electrons forming positive ions', whereas in electronegative gases the molecules absorb the 'free electrons forming negative ions'. Electronegative gases are more common in use. They are non-explosive and non-flammable.

Depending upon their nature, the dielectric gases can also be classified into following groups.

1. Pure gases such as hydrogen, helium, nitrogen
2. Complex (or mixture) gases such as air
3. Oxide gases such as sulphur dioxide (SO_2), carbon dioxide (CO_2)
4. Hydrocarbon gases such as ethane (C_2H_6), methane (CH_4), propane (C_3H_8) etc.

Amongst these, the pure nitrogen is most widely used dielectric after air. It is cheap and abundantly available in air.

10.7.1 Air

Air is a natural kind of gaseous dielectric. It is inherently present in the surrounding of transmission and distribution conductors. The atmospheric air is a mixture of a number of gases. Major constituents in it are nitrogen ($\approx$ 78% by volume) and oxygen ($\approx$ 20.9% by volume). Besides CO_2, Ne, He, CH_4, H_2 etc. as the other constituents in its composition; it also contains water and traces of argon (Ar), krypton (Kr), nitrous oxide (N_2O), xenon (Xe), carbon monoxide (CO), and ozone (O_3). For all practical purposes the air is considered to be an electropositive dielectric gas. It acts as insulation in overhead transmission lines, switches, plugs, and various electrical machines.

Air possesses excellent dielectric properties, important among them are the following.

1. Its dielectric loss is negligible, *i.e.* $\tan \delta \approx 0$.
2. Its dielectric strength is 3 to 5 kV/mm.
3. Its permittivity is 1.0006.
4. It is excellent for low voltage services.

Air is also used for thermal protection of structures. For that, the air-filled foam glass and foamed plastics are used. Air is not suitable for thermal insulation at cryogenic temperatures as it solidifies at 81.3 K (−191.7°C). Hence the insulating layers are kept free from air, and are sealed.

10.7.2 Nitrogen

After air, it is another widely used gaseous dielectric. Since nitrogen is chemically inert, it prevents oxidation and reduces deterioration of apparatus for which it is used. Nitrogen is generally used under pressure such as in some types of capacitors. It is also used in high-voltage *gas-filled pressure cables* in conjunction with *oil-treated paper insulation.*

10.7.3 Vacuum

Vacuum means a free space without air. As the creation of free space without air is almost impossible, a pressure of the order of 10^{-4} Torr* (1.333×10^{-2} Pa) or less can be treated as a vacuum for all practical purposes. The idea of vacuum as insulation is as old as 1897 when R.W. Wood investigated the production of X-ray tubes. According to recent investigations, the dielectric mechanism of vacuum is a complicated phenomenon. It involves an *alternative emission* mechanism and a complex *non-metallic emission* mechanism, which suggest a 'prebreakdown electron emission' in vacuum. Accordingly, when the voltage across a very small gap (a few mm) is sufficiently increased, a relatively steady current begins to flow. For longer gap spacings (> 10 mm), small pulse currents of millisecond

* 1 Torr = 1 mm of Hg

duration and charges of the order of micro-coulomb (μC) exist without a steady current.

Vacuum is an effective means of insulation. It finds applications in following major devices.

- Electronic valves
- Microwave tubes
- Photocells
- Electrostatic generators
- X-ray vacuum tubes
- Low-loss capacitors
- Circuit breakers
- Outer space systems
- Particle accelerators and separators
- Controlled nuclear fusion devices

The electrostatic generators, low-loss capacitors, and vacuum circuit breakers (instead of common oil circuit breakers) are used as high voltage apparatuses. These are recent developments.

10.7.4 Vacuum as Reflective Insulation

Vacuum is also used as thermal protection of structures and transporting systems using the technique of reflective insulation. In this technique, highly reflective surfaces are separated by a large width of air space (vacuum). Conduction and convection are minimum at about 20 mm width of air. It is employed in *tankers* transporting the following liquid fuels and explosives.

1. Liquid oxygen
2. Liquid hydrogen
3. Liquid nitrogen

Aluminium foils on paper, reflective aluminium surfaces separated by glass fibre lamina, and Mylar (aluminized plastic films) under high vacuum are used for this purpose.

10.8 Ferroelectric Materials

Ferroelectric materials (or ferroelectrics) are those dielectrics which have dielectric constant above 2000. They are analogous to ferromagnetics (see chapter 11) as all the electric dipoles are aligned in the same direction even

Table 10.7 Properties of some ferroelectrics

Material	**Spontaneous polarization (C/m^2)**	**Ferroelectric Curie temperature T_{fe}(°C)**
KH_2PO_4	0.05	-148
$BaTiO_3$	0.25	122
$KNbO_3$	0.30	437
$PbTiO_3$	0.50	492

in the absence of an electric field. When electric field is applied to ferroelectric materials, the ferroelectric domains get aligned in the direction of applied field. Hence, large and spontaneous polarization is observed in such solids. They also exhibit the property of *piezoelectricity.* A change in their structure and behaviour is seen at a critical temperature called *ferroelectric curie temperature* T_{fe}.Some ferroelectric crystals and their properties are listed in Table 10.7.

Examples and applications. Other ferroelectric materials are Rochelle salt (terahydrate of potassium tartrate), $SrTiO_3$, KH_2AsO_4, lead-zirconate ($PbZrO_3$), lithium niobate ($LiNbO_3$) and lithium tantalate ($LiTaO_3$). Ferroelectrics are used in applications such as electro-optic materials, sonar devices, strain gauges, microphones and miniature capacitors.

10.8.1 Anti-Ferroelectric Materials

Similar to antiferromagnetic materials (see chapter 11), these are ionic crystals having lines of ions spontaneously polarized in anti-parallel direction with respect to alignment of neighbouring ions. Examples of such materials are WO_3, $NaNbO_3$, $PbHfO_3$, $Ag_2H_3IO_9$, etc. Their curie temperatures are in the range of 227 K to 1010 K.

10.9 Barium Titanate: A Ferroelectric Ceramic

Earlier in art. 10.8, we have briefly discussed the ferroelectric materials. Barium titanate ($BaTiO_3$) is most important among them. It is a mixed oxide having perovskite structure as shown in Fig. 10.1a. It is an artificial ferroelectric material. As discussed in art. 10.8, its $\varepsilon_r > 2000$. The curie temperature of $BaTiO_3$ is 130°C at which, there occurs a change in its structure.

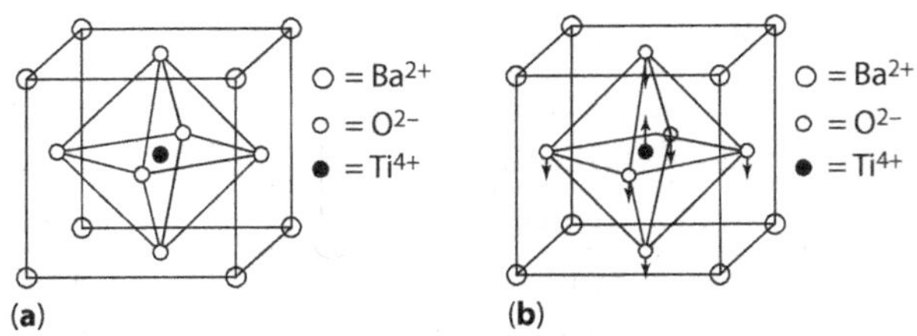

Figure 10.1 Structure of $BaTiO_3$ is (a) cubic above 130°C, and(b) tetragonal below 130°C.

Its structure is 'cubical' *above 130°C* and 'tetragonal' below it. When $BaTiO_3$ crystal is cooled below 130°C, the Ti^{4+} ion shifts on one side of the body centre and O^{2-} ions in opposite side as shown in Fig. 10.1b. Therefore, the centres of positive and negative charges do not coincide, and local dipoles are created throughout the crystal. Since the dipoles of neighbouring unit cells are all aligned, they induce a large polarization in solid.

The large polarization thus produced induces a large electric dipole moment. As there is displacement in the centres of positive and negative charges, the $BaTiO_3$ crystal elongates. Hence the geometry $a = b = c$ of cubic crystal becomes $a = b \neq c$ which characterizes a tetragonal crystal. The geometrical details of $BaTiO_3$ crystal in tetragonal form are summarized below.

- $a = b = 3.992$ Å
- $c = 4.036$ Å
- $c/a = 1.011$

10.9.1 Effect of Temperature on Structure of $BaTiO_3$

The behaviour of $BaTiO_3$ is peculiar under varying temperature. When cooled down to 5°C, its tetragonal structure changes to 'orthorhombic' form and on further cooling to -90°C, it transforms to 'rhombohedral' form. The same is true during heating also. These are summarized below.

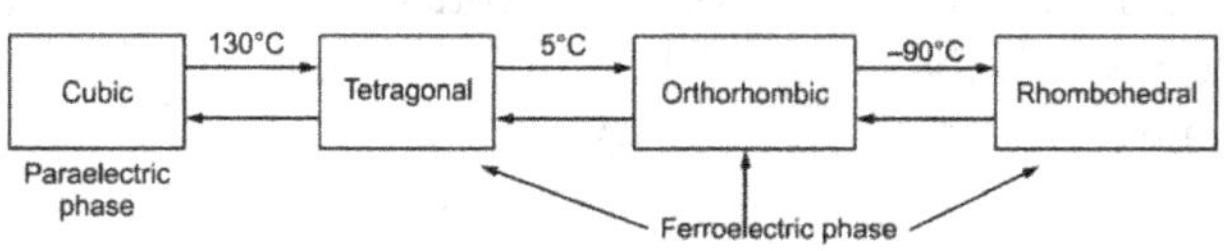

It also shows that the $BaTiO_3$ crystal remains in paraelectric phase above 130°C, but occupies ferroelectric phase below it.

Dielectric constant versus temperature. The dielectric constant of $BaTiO_3$ is affected considerably by temperature as shown in Fig. 10.2. It shows a rise with increasing temperature. There are abrupt and sharp rises at critical temperature viz. –90°C, 5°C and 130°C.

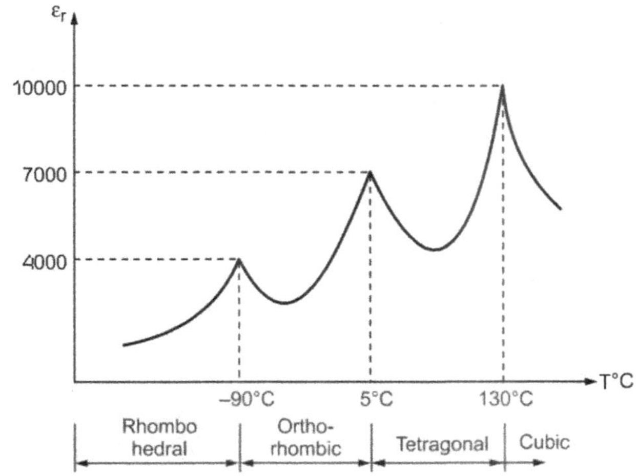

Figure 10.2 Variation in dielectric constant ε_r as a function of temperature *T*.

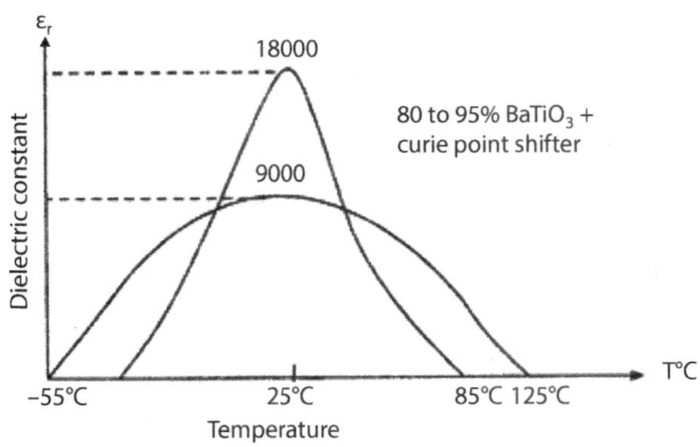

Figure 10.3 Curie point shifter removes the sharpness in ε_r vs *T* curve and smoothens it over a range of operating temperature

10.10 Modified Barium Titanate

To further enhance the characteristics of $BaTiO_3$, it *is* modified by adding curie point shifter. Besides changing the curie point to different operating temperature range, it also smoothens the profile of abrupt change in dielectric constant, Fig. 10.3. The shifting in curie point is accomplished by using some substitutions. Main among them are given below.

Table 10.8 Effect of different curie point shifters on curie temperature

Base titanate	+	Curie point shifter	Rate of change in curie temperature (dTc/dx)	Solid solubility (%)
$BaTiO_3$	+	$BaZrO_3$	−5.3°C	100%
$BaTiO_3$	+	$BaSnO_3$	−8°C	100%
$BaTiO_3$	+	$BaHfO_3$	−5°C	100%
$BaTiO_3$	+	$LaTiO_3$	−18°C	15%

***$BaTiO_3$ + $PbTiO_3$*.** This substitution results in formation of $(Ba_{1-x}Pb_x)TiO_3$ when $0 \leq x \leq 1$. On mixing them, the curie temperature rises at a rate of 3.7°C *i.e.* $dT_c/dx = 3.7°C$. The value of T_c for $BaTiO_3$ is 130°C and that for $PbTiO_3$ is 490°C.

***$BaTiO_3$ + $SrTiO_3$*.** At room temperature, it is not a ferroelectric. Its structure is cubic perovskite. On mixing them, the curie temperature lowers down by 3.7°C *i.e* $dT_c/dx = -3.7°C$.

***$BaTiO_3$ + $CaTiO_3$*.** This mixture has no effect on curie temperature, but it lowers the *tetragonal orthorhombic* transition temperature.

For other substitutions, the details are given in Table 10.8.

Applications. The ferroelectric materials are employed in following main applications.

- Matrix addressed memories
- Shift registers and switches (i.e. transpolarizers)
- Multiplate capacitors
- Thermistors
- Light deflectors, modulators and displays
- Holographic storage

10.11 PLZT as an Electro-Optic Material

Polycrystalline lanthanum modified lead zirconate titanate (PLZT) is an electro-optic (EO) material which has many advantages over single crystal EO materials. These include ease of manufacturing, low operating voltages and reduced long range strain effects. PLZT based devices, with compositions of 8.8–9.5/65/35*, are used in a variety of opto-electronic applications including high speed scanning dynamic lenses, optical switches, shutters and eye-protection devices.

Bulk PLZT can be highly scattering which has constrained the geometry of many devices by using thin wafers with interdigital surface electrode designs. These design geometries can require the application of very strong electric fields. For example, we consider a programmable phased array device using PLZT 9.0/65/35 with surface electrodes separated by 40 mm. To achieve an average phase shift of light passing through such a device, one requires applying the greater field strengths throughout the gap region. At these field strengths, we find that even thin wafers of PLZT become highly scattering. This scattering can significantly reduce the optical signal strength, can create off-axis noise and depolarize the transmitted light. Therefore, design geometries that reduce the operating electric field strength and the associated scattering, in them the depolarizing effects are critical for PLZT based devices. Computer aided design and modeling (or CAD/CAM) can be a very useful tool in the development of electro-optic (EO) device geometries for improved performance.

10.12 Piezoelectricity

Mechanical strains can polarize a crystal by displacing ions relative to one another provided that the crystal does not have a centre of symmetry. By a centre of symmetry, we mean a point in a crystal about which the lattice sites and atoms are symmetric. Figure 10.4a illustrates this condition; the centre of symmetry cancels out all possible polarizations. When the charge redistribution is non-symmetrical, Fig. 10.4b, distortion in the lattice induces displacement of positive and negative charges. This phenomenon is known as *piezoelectric effect*.

10.12.1 Characteristics and Uses

- In this effect, an electric field causes generation of mechanical strain in the dielectric crystal.
- Piezoelectric crystals such as quartz, CdS, ZnS, $PbZrO_3$ and ZnO do not have centre of symmetry, therefore they exhibit piezoelectricity.

---

* PLZT is generally designated by La/Pb $ZrO_3/PbTiO_4$ combination e.g. 8.8 to 9.5, 65% and 35% respectively.

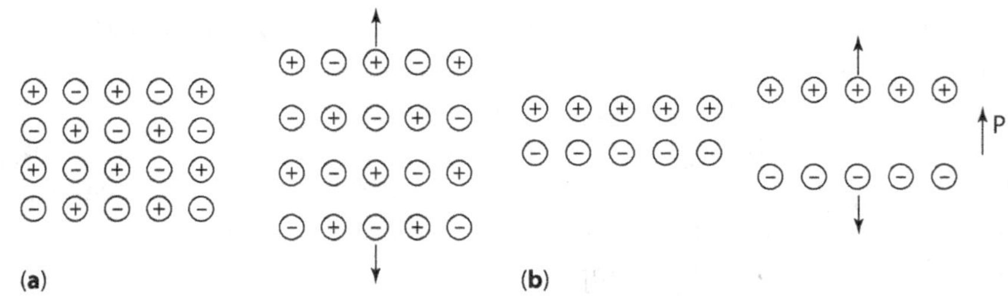

Figure 10.4 Piezoelectricity and crystal symmetry explains that in a crystal, lacking a centre of symmetry, an elastic strain induces a dipole moment (a) tensile strain of ionic crystal with centre of symmetry, no induced dipole moment and (b) tensile strain of ionic crystal with no symmetry centre, showing strain induced dipole moment.

- Quartz, a type of ceramic, is widely used as piezoelectric material.
- It is used as transducer for changing electrical signal into mechanical signal and vice-versa.
- It can transform mechanical movements, sound waves and vibrations into electrical potentials.
- A quartz crystal is used in the production of ultrasonic waves.

10.12.2 Mechanism of Piezoelectricity

The phenomenon of piezoelectricity is of great importance in electromechanical transducers. Let us consider a piezoelectric material which elongates along the direction of polarization. If this material is subjected to a compressive stress along the axis of polarization, and if only reversible processes are considered, the material can best respond to the compressive stress by polarizing itself at right angles. This will minimize the strain energy of the system and will lead to ionic displacements at right angles to the original direction of polarization and compressive stress. The result is the generation of a voltage along the direction perpendicular to the stress where there was previously no voltage. Therefore, the compressive stress has generated a change in voltage. This is the piezoelectric effect.

10.12.3 Inverse Piezoelectric Effect

The inverse effect is the change in strain along the axis of polarization caused by the application of an electric field. Suppose an electric field is applied perpendicularly to the direction of polarization. This will eventually result in a rotation of polarization into the field direction, and this causes a change in strain along the direction of the field. By the Poisson effect, there will also be a change in strain at right angles to this direction.

10.12.4 Piezoelectric Materials

Although the quartz was the first piezoelectric material to be used in practical devices, many more are used now-a-days. These are listed in Table 10.9.

10.12.5 Effect of Temperature on Piezoelectric Crystal

All ferroelectric materials are piezoelectric but all piezoelectric materials are not ferroelectric. Quartz is piezoelectric but not a ferroelectric material. When a piezoelectric crystal is heated or cooled, the interatomic distance increases or decreases asymmetrically. This causes a change in the polarization which induces a potential difference within the crystal. This effect *is* known as *pyroelectric effect.*

Table 10.9 Important piezoelectric materials

S. No.	Material	Formula	Piezoelectric coefficients (CN^{-1})	Relative permittivity (ε_r)
1.	Quartz	SiO_2	-2.25×10^{-2}	4.58
2.	Ammonium dihydrogen phosphate (ADP)	$NH_4H_2PO_4$	5×10^{-11} at 0°C	44.3
3.	Lithium tantalate	$LiTaO_3$	8×10^{-12}	53.5
4.	Lithium niobate	$LiNbO_3$	1.6×10^{-11}	85.2
5.	Potassium dihydrogen phosphate (KDP)	KH_2PO_4	Similar to ADP	44–45
6.	Polyvinylidene fluoride (PVDF)	$(CH_2\text{-}CF_2)n$	1.82×10^{-12}	160–200
7.	Lead zirconate titanate (PZT)	$PbTi_{0.48}Zr_{0.52}O_3$	-9.4×10^{-11}	730
8.	Rochelle salt	$NaKC_4H_4O_6 \cdot 4H_2O$	2.33×10^{-9}	3000

10.13 Piezoelectrics in Transducer Uses

Piezoelectric crystals are used in devices called transducers which convert electrical energy into mechanical energy. These are used in microphones, phonograph pickups, strain gauges, and sonar devices. Quartz which is piezoelectric and not ferroelectric, is often used in such devices, but it has low sensitivity and requires voltage amplification. Rochelle salt ($KNaC_4H_4O_6$), a ferroelectric is also used in transducer applications. It is readily attacked by moisture and can only be used in temperature range of −18°C to 24°C because it undergoes phase changes. Barium titanate has less piezoelectric sensitivity than Rochelle salt, but it can be used over wider range of temperature and is resistant to atmospheric attack. Barium titanate piezoelectric materials are fabricated by ceramic processes in different shapes. They are given a polarization treatment by cooling through the Curie temperature in a strong electrical field. They can be used in this state up to the temperature of about 70°C. For higher temperature use, the lead titanate ceramics are employed.

10.13.1 Working of Piezoelectric Transducer

Ultrasonic testing utilizes the high frequency acoustic (sound) waves for testing of surfaces and internal defects in metals and non- metals. The acoustic waves are generated by *Piezoelectric transducers* in the Frequency range of 1 to 10 MHz. Velocity of these waves depends on the nature of transmitting medium. It is 300 m/s in air at sea level but is lower in solid medium. Therefore in common test materials, the acoustic wavelengths λ (velocity = frequency × wavelength) are of the order of 1 to 10 mm. During the test, a highly directional sound beam is transmitted to the test piece, that propagates through the material but is dissipated or reflected by discontinuities/defects. These defects are recorded by a set of instruments and display systems using 'pulse-echo technique'. The ultrasonic testing provides information regarding the size, depth and location of discontinuities in materials/components.

In it the ultrasonic waves are produced by piezoelectric effect within the crystal probe. The probe is placed on the work-piece surface. The piezoelectric action is reversible interaction between elastic strain and electric field. If the piezoelectric crystal is slightly compressed, it produces an electric signal. Similarly, when this crystal is excited with an alternating current of ultrasonic frequency, the ultrasonic waves are introduced in the work-piece. Thus the crystal probe acts as receiver as well as transmitter.

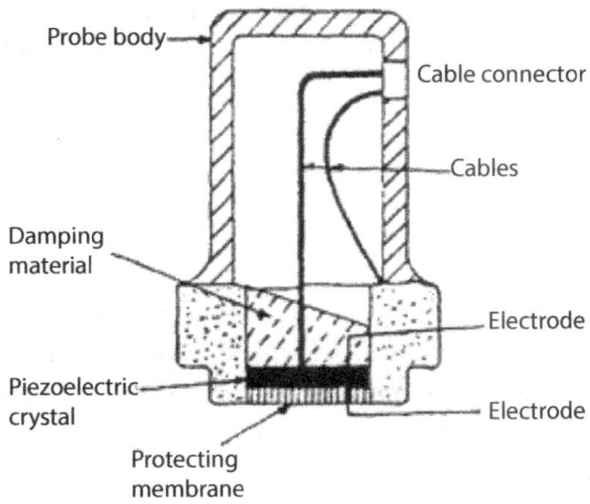

Figure 10.5 Details of an ultrasonic probe

Crystal probe. The transducer (piezoelectric crystal) may be made of the following materials.

- Quartz
- Lithium sulphate ($LiSO_4$)
- Lead niobate ($PbNbO_3$)
- Barium titanate ($BaTiO_3$)
- Lead zirconate titanate (PZT)

Amongst these the quartz is most commonly used material due to its excellent piezoelectric properties, mechanical and dielectric strength, and thermal stability. It can operate at high temperatures upto about 500°C. This crystal is mounted suitably in a probe before use. One such probe is shown in Fig. 10.5. It protects the transducer from mechanical damage and also the operator from electric shock.

10.14 Relation Between Young's Modulus and Electric Field in Piezoelectric Material

In a piezoelectric material, the electric field E produced by an applied stress σ remains proportional to it. Thus

$$E \propto \sigma$$

$$\therefore \quad E = \lambda\sigma \tag{10.2}$$

where λ is a constant, also known as voltage-output coefficient.

Piezoelectric modulus. Similarly, when a piezoelectric material is subjected to an electric field *E*, a strain ε is produced in it which is proportional to electric field. Thus

$$\varepsilon \propto E$$

or
$$\varepsilon = p_{coeff} E \tag{10.3}$$

where p_{coeff} is a constant, also known as *piezoelectric constant or piezoelectric coefficient or piezoelectric modulus.* Typical values of piezoelectric coefficients for some piezoelectric materials are given in Table 10.9.

We also know that the stress applied on a material is proportional to the strain produced *i.e.*

$$\sigma \propto \varepsilon$$

or
$$\sigma = Y\varepsilon \tag{10.4}$$

where Y is Young's modulus (or modulus of elasticity) of the material.

On correlating Eqns 10.2, 10.3 and 10.4; we get

$$\begin{aligned} E &= \lambda\sigma = \lambda(Ye) \\ &= \lambda(Y \times p_{coeff} E) \end{aligned}$$

$$Y = \frac{E}{\lambda p_{coeff} E} = \frac{E}{\lambda p_{coeff}} \tag{10.5}$$

These are important relations that are used to determine the values of σ, ε, *E* and Y for piezoelectric materials. These are mainly used for transducer applications.

Example 10.1 A barium titanate ($BaTiO_3$) wafer is of 0.15 mm thickness and modulus of elasticity = 70 GPa. If it is subjected to a compressive stress of 25 MPa, find the potential difference produced across it. Take piezoelectric constant for this wafer as 1.0×10^{-10} m/V.

Solution. Given are

$$\sigma = 25 \text{ MPa} = 25 \times 10^6 \text{ N/m}^2,$$

Thickness $t = 0.15 \text{ mm} = 0.15\text{x } 10^{-3} \text{ m}$,

Young's modulus $Y = 70 \text{ GPa} = 70 \times 10^9 \text{ N/m}^2$.

Since the electric field E produced by stress σ is related as

$$E = \lambda\sigma$$

and the modulus of elasticity as

$$Y = 1 / (\lambda t)$$

in which λ is a constant,

$$\therefore \qquad E = \sigma / (Y \times t)$$

Since the electric field E is equal to the potential difference V per unit thickness of the wafer,

Therefore
$$V = \frac{(25\times10^{6})\times(0.15\times10^{-3})}{(1.0\times10^{-10})\times(70\times10^{9})}$$

$$= \mathbf{535.7\ V}$$

Example 10.2 A quartz crystal has to be used to generate the vibrational frequency of 434 kHz along its thickness. If the Young's modulus along the appropriate direction is 80 GPa, and density is 2655 kg/m^3, determine the required thickness of the crystal.

Solution: The frequency of the thickness vibrations may be obtained from

$$f = \frac{n}{2t}\sqrt{\frac{E}{\rho}} \qquad (i)$$

where n = 1, 2, 3 ... denote the fundamental first overtone, second overtone etc.; t is thickness of the crystal, E is Young's modulus along appropriate direction, and ρ is density of the crystal. As the given data are

f = 434 kHz = 4.34×10^5 Hz, E = 80 GPa = 80×10^9 N/m^2,

ρ = 2655 kg/m^3, and let n = 1; then from Eqn. (i), we find

$$4.34\times10^{5} = \frac{1}{2t}\sqrt{\frac{80\times10^{9}}{2655}}$$

$$\therefore \quad t = \frac{1}{2\times 4.34\times 10^5}\sqrt{\frac{80\times 10^9}{2655}}$$

$$= (1.152 \times 10^{-6}) \times 5490.28$$

$$= 6.32 \times 10^{-3}\ \text{m} = 6.32\ \text{mm}$$

10.15 Electrostriction

When an electric field is applied to a dielectric, there is a change in its dimensions due to distortion of electronic configuration, or due to the rotation of permanent dipoles, Fig. 10.6a. This behaviour of dielectrics is known as *electrostriction.* Electrostriction is considerable at very high electric fields. Polarization does not occur in a symmetric dielectric crystal. The crystal shown in Fig. 10.6a is symmetrical.

10.16 Pyroelectricity

In accounting for the piezoelectric effect, it is necessary to consider the symmetry of the crystal. When under stress, the centres of gravity of the positive and negative charges are separated, forming an electrostatic dipole and hence the polarization of the crystal. There are many materials in which the symmetry is such that the centres of gravity of the positive and negative charges are separated even without a stress being applied. These

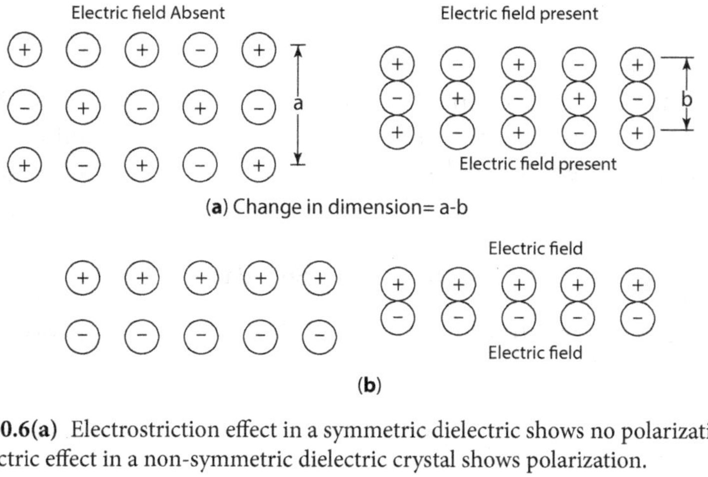

Figure 10.6(a) Electrostriction effect in a symmetric dielectric shows no polarization, (b) Piezoelectric effect in a non-symmetric dielectric crystal shows polarization.

exhibit spontaneous polarization, which means that there are permanent electrostatic charge on the surfaces of the crystal, with one face positive and another negative depending on the direction of the polarization vector.

10.16.1 Pyroelectric Effect

The spontaneous polarization is a strong function of temperature because the atomic dipole moments vary as the crystal expands or contracts. Heating the crystal will tend to desorb the surface neutralizing ions as well as changing the polarization, so that a surface charge may then be detected. Thus the crystal appears to have been charged by heating. This is called the *pyroelectric effect*. For the first time it was observed in the natural crystal tourmaline.

10.16.2 Pyroelectric Coefficient

The electric field developed across a pyroelectric crystal can be remarkably large when it is subjected to a small change in temperature. The pyroelectric coefficient C_{pyro} may be defined as the change in flux density in the crystal due to a change in temperature i.e.

$$C_{\text{pyro}} = \frac{\partial D}{\partial T} \tag{10.6}$$

Its unit is coulombs per square centimetre per degree. For example, a crystal with a typical pyroelectric coefficient of 10^{-8}C cm^{-2} K^{-1} and a relative permittivity of 50, develops a field of 2000 V cm^{-1} for 1 K temperature change.

10.16.3 Pyroelectric Devices

Pyroelectric devices can be used to detect any radiation that results in a change in temperature of the crystal, but are generally used for infrared detection. Because of its extreme sensitivity, a temperature rise of less than one-thousandth of a degree can be detected. The detector must be designed so that the heat generated in the crystal the radiation does not flow away too quickly. Such detectors are widely used in burglar alarms, which detect the thermal radiation from a human body. By using a pyroelectric as the sensitive screen in a television camera tube, infrared images can be formed from differing heat radiation from the scene being viewed, so that the operator can 'see' in the dark. Pyroelectric devices are used in a wide variety of satellite and military applications.

10.17 Lead Zirconate Titanate (PZT): A Piezoelectric Ceramic

It is a solid solution of lead zirconate ($PbZrO_3$) and lead titanate ($PbTiO_3$). Both can be mixed in different proportions to yield the properties of desired choices. Their compositions possess the following favourable qualities.

1. Wide range of dielectric constants.
2. High curie temperature that permits high temperature operations.
3. Electromechanical coupling coefficients better than the $BaTiO_3$.
4. They can be produced easily by sintering.

$PbZrO_3$ is an antiferroelectric material. At room temperature, its structure is cubic which transforms into orthorhombic above curie temperature of 230 °C. i.e.

$$\text{Cubic } (PbZrO_3) \xrightarrow{230^\circ C} \text{Orthorhombic } (PbZrO_3)$$

$PbTiO_3$ is a ferroelectric material. Below curie temperature ($T_c = 490$ °C), its structure is cubic but it changes to tetragonal structure above it. Thus

$$\text{Cubic } (PbTiO_3) \xrightarrow{490^\circ C} \text{Tetragonal } (PbTiO_3)$$

Its c/a ratio is 1.04. PZT is prepared by mixing the zirconate and titanate in a ball mill and then calcinising above 1000 °C.

10.17.1 Different Types of PZTs and Their Uses

The PZTs are of two types viz. hard PZT and soft PZT. Their details are compared as follows.

Description	Hard PZT	Soft PZT
Electric losses	Low	High
Permittivity	Less	More
Piezoelectric coefficients	Low	High
Making pole and dipole	Difficult	Easier
Doping ions	Na^+, K^+, $Al3^+$, $Fe3^+$	$La3^+$, $Sb5^+$, $Nb5^+$

Applications. Main applications of PZT in modern uses are the following.

- Ultrasound
- Sonar
- Buzzers
- Spark generators
- Sensors
- Transducers
- Actuators
- Filters

10.18 Lead Lanthanum Zirconate Titanate (PLZT)

It is a composition of lanthanum (La) with PZT. The lanthanum is added to modify the properties of PZT. Consequently, we find

1. Decreased curie temperature
2. Increased dielectric constant
3. Increased squareness of hysteresis loop
4. Higher electromechanical coupling coefficients
5. Decreased coercive field, and
6. Better optical transparency

The proportion of La in PZT plays an important role in deciding the property of PLZT. It is added in different proportions. If the piezoelectric properties of PLZT is to be enhanced, La should be added less than 5%. But if the PLZT has to be used in electro-optical applications, a proportion of more than 5% is required. PLZT is generally designated by La/$PbZrO_3$/$PbTiO_3$ combination e.g. 7/55/45, 2/65/35, 12/40/60 etc. Various properties of some PLZT composition are given below in Table 10.10 for a ready reference.

10.18.1 General Formula and Hysteresis Loop of PLZT

The general formula for PLZT is given by $(Pb_{1-x}La_x)\ (Zr_{1-y}Ti_y)_{1-x/4}\ V^B{}_{0.25x}\ O_3$ and $(Pb_{1-x}La_x)_{1-0.5x}\ (Zr_{1-y}Ti_y)V^A{}_{0.5x}O_3$. The first formula assumes

Table 10.10 Various Properties of Some PLZT

Composition	Dielectric constant ε_r	Loss tangent tan δ (%)	Curie temperature T_c (°C)	Coercivity H_c (kV/cm)	Specific gravity
2/65/35	650	2.5	320	13.7	7.98
6/65/35	1210	1.6	240	8.7	7.86
10/65/35	5100	5.4	−25	0.0	7.78
14/65/35	1450	2.3	–	0.0	7.45

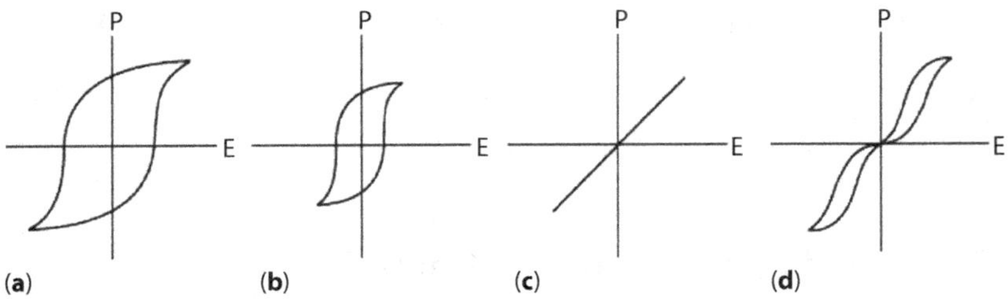

Figure 10.7 Hysteresis loops obtained for different ferroelectric compositions (*a*) tetragonal ferroelectric phase (F_T), (*b*) rhombohedral ferroelectric phase (F_R) (*c*) cubic relaxor ferroelectric phase (F_c), and (*d*) an orthorhombic antiferroelectric phase (A_o)

that La^{3+} ions go to the *A* site and vacancies (V^B) are created on the *B* site to maintain the charge balance. The second formula assumes that the vacancies are created on the *A* site. The actual structure may be due to the combination of *A* and *B* site vacancies.

The electro-optic applications of PLZT ceramics depends on the composition. Figure 10.7a-d shows the hysteresis loops for various PLZT compositions from the phase diagram. PLZT ceramic compositions in the tetragonal ferroelectric (F_T) region show hysteresis loops with a very high coercive field (E_c). Materials with this composition exhibit linear electro-optic behaviour for $E < E_c$. PLZT ceramic compositions in the rhombohedral ferroelectric (F_R) region of the PLZT phase diagram have loops with a low coercive field. These PLZT ceramics are useful for optical memory application.

PLZT ceramic compositions with the relaxor ferroelectric behaviour are characterized by a slim hysteresis loop. They show large quadratic electro-optic effects which are used for making flash protection goggles to shield them from intense radiation. This is one of the biggest applications of the electro-optic effect shown by transparent PLZT ceramics. The PLZT ceramics in the antiferroelectric region show a hysteresis loop expected from an antiferroelectric material. These components are used for memory applications.

10.19 Solved Examples

Example 10.3 How can the ferroelectric materials be classified on the basis of their chemical composition and structure ?

Solution. Based on chemical composition and structure, the ferroelectric materials may be classified into following groups.

1. Tartrate group such as Rochelle salt (KH_3PO_8), NH_4Rb.
2. Oxygen octahedron group such as $BaTiO_3$.
3. Phosphates and arsenates group of alkali metals such as KH_2PO_4.

Example 10.5 Compare the properties of ferroelectricity and piezoelectricity.

Solution. A comparison is given below.

S. No.	Description	Ferroelectricity	Piezoelectricity
1.	Source of polarization	Dipole interaction energy	By application of an external stress
2.	Type of crystals in which occurs	Few non-centrosymmetric crystals which provide a favourable axis of polarity	Non-centrosymmetric crystals
3.	Inter-relation	All ferroelectrics are piezoelectrics	All piezoelectrics are not necessarily ferroelectrics
4.	Value of piezo-electric coefficient	Very large	Comparatively less

Quick Revision Summary

Ageing of mineral insulating oil. It refers to degradation of physical as well as electrical properties of oils with time due to ingress of moisture, oxygen and heat. Ageing of oils leads to deterioration of their insulating properties such as loss of electric strength, change in viscosity, formation of sludge etc.

Asbestos is a fibrous silicate mineral. It is used as insulator in the sheet, powder and wool forms.

Black varnish. It is a resinous solution of bitumen and asphalt in oil. It is of black colour due to the presence of black coloured bitumen as base material. Black varnish is less hygroscopic and does not age quickly, but has poor heat resistance. It is used to manufacture the varnished cloth for use as impregnent in winding the armature coil.

Electro-optic effect is the behaviour of a material in which its optical isotropic nature changes to anisotropic nature on application of an electric field. This effect is seen in $LiNbO_3$, $LiTiO_3$, benzene etc.

Review Questions

1. Classify various insulating materials in different categories. How do the solid, liquid and gaseous insulating materials compare with each other?
2. Enlist different ceramic insulating materials and briefly explain each of them.
3. Distinguish between the following.
   a. Muscovita mica and phlogopite mica
   b. Low voltage porcelain and high voltage porcelain
4. What are different kinds of glass insulating materials? Write their properties and applications.
5. Describe various properties of any two types of ceramic insulating material, any two organic insulating materials, and any two fibrous insulating material.
6. Classify insulating materials from different viewpoints. Suggest applications of the following materials stating the reasons for such uses.
   a. Rubber b. Paper
   c. Porcelain d. Bakelite
7. Differentiate between thermosetting and thermoplast plastics. Quote their examples and briefly explain their utility as insulating materials.
8. Discuss the importance of following insulating materials.
   a. Transformer oil in power transformer.
   b. Gases as dielectrics in high-voltage gas-filled pressure cables.
   c. Synthetic resin in medical electronics.
   d. Vacuum as a dielectric in circuit breakers.
9. What are the requirements of a good liquid insulating materials? Write the characteristics of transformer oil.
10. What are ferroelectric materials? Write down their names, properties and applications.
11. Describe the salient features, structure and properties of barium titanate as a promising ferroelectric material. What is the effect of temperature on its structure?

12. What is meant by 'modified barium titanate'? Name different modifiers used for modifying the properties of $BaTiO_3$. What improvements are noticed by such modifications? Mention ε_r vs T curve in this regard.
13. What is PLZT? Write its properties, features and applications.
14. What is piezoelectricity? Enumerate different piezoelectric materials and compare their qualities. Write various applications based on piezoelectric effect.
15. Explain the phenomenon of electrostriction. What is its importance for transducer materials and devices?
16. Deduce expressions to correlate stress, strain, Young's modulus and electric field in a piezoelectric material.
17. How are the elastic and electric fields related to each other in a piezoelectric material? Define piezoelectric coefficient.
18. Explain pyroelectricity. Name different pyroelectric materials. Discuss the significance of pyroelectric effect on pyroelectric devices.
19. Distinguish clearly the phenomena of ferroelectricity, piezoelectricity and pyroelectricity. Name their specific applications also.
20. Discuss the following term :
    a. Pyroelectricity
    b. Pyroelectric devices
21. Differentiate between the following.
    a. Ferroelectric materials and anti-ferroelectric materials.
    b. Electrostriction effect and pyroelectric effect.
22. Write notes on the following.
    a. Air and nitrogen as insulating material
    b. Cable and capacitor paper as insulating material.
    c. Use of micanite and glass bonded mica in electrical insulation.

Numerical Problems

1. Piezoelectric coefficient for a material is 1.0×10^{-10} m/V and its Young's modulus is 70 GPa. A 0.20 mm thick wafer of it is subjected to a compressive stress of 30 MPa. Determine the potential difference produced across it.
2. A potential difference of 1.9 kV is to be produced across 't' thick slice of quartz, when it is subjected to a sress of 15 MPa.

Taking modulus of elasticity of quartz as 70 GPa and piezoelectric coefficient as 2.3×10^{-12} m/V, determine the thickness of the slice.

Objective Questions

1. In a semiconductor strain gauge, the change in resistance on application of strain is mainly due to the change in its

   a. length b. diameter
   c. resistivity d. length and diameter

2. Match list I with list II, and choose the correct answer from the codes given below the lists.

List I (Items)	*List II (Applications)*
A. Muscovita	1. Glass
B. Pyrex	2. Thermoplast
C. Bakelite	3. Synthetic rubber insulation
D. Butyl	4. Epoxy
	5. Mica

Codes:

	A B C D		A B C D
a.	5 1 4 3	b.	1 5 4 3
c.	5 1 3 4	d.	1 5 3 4

3. For high speed reading and storing of information in a computer, the use is made of

   a. ferrite b. piezoelectrics
   c. pyroelectrics c. ferromagnetics above 768°C

4. Which of the following pairs does *not* match?

a. Dielectric strength of air	3 to 5 kV/mm
b. Vacuum may be used for	reflective insulation of liquid O_2 tanks
c. Permissible water content	50 ppm in transformer oil is
d. Carbon-pile resistors	for insulation of oil-filled cables

11

Magnetic Materials: Properties and Behaviour

11.1 Origin of Permanent Magnetic Dipole

Magnetic properties of solids originate from the motion of their electrons, and their magnetic moments or dipoles.

- The electron orbits are always associated with magnetic moments.
- The orbital arrangements may be such that the atoms may or may not have a permanent magnetic moment.
- The magnetic moments, if present, interact in different ways.
- They may be aligned, parallel or anti-parallel, equal or unequal in magnitude.
- Permanent magnetic dipoles exist even in the absence of an applied field.

11.1.1 Spinning Electrons Acting as Extremely Small Magnets

Origin of magnetic dipoles can be understood well by knowing the electron configuration and their filling sequence in an atom. Since the electrons spin about their own axes and also revolve in well- defined orbits, hence they constitute a current loop and create their own magnetic fields. Thus a charged particle under an angular momentum contributes to a permanent dipole moment. Origin of permanent magnetic dipole is thus attributed to spinning electrons that act as extremely small magnets having their two ends as north and south poles.

Uses. Magnetic materials find use in following major fields.

i. Magnetic materials are of great use in electrical equipments such as given below.
   - Transformers
   - Alternators
   - Motors
   - Electromagnets.
ii. Magnetic tapes and thin films, ferrites, and metallic glasses find special application in the following areas.
   - Memory of computer cores
   - Magnetic shielding
   - Recording devices, and
   - Credit card

Recent advances. Ceramic magnetic materials such as $BaO.6Fe_2O_3$ and $CoO.Fe_2O_3$, Cunife (Cu + Ni + Fe) and Cunico (Cu + Ni + Co) ductile alloys, and powdered manganese bismuthide (MnBi) are the recent additions in magnetic materials.

11.2 Terminologies Defined

Magnetic dipole moment. An electron creates its own magnetic field. A charged particle under an angular momentum, contributes to permanent dipole moment. Magnetic dipole moment μ_m is due to the orbital angular momentum of electrons owing to quantum numbers, and is given by

$$\mu_{\mathrm{m}} = \frac{ep}{2m} \tag{11.1}$$

where e is electronic charge, m is the mass of an electron, and $p = m\omega r^2$ is its angular momentum. Here, ω is angular velocity and r the radius of electron motion.

Bohr magneton β is the unit of magnetic moment of an electron spin, and is expressed by

$$\beta = \frac{\pm eh}{4\pi m} = \pm 9.273 \times 10^{-24}\,\text{Am}^2 \quad (11.2)$$

where h is Planck's constant. Net magnetic moment of two electrons of opposite spins is zero but the atoms and molecules having unpaired electrons possess some net moment.

Magnetization M of a solid is the sum of magnetic moments per unit volume V. It is given by

$$M = \sum \frac{\mu_m}{V} \quad (11.3)$$

Magnetic flux density or magnetic induction B is the ability of a magnetic material to retain magnetism when a magnetic field H is applied on it. It is related as

$$B = \mu H = \mu_0 \mu_r H \quad (11.4)$$

where μ, μ_0 and μ_r are absolute permittivity, permittivity of free (vacuum) space, and relative permittivity of the solid respectively. The value of $\mu_r = 1$ for free space and is more than 1 for solids. The magnetic flux density in a solid may also be defined as

$$B = \mu_0 (M + H) \quad (11.5)$$

Susceptibility χ. The magnetization of a solid is proportional to the applied magnetic field. So

$$M \propto H$$

or $$M = \chi H$$

Therefore susceptibility is expressed as

$$\chi = \frac{M}{H} \quad (11.6)$$

It is a parameter to assess the magnetic ability of a material. Its value may be positive or negative. Higher the positive value, better is the magnetic quality of material. Susceptibility of some materials are given in Tables 11.1 and 11.2.

Table 11.1 Susceptibility of Some Diamagnetic Solids

Material	Si	Al_2O_3	Diamond	Au	Graphite
Susceptibility χ (10^{-5})	-0.3	-0.5	-2.1	-3.6	-12.0

Table 11.2 Susceptibility of Some Paramagnetic Solids

Compound	$NiSO_4$	Fe_2O_3	$MnSO_4$	$FeCl_2$
Susceptibility χ(10^{-3})	1.2	1.4	3.6	3.7

Curie constant C_c. Susceptibility is independent of magnetizing field but varies inversely with absolute temperature T. So

$$\chi \propto \frac{1}{T}$$

or

$$\chi = \frac{C_c}{T} \tag{11.7}$$

where the constant of proportionality C_c is called Curie constant. It is expressed by

$$C_c = \frac{N\mu\beta^2}{k} \tag{11.8}$$

where N is electron spins per m^3 in a solid, and k is Boltzmann constant. The value of N may be obtained from

$$N = N_p + N_a \tag{11.9}$$

in which N_p and N_a are the dipoles per m^3 parallel to, and antiparallel to the applied field respectively.

11.2.1 Relation between Relative Permeability and Magnetic Susceptibility

Relative permeability μ_r and magnetic susceptibility χ are two important parameters of a magnetic material. A relation between them can be derived as follows.

$\because$	$B = \mu_0\mu_r H$	(from Eqn. 11.4)
and	$B = \mu_0(M + H)$	(from Eq. 11.5)
$\therefore$	$\mu_0\mu_r H = \mu_0(M + H)$	(*i*)
Knowing that	$M = \chi H$	(from Eq. 11.6)

We can write Eq. (*i*) as

$$\mu_0\mu_r H = \mu_0(\chi H + H)$$
$$= \mu_0\chi H + \mu_0 H$$

or $\mu_0\mu_r H - \mu_0 H = \mu_0\chi H$

i.e. $(\mu_0\mu_r - \mu_0)H = \mu_0\chi H$

hence $\mu_0(\mu_r - 1) = \mu_0\chi$

or $\mu_r - 1 = \chi$

Thus $$\mu_r = 1 + \chi \quad (11.10)$$

Example 11.1 A magnetic field of 2400 A/m is applied to a material having a susceptibility of 1500. Determine (*a*) its relative permeability, (*b*) intensity of magnetization, and (*c*) the remanance.

Solution. (*a*) From Eq. 11.10, we find

$$\mu_r = 1 + \chi = 1 + 1500 = 1501$$

(*b*) Since $$\chi = \frac{M}{H}$$

$\therefore$ $$M = \chi H = 1500 \times 2400$$
$$= 3.6 \times 10^6 \text{ A/m}$$

(*c*) As $$B = \mu_0\mu_r H$$

$\therefore$ $$B = (4\pi \times 10^{-7}) \times 1501 \times 2400$$
$$= 4.52 \text{ T}$$ (*T* stands for tesla)

Example 11.2 Relative permeability of 'supermalloy' is 200 000. It has a magnetization of 6000 A/m. Determine the strength of the magnet thus produced. Take $\mu_0 = 4\pi \times 10^{-7}$ henry/metre.

Solution. Given are: $\mu_r = 200\ 000$, and $M = 6000$ A/m.

Since the strength of the magnet is expressed in Tesla (or Wb/m^2) for which we have to determine magnetic flux density B, hence using Eqns. 11.6 and 11.10, we write

$$M = \chi H$$

$$= (\mu_r - 1)H$$

So $$6000 = (200\ 000 - 1)H$$

or $$H = \frac{6000}{199999} = 0.03$$

Now on using Eqn. 11.4, we find as

$$B = \mu_0 \mu_r H$$

$$= (4\pi \times 10^{-7}) \times 200\ 000 \times 0.03$$

$$= 7.539 \times 10^{-3}\ \text{Wb/m}^2 = 0.007\ \text{tesla}$$

Example 11.3 Nickel has a magnetic moment of 0.6 times of Bohr magneton per atom. If its lattice constant is 0.35 nm, find the saturation magnetization.

Solution. Given are: $M = 0.6\,\beta$, β (Bohr magneton) $= 9.27 \times 10^{-24}$ Am^2,

and $$a = 0.35\ \text{nm} = 0.35 \times 10^{-9}\ \text{m}.$$

The structure of Ni unit cell is FCC, and effective number of atoms/unit cell $N_e = 4$. Therefore

$$M_s = \frac{N_e M}{a^3}$$

$$= \frac{4}{(0.35 \times 10^{-9})^3} \times 0.6 \times 9.27 \times 10^{-24}\ \text{A/m}$$

$$= 0.51 \times 10^6\ \text{A/m}$$

11.3 Classification of Magnetic Materials

The magnetic materials are classified into following types based on the arrangement of their magnetic moments.

1. Diamagnetic materials,
2. Paramagnetic materials,

3. Ferromagnetic materials,
4. Antiferromagnetic materials, and
5. Ferrimagnetic materials (or ferrites).

11.3.1 Distribution of Magnetic Moments

The distribution of magnetic moments and their orientations for these materials are shown in Figs. 11.1*a-e*.

- The permanent dipoles are absent in diamagnetic materials, Fig. 11.1*a*, therefore they cannot be made magnets.
- The permanent moments are randomly oriented in paramagnetic materials, Fig. 11.1*b*. They align themselves parallel to the applied field feebly and become weak magnets.
- Figure 11.1*c* depicts a ferromagnetic material in which the magnetic moments are aligned, and are equal in magnitude. They are very good magnetic materials.
- Antiferromagnetic material and ferrimagnetic materials are sub-classes of ferromagnetic materials. In antiferromagnetic materials (Fig. 11.1*d*), the magnetic moments of equal magnitude are aligned in opposite directions.
- In ferrimagnetic materials, (Fig. 11.1*e*), the magnetic moments of unequal magnitude are aligned in opposite directions.

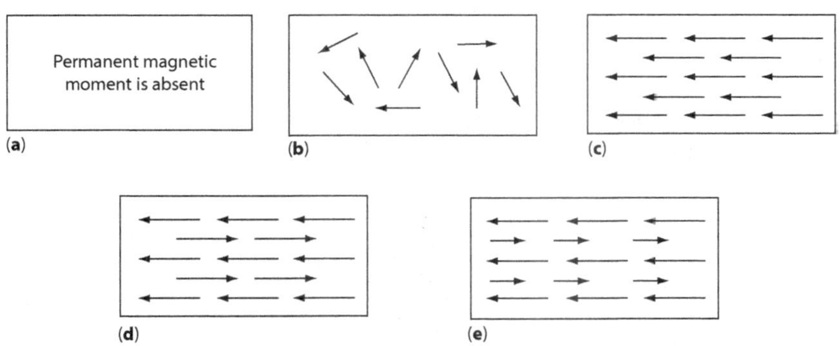

Figure 11.1 Behaviour of materials under the influence of magnetic field shows (*a*) diamagnetic (*b*) paramagnetic, (*c*) ferromagnetic (*d*) antiferromagnetic, and (*e*) ferrimagnetic nature.

11.4 Diamagnetism and Diamagnetic Materials

On application of a magnetic field to a diamagnetic solid, the orbital motion of electrons in the atoms modify, and very weak magnetic moments are induced. The interaction between neighbouring dipoles is absent and permanent magnetic moments do not exist. The diamagnetic substances position themselves in a direction perpendicular to the applied field. This behaviour shown by them is referred as diamagnetism. Gold, silicon, diamond, ionic solids and molecular solids are diamagnetic materials.

11.4.1 Negative Susceptibility

Diamagnetic solids repel the magnetic lines of force when subjected to an applied magnetic field, Fig. 11.2*a*. Their susceptibility is negative and of the order of 10^{-5}. Diamagnetic susceptibility of some materials are shown in Table 11.1. This susceptibility is related to the relative permittivity at all temperatures as

$$\chi = \mu_r - 1$$

11.5 Paramagnetism and Paramagnetic Materials

Atoms and ions possessing odd number of electrons give rise to residual permanent magnetic moment. If some inner orbits are incomplete in an atom having an even number of electrons, the solid may behave as paramagnetic.

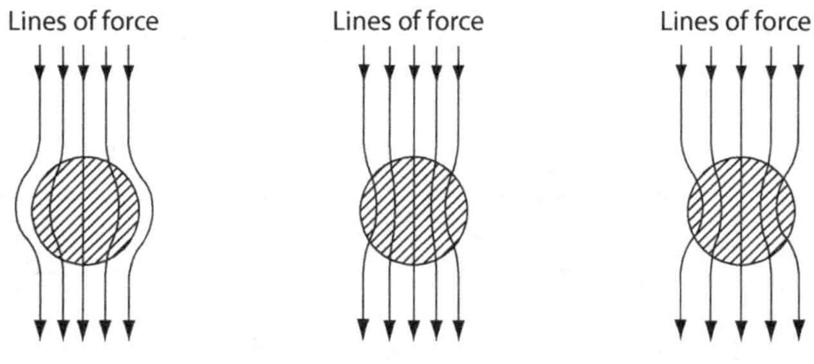

Figure 11.2 Magnetic lines of force are (*a*) repelled by diamagnetic solids, (*b*) feebly attracted by paramagnetic solids, and (*c*) strongly attracted by ferromagnetic solids.

- These randomly oriented magnetic moments as a whole have negligible net magnetic moment in the solid.
- When a magnetic field is applied on them, the random magnetic moments align themselves in the direction of the field.
- This results in a feeble magnetization, and is referred to as *paramagnetism.*
- This nature is found in the oxides, chlorides and sulphates of some metals.

11.5.1 Rare-earth Based Paramagnetic Salts

Paramagnetic salts which are rare-earth based compounds, such as $ErNi_2$, EuS, Gd_5 Si, DAG ($Dy_3Al_5O_{12}$) and GGG* ($Gd_3Ga_5O_{12}$), also fall in the category of paramagnetic materials. The magnetic lines of force are weakly attracted by them as shown in Fig. 11.2*b*. Their susceptibility is positive and of the order of 10^{-3}. Paramagnetic susceptibility of some solids is shown in Table 11.2.

Equation 11.10 also holds good for paramagnetic materials.

11.6 Ferromagnetism and Ferromagnetic Materials

Spontaneous magnetic moments exist in ferromagnetic solids even in the absence of applied magnetic field. The strong interactions between the atoms in a solid hold magnetic moments of neighbouring atoms parallel to each other. This results in ferromagnetism. Iron, nickel and cobalt are very good ferromagnetic materials. Gadolinium possesses this property at low temperatures (below 290 K). Ferromagnetism is pronounced in some oxides and carbides also. Europium oxide (EuO) is one such example.

The magnetic lines of force are strongly attracted by such solids as shown in Fig. 11.2*c*. Their susceptibility is positive and of the order of 10^2 to 10^5.

11.7 Antiferromagnetism and Antiferromagnetic Materials

The tendency for parallel alignment of electron spins is due to the quantum mechanical exchange forces. When the distance between interacting atoms is small, the exchange forces produce a tendency for antiparallel

* GGG stands for gadolinium-gallium- garnet.

alignment of electron spins of neighbouring atoms. Consequently, there is no net magnetic moment in the solid. The atoms in the first transition metals, such as copper and zinc, in the solid state are said to be anti-ferromagnetically coupled. Oxides of manganese, iron and cobalt; and compounds of nickel and manganese are the examples of this class of magnetic materials.

11.7.1 Maximum Susceptibility and Neel Temperature

The susceptibility of antiferromagnetic materials is a function of temperature. Figure 11.3 shows variation of susceptibility of MnF_2 with temperature. The maximum susceptibility χ_{max} occurs at a particular temperature known *as Neel temperature* T_N. Neel temperature of some compounds are given in Table 11.3. It can be observed that this temperature is much lower than the room temperature.

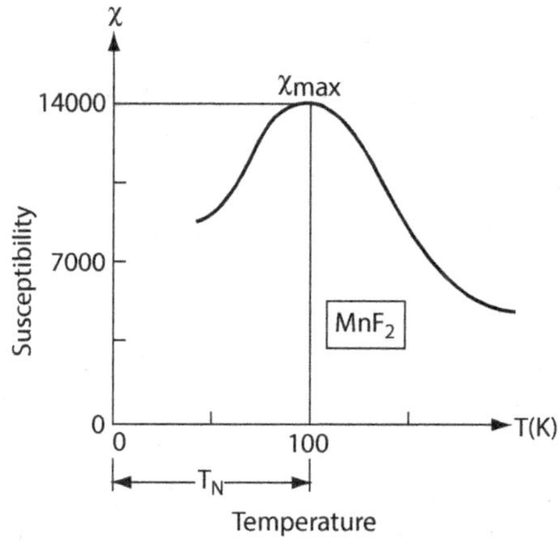

Figure 11.3 Susceptibility of antiferromagnetic MnF_2 as a function of temperature. Maximum susceptibility occurs at Neel temperature

Table 11.3 Neel Temperature of Some Antiferromagnetic Materials

Compound	MnF_2	MnO_2	MnS	FeO	NiF_2	CoO
Neel temperature T_N(K)	72	84	165	198	73	292

11.8 Ferrimagnetism and Ferrites

Some compounds have unequal magnetic moments although their atoms are antiferromagnetically coupled. The magnitude of magnetic moment is more in one direction than in the other (Fig. 11.1*e*). This behaviour of a material is called *ferrimagnetism*. *Garnets* and *ferrites* show this behaviour.

Ferrites are compounds of two metallic oxides of which one is invariably an iron oxide. The other metallic oxides may be bivalent elements such as Ni, Mn, Zn, Cu and Fe.

- Symbolically, ferrites may be designated as ($Me_tO.Fe_2O_3$) in which 'Met' stands for metal and expresses the above elements.
- Magnetite (Fe_3O_4) is a compound of ferrous oxide (FeO) and ferric oxide (Fe_2O_3) and is expressed as ($FeO.Fe_2O_3$).
- Its structure is cubic.
- More common ferrites are manganese ferrite ($MnFe_2O_4$), nickel ferrite ($NiFe_2O_4$), magnesium-manganese ferrite, zinc-manganese ferrite and zinc-nickel ferrite.

11.8.1 Properties of Ferrites

Ferrites are ceramic materials possessing following favourable properties.

1. Very high electrical resistivities.
2. Low power loss at high frequencies.
3. Suitable for temporary as well as permanent magnetic applications due to their spontaneous magnetization.
4. Conductivity behaviour like those of semiconductors.

However, they suffer from poor machinability and brittleness. The magnetic hysteresis loop (art. 11.11) of ferrites may vary from very narrow to very wide. Their shape is almost rectangular as shown in Fig. 11.7*b* (see Example 11.4).

11.8.2 Applications of Ferrites

Ferrites with narrow hysteresis loop form soft magnets. They are used for the applications such as given below.

- Audio transformers
- Television transformers
- Gyrators
- Inductance cores.

The Mn-Mg ferrite with almost rectangular hysteresis loop is employed to make the following components.

- Memory cores of computers.

Barium ferrite ($BaO.6Fe_2O_3$) is used to make the following items.

- Permanent magnets.

11.8.3 Soft and Hard Ferrites

Ferrites exhibit hysteresis during magnetization. The hysteresis loops of Cu-Mn ferrite, Mg-Mn ferrite and Zn-Mn ferrite are square, and they are used as memory cores in computers. The ferrites may also be classified as follows.

1. Soft ferrites
2. Hard ferrites
   - Lithium ferrites are soft ferrites. They possess square hysteresis loop and low dielectric losses. These serve as low mobility semiconductors. Due to their high curie temperature, they are used for applications at microwave frequencies. Hard ferrites such as barium, strontium and lead ferrites show semiconductor behaviour.
   - A general purpose Ni-Zn ferrite has relaxation time of the order of 10^{-9}s at microwave frequencies whereas this relaxation time for Co-Zn ferrite is of the order of 10^{-10} s. They are made by powder compaction technique. Their relative permeability is typically of the order of 10^4, and resistivities of 10^7 ohm m.

11.8.4 Spinel, Garnet and Magnetoplumbite

Ferrites have three different types of crystals viz. spinel, garnet, and magnetoplumbite. They are, therefore, classified accordingly as given below.

Type	Structure	General formula	Example
• Spinel • Garnet • Magnetoplumbite	Cubic Cubic Hexagonal	MFe_2O_4 MFe_2O_{12} MFe_2O_{19}	M = Zn, Mg, Co, Cd, Cu, Mn, Ni M = Sm, Tb, Ho, Er, Lu, Tm, Y, Eu, Gd, Dy M = Ba

The spinel ferrite unit cell contains 8 × MFe_2O_4 units where *M* is a divalent metal ion. This is similar to $MgAl_2O_4$, which is a mineral spinel. The iron and metal ions occupy octahedral and tetrahedral sites of the spinel lattice respectively. This distribution imparts interesting magnetic and electrical properties to ferrites.

11.8.5 Normal and Inverse Spinel

The spinel can be of two types (*i*) Normal spinel, and (*ii*) Inverse spinel. Zinc ferrite ($ZnFe_2O_4$) is a *normal spinel* while magnetite (Fe_3O_4) is an *inverse spinel.* The curie temperature of spinel ferrites ranges between 700 K to 860 K. Their conductivity varies between 10^2 to 10^{-11} ohm-cm.

11.8.6 Garnets

This is the name for a class of compounds crystallizing in a certain crystal structure. As far as magnetic properties are concerned, their most interesting representative is yttrium-iron garnet ($Y_3Fe_5O_{12}$), which happens to be ferromagnetic for a curious reason. The spin of the yttrium atoms is opposite to the spin of the iron atoms so the magnetic moments line up alternately if the orbital magnetic moments are small. But for yttrium the orbital magnetic moment is larger than the spin, and is in the opposite direction. Hence the total magnetic moment of yttrium atom is in the same direction as that of iron, making the compound ferromagnetic.

11.8.7 Ferrites in Memory Devices

We have already discussed the basic aspects of ferrites. Some more information and the practical/commercial aspects will be discussed now. For a typical ferrite the properties are as follows.

- Coefficient of linear thermal expansion ≈ 10^{-5}/°C
- Thermal conductivity ≈ 5 × 10^2 W/m°C

- Thermal capacity ≈ 0.17 kcal/kg°C
- Density = 3000 to 5000 kg/m^3

Ferrites are suitable for use in those devices which require

i. high speed response,
ii. very low power loss, and
iii. high frequency.

This makes them useful for the purpose of memory devices in computers. In memory devices, they are formed as miniature ring cores which are fitted on the current carrying conductor in the circuit. Consequently, they are magnetized each time if the phase of the current changes. In this way it memorizes the possible states.

11.9 Curie Temperature

The aligned magnetic moments tend to be random due to thermal energy. Therefore, ferromagnetic solids behave as paramagnetic solids at high temperatures. A critical temperature at which the alignment of magnetic moments vanish is called *Curie point* or *Curie temperature.* The effects of curie temperature on various magnetic properties are as follows.

- The susceptibility of a ferromagnetic material decreases with an increase in the curie temperature.
- Magnitude of exchange energy has an effect on the value of curie temperature.
- Higher the energy, higher will be the curie temperature.

Curie temperature is different for ferromagnetic and paramagnetic materials. It is, therefore, referred to as *ferromagnetic curie temperature* T_{fe} *and paramagnetic curie temperature* T_{pe}. These values for some materials are shown in Table 11.4.

Figure 11.4 shows the variation of saturation magnetization as a function of critical temperature. The saturation magnetization M_s becomes zero at T_c, and the material practically becomes nonmagnetic.

Table 11.4 Ferromagnetic curie temperature and paramagnetic curie temperature of some materials

Material	Ferromagnetic Curie temperature T_{fe}(K)	Paramagnetic Curie temperature T_{pe}(K)
Fe	1045	1095
Co	1395	1430
Ni	630	650
Gd	289	—
MnF_2	—	115
MnO_2	—	315
MnS	—	530
FeO	—	570
NiF_2	—	115
CoO	—	280

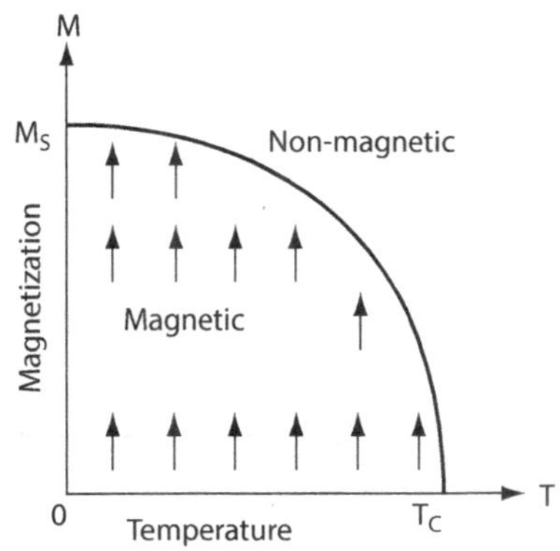

Figure 11.4 Saturation magnetization as a function of critical temperature demarcates magnetic and nonmagnetic regions.

11.9.1 Requirement of Higher and Lower Curie Temperature for Different Applications

To maintain magnetic properties in transformers, motors, alternators, relays and permanent magnets; materials used must have curie temperature higher than their maximum operating temperatures.

A lower curie temperature than the lowest operating temperature is required in the spectrometers, cathode-ray tubes, and in compass housings. Non-magnetic quality is desired in these cases. Therefore, curie temperature of a material needs to be varied to get the desired results. It is varied by suitable alloying.

11.10 Laws of Magnetic Materials

The variation of susceptibility χ of different materials at ferromagnetic curie temperature T_{fc}, and paramagnetic curie temperature T_{pc} follows certain established laws. These laws are

1. Curie law,
2. Curie-Weiss law, and
3. Neel law

Curie law. Curie law applies to the elements in pure state, oxides and salts of paramagnetic nature. It is given by

$$\chi = \frac{C_C}{T} \tag{11.11}$$

where T is the absolute temperature of the materials.

Curie-Weiss law. Several paramagnetics in bulk state obey Curie-Weiss law expressed as

$$\chi = \frac{C_C}{T - T_{pc}} \tag{11.12}$$

This law also holds good for ferromagnetic materials at $T > T_{fc}$.

Neel law. Antiferromagnetic materials are governed by Neel law, given as

$$\chi = \frac{C_C}{T + T_{pc}} \tag{11.13}$$

These laws are illustrated in Fig. 11.5 on the reciprocal of susceptibility versus temperature plot.

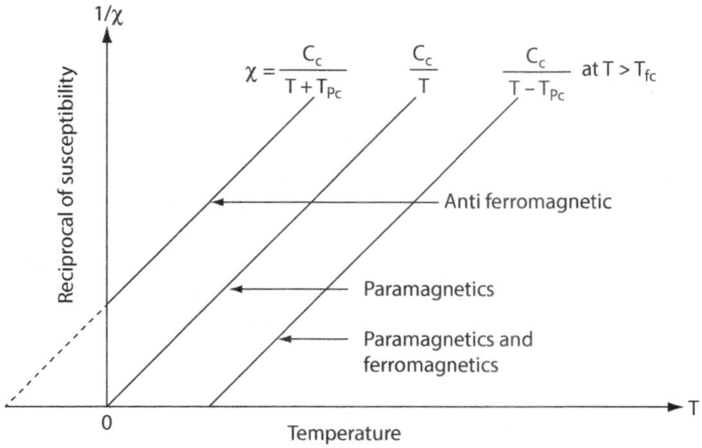

Figure 11.5 Illustration of different laws obeyed by magnetic materials.

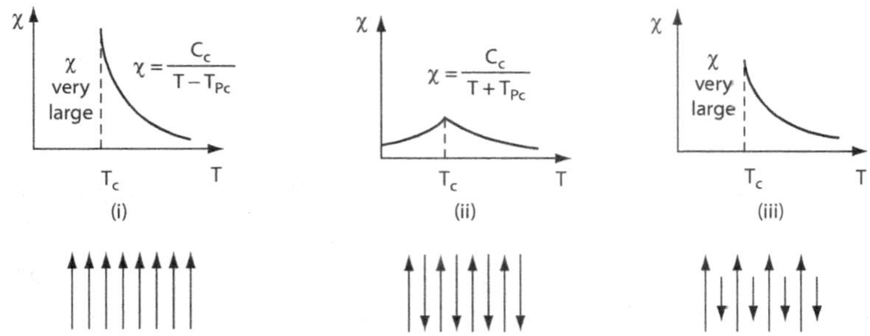

Figure 11.6 Magnetic susceptibility versus T for (*i*) ferromagnetic, (*ii*) antiferromagnetic, and (*iii*) ferrimagnetic materials. Magnetic moment alignment is also indicated for each case

11.10.1 Effect of Temperature on Magnetic Susceptibility

The effect of temperature on magnetic susceptibility may be understood from the $1/\chi$ vs T plot of Fig. 11.5, but for more clarity it is further shown as χ vs T in Fig. 11.6. The details are self explanatory.

11.11 Magnetization Curve, and Initial and Maximum Permeability

The relationship between changing magnetic field H and its influence on the magnetic flux density B can be represented by a curve. This is called *magnetization curve* and is shown by *OACD* path in Fig. 11.7*a*. This curve has three regions.

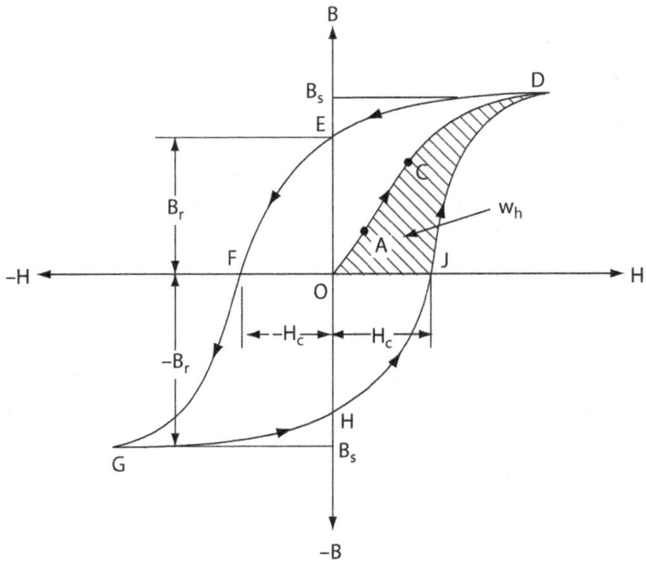

Figure 11.7 (*a*) Magnetization curve *OACD*, and magnetic hysteresis loop *ODEFGHJD* showing variation of magnetic induction and the magnetic field

In the first region *OA*, the curve rises up. The second region *AC* has almost a constant slope which starts bending near *C*. The third region *CD* has a decreasing slope, and becomes almost horizontal at *D*. At this part the material has reached to a saturation flux density *B*. If *H* is enhanced further, there will be no useful increase in the value of flux density. The slopes of the first and second regions are known as *initial permeability* and *maximum permeability* respectively.

11.11.1 Magnetic Hysteresis Loop (or Cycle Magnetization)

Remanence. The changes in magnetization in a ferromagnetic material lag behind the variations of the magnetic field applied on it. In Fig. 11.7*a*, when the magnetic field is decreased and brought back to zero, the curve assumes *DE* path. Even at $H = 0$, the magnitude of magnetic flux is equal to $OE = B_r$. This is residual magnetic flux density, and is also known as *retentivity* or *remanence* of ferromagnetic material. Higher this value, better is the quality of ferromagnetic material.

Coercive force. When direction of the *field H* is reversed, the curve follows *EF path. F* is another important point. Here *OF* indicates *demagnetization*

force or *coercive force*. It means that a ferromagnetic material requires magnetic field of $H = -H_c$ to bring down its flux density to zero. A strong ferromagnetic material will need a higher value of $-H_c$. On further increasing the magnetic field, the curve follows *FG* path and reaches to saturation flux density $-B_s$ in the opposite direction. Thus one cycle is complete.

Now the magnetic field is applied again on the ferromagnetic material. This time the *B-H* curve follows *HJD* path. The curve does not take on the original *OACD* path. The curve *DEFGHJD* is called *magnetic hysteresis loop*. It should be clear that $OB_s = -OB_s$, $OF = OJ$, and $OE = OH = B_r$.

Example 11.4 The B-H curve for a magnetic material (ferrite) is shown in Fig. 11.7*b*. Estimate the hysteresis loss per cycle for it. Assume the loop to be roughly a rectangle.

Solution. From the given Fig. 11.7*b*, we observe that

The width of the loop $= (OA + OB) = 80 + 80 = 160$ A/m

Height of the loop $= OC + OD$

$= 0.15 + 0.15 = 0.30$ Wb/m^2

$\therefore$ Area of the loop $= (160 \times 0.3)$ Wb/m^2 A/m

$= 48$ T A/m $= 48$ J

This is the energy loss per unit volume of the magnetic material during one cycle.

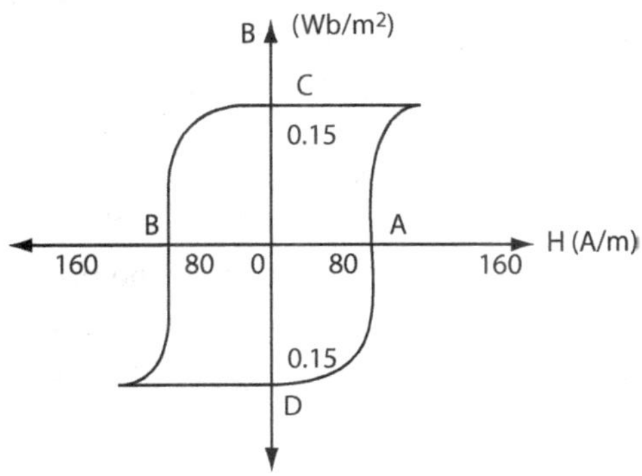

Figure 11.7b B-H curve of a magnetic material, generally ferrite, is almost rectangular.

11.12 Hysteresis and Eddy Current Losses

The enclosed area *OACDJO* in magnetic hysteresis loop is called *hysteresis loss* W_h. In a nonlinear *B-H* curve, *B* always lags behind *H*. When the applied magnetic field varies cyclically in subsequent 2nd, 3rd,... *n*th cycles, it causes eddy current loss W_e. This is an undesired effect.The work required to go round the hysteresis loop is proportional to the area enclosed. The sum of hysteresis loss W_h and the eddy current loss W_e is known as *core loss* W_c.

Thus the total power loss is given by

$$W_c = W_h + W_e \tag{11.14}$$

where
$$W_h = \eta B_{\max}^2 fV \text{watt} \tag{11.15}$$

and
$$W_e = \lambda B_{\max}^2 f^2 t^2 V \text{watt} \tag{11.16}$$

Here
$$\text{loop area } A = \eta B_{\max}^{1.6} \tag{11.17}$$

in which η is Steinmetz coefficient, λ is eddy current constant of material, $B_{\max} = B_s$ is the saturation flux density, *t* is thickness of the material, *f* is the frequency of reversal of magnetic field and *V* the volume of material.

11.12.1 Eddy Current Loss per Unit Volume*

The term eddy current is applied to an electric current which circulates within a mass of conductor material, when the material is situated in a varying magnetic field. These eddy currents result in a loss of power. Eddy current losses together with the hysteresis losses cause heating of magnetic materials. To reduce these losses, not only are the materials with low Steinmetz coefficients chosen but the magnetic core is also made in the form of laminations to reduce the eddy current loss.

The total power loss in a plate of thickness $t = \dfrac{4}{3}\dfrac{B_m^2 f^2 t^3 dlk^2}{\rho}$ (for derivation, see the source reference)

where *t* is thickness of lamination, *l* is length of the plate, *d* is depth of the plate, 2*d* is length of eddy current path, *k* is form factor of wave (it is 1.11 for sinusoidal waveform) and ρ is resistivity of the material. As the volume of the plate = *d.l.t*,

*Source: Electromagnetic Theory by Professor J. R. Lucas.

$$\text{Eddy current loss per unit volume} = \frac{4}{3}\frac{B_m^2 f^2 t^2 k^2}{\rho} \tag{11.18}$$

It is thus seen that the eddy current loss per unit volume is proportional to the square of the thickness of the individual plates. Thus for a given volume, if the thickness of the laminations is made very small, the eddy current losses can be minimized.

Comparison of properties of eddy current and hysteresis losses. These are

1. Eddy current loss is proportional to the square of the frequency, while the hysteresis loss is directly proportional to the frequency.
2. Eddy current loss is proportional to the square of the peak flux density, while the hysteresis loss directly is usually proportional to the 1.6th power of the peak flux density.
3. Eddy current loss is proportional to the square of the thickness of the laminations, while the hysteresis loss does not depend on thickness of laminations.
4. Eddy current loss is dependent on the resistivity of the material, while the hysteresis loss is dependent on the Steinmetz constant of the material.

Example 11.5 Determine the power loss due to hysteresis in a transformer core of 0.01 m³ volume at 50Hz frequency. The are of the loop is 600 J/m².

Solution. Given data are

loop area A = 600 J/m², f = 50 Hz and V = 0.01 m³

Using Eq. 11.17,

$$A = \eta B_{\max}^{1.6}$$

in Eq. 11.15, we get

$$\begin{aligned} W_h &= \eta B_{\max}^{1.6} fV \\ &= (600\text{J}/\text{m}^2)\times 50\text{Hz}\times 0.01\text{m}^3 \\ &= 300\,\text{Watt} \end{aligned}$$

Example 11.6 In a reactor, the hysteresis loss is 300 W when the maximum flux density is 0.9 Wb/m² and frequency 50 Hz. Find the loss at 40 Hz if the maximum flux density is raised to 1.1 Wb/m². Assume the hysteresis loss to varry as $(B_{max})^{1.7}$.

Solution. Using Eq. 11.15, the hysteresis loss is given by

$$300 = \eta \times (0.9)^{1.7} \times 50 \times 10^{-7} \times V \quad \text{(i)}$$

and

$$W_{h2} = \eta \times (1.1)^{1.7} \times 40 \times 10^{-7} \times V \quad \text{(ii)}$$

On dividing (ii) by (i), we get

$$\frac{W_{h2}}{300} = \frac{\eta \times (1.1)^{1.7} \times 40 \times 10^{-7} \times V}{\eta \times (0.9)^{1.7} \times 50 \times 10^{-7} \times V}$$

$$\therefore \quad W_{h2} = \frac{300 \times (1.1)^{1.7} \times 4}{(0.9)^{1.7} \times 5}$$

$$= 240 \times 1.22 = 293\text{W}$$

Example 11.7 Calculate the loss per kg in a specimen of alloy steel for a maximum flux density of 1.1 Wb/m² and a frequency of 50 Hz using 0.5 mm thich sheets. The resistivity of alloy steel is 30×10^{-8} ohm-m. The density is 7800 kg/m³. Hysteresis loss in each cycle is 380 W-s/m³. Assume the waveform to be sinusoidal.

Solution: Given data are: $B_m = 1.10$ Wb/m², $f = 50$ Hz, $t = 0.5$ mm = 5×10^{-4} m, $\rho = 30 \times 10^{-8}$ ohm-m, density $\rho' = 7800$ kg/m³, $m = 1$kg, $W_h = 380$ W-s/m³, and $k = 1.11$.

$$\because \quad W_e = \frac{4}{3}\frac{B_m^2 f^2 t^2 k^2 V}{\rho}$$

$$\therefore \quad W_e = \frac{4}{3}\frac{(1.1)^2 (50)^2 (5 \times 10^{-4})^2 \times (1.11)^2}{30 \times 10^{-8} \times 7800}$$

$$= \frac{1.33 \times 1.21 \times 625 \times 1.232}{2340}$$

$$= 0.5309 \text{ watt / kg}$$

Example 11.8 Calculate the loss of energy per hour due to hysteresis in an iron piece of 50 kg mass, 7500 kg/m³ density and subjected to 50 Hz frequency; if its hysteresis loop area is found to be 150 m². The scale factors on ordinate and abscissa are 1 cm = 0.008 Wb/m² and 1 cm = 20 A/m, respectively.

Solution. Given data are: Frequency $f = 50$ Hz, Mass $m = 50$ kg, Hysteresis loop area = 150 m², scale factors: 1 cm = 0.008 Wb/m², 1 cm = 20 A/m, Density of iron $\rho = 7500$ kg/m³

$$\therefore \qquad \text{Volume } V = \frac{\text{Mass}\,(m)}{\text{Density}\ \rho} = \frac{50}{7500}$$

As the energy lost during each cycle = Area of loop

$$= 150 \times 0.008 \times 20 \times 10^4 \qquad (\therefore\ 1\ \text{m}^2 = 10^4\ \text{cm}^3)$$

$$= 24 \times 10^4\ \text{J/m}^3$$

and power loss due to hysteresis

$$= \text{Area of loop} \times \text{frequency} \times \text{volume}$$

$$= (24 \times 10^4) \times (50) \times \frac{50}{7500}$$

$$= 8 \times 10^4\ \text{watt} = 8 \times 10^4\ \text{J/s}$$

$\therefore$ Energy lost in one hour = Power × Time

[Time = 1 hour = 60 × 60 seconds]

$$= (8 \times 10^4) \times (60 \times 60)$$

$$= 288 \times 10^6\ \text{Joule} = 288\ \text{MJ}$$

11.13 Domain Theory

According to domain theory, the ferromagnetic materials have sub-microscopic regions. These regions marked A, B, ..., G, H in Fig. 11.8*a* are called *domains*. The magnetic moments are aligned in each domain at room temperature even in the absence of an applied magnetic field. Considering all the domains of the material in totality, the magnetic moments are oriented randomly as shown in Fig. 11.8*b*. Their resultant effect is to produce zero magnetization.

When a magnetic field H is applied on the material, the magnetic moments in each domain align themselves in the direction of the field as shown in Fig. 11.8*c*. This results into a large magnetization of the solid. It

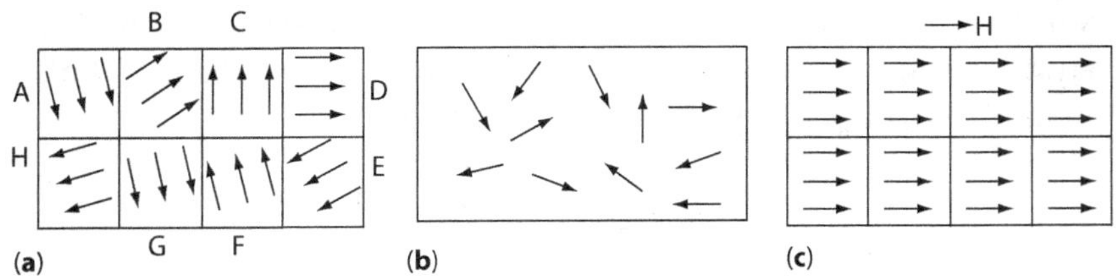

Figure 11.8 (*a*) Magnetic moments are aligned in each domain in the absence of magnetic field, (*b*) net magnetization in material is zero due to random orientation, and (*c*) magnetic moments in each domain align in the direction of applied magnetic field.

is due to the alignment of different domains which are already ferromagnetic. Above explanation is known as *Weiss domain theory,* after the name of its originator.

11.14 Magnetostriction

Magnetostriction is the behaviour of ferromagnetic materials in which their dimensions and shapes are changed when they are magnetized. The dimensional changes occur in their length, width, and volume. The shape change occurs mainly in materials having hexagonal structures such as Co, Mg etc. Non-cubic crystals such as MnO deform to trigonal shape. Materials having cubic crystals change in their dimensions only, and not in their shapes. Depending upon the nature of changes, the types of magnetostriction may be called as follows.

1. Linear magnetostriction
   i. longitudinal magnetostriction
   ii. transverse magnetostriction
2. Volume magnetostriction.

The linear change in dimension may be positive (*i.e.* elongation or increase) or negative (*i.e.*contraction or decrease). The amount of change depends upon the intensity of magnetic field.

11.14.1 Salient Features of Magnetostriction

Salient features of magnetostriction effect are the following.

1. This effect is analogous to electrostriction that occurs in polarized dielectrics (see chapter 10 for details).

2. It has an inverse effect also in that, if the physical dimensions of magnetostrictive materials are changed by application of external force, a change in magnetization also occurs.
3. A very small change in volume occurs at ordinary temperatures also, if a very strong magnetic field is applied on the magnetic material.
4. The change in dimension takes place in the direction of magnetization.
5. Internal strains within the material influence the direction of magnetization. However, it is insensitive to change in direction of magnetization by 180°, when strained.
6. Internal strains greatly influence the shape of hysteresis loop of material. The loop is generally broadened.

11.14.2 Mechanism of Magnetostriction

The causes of magnetostriction in materials are several interactions, inter-atomic spacing, and exchange energies. When dipole moments of atoms are rotated, the bond length between the atoms changes. It affects upon the atomic spacing which may increase or decrease, depending upon whether the dipoles are of attracting or repelling nature with each other. For example, Ni contracts in the direction of magnetization and expands in the transverse direction by about 40 ppm at saturation magnetization.

The dimensional changes, whether positive or negative, are a function of applied field and they differ in different materials. For example in a low field, the iron elongates in the direction of magnetization whereas nickel and cobalt contract. But under high magnetic fields, all the above three metals contract. This behaviour is shown in Fig. 11.9 on a plot between applied field H and coefficient of magnetostriction $\lambda_{m.}$

11.14.3 Magnetostrictive Materials

The property of magnetostriction is observed in iron, nickel, cobalt, and gadolinium. It is also observed in some alloys of iron with chromium, cobalt and aluminium. However, this is most marked in pure nickel and nickel-iron alloys. An addition of nickel in iron has a considerable effect on magnetostriction characteristic of the alloys. For example,

1. Pure iron shows either a positive or negative magnetostriction on magnetization, which on adding nickel exhibits only positive magnetostriction for all the field intensities.

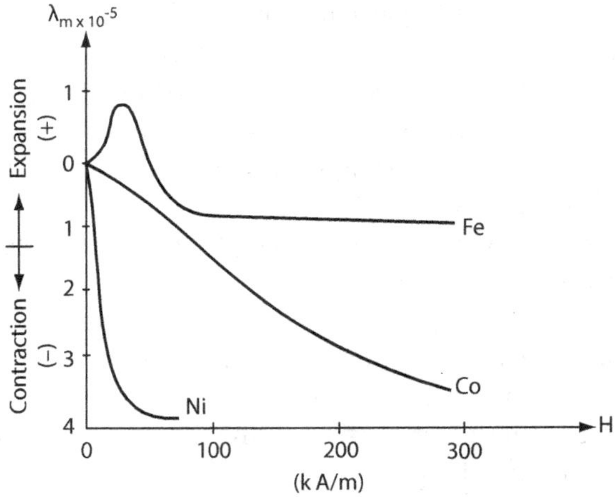

Figure 11.9 Magnetostriction behaviour of iron, nickel and cobalt under the influence of applied magnetic field.

2. Longitudinal magnetostriction diminishes to zero on adding 30% Ni in iron.
3. Volume magnetostriction becomes zero on adding 36% Ni in iron. (A 36% Ni-iron alloy is called INVAR and possesses almost zero value of coefficient of thermal expansion). It is because the volume contraction due to loss of magnetization on heating upto curie temperature, compensates for very small value of thermal expansion.
4. Metglass, an amorphous alloy, having a composition of $Fe_{81}B_{13.5}\ Si_{3.5}C_2$ is available in the form of thin ribbon of around 25 μm thickness.
5. Terfenol, an alloy of terbium and dysprosium with steel, having a composition of $Tb_{0.3}Dy_{0.7}Fe_2$.

The magnetostrictive materials find use in the following systems.

- Audio-frequency oscillators at sonic and ultrasonic frequencies.
- Magnetostrictive transducers.

11.15 Ferromagnetic Anisotropy

Anisotropy (*i.e.* non-isotropy) is a direction dependent property of materials. In it, the nature of properties is dissimilar in all the directions. Similar to mechanical anisotropy in which the mechanical properties are dissimilar in different directions, if the magnetic properties are different in different directions, it is known as *magnetic anisotropy.* Thus the 'ferromagnetic anisotropy' refers to such a property of materials in which the magnetization and saturation behaviours are different in different directions of a crystal, under the influence of applied magnetic field.

11.15.1 Anisotropy in Single Crystal

The *B-H* curve shown in Fig. 11.7 is drawn for an arbitrary direction. For a single crystal of ferromagnetic materials, the path of *B-H* curve differs in different directions relative to the applied field. Figure 11.10 shows such paths for iron and nickel along [100], [110] and [111] directions. It is observed that the magnetization and occurrence of saturation are different in different directions. For iron (BCC crystal), the direction of 'easy magnetization' is [100] and in nickel (FCC crystal) it is [111], but the direction of hard 'magnetization' in iron is [111] and in nickel is [100]. Similarly, the direction of easy magnetization in cobalt (HCP crystal) is along the *c*-axis.

11.15.2 Soft and Hard Directions

Ferromagnetic materials exhibit directional nature in their behaviour. They get magnetized differently in different directions when subjected to an applied magnetic field. Some magnetic materials offer high permeability along certain directions and are easily magnetised. Such directions are called *soft directions.* But these materials are difficult to magnetise in some other directions which are called *hard directions.* These directions are different in different materials.

11.15.3 Methods of Inducing Magnetic Anisotropy

Materials are generally available in their bulk form. To induce magnetic anisotropy (i.e. directional property) in them, they are treated by the following methods.

i. Cold rolling
ii. Magnetic annealing
iii. Magnetic quenching.

The magnetic annealing and quenching are specialized heat treatment processes that are performed in a magnetic field. Annealing requires slow cooling but the quenching is done rapidly below curie temperature.

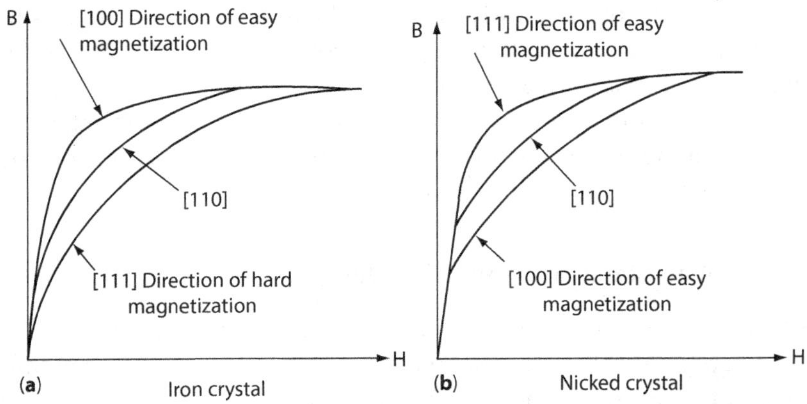

Figure 11.10 Magnetization curve follows different paths in different directions due to magnetic anisotropy in iron, and nickel.

Magnetic annealing is a heat treatment process which is performed to improve ductility, enhance machinability, refining the grains, and softening the metals. When the process is performed in absence of magnetic field, it is known as mechanical annealing (or simply annealing). And if the process is performed in presence of an applied magnetic field, it is called 'magnetic annealing'. Its treatment alters the mechanical properties of materials. For example, the demagnetizing force and magnetization of AlNiCo increases considerably on magnetic annealing.

11.16 Domain Growth and Domain Wall Rotation

A ferromagnetic material consists of a large number of domains of different structures. The number of domains depends on the size of crystal, while the structure depends upon the shape of crystal. Adjacent domains which are magnetized in different directions, are separated by a transition layer called as block wall or domain wall, Fig. 11.11*a*. When an external field *H* is applied to the crystal, the domain walls start moving. This movement is of rotational nature as shown in Fig. 11.11*b*. The domains in favourable direction of spontaneous magnetization grow at the expense of those domains that are magnetized in unfavourable directions, by virtue of motion (rotation) of domain walls. Thus, some domains grow (marked *A* in Fig. 11.11*b*) while the others shrink (marked *B* in Fig. 11.11*b*).

Initially there is neither much rotation in domain walls nor substantial growth in the domains, but with increase in applied field the rotation and growth (accompanied with shrink in less favourably oriented domains)

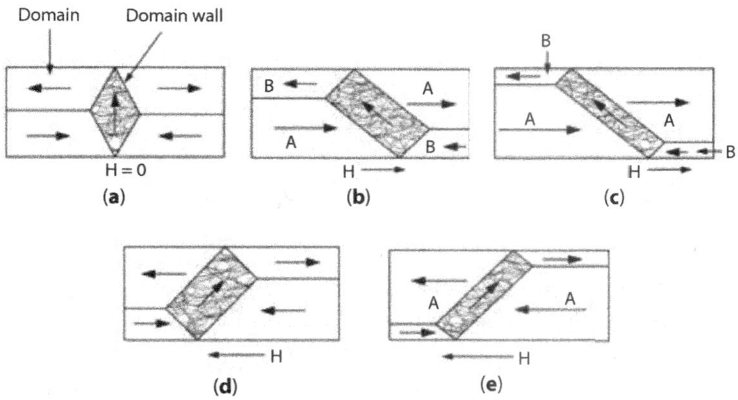

Figure 11.11 Explanation of domain growth and domain wall rotation (*a*) no effect as $H = 0$, (b) favourably oriented domains '*A*' grow while the unfavourably oriented domains '*B*' shrink, and domain wall rotates, (c) growth and rotation effects are more at larger '*H*', *(d)* and *(e)* the growth and rotation reverses when the applied field '*H*' is reversed.

increases as shown in Fig. 11.11*c*. However, the rotation and growth stops when the flux density reaches its saturation level. And when reversed field is applied, the rotation and growth takes opposite turn, Figs. 11.11*d-e*. The phenomena of domain wall rotation and domain growth are important in the sense that the behaviour of *B-H* curve for ferromagnetic materials largely depends upon them.

11.17 Derivation of Langevin's Theory of Diamagnetism and Expression For Diamagnetic Susceptibility

Let we consider a single electron moving in a circular orbit around the nucleus in the atom as shown in Fig. 11.12a. The revolving electron is kept in circular orbit by a force of attraction F_{att} due to central positive nucleus balanced against the centrifugal force. This gives

$$F_{\text{att}} = \frac{mv^2}{r} = m\omega^2 r \tag{11.19}$$

where m is mass, v and ω are linear and angular velocities, and r is radius of the orbit.

Let the intensity of applied magnetic field is H *which* is perpendicular to the plane of the orbit. If the electron is moving in clockwise direction, a force equal to Hev will be directed outwards as shown in Fig. 11.12a.

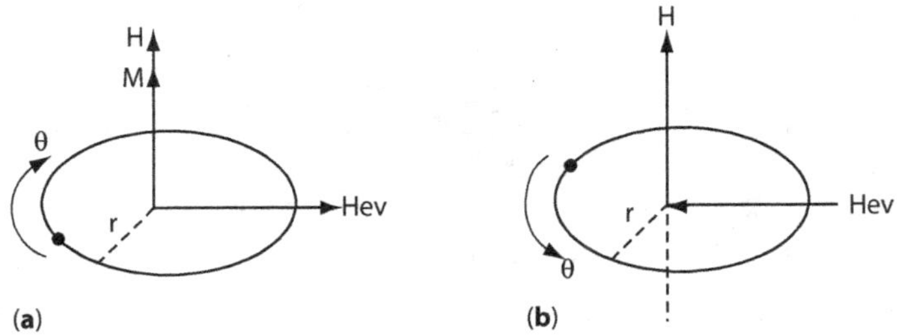

Figure 11.12 The force acts (a) outwards when the electron moves clockwise, and (b) inwards when the electron moves in anticlockwise direction.

However, if the electrons move in anticlockwise direction, the force *Hev* will act inwards (Fig. 11.12b).

11.17.1 Larmor Precession

According to Larmor, the effect of applied field is to cause a precessional motion of the orbit which gives rise to a change in angular velocity $\delta\omega$. Then for clockwise motion of the electron, the condition of stable motion in the same orbit gives

$$F - Hev = m(\omega + \delta\omega)^2 r \tag{11.20}$$

i.e.

$$m\omega^2 r - He(\omega r) = m\omega^2 r + 2m\omega\delta\omega r + m\delta\omega^2 r$$

Assuming that $\delta\omega$ is small as compared to ω, the higher order $\delta\omega^2$ is neglected.

So

$$\delta\omega = -\frac{He}{2m} \tag{11.21}$$

Here, the –ve sign indicates a change in angular velocity due to Larmor precession in a sense opposite to that of the electron.

Change in magnetic moment. The change in angular velocity gives rise to a change in magnetic moment *M*, which is given by

$$\begin{aligned} \delta M &= e\pi r^2\left(-\frac{He}{4\pi m}\right) \\ &= -\frac{He^2 r^2}{4m} \end{aligned} \tag{11.22}$$

where e is the charge revolving in a circle of radius r. The sense of this change in magnetic moment is always opposite to that of the field, whether the electron moves in clockwise or in anticlockwise direction.

11.17.2 Determining the Diamagnetic Susceptibility

For an atom of atomic number Z, there are Z electrons and Z orbits to consider. It is possible that all the electrons may not rotate in the same circular orbit, therefore r^2 in Eqn. 11.22 may be replaced by a mean value R^2. And also the orbits may not all be perpendicular to the direction of field (may be inclined at different angles). Hence, assuming that the orbits have all possible orientations with respect to the field and considering the spherical symmetry of the diamagnetic atom, r^2 in Eqn. 11.22 may be replaced by (2/3) R^2. Hence total magnetic moment of the atom is given by

$$M_{\text{total}} = -\frac{ZHe^2}{4m}\frac{2}{3}R^2 = -\frac{1}{6}\frac{N_A ZHe^2 R^2}{m} \tag{11.23}$$

where N_A is Avogadro's number.

Since the intensity of magnetization I is the magnetic moment per unit volume, therefore

$$I = \frac{M}{V} = \frac{-\frac{1}{6}\frac{N_A ZHe^2 R^2}{m}}{\frac{A_W}{\rho}} \tag{11.24}$$

where A_w is atomic weight and ρ the density of material.

The Volume susceptibility χ_{voi}, mass susceptibility χ_{mass} and atomic susceptibility χ of a diamagnetic material may be found as

$$\chi_{\text{vol}} = \frac{I}{H} = -\frac{1}{6}\frac{N_A Ze^2 R^2}{m}\frac{\rho}{A_W} \tag{11.25a}$$

$$\chi_{\text{mass}} = \frac{I}{\rho H} = -\frac{1}{6}\frac{N_A Ze^2 R^2}{mA_W} \tag{11.25b}$$

$$\chi = \frac{\delta M}{H} = -\frac{1}{6}\frac{N_A Ze^2 R^2}{m} \tag{11.25c}$$

Using Eqns. 11.1 to 11.9, the above equation can be converted to any other form of convenience.

11.18 Derivation of Langevin's Theory of Paramagnetism and Expression for Paramagnetic Susceptibility

This theory is based on the kinetic theory and the relation for a paramagnetic gas. The molecules of paramagnetic gas are assumed small permanent magnets on account of the circulating electrons. If M is the molecular magnetic moment acting on all molecules and H is uniform intensity of the applied field, the magnetic potential energy of each molecular magnet will be given by $-MH\cos\theta$, where θ is the angle between the axis of molecular magnet and the direction of the field.

Now considering unit volume of the gas containing N numbers of molecules and if dN number out of them have their magnetic axes inclined at θ angle with the direction of the field, then on the basis of kinetic theories we may write

$$dN = Ce^{(MH\cos\theta/kT)}\sin\theta d\theta \tag{11.26}$$

where C is a constant. Since the possible orientations are contained between θ and π, we write

$$N = \int_0^{\pi} dN \tag{11.27}$$

On simplifying Eqn. 11.26 and 11.27, we obtain

$$C = \frac{\dfrac{NMH}{kT}}{e^{\left(\frac{MH}{kT}\right)} - e^{-\left(\frac{MH}{kT}\right)}} \tag{11.28}$$

11.18.1 Determining the Intensity of Magnetization

As the resolved component of magnetic moment of dN molecules in the field direction is $M\cos\theta$, the total magnetic moment due to all the N molecules contained in unit volume is ($I = M/V$)

$$I = \int_0^{\pi} M\cos\theta dN$$

Its evaluation, on substituting the value of constant *C*, gives

$$I = MN\left[\coth\left(\frac{MH}{kT}\right) - \frac{1}{\left(\frac{KH}{kT}\right)}\right] \tag{11.29}$$

11.18.2 Langevin Function

In Eqn. 11.29, *coth(MH/ kT)* is a hyperbolic function and *MN* represents the magnetic moment per unit volume when all the molecules are turned in the direction of magnetizing field. Hence, it gives the saturation value of *I* i.e. *MN* = *I*. Therefore

$$\frac{I}{I_s} = \left[\coth\left(\frac{MH}{kT}\right) - \frac{1}{\left(\frac{MH}{kT}\right)}\right] = L\left(\frac{MH}{kT}\right) \tag{11.30}$$

This term *L(MH/kT)* is called Langevin function. For small value of '*MH/kT*', which corresponds to all practical purposes, the value of Langevin function is

$$\frac{I}{I_s} = \frac{1}{3}\left(\frac{MH}{kT}\right), \quad \text{or} \quad I = \frac{I_s}{3}\left(\frac{MH}{kT}\right) \tag{11.31}$$

11.18.3 Determining the Paramagnetic Susceptibility

Now the expressions for susceptibilities may be derived as follows.

$$\chi_{\text{volume}} = \frac{I}{H} = \frac{I_s}{3H}\left(\frac{MH}{kT}\right) = \frac{MN}{3}\frac{M}{kT} = \frac{M^2N}{3kT} \tag{11.32a}$$

$$\chi_{\text{mass}} = \frac{\chi_{\text{volume}}}{\rho} = \frac{M^2N_m}{3kT} \tag{11.32b}$$

where N_m *is* the number of molecules in unit mass.

$$\chi_{\text{molecular}} = \frac{M^2 N_M}{3kT} \tag{11.32c}$$

where N_M is the number of molecules in a gram molecule (Avogadro's number).

11.19 Solved Examples

Example 11.9 How can the susceptibility and permeability of a diamagnetic and a paramagnetic material be determined?

Solution. These can be determined by measuring the force exerted upon the material in an inhomogeneous field, by using a 'Gouy balance'. Since, $(\mu_r - 1) = \chi$, hence determination of any one parameter will yield the value of other parameter.

Example 11.10 What are paramagnetic salts and where are they used?

Solution. Paramagnetic salts are the rare-earth based compounds and are used to obtain very low temperatures (< –272°C) by adiabatic demagnetization. They are used in solid state MASER.

Example 11.11 Cu, Mn and Al each as a separate metal does not exhibit ferromagnetism but their alloy Cu + Mn + Al (commonly known as Heusler alloy) exhibits ferromagnetism-Why ?

Solution. A metal or an alloy exhibits ferromagnetism below its ferromagnetic curie temperature T_{fc}. The T_{fc} of a metal or alloy can be altered (moved up or down) by addition of other elements. The T_{fc} of Fe is (772°C), of Ni is (357°C) and of Co is (1122°C). But an addition of Cu of melting point 1083°C, Mn of melting point 1245°C and Al of melting point 660°C shifts the T_{fc} of Heusler alloy, hence it exhibits ferromagnetism.

Example 11.12 What are the reasons of following occurrences?

a. Although the manganese (Mn) has five unpaired electrons per atom, yet it is not in the list of ferromagnetic materials –Why?
b. Curie temperature of Si-steel, a soft magnetic material, is 745°C. What changes will occur in the properties of this material below and above 745°C?

Solution.

a. The magnetic moments in neighboring atoms of Mn cancel out each other, that is why it is not a strong magnetic material (i.e. ferromagnetic). It is worth to recall that Mn is anti-ferromagnetic material.

b. Below curie temperature (745°C), the material will show highly magnetic character i.e. ferromagnetic. It will be however, a weak magnetic material (i.e. paramagnetic) above this temperature.

Example 11.13 Prove that the values of magnetic moments (in Bohr magneton) in iron, cobalt, and nickel are 4, 3 and 2 respectively.

Solution. The atomic number of iron, cobalt and nickel respectively are 26, 27 and 28. The electronic configuration of iron is $1s^2\ 2s^2\ 2p^6\ 3s^2\ 3p^6\ 3d^6\ 4s^2$. The configuration of cobalt and nickel are the same except for the arrangement in 3d-suborbit. Instead of $3d^6$ in Fe, it is $3d^7$ for cobalt and $2d^8$ for nickel. Thus

Element	Number of electrons in 3d-suborbit	Arrangement of spin moment	Value of magnetic moment in Bohr magneton = (upward moment - downward moment)
Fe	6	↑↑↑↑↑↓	5 - 1 = 4
Co	7	↑↑↑↑↑↓↓	5 - 2 = 3
Ni	8	↑↑↑↑↑↓↓↓	5 - 3 = 2

Quick Revision Summary

Antiferromagnetism. It is the opposite alignment of adjacent atomic magnetic moments in a solid by the exchange interaction.

Coercive force. It is the applied field required to reduce the induction of a magnetized material to zero.

Curie constant. It is the proportionality constant between χ_m and $1/T$ in the Curie law.

Curie temperature. It is the temperature above which, if a ferromagnetic material is heated, it becomes paramagnetic.

Domain. It is a region in a ferromagnetic or ferrimagnetic material where all the moments are aligned.

Eddy current loss. It is the power loss of magnetic materials in alternating fields due to induced currents in the material.

Ferrimagnetism. It is a special case of antiferromagnetism, where the opposed moments are of different magnitudes and thereby results in a large net magnetization.

Ferromagnetism. The appearance of a very large magnetization due to the parallel alignment of neighboring magnetic moments by an exchange interaction is known as ferromagnetism.

Hard magnetic material. It is a material having high coercive force and high saturation magnetization.

Hysteresis. It is the irreversible *B-H* characteristic of ferromagnetic and ferrimagnetic materials.

Hysteresis loss. It is the work dissipated in tracing a *B-H* loop, essentially the work needed to move the domain boundaries during a single magnetization cycle.

Initial and maximum permeability are the slopes of first and second regions of magnetization curve, (see Fig. 11.7a).

Magnetic annealing is a heat treatment process which is performed to improve ductility, enhance machinability, refining the grains and softening the metals. When the process is performed in absence of magenetic field, it is known as mechanical annealing (or simply annealing), and if the process is performed in presence of an applied magnetic field, it is called 'magnetic annealing'. Its treatment alters the mechanical properties of materials. For example, the demagnetizing force and magnetization of ALNICO increases considerably after magnetic annealing.

Magnetostatic energy and isotropic exchange energy. Magnetostatic energy is the energy of dipole moments of a material, whereas the *isotropic exchange energy* is the energy of interaction between the dipoles. Due to this energy, the dipoles line-up together.

Magnetostriction. It is the change in length along the direction of magnetization of a multi-domain solid.

Magnetostrictive energy. It is the energy due to the mechanical stresses generated by magnetostriction in the domains.

Neel temperature. It is the temperature at which the susceptibility is maximum in antiferromagnetic material.

Paramagnetism. The small positive susceptibility due to the weak interaction and independent alignment of permanent atomic and electronic magnetic moments with the applied field, is known as paramagnetism.

Reluctivity. It the reciprocal of permeability and is expressed as $1/\mu_r$ for a medium. Its value is little less than one for diamagnetic materials, little more than one for paramagnetic materials, and a small fraction of one for ferromagnetic materials. For steel it is 1/2000.

Remanence. It is the value of B in the specimen when H is reduced to zero.

Saturation. It is the maximum value of magnetization M_s or magnetic induction B_s for a ferromagnetic material.

Soft magnetic material. It is a higher permeability material having low hysteresis and low coercive force.

Weiss field. It is a hypothetical internal magnetic field, strong enough to make the spin magnetic moments in a solid line up, despite the effect of thermal energy.

Review Questions

1. Explain the mechanism of origin of permanent magnetic dipole. Describe all possible applications of magnetic materials. What is Cunife and how is it different from Cunico?
2. What are the effects of dipole moments on the magnetic behaviour of materials?
3. Define and explain the following.
   a. Magnetization
   b. Susceptibility
   c. Bohr magneton, and
   d. Curie constant
4. In magnetic material, prove that $\mu_r = 1 + \chi_m$, where χ_m is magnetic susceptibility and μ_r is relative permeability.

5. Classify magnetic materials. Write examples, salient features and applications of each of them.
6. Describe the following with the help of suitable diagram,
   a. Neel temperature b. Antiferromagnetism.
7. Differentiate between the following.
   a. Diamagnetism and paramagnetism
   b. Ferromagnetism and ferrimagnetism
8. Explain the magnetic hysteresis curve. Why is it called hysteresis curve? Discuss coercive force and retentivity with the help of a B-H curve.
9. Explain the following terms.
   a. Hysteresis loss
   b. Curie temperature
   c. Paramagnetic curie temperature
10. Discuss different laws being followed by various magnetic materials. What is the effect of temperature on magnetic susceptibility?
11. What are ferrites? Discuss their salient characteristics. Name different types of ferrites and write their applications.
12. Explain the properties of soft and hard magnetic material along with initial and maximum permeability.
13. Compare hard and soft magnetic materials. Why is a soft magnetic material preferred over a hard magnetic material for use in the transformer core?
14. Discuss the phenomenon of domain theory. Steel is ferromagnetic but is not a magnet in the absence of magnetic field-Why?
15. Write notes on the following.
    a. Soft and hard ferrites
    b. Domain theory
    c. Garnets
    d. Core losses
16. What are the effects of a large magnetic moment on curie temperature of materials?
17. What is magnetostriction? Discuss its mechanism and salient features. Name some magnetostrictive materials.
18. Explain the phenomenon of magnetic anisotropy. Discuss its importance for magnetic materials. Suggest various methods of inducing magnetic anisotropy.
19. Describe the occurrence of domain growth and domain wall rotation with the help of suitable diagrams and illustrations.

20. Discuss the following :
    i. Magnetocrystalline anisotropy
    ii. Domain wall motion
21. Derive the Langevin's theory of diamagnetism stating the assumptions and approximations made therein. Also write the expressions for diamagnetic susceptibility.
22. Derive the Langevin's theory of paramagnetism stating the assumptions and approximations made therein. Also write the expressions for paramagnetic susceptibility.
23. Discuss the Larmor precession and Langevin function.

Numerical Problems

1. Diamagnetic susceptibility of gold is -0.6×10^{-5}. Calculate the magnetic flux density and the magnetization for an applied field of 200 kA/m.
2. Determine the magnetization and the magnetic induction in a material of magnetic susceptibility 1.3×10^{-4} lying in a magnetic field strength of 7.8×10^{4} A/m. Take $\mu_0 = 4\pi \times 10^{-7}$ henry per metre.
3. Determine the magnetic induction and the magnetization in a gold crystal under an applied field of 200 kA/m. The diamagnetic susceptibility of gold is -0.6×10^{-5}.
4. The average magnetic moment along the direction of applied field in a paramagnetic material is 2.79×10^{-3} Bohr-magneton per spin. Determine the applied field at room temperature.
5. The iron loss in a transformer core at 25 Hz is 245 W and at 60 Hz is 900 W. Estimate (*a*) the eddy current loss, and (*b*) the hysteresis loss at 60 Hz.
6. The electrical resistivity of 5% silicon-iron alloy is 7 times to that of a pure iron metal. Determine the decrease in eddy current loss for the above alloy as compared to pure iron.
7. On alloying the iron with 5% silicon, its electrical resistivity increases with a corresponding decrease in eddy current losses. If the resistivity increases by 700%, how much will be the decrease in eddy current losses?
8. During a test of transformer core, the iron loss of 245 watts and 900 watts were observed at 25 Hz and 60 Hz respectively. Compute the hysteresis loss at 60 Hz, and the eddy current loss.

9. A cobalt specimen of 76 kg mass and specific gravity 8.9 is given a flux density of 1.5 tesla at 60 Hz. Assuming Steinmetz constant as 740 J/m^3, calculate the energy loss due to hysteresis in one hour. Estimate the area of magnetic hysteresis loop also which is drawn to the following scale.

15 AT/m = 1 cm on abscissa, 0.2 tesla = 1 cm on ordinate.

10. Magnetic field strength in a piece of copper is 1 MA/m. If the magnetic susceptibility of copper is -0.5×10^{-5}, determine the flux density and magnetization in copper.
11. Saturation value of magnetization in an iron piece is 1.75×10^6 A/m. The iron has BCC structure having lattice dimension 2.86 Å. What will be the average number of Bohr magnetons contributed to magnetization per atom.
12. An iron specimen weighing 50 kg is subjected to magnetization at 50 c/s. The area of its hysteresis loop is equivalent to 0.25 kJ/ m^3. If the weight density of iron is 7500 kg/m^3, determine the loss of energy per hour in the specimen. [**Hint :** Hysteresis loss/m^3 = 0.25 kJ/cycle]
13. A 300 kW dynamo having 8 poles is driven at 375 rpm. Volume of its armature core made of a material of specific gravity 7.8, is 7.63×10^{-2} m^3. It is subjected to a maximum flux density of 1.2 tesla. Determine the core loss in it. Take the value of hysteresis coefficient as 0.002.
14. What is (*a*) the intensity of magnetisation, and (*b*) the flux density produced, when a magnetic field of strength 50 A/m is applied to a Fe-3% Si alloy with a relative permeability of 7000?
15. Figure 11.13 shows the hysteresis loop for a ferromagnetic material. Estimate from the graph the values of (*a*) the remanence, (*b*) the coercive field, (*c*) the saturation flux density, (*d*) the energy loss due to hysteresis during one cycle, and (*e*) $(BH)_{max}$.
16. A transformer has an eddy current loss of 12 W when operating at a frequency of 100 Hz. What will be the eddy current loss when it is used at 50 Hz, the flux density remaining unchanged?
17. When the maximum flux density in an inductor is 1.2 T, the hysteresis loss is 4.0 W. What will be the loss when the maximum flux density is 1.6 T? The Steinmetz constant is 1.6.
18. The iron losses of a transformer are measured at different frequencies and a graph plotted of (iron loss/frequency) against the frequency. If the graph has an intercept with the (iron loss/

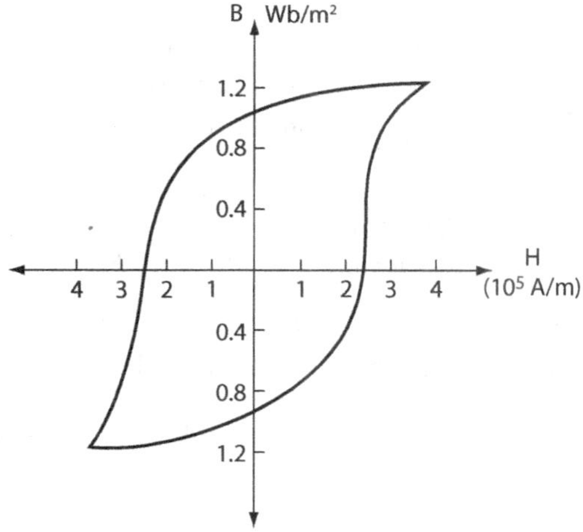

Figure 11.13

frequency) axis of 0.20 W/Hz and a slope of 0.010 W/Hz, what will be the hysteresis and eddy current losses at 100 Hz?

19. The magnetic field strength in a piece of copper is 10^6 ampere m^{-1}. Given that the magnetic susceptibility of copper is -0.5×10^{-5}, find (*a*) the flux density, and (*b*) the magnetization in the copper.
20. The magnetic field strength in a piece of Fe_2O_3 is 10^6 ampere m^{-1}. Given that the susceptibility of Fe_2O_3 at room temperature is 1.4×10^{-3}, find the (*a*) flux density, and (*b*) the magnetization in the material, (*c*) What is the magnetization at the temperature of liquid nitrogen?

Objective Questions

1. Consider the following statements.
   The coercive force can be increased by
   1. adding cobalt because it is ferromagnetic material.
   2. adding gold because it is diamagnetic material.
   3. adding supermalloy.
   4. adding copper.

   Of these statements,

   a. 4 alone is correct b. 1 alone is correct
   c. 2 and 3 are correct d. 1, 2 and 4 are correct

2. Consider the following statements.
   A : Domain walls can be easily moved in a magnetic field if the material possesses high permeability.
   R : Highly permeable materials have their permeability less than 1.
   Of these statements,

   a. Both A and R are true b. Both A and R are wrong
   c. A is true but R is wrong d. R is true but A is wrong

3. High initial permeability in Fe-Ni alloys helps in (reducing/increasing/maintaining) the area under hysteresis loop.

4. For high speed reading and storing of information in a computer, the use is made of

   a. ferrites b. piezoelectrics
   c. pyroelectrics d. ferromagnetics above 768°C

5. The poles of alternators are usually made of

   a. wrought iron b. permalloy
   c. CdS d. alnico

6. The residual magnetic flux density is more in case of

   a. metallic magnets b. ceramic magnets
   c. graphite d. iron oxide

7. A suitable material for VHF (very high frequency) application is

   a. silicon steel b. alnico
   c. cobalt steel d. ferrite

8. In general, the effect of temperature on relative permittivity of a material is to

   a. reduce it b. increase it
   c. maintain it d. all of the above

9. Bohr magneton of elements depends on their

   a. atomic number b. electrons in the outermost orbit
   c. atomic weight d. unbalanced electron spin in $3d$-state

10. Air exhibits

    a. ferromagnetism b. paramagnetism
    c. antiferromagnetism d. ferrimagnetism

11. The structure of a garnet is

    a. cubic b. hexagonal
    c. spinel d. ferrox cube

12. Recent example of a hard magnetic material is

    a. alnico b. hycomax
    c. CrO_2 d. invar

13. The permeability of iron can be increased by

    a. alloying it with nickel b. purifying it
    c. deposition of carbon d. electroplating

14. Match List 1 with List II, and choose the correct answer from the codes given below the lists.

	List I		*List II*
A.	Contraction and expansion	1.	[100] in nickel and [111] in iron
B.	Direction dependent behaviour	2.	magnetic annealing
C.	Direction of hard magnetization	3.	magnetostriction
D.	Domain wall rotation and domain growth	4.	ferromagnetic materials
		5.	magnetic anisotropy

Codes:

	A	B	C	D		A	B	C	D
a.	3	5	1	4	b.	1	3	5	2
c.	2	1	4	5	d.	5	2	1	3

12

Magnetic Materials: Types and Applications

After discussing the properties and behaviour of magnetic materials in chapter 11, we shall now study details of their different types and applications.

12.1 Types of Magnetic Materials

Depending upon the arrangement of magnetic moments, structures, texture, composition and properties; the magnetic materials may be classified as follows.

I. **On the Basis of Arrangement of Magnetic Moments**
   1. Diamagnetic materials
   2. Paramagnetic materials
   3. Ferromagnetic materials
   4. Antiferromagnetic materials
   5. Ferrimagnetic materials (or ferrites)

II. **On the Basis of Easeness in Magnetization**
   1. Hard (or high coercive) magnetic materials
   2. Soft (or high permeable) magnetic materials

III. On the Basis of Structure of Solids
1. Metallic magnetic materials
2. Ceramic (or oxide, or amorphous) magnetic materials

IV. On the Basis of Composition
1. Ferrous alloyed magnetic materials
2. Non-ferrous alloyed magnetic materials

V. On the Basis of Texture
1. Powdered magnetic materials
2. Textured magnetic materials
   i. randomly oriented
   ii. preferred (favourably) oriented

VI. On the Basis of Different Specialties
1. Metallic glass
2. Magnetic bubble
3. Magnetodielectric magnetic materials
4. Magnetostrictive magnetic materials.

Amongst these, the ferromagnetic materials are most widely used. Antiferromagnetic materials and ferrites are its sub-classes.

Ferrites, now-a-days, have emerged as an outstanding magnetic materials. They are getting increasing use in computer components. Metallic magnets particularly in alloy form and ceramic magnets are equally popular. Metallic glass and magnetic bubble are the recent magnetic materials. Since the ceramics are metallic oxides, they are also known as *oxide magnets.* We have already discussed that the ceramics have non-crystalline (or amorphous) structures, therefore ceramic magnets are also called as *amorphous magnets.*

12.2 Magnetic Materials

Magnetic materials are available in different forms and magnetization ranges. Each form possesses a special property that suits a specific application. However, the magnetic materials are broadly classified into *soft magnetic* materials and *hard magnetic* materials. Ferrites are modern materials with interesting characteristics. Metallic glass (or metal-ceramic) and magnetic bubbles are among the list of latest materials. All these are described in subsequent articles.

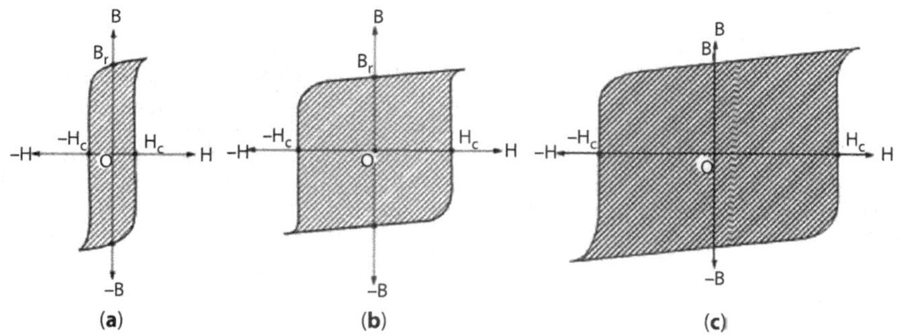

Figure 12.1 Hysteresis loop of (*a*) a soft magnetic material gives small loop (hatched) area; (*b*) a hard magnetic material gives medium area, and (*c*) large hatched area

12.3 Soft Magnetic Materials

These materials are also known as permeable magnetic materials as they possess high permeability. The *B-H* curve of a soft magnetic material is shown in Fig. 12.1*a*. Due to very low coercive force, the hysteresis loop gives small area under the *B-H* curve. Due to smaller area of hysteresis loop, the power losses in such materials are low, and they can be used as magnets at high frequencies.

12.3.1 Transformer Purpose Sheet Form Magnetic Material

Soft magnetic materials are suitable for use in transformer cores where the frequency of applied magnetic field reverses frequently. These materials help in reducing the eddy current losses if the thickness of the core material is very thin. Thin sheets of permeable metal or alloy of 0.025 to 1 mm thickness are assembled together to form a laminated core. These sheets are insulated to minimize eddy current.

12.3.2 Powder Form Magnetic Material

Powdered material cores produced from very fine metallic powders of 1 to 10 μm sizes are more recent advancements. They are superior than thin laminated cores, and are suitable for application as in very high frequency coils. Properties of some soft magnetic materials are listed in Table 12.1.

Table 12.1 Properties of soft magnetic materials

Material	Relative permeability μ_r		Hysteresis loss		Saturation flux density B_s	
	Initial	Maximum	(J/m^3)	(J/kg-cycle)	(Tesla)	(MA/m)
• Iron	250	—	500	—	2.2	—
• 4% Si iron (random-oriented)	500–1500	7000	95	0.02	2.0	1.55
• Permalloy (45% Ni)	2700	—	120	—	1.6	—
• Permalloy (78% Ni)	—	100000	—	0.0005	—	0.87
• Supermalloy (79% Ni, 5% Mo)	100000	1000000	21	0.0001	0.8	0.70

12.4 Hard Magnetic Materials

Hard or permanent magnetic materials have an ability to retain magnetic field. They are characterized by large coercive force and sufficient permeability. Therefore, area below their *B-H* curves is large, and is typically as shown in Figs. 12.1*b* and *c*. Material of Fig. 12.1*c* is a better hard magnetic material due to its larger hatched area than that of Fig. 12.1*b*. They are used to make permanent magnetic poles for alternators and motors. Commonly used hard magnetic materials are the following.

i. High carbon steel, Alnico (Al + Ni + Co + Fe), Cunife (Cu + Ni + Fe) and Cunico (Cu + Ni + Co) etc.
ii. Powdered iron-based magnetic materials of manganese bismuthide (MnBi) develop very high coercive forces.
iii. Powdered ceramic permanent magnetic materials such as $CoO.Fe_2O_3$ and $PbO.6Fe_2O_3$ exhibit large energy product.
iv. Alnico retains about 70% magnetization at 600°C.

Properties of some hard magnetic materials are given in Table 12.2.

Table 12.2 Properties of hard magnetic materials

Material	Coercive force H_c (kA/m)	Residual flux density B_r (Tesla)	Energy product B_rH_c(kJ/m^3)
• Iron-fine iron powder	0.08–40	—	—
• High carbon steel	3.98	0.90	3.58
• 4% Cr-steel	5.17	0.95	4.91
• 5% W-steel	5.57	1.05	5.85
• 36% Co-steel	18.31	0.95	17.40
• Al-Ni-Co alloy steel	60–120	0.8–1.2	50–145
• Rare earth alloyed cobalt, $SmCo_5$, $PrCo_5$	200	1.0	200
• $Nd_2Fe_{14}B$	—	—	400

An approximate estimate provides the following relations.

i. For metallic magnets,

$$B_r > 2H_c \qquad \text{and} \qquad BH_{max} = ½B_rH_r \tag{12.1}$$

ii. For ceramic magnets

$$B_r \leq H_c \qquad \text{and} \qquad BH_{max} = ¼\, B\, r^2 \tag{12.2}$$

Recent advances. Aclomax (55% Fe, 11% Ni, 22% Co, 8% Al, 4% Cu) and hycomax (50% Fe, 21% Ni, 20% Co, 9% Al) are recently developed hard magnetic materials.

12.4.1 Soft Magnetic Materials Versus Hard Magnetic Materials

A comparison between soft magnetic materials and hard magnetic materials is given below for a quick understanding.

Description	Soft magnetic material	Hard magnetic material
• Area of hysteresis loop	Smaller	Larger
• Hysteresis loss	Less	More
• Permeability	Larger	Smaller

(*Continued*)

Description	Soft magnetic material	Hard magnetic material
• Magnetic reluctance	Low	High
• Susceptibility	Low	High
• Retentivity	Smaller	Larger
• Coercivity	Smaller	Larger
• Magnetization	Easier	Difficult
• Demagnetization	Easier	Difficult
• Resistivity	Higher	Lower
• Eddy current loss	Lesser	Greater
• Magnetostatic energy	Smaller	Larger
• Magnetic saturation requirement	Less magnetic force is needed	High magnetic force is needed
• Domain wall movement	Considerable	Negligible
• Effect of changes in magnetic field	Large change in magnetization	Comparatively less change
• Mechanical hardness	Comparatively less	More

12.5 High Energy (Product) Hard Magnetic Materials (HEHMMs)

Those hard magnetic materials whose energy products $(B_rH_c)_{max}$ are in excess of about 80 kJ/m^3 are called high energy HMMs. It is worthwhile to remember that the conventional HMMs generally possess an energy product of about 3 to 80* kJ/m^3. The high energy HMMs are recently developed materials having the structure of intermetallic compounds. They are available in different compositions. Common examples of such materials are rare earths and some alloys, but main among them having greater commercial utilities are the following.

1. Samarium-Cobalt rare earth ($SmCo_5$), and
2. Neodymium-Iron-Boron alloy ($Nd_2Fe_{14}B$)

12.5.1 Samarium-Cobalt Rare Earth

$SmCo_5$ belongs to the family of those alloys which are formed by combining iron or cobalt with a rear earth element. A number of alloys of this

* The demarcating value of 80 kJ/m^3 is arbitrary and most acceptable.

family possess high energy product but even more higher energy product of $SmCo_5$ makes it most suitable for engineering applications. These are produced by sintering method of powder metallurgy technique. Its magnetic properties and applications are illustrated in Table 12.3.

12.5.2 Neodymium-Iron-Boron Alloy

$Nd_2Fe_{14}B$ alloy is more useful high energy HMM and is used to make magnets for diversified applications. They are cheaper than samarium (Sm) which is costlier and also a rare material. Energy product of $Nd_2Fe_{14}B$ alloy is about 1.5 times greater than that of samarium-cobalt rare earth and coercivity is also substantially higher. It is fabricated by either sintering method (powder metallurgy) or by rapid solidification (melt spinning) technique. Its magnetic properties are illustrated in Table 12.3 alongwith those of $SmCo_5$ and Alnico, showing a comparison between these materials.

Table 12.3 Magnetic properties of high energy HMMs and their comparison with a conventional (non-high energy) hard magnetic material

Property	**Samarium-cobalt rare earth**	**Neodymium-iron-boron alloy**	**Alnico 8**
Composition/ formula	$SmCO_5$	$Nd_2Fe_{14}B$	7% Al, 15% Ni, 35% Co, 4% Cu, 5% Ti, rest Fe
Maximum energy product (kJ/m^3)	170	255–400	36
Remanence (tesla)	0.92–1.0	1.16	0.76
Coercivity (kA/m)	200	220–345	125
Resistivity (ohm m)	5×10^{-7}	16×10^{-7}	—
Curie temperature (°C)	725	310	860
Fabrication technique	Sintering	Sintering, or rapid solidification	Sintering
Cost	Costlier	Cheaper	Cheaper
Main applications	As magnets for fractional horse power motors	As magnets for fractional horse power motors	As magnets for large horse power motors

12.6 Commercial Grade Soft Magnetic Materials

Commercial grade soft magnetic materials are available in different compositions. Each of them suits to some specific application. They are generally hard and brittle. A brief description of important among them is given below.

Alsifer. It is an Fe-Si-Al alloy having an approximate composition of 9.5% Si + 5.6% Al + rest Fe. Its relative permeability (initial to maximum) lies between 10 000 to 35 000, coercivity is 0.02 oersted, and resistivity is about 0.80 ohm m. It is used in capacitors.

Permalloy. It is a Fe-Ni alloy having 40 to 80% nickel contents. Based on the basis of nickel content, it can be divided into two groups.

i. Low Ni permalloy
ii. High Ni permalloy

If the nickel content is upto about 50%, it is called as low nickel permalloy. But if this percentage is above 50%, it is termed as high nickel permalloy. Both types are alloyed with small amounts of manganese, chromium and molybdenum also. Their properties are already shown in Table 12.1.

Supermalloy. This is also an Fe-Ni alloy containing a high percentage of nickel (about 79%) alongwith about 5% molybdenum. It possesses a very high relative permeability, both initial and maximum, which is displayed in Table 12.1 along with other magnetic properties.

Mumetal. This is an Fe-Ni alloy having some copper and chromium contents also. Its typical composition is 75% Ni + 5% Cu + 2% Cr + rest Fe. Its relative permeability lies in between the permeabilities of high- nickel permalloy and supermalloy. The resistivity of mumetal is high, therefore the eddy current losses are lower. Its magnetic, thermal, physical and other properties are given below for a ready reference.

- Remanence = 0.47 tesla
- Coercive force = 0.2 A/m
- Resistivity = 0.6×10^{-7} ohm m
- Coefficient of linear expansion = 13×10^{-6} / °C
- Specific heat = 0.10 kcal/kg°C
- Weight density = 8800 kg/m^3
- Relative permeability = 20 000 to 130 000

Perminvar. It is an Fe-Ni alloy having Co as another constituent. Its permeability is independent of field strength. It is used to make transformers and chokes.

12.7 Commercial Grade Hard Magnetic Materials

Commercial grade hard magnetic materials are available in different compositions. They are generally less ductile and difficult to shape. They are usually casted and finished by grinding process. Each of them fulfils the requirements of certain specific applications. A brief description of important among them is given below.

Permendur. It is an alloy of iron with cobalt and vanadium having a composition of 49% Co +2% V + rest Fe. This is capable of setting a much greater flux density (a strong magnetic field) in air gaps such as in microphones, oscillographs, and electromagnets. Therefore, it is used to make cores and poles of magnetic circuits for them. This alloy is superior to electrical steel in many respects. Thermal, physical and other properties of permendur are given below.

- Remanence = 1.5 tesla
- Coercive force 160 A/m
- Resistivity = 0.47×10^{-6} ohm m
- Coefficient of linear expansion = $9 \times 10^{-6}/^{\circ}C$
- Weight density = 8050 kg/m^3

Alnico. It is basically an alloy of Al, Ni and Co with Fe. However, Cu is also added in some brands. Depending upon the composition and properties, the alnico is available in different grades. Their composition and properties are listed in Table 12.4.

Cunife. These are recently developed copper-nickel and copper-nickel cobalt alloy steels. They are ductile and can be cold worked by ordinary means. They are available in two different grades whose details are given in Table 12.5.

Other Commercial Grades: Some more names in this list are the following.

- **Vectolite** containing 44% Fe_3O_4 + 30% Fe_2O_3 + 26% Co_2O_3. Its remanance is 0.16 tesla, coercivity 70 000 ampere-turn/meter, and energy product as 4 kJ/m^3.
- **Cobalt-chrome steel** containing 16% Co + 9% Cr + 0.3% Mn +1% C + rest Fe.

Table 12.4 Composition and properties of different grades of alnico

Type/grade	Composition (%)	Coercivity (AT/m)	Retentivity (tesla)	Energy Product B_rH_c (Wb/m^2 . A/m)	Density (kg/m^3)
Alnico 1	12 Al, 21 Ni, 5 Co, rest Fe	35000	0.72	11200	6100
Alnico 2	10 Al, 17 Ni, 12.5 Co, 6 Cu, rest Fe	44000	0.72	12800	7100
Alnico 2 (sintered)	10 Al, 17 Ni, 12.5 Co, 6 Cu, rest Fe	41500	0.69	11200	6900
Alnico 4	12 Al, 28 Ni, 5 Co, rest Fe	56000	0.55	10400	7000
Alnico 5	8 Al, 14 Ni, 24 Co, 3 Cu, rest Fe	44000	1.25	36000	7400
Alnico 12	6 Al, 18 Ni, 35 Co, 8 Ti, rest Fe	72000	0.57	14400	7200

Table 12.5 Composition and properties of different grades of cunife

Type/ grade	Composition (%)	Coercivity (AT/m)	Retentivity (tesla)	Energy product (Wb/m^2 . A/m)	Density (kg/m^3)
Cunife 1	60 Cu, 20 Ni, rest Fe	44000	0.54	12000	8600
Cunife 2	≈ 70 Cu, 10 Ni, rest Fe	20500	0.73	6400	8600

- **Comol** containing 17% Mo + 12% Co + rest Fe. Its properties are
  i. Coercivity = 20 000 AT/m
  ii. Retentivity=1.05tesla
  iii. Energy product = 8800 Wb/m^2 . A/m
  iv. Density = 8400 kg/m^3

- **Vicalloy** containing 52% Co + 14% V + rest Fe. Its properties are
  i. Coercivity = 32 000 AT/m
  ii. Retentivity = 0.96 tesla
  iii. Energy product = 22 400 Wb/m^2 . A/m
  iv. Density = 8100 kg/m^3

  It is also available in other compositions of Co and V.

- **Platinum-cobalt alloy** containing 77% Pt + 23% Co. Its properties are
  i. Coercivity = 208 000 AT/m
  ii. Retentivity = 0.45 tesla
  iii. Energy product = 30 400 Wb/m^2 . A/m
  iv. Density = 1725 kg/m^3
- **Silmanal** containing about 87% Ag + 8.5% Mn + 4.5% Al. Its properties are
  i. Coercivity = 47 500 AT/m
  ii. Retentivity = 0.57 tesla
  iii. Energy product = 680 Wb/m^2 . A/m
  iv. Density = 9000 kg/m^3

12.8 Ferrites in Memory Devices

We have already discussed the basic aspects of ferrites in chapter 11. Some more information and the practical/commercial aspects will be discussed now. For a typical ferrite the properties are as follows.

- Coefficient of linear thermal expansion = 10^{-5}/°C
- Thermal conductivity = 5×10^2 W/m°C
- Thermal capacity = 0.17 kcal/kg°C
- Density = 3000 to 5000 kg/m^3

Ferrites are suitable for use in those devices which require (i) high speed response, (ii) very low power loss, and (iii) high frequency. This makes them useful for the purpose of memory devices in computers. In memory devices, they are formed as miniature ring cores which are fitted on the current carrying conductor in the circuit. Consequently, they are magnetized each time if the phase of the current changes. In this way it memorizes the possible states.

Magnetic properties and main applications of some commercial grade soft ferrites are displayed in Table 12.6.

Table 12.6 Magnetic materials of soft ferrites

Name of ferrite	Composition	Relative permeability (μ_{ri}, $\mu_{r\,max}$)	Magnetic flux density B_s, B_r (Wb/m^2)	Hysteresis loss (J/m^3)
• Mn-Zn ferrite	$Mn_{0.5}Zn_{0.5}Fe_2O_4$	—, 2500	—, —	≈ 40
• Ferroxcube A	48% MnO. Fe_2O_3 + 52% ZnO. Fe_2O_3	1200, —	0.33, —	—
• Ferroxcube B	36% NiO. Fe_2O_3 + 64% ZnO. Fe_2O_3	650, —	0.36, —	—
• Ni-Zn ferrite	$Ni_{0.5}Zn_{0.5}Fe_2O_4$	—, 1000	—, —	≈ 30
• Nickel ferrite	NiO.Fe_2O_3	17, —	0.23, —	—
• Yttrium-iron garnet	$Y_3Fe_5O_{12}$	—, —	—, —	Very low
• Ferrimagnetic garnet	(Sm,Gd,Eu). Fe_5O_{12}	—, —	—, 1	—

12.9 Magnetic Storage

Modern hi-tech era is of information storage. Magnetic reading has almost become universal for storage of electronic information. The overwhelming use of computer hard disk, floppy disk, audio and video tapes, ATM (Automatic Teller Machine) and credit cards etc. can be seen in this perspect. Magnetic disks are used as secondary memory (storages) in computers because of their capability of storing large information at lower cost. And also

- In a magnetic storage medium the sound or visual images, computer bits and bytes are recorded in the form of electrical signals on a very small segment of magnetic material.
- Transfer and retrieval of information from disk or tape is accomplished by an 'inductive read-write head'.
- This head consists of a wire-coil wound around a magnetic material core in which a gap is cut.
- The data information is written within the coil by electrical signal, which then generates a magnetic field across the gap.
- In turn, this field magnetizes a very small area of disk or tape within the proximity of the head.

- Upon removal of field, the magnetization remains i.e. the signal has been stored.

The magnetic materials for storage purposes may be both: hard magnetic materials and soft magnetic materials. They are used in various forms such as tapes, films, bubbles etc. Important among these are briefly discussed in following sections.

12.9.1 Magnetic Tapes and Films

Extremely rapid magnetization is possible by thin magnetic (metallic) films produced on a non- magnetic substrate. The process of electroplating, evaporation, or sputtering are employed for this purpose. The coercive force can be enhanced by adding cobalt. Magnetostriction (see art. 11.14) effect becomes zero in an iron alloy with 81.5% Ni. Metallic films must be as thin as possible to minimize eddy-current losses. Magnetic tapes and films are also made of FeCoNi, $Gd_3Fe_5O_{12}$ and CrO_2.

12.10 Metallic Glasses

Most metals and alloys are crystalline i.e. their atoms are arranged in a regular, ordered pattern that extends over long distances (hundreds or thousands of atoms). These regions of ordered atomic arrangement are crystals. The regular arrangement of atoms in a crystalline material can be directly viewed using a transmission electron microscope (TEM). Many of the important properties of engineering alloys can be explained in terms of this sort of crystalline order, or in many cases in terms of defects in the crystal structure.

Metallic glasses, in contrast, are alloys that are noncrystalline or amorphous. There is no long range atomic order. In this case, the atoms are more or less randomly arranged. This tells us that there are none of the long rows of atoms and the material is indeed amorphous.

12.10.1 Interesting Amorphous Material

Making amorphous solids is nothing new. Many common materials including oxide glasses such as ordinary window glass and most polymers, are amorphous. It is quite unusual, however, for a metallic material to be amorphous. The technique of making a metallic glass is to cool down a metallic liquid (which has a disordered structure as well) so rapidly that there is not enough time for the ordered, crystalline structure to develop. In the original metallic glasses the required cooling rate was quite fast, as much as a million degrees Celsius per second ! More recently, new alloys have been developed that form glasses at much lower cooling rate, around 1-100 degree per second. While still fairly rapid, it is slow enough that we can cast bulk ingots of these metallic alloys, and they will solidify to form glasses.

From an engineering point of view, our interest in metallic glasses stems from their unique structure. Since the structure of a material determines its properties, one might expect that a material with an unusual structure might have interesting properties. This is certainly true of metallic glasses. For instance, metallic glasses can be quite strong yet highly elastic, and they can also be quite tough (resistant to fracture). Even more interesting are the thermal properties. For instance, just like an oxide glass, there is a temperature (called the 'glass transition temperature') above which a metallic glass becomes quite soft and flows easily. This means that there are lots of opportunities for easily forming of the metallic glasses into complex shapes.

12.10.2 Unusual Properties of Metallic Glasses

These bulk metallic glasses (BMGs) have unusual properties. They are typically much stronger than crystalline metal counterparts by a factor of 2 to 3. They are quite tough (much more so than ceramics), and have very high strain limits for Hooke's law obeying materials as shown in Fig. 12.2. Metallic glasses possess several other superior properties as compared to crystalline (conventional) metals. These are as given below.

i. High strength, about 3 times stronger than steel.
ii. High springing nature, about 10 times to that of a best quality industrial steel.
iii. High strength to weight ratio
iv. High hardness
v. Becomes soft on heating, therefore are easily malleable
vi. Extreme springiness and rebound characteristics
vii. Good acoustic damping properties
viii. Highly elastic
ix. Highly tough

12.10.3 Fabrication of Metallic Glass

Unlike conventional metals, which are usually cooled slowly until they fully solidify, metallic glasses are cooled very rapidly and very uniformly to freeze their random atomic pattern in place, before crystallization occurs due to the nucleation and growth of crystal grains. Earlier, the only way to

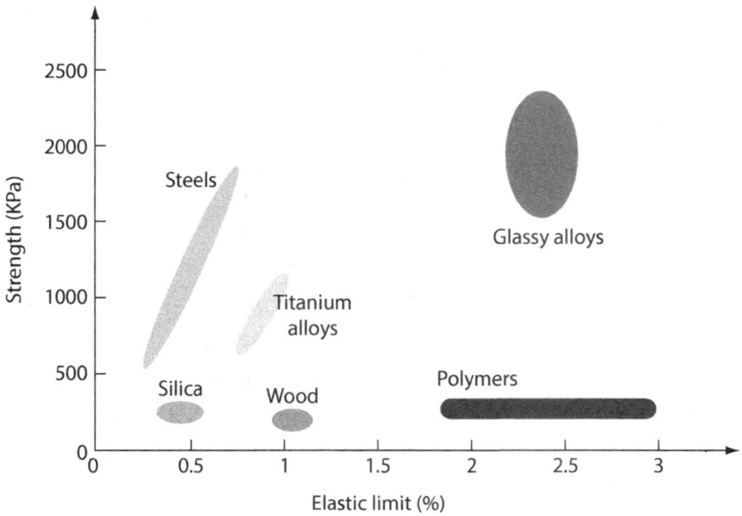

Figure 12.2 Typical strengths and elastic limits for various materials show that the glassy alloy i.e. metallic glasses are unique.

extract heat fast enough to maintain the metal's random state was to keep the metastable material very thin through special techniques such as splat cooling, in which the droplets of molten metal was quickly frozen on a cold surface. Continuous amorphous metal ribbons less than 0.1 mm thick could also be formed at a cooling rate of 1 million °C per second, by pouring molten metal onto a cold spinning wheel.

Rapid quenching and supercooling. Metallic Glasses (MG) are obtained by rapidly quenching of the melt at a quenching rate which is high enough to prevent crystallization. During the quench, the supercooled liquid becomes more and more dense and the atoms in the liquid become trapped in cage formed by their neighbours. The viscosity then increases dramatically and the supercooled liquid becomes an amorphous solid. The process is known as the glass transition. At temperatures well below the glass transition temperature Tg, the atomic mobility is too low for crystallization to take place.

Recent developments. A renewed interest in metallic glasses has developed in the last few years due to the discovery of a number of alloys that can be amorphised at cooling rates as low as 1 K/s. Consequently, the material of which the mechanical strength is one of the advantageous properties, can now be made in dimensions of the order of millimeters instead of 10^{-6} m thick melt spun ribbons.

Table 12.7 Example of some metallic glass (alloys) showing their constituents and specialities

Material system	Specialities
• Cu-Zr	Binary alloy
• Cu-Zr-Al	Ternary alloy
• Cu-Zr-Al-Y	Quarternary alloy
• Ta-Cu-Ni-Al	Plastic strain $\varepsilon_p = 4.5\%$
• Pt-Cu-Ni-P	Platic strain $\varepsilon_p = 20\%$
• $Pd_{40}Ni_{40}P_{20}$	Palladium-nickel based alloy
• $Fe_{40}Ni_{40}P_{14}B_6$	Iron-nickel based alloy
• $W_{60}Ir_{20}B_{20}$	Tungsten-iridium based alloy
• $Zr_{41.2}Ti_{13.8}Cu_{12.5}Ni_{10.0}Be_{22.5}$	Vitreloy (or liquimetal) used for golf club heads
• $Al_{90}Fe_5Ce_5$	Aluminium-iron-cesium based alloy
• $La_{55}Al_{25}Ni_{20}$	Lanthanide based alloy

12.10.4 Materials System and Salient Applications of Metallic Glasses

Various material systems of metallic glasses and their specialities are given in Table 12.7. The applications are also given below.

12.10.5 Applications

Salient applications of metallic glasses are the following

- Transformers
- Magnetic head recorder
- Metallic glass ribbon for different uses
- Wear resistant coatings
- Magnetic shielding
- Electric motors
- Golf club head
- Engine parts
- Military applications such as armor-piercing projectiles

12.10.6 Metallic Glasses in Electronic Uses

Metallic glasses are excellent ferromagnets possessing high magnetic moments,very high permeability, and zero magnetostriction. They are hard and corrosion resistant, therefore they are very much suitable for use as magnetic head recorder. As these can be easily magnetized, hence they also find applications in magnetic shielding, motors, transformers etc. Cobalt - based glass alloys have established their superiority over super-malloys at frequencies above 0.1 MHz.

The absence of grain boundaries, high mobility of domains, high electrical resistivity, wear resistance and high flux density make the metallic glass a superior magnetic material.

12.11 Magnetic Bubbles

Magnetic bubbles are memory devices having storage material of either amorphous magnetic layer or magnetic garnet. The iron garnet materials are grown on GGG or $Sm_3Ga_5O_{12}$. The bubble diameter of garnet system is proportional to $1/M^2$ where M is the saturation magnetization. These are made of soft magnetic materials of few micron diameter. Being embedded in the matrix of different spin orientations, they are separated by an equal distance among themselves. Magnetic bubble memory chips of 4 MB capacity has already been developed (before the year 2000). To increase the memory, the bubble diameter and the film thickness have to be reduced. Techniques of lithography and film growth are employed for the same. Bubble memories are suitable for small sizes, low power dissipation and highly reliable applications. They are employed in memory devices of computers.

12.11.1 Working Principle

This device works on the principle that the small regions of magnetic materials can have different magnetic alignments within a uniform physical shape. The technique is to grow very thin film of either ortho-ferrites or garnets, epitaxially on a suitable substrate. The usual material is garnet with the general formula $R_3Fe_5O_{12}$, where R represents yttrium or a combination of rare earth ions. Sometimes gallium or aluminium is substituted for some of the iron, to lower the saturation magnetization. A typical composition thus formed is $Eu_1Er_2Ga_{0.7}Fe_{4.3}O_{12}$. The film is only a few micrometers thick. All the domains can be aligned in a weak magnetic field normal to the film. Then by applying a stronger localized field in opposite direction, it is possible to produce a cylindrical domain, called a 'magnetic bubble', with its magnetic axis inverted (Fig. 12.3*a*).

When the strong field that created it, is removed, it turns out that with suitable materials the domain wall coercivity is great enough to produce a

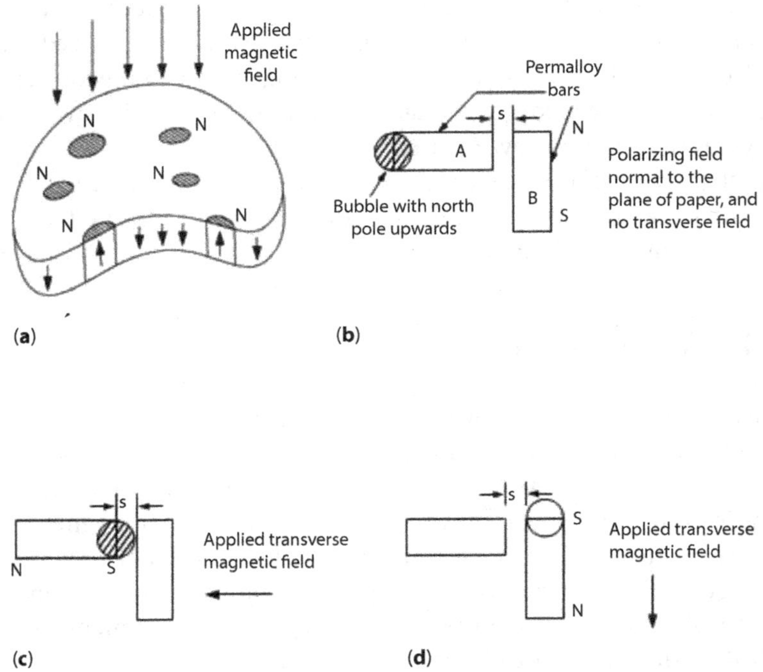

Figure 12.3 Magnetic bubble domains (*a*) applied field for stable bubbles (*b-d*) illustrate the movement of bubble.

stable bubble. The most stable bubble size will be obtained when the radius is approximately equal to the garnet film thickness of a few micrometres.

12.11.2 Moving the Magnetic Bubble

One way of achieving the controlled motion is by printing a pattern of small permalloy bars on the surface. The usual manufacturing technique for this is photoengraving using a photoresist material. To show the principles of how the bubbles can be persuaded to move from one place to another, we consider two typical permalloy bars A and B on the surface and assume the presence of a bubble with its north pole upwards, as shown in Fig. 12.3b.

In the absence of a magnetic field the permalloy bars are unmagnetized and have no effect upon the bubble. However, if a magnetic field as shown in Fig. 12.3*c* is applied, then bar becomes magnetized and the north pole of the bubble will move to the south pole of bar *A*. To move the bubble to bar *B*, we need to change the direction of magnetic field as shown in Fig. 12.3*d*. Then bar *B* becomes magnetized, and the bubble moves to the south pole of bar *B*.

12.11.3 Information Storage Density

The bubbles, by their presence or absence, may be used for storing the binary information and that the information can be read out. The advantage of using magnetic bubble is mainly, the density. With presently available photoengraving techniques, the contiguous structure may lead to a density of 10 million bits per cm^2. This is of course not a random access memory, the information has to be read out serially; the achievable speed may be a few hundred kbit/s. It could be used at an advantage when large blocks of data need to be transferred to the main memory of a computer for processing.

12.12 Effects of Alloying Elements on Magnetic Properties

Effects of different alloying elements are widely varying. Even different composition of the same element has different effects. These effects are briefly illustrated as follows.

Alloying element	Effects on magnetic properties
• Nickel	Improves initial and maximum relative permeabilities, reduces hysteresis loss and saturation flux density.
• Silicon	Improves initial and maximum relative permeabilities.
• Copper	Reduces the net magnetization.
• Zinc	Reduces the net magnetization, considerably.
• Aluminium	Reduces the net magnetization, considerably.
• Cobalt	Enhances the coercive force and energy product, and imparts magnetization at high temperatures.
• Manganese	Increases coercivity and hardness.
• Tungsten	Increases remanence and abrasion resistance.

12.13 Textured Magnetic Materials

Textured magnetic materials (also called grain-oriented materials) are made by orienting the crystals/grains in a preferred direction. The orientation is mainly accomplished by rolling process. By texturing the grains, the material exhibits improved magnetic properties in the direction of aligned orientation than in non-aligned directions. They can be easily magnetized also with a smaller magnetic field.

Cold rolled sheet steel used for making transformer core is an example of textured (or grain- oriented) magnetic material. It has a lower hysteresis loss due to texturing. Also by orienting the grains parallel to electric flux path, a high permeability is obtained which is a desired property for transformer core.

12.14 Amorphous (or Oxide) Magnetic Materials

Non-crystalline solids are called amorphous materials. They may be oxides, carbides, halides, nitrides etc. Most ceramics are metallic oxides, hence amorphous. Thus, the ceramic based magnetic materials are called amorphous or oxide magnetic materials. They may be amorphous paramagnetics such as Fe_2O_3, DAG($Dy_3Al_5O_{12}$); amorphous antiferromagnetics such as MnO_2, FeO, CoO etc., and amorphous ferrites such as $PbO.6Fe_2O_3$, $BaO.6Fe_2O_3$ etc.

12.15 Powder Magnetic Materials

These are hard magnetic materials composed of fine particles, bonded together by organic resins or non-magnetic metal binders such as lead. The fine particles in the form of powder are of 1 to 10 μm sizes and are manufactured by powder metallurgy techniques. These can be produced as Elongated Single Domain (ESD) magnets also by precipitating on liquid-mercury cathode. Powdered magnetic materials are used to make miniature magnets for fractional horsepower motors for use in the following applications.

- Hearing aids,
- Car windshield wiper-motor,
- Earphones,
- Clocks etc.

12.16 Solved Examples

Example 12.1 Suggest suitable magnetic materials, soft or hard type, for the following applications stating the reasons of such selection,

a. A magnetic door latch
b. A compass needle
c. Door seals in refrigerators
d. Computer disc drive

Solution.

a. A hard magnetic material is required. Since the door latch has to retain its magnetism, hence it should have permanent magnetism.
b. A hard magnetic material is required. Since the compass needle has to retain its magnetism, hence it should have permanent magnetism.
c. A hard ferrite magnet is required as it has higher coercivity, resistance to many chemicals, and being an oxide; it is ideal for use in wet or damp environments such as found in refrigerators.
d. High grade high energy hard magnetic materials like $Nd_2Fe_{14}B$ made by sintering process is suitable as only a small volume of magnetic material is required. It also decreases the access time.

Quick Revision Summary

Alnicos. A most widely used class of permanent magnet alloys, which contain Fe, Al, Ni, Co, and Cu or Ti.

Directions of easy magnetization. The crystallographic direction in a ferromagnetic single crystal which is most easily magnetized.

Energy product. The highest magnitude of the product of B_r and H_c to be found on the demagnetization curve of a hard magnetic material.

ESD (Elongated Single-Domain) magnet. A magnet made of aligned fine particles which are below the domain wall size and elongated in their direction of easy magnetization.

Ferrite. Material whose formula is $MO.Fe_2O_3$, where M is a divalent metal, whose structure is of the spinel family. The ferrimagnetic ferrites have the inverted spinel structure.

Garnet. A group of oxides similar to the spinel. The ferrimagnetic garnets have the formula $2M_2O_3.5Fe_2O_3$, where M is a trivalent rare- earth metal.

Inverted spinel. It has the same oxygen lattice as a spinel, with half the trivalent ions on tetrahedral sites. The ferrimagnetic ferrites are inverted spinels.

Iron group of elements. Those transition elements in which the 3d state are party empty are called 'iron group of elements'. Their atomic numbers are from 21 to 29. Fe, Ni, Co, Gd are such examples. Due to incompletely filled inner states, these elements influence the magnetic properties of materials considerably.

Magnetic anneal. It is the heat treatment of a magnetic material in a magnetic field which induces an anisotropy by aligning various features of the microstructure with the field, e.g., directional ordering in Permalloys, the shape of the α phase in Alnico 5, and the alignment of the powder particles in an ESD magnet.

Permalloys. These are the Ni-Fe alloys with high initial permeabilities and are used where high sensitivity is needed.

Preferred orientation (Texture). Crystallographic alignment of the grain structure in a polycrystalline solid.

Sintering is a fabrication process in which the agglomeration of powdered metal is done by the application of pressure and temperature. This process is employed to prepare the intricate and tiny sized products such as miniature magnets.

Skin effect. When the thin laminated cores (sheet) of apparatuses like transformers are subjected to alternating current, the alternating magnetic flux should normally get distributed over the entire cross- section of the sheet. But normally it is not so. The flux is forced outwards by the outer skin of magnetic sheet. This behaviour of sheet is called 'skin effect'. This is an undesired effect and is caused due to demagnetizing effect of eddy current.

Soft and Hard directions. Ferromagnetic materials exhibit directional nature in their behaviour. They get magnetized differently in different directions when subjected to an applied magnetic field. Some magnetic materials offer high permeability along certain directions and are easily magnetized. Such directions are called *soft directions.* But these materials are difficult to magnetise in some other directions which are called *hard directions.* These directions are different in different materials.

Spinel ($MgAl_2O_4$). A compound whose Al (trivalent) atoms and Mg atoms occupy tetrahedral sites in an FCC oxygen lattice. Other trivalent and divalent metal ions may be used in place of the Al and Mg respectively. $ZnFe_2O_4$ is an example.

Spinels are the compounds having general formula XY_2O_4, where X and Y are the cations, and oxygen is the anion. The cations X are in the four tetrahedral voids and cations Y in the eight octahedral voids for every four FCC packed oxygen anions.

Splat cooling refers to extremely fast cooling rate, generally exceeding one million degrees per second, and is used in the production of *metallic glass* by transforming liquid state into a solid state.

Superlattice alloys. These are alloys of platinum formed with iron or cobalt. Since structure of Pt is of FCC type, of Fe is BCC and Co is HCP; therefore the alloys formed by them are of special characters. When heated for hardening and then cooled, they form a superlattice of FCC and BCC, or FCC and HCP configurations. These are used to make hard magnetic materials.

Review Questions

1. Enlist different types of magnetic materials and briefly elaborate them stating their salient features and applications.
2. Explain the properties of soft and hard magnetic material along with initial and maximum permeability.
3. Compare hard and soft magnetic materials. Why is a soft magnetic material preferred over a hard magnetic material for use in the transformer core ?
4. Name different commercial grade soft magnetic materials and describe them stating their salient applications.
5. Name different commercial grade hard magnetic materials and describe them stating their salient applications.
6. Explain the properties, applications, and compositions of the following magnetic materials.
   a. Mumetal
   b. Supermalloy
   c. Ferroxcube
   d. Electrical sheet steel
7. What are the high energy hard magnetic materials? Describe their different types.
8. Compare the properties and applications of Samarium based and Neodymium based high energy hard magnetic materials.
9. Enumerate different kinds of magnetic materials for soft ferrites purposes. Write their composition, properties and applications.
10. Which materials are used for making magnetic tapes and films? Discuss their characteristics.

11. What are metallic glasses? How are they produced? Write their properties, examples and applications.
12. Explain the magnetic bubble. Discuss its working principle and write applications.
13. Discuss the effects of following alloying elements on magnetic properties of alloys magnets.
    a. Ni b. Si c. Co d. Mn
14. How do the addition of Cu, Zn, Al and W influence upon the properties of magnetic materials?
15. Write notes on the following.
    a. Oxide magnetic material
    b. Textured magnetic material
    c. Powder magnetic material
    d. Information storage density
16. Suggest suitable materials for the following applications.
    a. Transformer cores
    b. High frequency coils
    c. Audio frequency oscillators
    d. Fractional horsepower motors.
17. Explain reasons for the following.
    a. A soft magnetic material is preferred for use in transformer core.
    b. A powdered magnetic material is preferred for miniature size magnets such as for use in hearing aid system.
    c. Core of an alternating current machine becomes hot.
    d. Ferrites are used for data storage in computers.

Objective Questions

1. Match list I (memory elements) with list II (properties) and select the correct answer using the codes given below the lists.

List I	*List II*
A. Semiconductor memory	1. Destructive read out
B. Ferrite core memory	2. Combinational logic
C. Magnetic tape memory	3. Volatile

Codes:

	A B C		A B C
a.	1 2 3	b.	2 3 4
c.	3 4 1	d.	4 1 2

2. Match list I with list II and select the correct answer using the codes given below the lists.

List I	*List II*
A. Ferromagnetism	1. d.c. electrical resistivity vanishes at the critical temperature T_c
B. Semiconductor	2. Doping with impurity increases the electrical conductivity
C. Optical property of solid	3. An internal molecular field B_M which is proportional to magnetisation M exists at each dipole and aligns it parallel to other dipoles
D. Superconductivity	4. Above the Neel temperature, the dipoles become randomly oriented
	5. The conductivity of crystalline solid, semiconductors and dielectrics increases by radiation incident

Codes:

	A B C D		A B C D
a.	3 5 2 1	b.	3 2 5 1
c.	4 5 2 3	d.	4 2 5 3

3. High frequency transformer cores are generally made of
   a. Cast iron b. Mumetal
   c. Ferrite d. Graphite

4. For making tapes and films, the use is made of
   a. $Gd_3Fe_5O_{12}$ b. piezoelectrics
   c. pyroelectrics d. ferromagnetics above 768°C

5. The magnetic head recorders are usually made of
   a. wrought iron b. permalloy
   c. CdS d. metallic glass

6. The residual magnetic flux density is more in
   a. metallic magnets b. ceramic magnets
   c. graphite d. iron oxide

7. Manganese ferrite is a 1 : 1 mixture of
   a. MnO and CdO b. MnO and Fe_2O_3
   c. $MnCl_2$ and Fe_2O_3 d. $MnC1_2$ and CdO

8. Match list I with list II and select the correct answer using the codes given below the lists.

List I	*List II*
A. Mumetal	1. Very high magnetic permeability
B. Samarium/cobalt	2. Very high magnetic remanance
C. Sapphire	3. Very high thermal conductivity
D. Ferrites memory	4. Very stable magnetic permeability

Codes:

	A B C D		A B C D
a.	1 2 3 4	b.	2 3 4 1
c.	3 4 1 2	d.	4 1 2 3

9. Memory of a computer can be increased by
   a. increasing the diameter of magnetic bubble.
   b. decreasing the diameter of magnetic bubble.
   c. using alcomax, a recent hard magnetic material.
   d. using strontium titanate ($SiTiO_3$).

10. Consider the following statements.
    A. The immobile domain walls are responsible for large coercive force in hard magnetic materials.
    B. Perminvar is a magnetodielectric material.
    C. Alnico retains about 70% magnetization at 600°C.
    D. Ferroxcube is an example of hard ferrite.

    Of these, the correct statements are
    a. A and C b. B and D
    c. A, B and C d. B, C and D

11. Magnetic films are thinner than magnetic tapes, and are used for frequencies of 1 MHz or more. *(T/F)*

13

Superconductive Materials

13.1 Concept of Superconductors

Superconductors are those elements, compounds and alloys of metals and non-metals which exhibit extraordinary magnetic and electrical behaviour at extremely low temperatures (near absolute zero). Such low temperatures are not practically favourable for wide applications. The amazing property of superconductivity is, infact, desired at room temperature. Efforts are on to develop new superconducting materials at practically feasible temperatures. A breakthrough in this area is poised to cause a revolution in future life style.

13.1.1 Meaning of the Phenomenon of Superconductivity

Superconductivity is a phenomenon observed in superconducting materials. Superconductors are not the conductors of super qualities as can be misunderstood sometimes. These materials exhibit abrupt and sudden changes in their resistance and other behaviours at extremely low temperatures, generally below 10 K. They may belong to the categories of metals, non-metals, ceramics, composites, metallic compounds and alloys. Superconductors do not obey Ohm's law below a critical temperature T_c. The peculiar behaviour

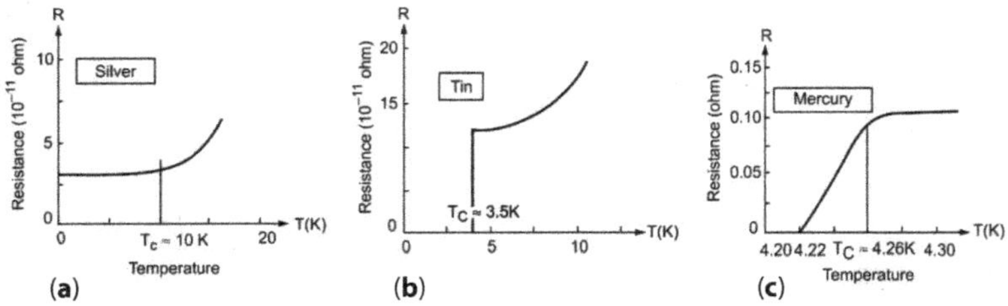

Figure 13.1 Variation of resistance as a function of temperature for (*a*) silver, (*b*) tin, and (*c*) mercury, shows that the materials do not obey Ohm's law below cirtical temperture.

of silver, tin and mercury are shown in Figs. 13.1*a-b-c* respectively. These curves show an unusual drop in their resistance below a certain critical temperature T_c.

Examples. Good conductors at room temperature e.g. gold, silver, copper, aluminium etc; and ferromagnetic materials such as iron, nickel and cobalt are absent from the list of superconductors. Superconducting elements are relatively poor conductors at room temperature. The critical temperature of some elements, compounds and alloys are illustrated in Tables 13.1 and 13.2.

13.2 Properties of Superconductors

Superconducting materials exhibit the following extraordinary properties below their critical temperatures.

i. The magnetic flux density, $B \approx 0$
ii. The relative permeability, $\mu_r \approx 0$
iii. The specific resistance, $\rho \approx 0$
iv. The magnetic susceptibility, $\chi \approx -1$
v. The power (copper) loss $I^2R \approx 0$

13.2.1 Meissner Effect

When a magnetic field is applied, these materials repel back all the lines of force as shown in Fig. 13.2. This is perfectly diamagnetic behaviour of a superconductor and is known as *Meissner effect.* It is helpful in protecting the devices from the undesired magnetic fields.

Table 13.1 Critical Temperatures of Some Superconducting Materials

Type	Symbol	Critical temperature T_c (K)	Symbol	Critical temperature T_c (K)
Element				
	Al	1.19	V	5.13
	Sn	3.50	Zn	0.79
	Hg	4.26	Ti	0.53
	Ag	10.00	Pb	7.17
Compound				
	CuS	1.6	Nb_3Sn	18.3
	SuSb	3.9	Nb_3Ge	23.2
	Pb_2Au	7.0	NbN	14.7
	MoN	12.0	V_3Si	17.0
Alloy				
	La-Ba-Cu-O	34.0	$YBa_2Cu_3O_{7-x}$	90.0
	$Pb_2Sr_2(Y,Ca)Cu_3O_8$	76.0	$Bi_2Sr_2Ca_2Cu_3O_{10}$	109.0
	$AuBa_2Ca_3Cu_4O_{11}$	99.0	$TlBa_2Ca_2Cu_3O_{9+}$	123.0

Example 13.1 Prove that the susceptibility of a superconductor is –1 and relative permeability is zero.

Solution. Recalling the relation

$$B = \mu_0(M + H) \quad (i)$$

and knowing that $B = 0$ for a superconductor,

$$\therefore \quad 0 = \mu_0(M + H)$$

$$\text{or} \quad M = -H \quad (ii)$$

$$\because \quad M = \chi H \quad (iii)$$

$\therefore$ On putting Eq. (*ii*) in Eq. (*iii*), we get

$$-H = -\chi H$$

$$\text{or} \quad \chi = -1$$

$$\text{And also because} \quad \chi = \mu_r - 1$$

$$\therefore \quad \mu_r = -1 + 1 = 0$$

Table 13.2 Critical Magnetic Fields and Critical Temperatures of Some Superconducting Materials

Material	**Critical temperature T_c(K)**	**Critical magnetic field strength H_c(A/m)**	**Remark**
Elements			
Al	1.19	0.0099	Type I,
Pb	7.17	0.0803	at 0 K,
Mo	0.93	0.0096	H_c
Ta	4.48	0.0829	
Ti	0.53	0.0056	
Compounds			
Nb_3Sn	18.3	21.0	Type II,
Nb_3Al	19.9	22.0	at 4.2 K,
Nb_3Ge	23.2	37.0	Hc2
V_3Ga	15.4	22.0	
$Pb\ Mo_6\ S_8$	15.0	60.0	
$La_{80}\ Au_{20}$	8.7	20.0	Metallic glass
Alloys			
Nb-Ti	10.2	11.0	47% Ti
Nb-Zr	10.8	8.5	33% Zr
Nb-Zr-Ti	10.3	10.5	42% Zr, 6% Ti

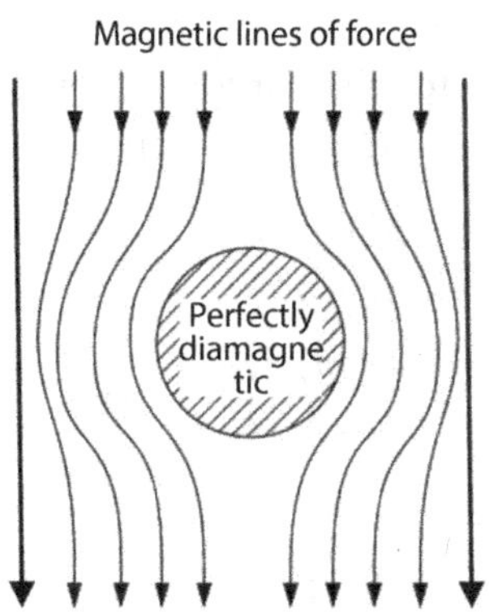

Figure 13.2 A superconductor is perfectly diamagnetic having magnetic susceptibility $\chi = -1$. It repels back the magnetic lines of force.

13.3 Types of Superconductors

Superconductors are classified into the following categories owing to different considerations:

I. **On the Basis of Working Temperature**
   1. Low temperature superconductors (LTS)
   2. High temperature superconductors (HTS)

II. **On the Basis of Kind of Material**
   1. Metallic superconductors (MS)
   2. Intermetallic compound superconductors
   3. Ceramic superconductors (CS)
   4. Alloy superconductors

III. **On the Basis of Application**
   1. Magnetic grade superconductors
   2. Non-magnetic grade superconductors

IV. **On the Basis of Penetration of Magnetic Lines of Force**
   1. Type I or ideal superconductors
   2. Type II or hard superconductors

LTS and HTS. Low temperature superconductors are those having their critical temperatures less than 25 K ($T_c < 25$ K). Superconductors, whose $T_c > 25$ K, are known as high temperature superconductors. LTS are mostly metallic while HTS are of ceramic type. Nb-Ti, Nb-Zr, Nb_3Sn, Nb_3Al, V_3Ga etc. are LTS. These are suitable for magnet applications. HTS is observed in perovskite oxides which, infact, are ceramics.

Type I and Type II. Type I superconductors such as Hg, Pb, In, etc. are unsuitable as magnets because of their low H_c values ($H_c < 10^5$ A/m corresponding to about 0.1 tesla). Type II superconductors are of magnetic grade such as Nb-Ti alloy and intermetallic compound Nb_3Sn. The bismuth lead strontium calcium copper oxide (Bi Pb Sr Ca CuO) tape is a type II superconductor. It is silver-sheathed.

13.3.1 Effect of Pressure and Temperature on Superconductivity

Pressure and temperature also have a meaning with superconducting magnets. A typical 10 T (tesla) magnet is subjected to an equivalent magnetic pressure of 40 MPa (400 atmosphere) whether it is superconducting and operating at 4.2 K (temperature of cooled liquid helium), or 77 K (cooled liquid nitrogen), or room temperature (cooled water).

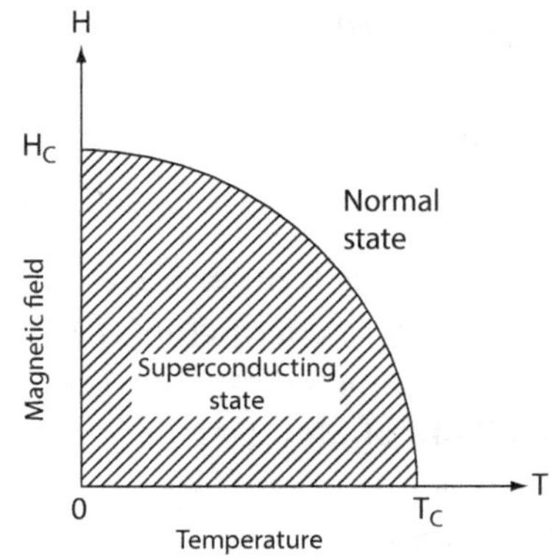

Figure 13.3 Variation of critical magnetic field H_c as a function of temperature demarcating the superconducting and normal states.

13.4 Critical Magnetic Field and Critical Temperature

As described earlier, the critical temperature T_c is the transition temperature at which the normal state of a material changes to a superconducting state. This critical temperature can be varied by an application of magnetic field H. Variation of T_c with the change in magnetic field is shown in Fig. 13.3. The material is in normal state for $H > H_c$ and in the superconducting state for $H \leq H_c$, where H_c is the critical magnetic field. The two critical values T_c and H_c are related by

$$H_c = H_0\left(1 - \frac{T^2}{T_c^2}\right) \tag{13.1}$$

where H_0 is a material constant. It can be determined from

$$H_0 = T_0\sqrt{\frac{\gamma_c}{2\mu_0}} \tag{13.2}$$

where γ_0 is a constant that accounts for electronic heat capacity in the normal state. The value of $\mu_0 H_0 = 0.06$ tesla for Pb and Cd; $H_0 = 30$ gauses and $T_c = 0.52$ K.

13.4.1 Effect of Isotopic Mass on Critical Temperature

The critical temperature is different for different materials as it depends on their isotopic mass. Both the above quantities are related as

$$T_c\sqrt{M} = \text{constant} \tag{13.3}$$

where M is the isotopic mass. It is, therefore, clear that the critical temperature decreases with increasing isotopic mass.

13.4.2 Silsbee Rule

The magnetic field which causes a superconducting state to behave as normal state, is not necessarily an externally applied field. A flow of electric current in a conductor can also cause this effect. *Silsbee* has shown that a critical current I_c flowing in a long circular wire of radius r induces

$$I_c = 2\pi r H_c \tag{13.4}$$

It indicates that the superconductivity in a long circular wire will be destroyed when $I > I_c$. As current needed for producing strong magnetic fields is high, therefore $I > I_c$ prevents the use of superconductors as coils.

Example 13.2 Critical temperature of Pb in superconducting state is 7.17 K under zero magnetic field. The value of critical field for it is 0.0803 A/m at 0 K. Determine its critical field at 3 K and 10 K.

Solution. Given are: $H_0 = 0.0803$ A/m, $T = 3$ K and 10 K,

$T_c = 7.17$ K. To determine H_c.

On substituting the appropriate values in Eqn. 13.1, we find

For $T=3K$; $\quad H_c = 0.0803[1-3^2/7.17^2]$

$= 0.803[1 -0.175] = 0.06624$ A/m

For $T = 10K$; $\quad H_c = 0.0803[1-10^2/7.17^2]$

$= 0.0803[1 - 9.45] = -0.0758$ A/m

Discussion. The value of H_c at 10 K is –ve. It should not be so for a superconductor. In fact, occurrence of a –ve value for H_c indicates that the material is in normal state (and not in superconducting state) at this temperature.

Example 13.3 Calculate the critical current which can pass through a long thin superconducting wire of aluminium of diameter 1 mm. The critical magnetic field for aluminium is 7.9×10^3 A/m.

Solution. The critical current in superconducting state may be determined by Silsbee's rule. Hence using Eq. 13.4, we get

$$I_c = 2\pi r\, H_c$$
$$= 2\pi \times (1 \times 10^{-3}) \times 7.9 \times 10^3$$
$$= 49.65 \text{ A}$$

13.4.3 Important Terminologies

Depth of penetration d_p. This term is related to as how deeply a magnetic field penetrates through the surface of a superconductor. An external magnetic field impinged upon a type II superconductor decays exponentially into the surface of the superconductor.

Coherence length l_c. It is the size of a Cooper pair that represents the shortest distance over which the superconductivity can be established in a material. Its value is typically of the order of 1000 Å, although it can be as small as 30 Å in copper oxides.

13.5 Ideal and Hard Superconductors

Based on the characteristics of penetration of magnetic lines of force into the material, the superconductors are classified as

1. Type I or ideal superconductors, and
2. Type II or hard superconductors.

As illustrated in Figs. 13.4*a-b*, type I superconductors show full penetration of magnetic field at H_c, while the type II allow gradual penetration. It starts from a lower critical field H_{c1} and finishes at the upper critical field H_{c2}. The higher critical magnetic field is due to its ability to trap the magnetic flux inside the material. The critical magnetic fields alongwith critical temperatures for Type I, Type II, metallic glass and other superconducting materials are given in Table 13.2.

13.6 Mechanism of Superconduction

The phenomenon of superconductivity is believed to occur due to the interaction of pairs of electrons. According to electron theory, the resistance of a metallic conductor should vary continuously with temperature, and it should completely disappear only at absolute zero (zero kelvin). But the

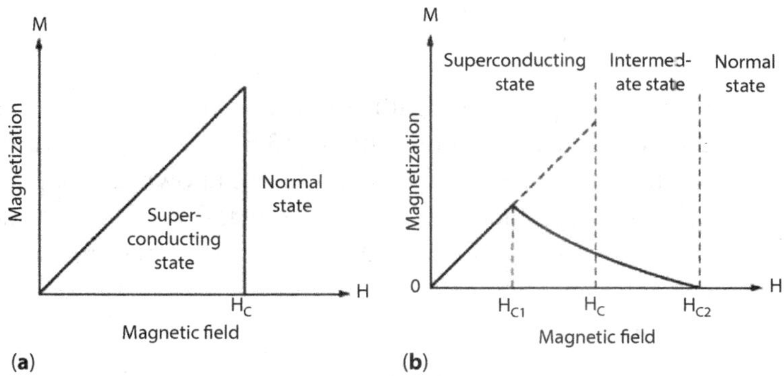

Figure 13.4 (*a*) Sudden change of magnetization in Type I superconductors, and (*b*) gradual penetration of magnetization in Type II superconductors

phenomenon of superconductivity at very low temperatures offers a challenge to the existing electron theory. Various theories in the past (Kapitza additional resistance theory, Thomson's electric doublet theory etc.) have failed to explain its phenomenon satisfactorily.

13.6.1 Various Theories of Superconductivity

Following theories are generally considered authentic for different types of superconductors.

1. London theory for type I superconductors
2. GLAG theory for type II superconductors
3. BCS theory incorporating the concept of super-electrons and supercurrent

13.7 London's Theory For Type I Superconductors

The theory of superconductivity for Type I superconductors was given by two brothers F. London and H. London. This electromagnetic theory accounts for Meissner effect. According to this theory, the critical current density in a superconductor may be given by

$$I_{cd} = n_e ev \tag{13.5}$$

where n_e is the electron concentration, e is electron charge, and v is the velocity of sound. The n_e is related as

$$n_e = \frac{2\rho N_A}{A_W} \tag{13.6}$$

where ρ is the density, N_A is the Avogadro's number and A_w is atomic weight of the superconductor. The numeral 2 in Eq. 13.6 accounts for two *superelectrons* for each atom. The two superelectrons are known as Cooper pair. The shielding of superconductor from external magnetic field is achieved by a supercurrent. The depth of penetration d_p of this current into the surface of superconductor can be determined from

$$d_p = \left(\frac{m}{\mu_0 n_e e^2} \right)^{1/2} \tag{13.7}$$

where m is the mass of electron, and μ_0 is the permeability of free space.

13.8 GLAG Theory For Type II Superconductors

This theory describes the relationship among penetration depth, mixed state and upper critical field. Type II superconductor may be treated as a mixture of finely divided Type I superconductor and a normal conductor. This mixture can be modelled into two kinds of Type II superconductors, viz.

1. Lamina model, and
2. Vortex or island model

The lamina model assumes superconducting laminae separated by the normal laminae. The vortex model considers normal state islands of hexagonal arrangement floating in a superconducting surrounding.

13.8.1 Distinguishing Between the Type I and Type II Superconductors

According to GLAG theory, a superconductor will be known as

Type I if $l_c \le \sqrt{2} d_p$ (13.8)

Type II if $l_c > \sqrt{2} d_p$ (13.9)

where l_c is *coherence length* and d_p is the depth of penetration of supercurrent into the surface. The effect of alloying is to shorten l_c and hence

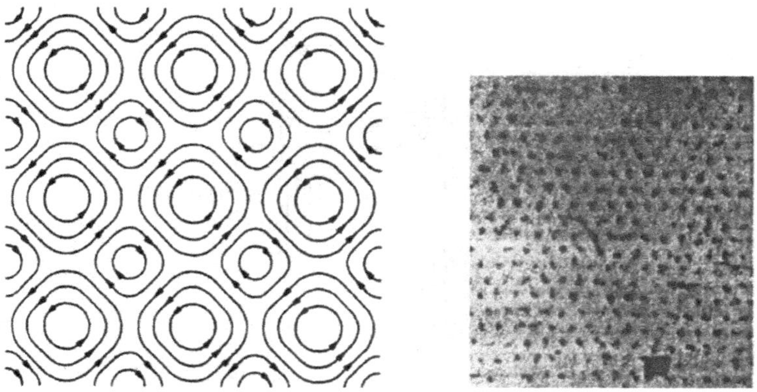

Figure 13.5 (*a*) The lines of current flow for a type II superconductor shows the magnetic field, which is maximum in the centres of the current vortices. (*b*) Triangular vortex structure on the surface of a superconducting material.

the mean free path of the normal electron. l_c is inversely proportional to normal-state electrical resistivity of the material.

- The critical current density of Type II superconductor is given by

$$I_{cd} = \frac{\lambda}{H + H_0} \tag{13.10}$$

where λ and H_0 are the constants. The value of H_0 is 14.5 tesla for Nb-Ti, and 27.9 tesla for Nb-Sn.

13.8.2 Variation of Magnetic Field

A two-dimensional analysis of type II superconductor shows that the intensity of magnetic field varies in a periodic manner with well-defined maxima as shown in Fig. 13.5*a*. It is quite clear that the role of the current is either 'not to let in' or 'not to let out' the magnetic field.

Supercurrent vortex. The density of superconducting electrons is zero at the maxima of the magnetic field. Thus, in a somewhat simplified manner, we may say that there is a normal region surrounded by a supercurrent vortex. There are lots of vortices; their distance from each other is about 1 μm. The vortex structure on the surface of a Pb-In rod at 1.1 K is shown in Fig. 13.5*b*. It has a triangular structure.

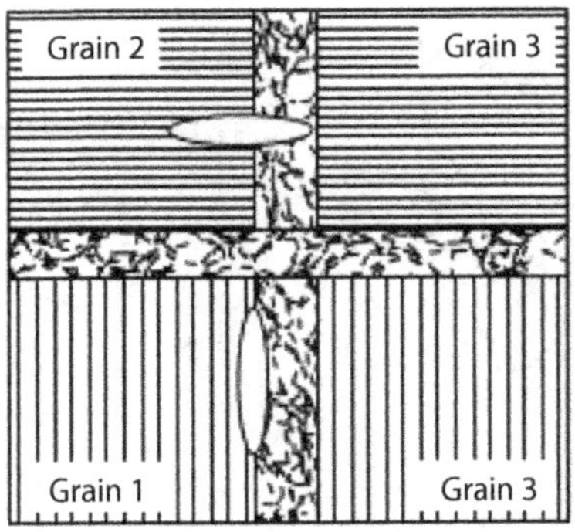

Figure 13.6 One disc-shaped Cooper pair meeting a grain boundary edge-on, and another face-on.

13.9 BCS Theory

Features. This is the most widely accepted theory to explain superconductivity. The theory asserts that, as the electrons pass through a crystal lattice, the lattice deforms inward towards the electrons generating sound packets known as 'phonons'. These phonons produce a trough of positive charge in the area of deformation that assists subsequent electrons in passing through the same region, in a process known as 'phonon-mediated coupling'.

13.9.1 Supercurrent

The BCS theory explains most of the phenomena associated with superconductivity in a natural manner. It explains electromagnetic and thermodynamic properties of LTS. It incorporates the hypothesis that the special pairs of electrons, called *Cooper pairs* (Fig. 13.6), carry the supercurrent.

13.9.2 Phonon and Photon

Presently *Bardeen-Cooper-Schreiffer (BCS)* theory is acceptable. This theory holds a three-way interaction between two electrons and a phonon responsible for superconducting effect. Phonon is different from *photon. Phonon is a quanta of energy in the elastic wave, while the photon is a quanta of energy in electromagnetic wave.* The movement of electrons without scattering produces a weak interaction which is experienced at extremely low temperatures due to negligible thermal excitation.

13.9.3 Electron-Phonon Interaction

The electron travelling in a solid interacts with the lattice vibrations. This interaction called *electron-phonon Interaction* leads to scattering of electrons and causes electrical resistivity. Since the number of phonons increases with temperature, the resistivity is a sensitive function of temperature particularly in the low temperature region, where it varies as T^5.

The BCS theory assumes that the electron- phonon interaction produces an attractive interaction, V_{ph}, between two electrons whose momentum and spin are related as $k\uparrow$ and $-k\downarrow$ i.e. between the electrons whose spin and the wave vector are equal in magnitude but opposite in sign. These are called *Cooper pairs* (or *superelectrons*). These electrons due to their charge also repel each other with an interaction V_c. But Bardeen, Cooper and Schreiffer assume that in the superconducting phase, V_{ph} dominates over V_c and the net interaction V is attractive. It is given as

$$V = (-V_{ph} + V_c) < 0 \tag{13.11}$$

13.9.4 Reason of Two Electrons Forming a Pair

The basis of BCS theory is that there is an attractive force between two electrons having nearly the same kinetic energy. This force in the right circumstances, causes them to move in pairs. Superconductivity occurs only if this attraction exceeds the electrostatic repulsion that always occurs between two electrons. The attractive forces arise because, in a crystal lattice of positive ions, an electron produces a small distortion of the lattice by attracting the positive ions towards itself, making the lattice slightly more dense in its vicinity. To a passing electron, this distortion will behave like a local increase in positive charge density and will attract the second electron towards it. Thus the end electrons tend to form pairs through their electrostatic interaction with the lattice. The attractive force is maximum for electrons having opposite spins, and moving with equal velocities in opposite directions.

- The attractive force makes the two electrons to move in a coordinated, but not identical way. The distance over which the two electron wave functions are coordinated is known as the *coherence length.*

13.9.5 Cause of Resistance for Superconductor being Zero

In terms of the energy level concept, let ΔE represents a gap in the allowed energy states for electrons, a so-called 'energy gap', lying at the Fermi energy level. Unless the electrons gain sufficient energy to cross the gap (and this is not possible at temperatures below T_c), there is no possibility of electrons changing their state of motion, because there are no vacant energy states that they can move to. Thus the normal mechanism of resistance to current flow cannot operate and the resistance of the material will be zero. This will remain true until the drift velocity of the electrons that carry the current exceeds a critical value, i.e. until the critical current is exceeded. The critical velocity corresponding to this critical current density is approximately given by

$$v_c = \frac{\lambda_F \Delta E}{2h} \tag{13.12}$$

where λ_F is the electron wavelength for an electron at the Fermi level.

13.9.6 Relationship between Coherence Length and Energy Gap

The coherence length l_c can be related to the energy gap through the approximate relationship given as

$$l_c = \frac{h^2}{m\pi(\Delta E)\lambda_F} \tag{13.13}$$

and is of the order of 1 μm in most elemental superconductors. This means that the electron pairing is extended over thousands of atomic spacings so that the atomic scale defects and impurities are too small to have much effect on superconducting behaviour.

13.9.7 Debye Temperature

Bardeen, Cooper and Schreiffer have solved the problem of electron energy states when the attractive interaction V is present. They found that there is an energy gap in the excitation spectrum as discussed earlier. This theory gives the following relation for critical temperature.

$$T_c = 1.14\theta_D e^{(-1/N(0)V)} \tag{13.14}$$

where θ_D is the debye temperature, $N(0)$ is the density of electron states at the Fermi surface and V is the net attractive interaction. Equation 13.14 accounts for isotopic effect because θ_D is proportional to $M^{-1/2}$. BCS theory also explains the existence of Meissner effect and the *coherence length.*

Example 13.4 An infinitely long superconducting lead (Pb) rod of circular cross-section is subjected to an uniform external magnetic field perpendicular to its axis. Calculate (*a*) electron density, (*b*) critical current density, and (*c*) depth of penetration at the superconductor's surface. For lead, the specific density is 11.4, atomic weight is 207.2 kg/kg-mole, and the velocity of sound in it is 1200 m/s.

Solution. The given data are:

$$\rho = 11.4 \times 10^3 \text{ kg/m}^3, A_w = 207.2 \text{ kg/kg- mole, and}$$

$$v = 1200 \text{ m/s} = 1.2 \times 10^3 \text{ m/s}$$

The standard values of other quantities are

$$N_A = 6.023 \times 10^{26} \text{ particle/kg-mole}$$

$$e = 1.6 \times 10^{-19} \text{ C/electron}$$

$$m = 9.1 \times 10^{-31} \text{ kg, and}$$

$$\mu_0 = 4\pi \times 10^{-7} \text{ H/m}$$

Lead is Type I superconductor, therefore London's theory of superconductivity will be applicable.

Hence, using Eq. 13.6

$$n_e = \frac{2\rho N_A}{A_W}$$

$$= 2(11.4 \times 10^3)(6.023 \times 10^{26})/207.2$$

$$= 6.62 \times 10^{28} \text{ electron/m}^3$$

Using Eq. 13.7, we obtain the critical current density as

$$I_{ed} = n_e ev = (1.6 \times 10^{-19}) \times (6.62 \times 10^{28}) \times (1.2 \times 10)$$

$$= 1.27 \times 10^{13} \text{ A/m}^2$$

The depth of penetration at the surface of lead, within which the superconducting current flows, may be found from Eq. 13.8.

Thus $$d_p = \left(\frac{m}{\mu_0 n_e e^2}\right)^{1/2}$$

$$= \frac{(9.1 \times 10^{-31})}{(4\pi \times 10^{-7})(6.62 \times 10^{-28})(1.6 \times 10^{-19})}$$

$$= 2.1 \times 10.8 \text{ m} = 210 \text{ Å}$$

13.10 Current Applications and Limitations

Currently, superconductors are used in the following important applications.

1. Switching and memory elements
2. Flip-flop devices
3. Small cryotrons
4. Very strong (or high field) magnets
5. Magneto-hydro-dynamic (MHD) generators
6. Cryogenic engines and heat valves
7. Magnetic detectors and radiation detectors
8. Magnetic resonanace imaging (MRI) for which the generation of high magnetic fields using superconducting solenoid are required.
9. High-resolution detection of magnetic flux using SQUID (Superconducting Quantum Interference Device) magnetometers.
10. Thin films of HTS for SQUIDS and microwave devices, both these require low power.
11. Miniature, low-power electronic devices, mostly based on Josephson effect.
12. Magnetometers.

13.10.1 Limitations of Superconductors

The only limitation of superconductivity is its occurrence at very low temperatures. Although its useful properties may be utilized at lower temperatures such as in cryogenic engines, but we desire the favourable properties of superconductivity at room temperature ($\approx$ 300 K) so that their beneficial aspects may be utilized in day-to-day life.

13.10.2 Likely Futuristic Scenario

Research is in full swing, everywhere in the world, to achieve this wonderful phenomenon at room temperature. We may achieve the goal within a year, or in a few years, or perhaps never. If the breakthrough comes into being, the likely applications may be conceived as follows :

1. Huge digital computers and supercomputers may be built which will consume fraction of a watt power only, that also after prolonged use. It will become a reality since materials in superconducting state require almost no power.
2. The electric transmission conductors of hair dimensions will be needed. Hence, the distance between two supporting towers may be many kilometers.
3. The transformer may become an object of museum. It is because the voltage drop will be negligible as I^2R loss ≈ 0. As there will be no need to step-up the voltage, hence use of transformer will not be required.
4. The tariff of electrical power consumption may be 0.123 units for 5 years in a house utilizing fans, bulbs, refrigerator, geyser, cooler and other electrical appliances. However, the cost of one unit may be a guess work.
5. The lifestyle of the society may see a revolutionary change more than the one that took place when vacuum tubes were replaced by transistors.

13.11 Milestones in Research and Development of Superconductors

The important developments in the field of superconductor technology may be summarized as below :

1. Onnes observed in 1911 that mercury when cooled to 3 *K*, possesses resistance equal to 1/10 million of its resistance at melting point.
2. Gorter and Casimir in 1934 proposed a two-fluid model to explain the properties of superconductors.
3. Two brothers *F.* London and *H.* London in 1935 proposed classical theory of superconductivity for type I superconductors.
4. GLAG in 1950 explained the magnetic behaviour of type II superconductors. Origin of word GLAG comes from the first letters of the investigators Ginzburg V., Landau L., Abrikosov A., and Gorkov L.

5. Bednorz, Cooper and Schreiffer in 1957 propagated a microscopic theory of superconductivity.
6. In 1960, it was demonstrated that A75 *compound* e.g., niobium tin (Nb_3Sn) can carry 100 kA/cm^2 in a field of 8.8 tesla. Its T_c was about 20 K.
7. In 1960s, superconducting generators and multi-filamentary Nb-Ti superconductors were developed.
8. In 1970s, multi-filamentary Nb_3Sn superconductors with $I_c >$ 100 A, cable-in-conduit (CIC) conductors, and hybrid magnets generating 30 T (tesla) were produced. Hybrid magnets keep the hysteresis loss W_h as a minimum.
9. The highest T_c till 1986 was recorded as 23 K in the Nb_3Ge alloy.

13.11.1 Modern Ceramic Superconductors

1. Bednorz and Muller in 1986, reported a favourable critical temperature T_c = 34 K in lanthanum barium copper oxide ($La_{1.85}\ Ba_{0.15}\ CuO_4$) superconductor. They received Nobel prize for this discovery.
2. Zhao in 1987 developed La Ba CuO superconductor which was superconducting at 70 K.
3. In 1988, it was discovered that the yttrium barium copper oxide i.e. *YBCO* ($Y\ Ba_2\ Cu_3\ O_{7-x}$) superconductors at 90 K. Critical current densities in excess of 10^8 A/cm^2 was achieved in it.
4. In the same year, bismuth and thallium superconducting compounds were discovered whose characteristics were in line with the yttrium compounds.
5. In 1988, the 100 K barrier was also crossed due to the development of bismuth strontium calcium copper oxide i.e. BSCCO (Bi Sr Ca CuO) showing T_c = 106 K. Zero resistance at 85 K was observed in it.
6. In 1988, Parkin developed thallium barium calcium copper oxide i.e. TBCCO (Th Ba Ca CuO) superconducting at 125 K, although its practical application is doubtful due to highly poisonous nature of thallium.
7. In 2005, the value of T_c = 115 K has been reported in $Sn_2\ Ba_2\ (Ca_{0.5}\ Tm_{0.5})\ Cu_3O_x$.
8. In the same year, the value of T_c = 127-128 K has been found with material system $Tl_2Ba_2Ca_2Cu_3O_{10}$.

9. The highest value of critical temperature $T_c = 138$ K reported so far is in $Hg_{0.8}Tl_{0.2}Ba_2Ca_2Cu_3O_{8.33}$.

The development is still to go a longway to achieve the superconductivity at room temperature *i.e.* 300 K.

13.12 Present Scenario of the Main Applications of High Temperature Superconductors

More applications of high temperature superconductors are found in the following areas.

1. Energy production
2. Energy conversion
3. Leviated trains that ride on a magnetic field
4. Powerful magnets for nuclear fusion reactors
5. Medical imaging machines
6. Detection of infrared radiation
7. High speed electronic signal processing
8. Energy storage

Brief description of some systems is given below.

Josephson Junction. It is a device having two superconducting plates separated by an oxide layer. Such junctions are used in SQUID (Superconducting Quantum Interference Device), computer memories and other microelectronic devices. Josephson junction works on the principle of Josephson effect. This quantum effect is characterized by the tunnelling of superelectrons through the insulator of Josephson junction, and is observed as flow of current without any driving potential.

Maglev or Magnetic Levitation. This is a levitation phenomenon caused by the opposing magnetic fluxes. The use of this phenomenon is in wheelless high speed trains that are equipped with superconducting magnets. The Nb_3 Sn and Nb-Ti are used as magnet grade superconductors.

Magnetohydrodynamics (MHD). MHD deals with the motion of an electrically conducting fluid under the influence of magnetic field. This principle is employed in

1. MHD power generation,
2. MHD ship propulsion, and
3. MHD magnets.

In MHD power generation, the thermal energy of hot ionized gas is converted into d.c. electric power. In MHD ship propulsion, the sea water is accelerated by magnetic field. In both these applications, the superconducting magnets are required. The Japanese experimental ship Yamato sailed in 1992 utilized MHD principle.

Magnetic resonance imaging (MRI). In this technique, the visual images of the brain and other body parts can be created. This helps in diagnosis of probable illness much earlier than they occur. Superconducting magnets above 2 T are suitable for this purpose.

Suspension systems and motors. Frictionless suspension systems may be realized by the interaction between a magnetic flux produced externally and the currents flowing in a superconductor. If the superconductor is pressed downwards it tries to exclude the magnetic field, hence the magnetic flux on which it rests is compressed and the repelling force is amplified. Since it is possible to impart high speed rotation to a suspended superconducting body, and that all the conductors in the motor are free of resistance, it is obvious that the idea of a 100% efficient motor can be closely approximated.

Radiation detectors. The operation of these devices is based on the heat provided by the incident radiation. The superconductor is kept just above its critical temperature, where the resistance is a rapidly varying function of temperature. The change in resistance is then calibrated as a function of the incident radiation.

13.13 Producing the Superconducting Solenoids and Magnets

Nb-Sn superconducting wire. One of the principal applications of superconductors is in the generation of high-intensity, high-stability magnetic fields for both scientific investigations and medical applications such as magnetic resonance imaging (MRI). In this case, the superconducting wires need to be fabricated to form the coils of the solenoid. These wires usually consist of niobium- titanium or less often niobium-tin in a matrix of copper which is extruded into a wire about 0.5 mm in diameter. These multi-filament wires are shown in Fig. 13.7*a-b*.

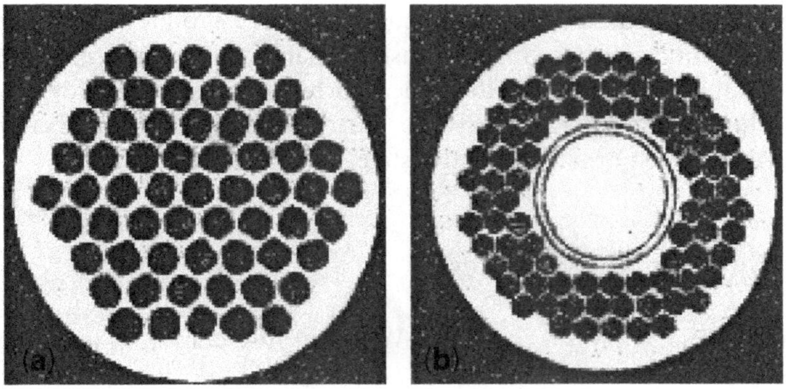

Figure 13.7 Cross-section configurations of a superconducting wire shows that the dark regions are the superconducting material, and the light regions are a matrix of normal conducting material (*a*) superconducting filaments are distributed throughout the matrix, and (*b*) a core of normal material is surrounded by superconducting filaments

13.14 MRI for Medical Diagnostics

In magnetic resonance imaging for medical diagnostics, the magnetic moment of the nucleus can be detected through nuclear magnetic resonance. The resonant frequency of a particular nucleus is dependent on its mass, nuclear magnetic moment and the field strength to which it is subjected to. The resonance can be caused by using a radio-frequency coil with an adjustable excitation frequency and detected by using a pick-up coil.

13.14.1 Magnet as a Critical Component

A field gradient is normally used to determine the spatial locations of the nuclei. The measured resonance frequency indicates the field strength which a particular nucleus experiences. From a three-dimensional map of the field strength over a given volume, the location of the nucleus within that volume can be found. In this application, it is essential to have a strong magnetic field which is stable and precisely controllable over the working volume. A high field strength is advantageous because it gives a higher resonant frequency and therefore a stronger signal to noise ratio. In this type of instrument, the magnet is therefore a critical component.

13.14.2 Superconducting Magnet

Superconducting magnet systems which can generate magnetic flux densities upto $B = 15$ tesla ($H = 12 \times 10^6$ Am^{-1} in free space) are available. The wires which are used to make the coils of the superconducting solenoid can carry much higher current densities than conventional conductors. For example in niobium-tin, the critical current density which provides an upper limit to the current density the material can sustain before making a transition to the normal or resistive state, is typically 10 MA cm^{-2}.

13.15 Solved Examples

Example 13.5 What is a cryotron ? What purpose does it serve and how does it work?

Solution. A cryotron is an assembly of a superconducting wire (core material) A surrounded by a coil of another superconducting material B. When temperature of a system in which the cryotron is used, is below the transition temperature of two materials, both A and B are superconducting; otherwise they attain a normal state. Based on this fact, a cryotron is used as an element in control devices such as **flip-flop** in a computer. Its working may be understood as follows.

The current I_A in the wire is controlled by the current I_B in the *coil.* It is because of the fact that magnetic field produced by the coil normally exceeds the critical field of the wire (core material) at operating temperature. The intensity of controlling current I_B required to make the core in normal state, depends upon d.c. current flowing through the core. It is so as the core current also produces a magnetic field.

Example 13.6 Suggest suitable materials for making a cryotron.

Solution. *Tantalum* is a suitable material for making wire (or core), if the operating temperature is extremely low ($\approx$ 4.2 K which is liquification temperature of helium). Since the coil has to be superconducting even if the control current flows, hence *niobium* or *lead* are suitable coil materials.

Example 13.7 What are supercooled coils ?

Solution. *Supercooled* coils are such electrical systems which can produce a flux density of 10 tesla or more in superconducting state. This flux density is 'many ten' times more than the flux density produced by normal coils at room temperature.

Example 13.8 What are cryogenic temperature, cryogenic industries, and cryogenic engine ?

Solution. 115.5 K (–157.5°C) is known as *cryogenic temperature.* Those industries in which the operations are performed below this temperature are called *cryogenic industries.* Industries involved in manufacturing of liquid oxygen (at –183°C), liquid nitrogen (at –196°C), liquid ammonia in fertilizer industry (at –190°C) etc. are such examples. Engines using liquid fuels at so low temperatures are called cryogenic engines.

Example 13.9 Critical magnetic field at zero kelvin and critical temperature for Pb are 65 kA/m and 7.18 K respectively. Determine the critical current density at 4.2 K in a lead wire of 1 mm diameter. Consider a parabolic dependence of H_c on temperature.

Solution. Given are: $H_0 = 65$ kA/m $= 65 \times 10^3$ A/m, $T_c = 7.18$ K,

$T = 4.2$ K, and $r =$ dia/2 = ½ mm $= 0.5 \times 10^{-3}$ m.

The parabolic dependence of H_c on T is expressed by Eqn. 13.1 which is given as

$$H_c = H_0\left(1 - \frac{T^2}{T_c^2}\right)$$

$$H_c = 65\times10^3\left(1 - \frac{4.2^2}{7.18^2}\right)$$

$$= 4.28 \times 10^4 \text{ A/m}$$

Substituting it in Eqn. 13.4, we find the critical current as

$$I_c = 2\pi r H_c$$

$$= 2\pi \times (0.5 \times 10^{-3}) \times 4.28 \times 10^4$$

$$= 134.392 \text{ A}$$

Critical current density J_c is found from

$$J_c = \frac{I_c}{A} = \frac{2\pi r H_c}{\pi r^2} = \frac{134.392}{\pi(0.5\times10^{-3})^2}$$

$$= 1.71 \times 10^8 \text{ A/m}^2$$

Quick Revision Summary

A-15 Superconductors are intermetallic compounds such as Nb_3Sn, Nb_3Al and V_3Ga having extremely low critical temperature. Their structure is cubic crystalline, which are also known as beta-tungsten (β-W) structure.

Josephson Junction is a thin insulating layer between two superconducting solids having superb memory, logic and storage functions. The switching time is of the order of 10^{-11} s.

Review Questions

1. What is superconductivity? How is it different from conductivity and semiconductivity?
2. What are the salient features and properties of superconductors? Why don't they obey Ohm's law below their critical temperatures? Hence, explain the mechanism of superconduction.
3. Explain the effect of critical magnetic field, critical current and the isotopic mass on critical temperature of a superconducting material.
4. Classify the superconductors from different viewpoints. Differentiate between the ideal superconductors and hard superconductors. Write the applications of superconductors stating their limitations.
5. Explain the following in a superconductor.
   i. Meissner effect
   ii. Type II superconductor
6. Distinguish between Type I and Type II superconductors with the help of suitable diagrams.
7. Bring-out differences between the following,
   i. Silsbee effect and Meissner effect.
   ii. LTS and HTS.
8. Explain the mechanism of superconduction. What are various theories in this regard?
9. Explain London's theory and GLAG theory for superconduction.
10. Describe BCS theory. In this regard, explain Cooper pair, super-current and superelectrons. How is phonon responsible for superconducting effect?

11. What is the present trend of development of superconducting materials? How will these be affecting the life style in future?
12. Describe the scenario of the up-to-date progress made in modern ceramic superconductors.
13. Explain the depth of penetration, coherence length, and MRI.
14. How are the superconducting materials produced into high field superconducting magnets?
15. Write notes on the following.
    a. Maglev.
    b. High temperature ceramic superconductors.
    c. MHD ship propulsion.
    d. Cryogenic engine.

Numerical Problems

1. Critical temperature of Pb in superconducting state is 7.17 K under zero magnetic field. The value of critical field for it is 0.0803 A/m at 0 K. Determine its critical field at 3 K and 10 K.
2. Critical magnetic field at zero kelvin and critical temperature for Pb are 65 kA/m and 7.18 K respectively. Determine the critical current density at 4.2 K in a lead wire of 1 mm diameter. Consider a parabolic dependence of H_c on temperature.
3. The Nb_3Sn is to be used as magnet grade superconductor to make the rail for a wheel-less train. Its critical field at 7 K and 14 K are 21 A/m and 10 A/m respectively. Determine its transition temperature and critical field at 0 K and 4.2 K for design requirements.
4. The depths of penetration for a Type I superconducting material has been found to be 39.6 nm at 3 K, and 1730 Å at 7.1 K. Estimate its (*a*) critical temperature, and (*b*) depth of penetration at absolute zero.

Objective Questions

1. Magnetic susceptibility of a material is -1. It belongs to the kind of solid known as

    a. ferrite　　b. superconductor, type I
    c. antiferromagnetic　　d. garnet

2. As compared to the elements, the compounds and alloys can be made superconductors at

   a. higher temperatures b. lower temperatures
   c. the same temperature d. much lower temperature

3. The magnetic lines of force are repelled back completely due to an effect called

   a. BCS effect b. Silsbee effect
   c. Meissner effect d. Bednorz effect

4. Phonon is a quanta of energy in the

   a. electromagnetic wave b. ultrasonic wave
   c. infra-red wave d. elastic wave

5. Yttrium-Ba-Cu oxide superconductor superconducts at -183°. It is a

   a. high temperature superconductor
   b. metallic superconductor
   c. type I superconductor
   d. poisonous superconductor

Fill in the Gap Type Questions

6. BCS theory holds good for ....
7. Critical current density in a superconductor is given by according to .... theory.
8. According to GLAG theory, the ratio of (l_c/d_p) is .... than 1.414.
9. Visual image of brain can be created by .... technique.
10. The levitation phenomenon will help in the development of fast moving .... train.
11. The conductivity of a conducting material on being subjected to critical field changes to

    a. normal state b. unstable state
    c. temperature-independent state d. temperature-dependent state

12. Match List I with List II, and choose the correct answer from the codes given below the lists.

	List I		*List II*
A.	Debye temperature	1.	Two number per atom
B.	Superelectrons	2.	Type II superconductors
C.	Coherence length	3.	Ideal superconductors
D.	Flip-flop	4.	$\propto$ (isotopic mass)$^{-1/2}$
		5.	Cryotron

Codes:

	A B C D		A B C D
a.	4 1 2 5	b.	5 4 3 2
c.	2 5 3 1	d.	1 2 4 3

14

Passive Components (Resistors)

14.1 Passive and Active Components

Electronic components/devices may be grouped into following two specialties.

1. Active devices, and
2. Passive devices

The *active devices* are those whose basic characteristics get changed when an electrical signal is applied on it, whereas the *passive devices* do not change their basic character when an electric signal is applied on it. The examples of active and passive devices are listed below.

Active devices
Lasers LEDs VCSELs (vertical cavity surface emitting laser) Amplifiers Sensors Terahertz devices

Passive devices
Resistors Capacitors Inductors Combination of resistors, capacitors and inductors Waveguides Multiplexers

Passive devices
Photonic crystals Filters Microlenses Isolators Polarizers Diffractive optical elements Branching and mixing components

Advanced structures such as given below are also considered to be the passive devices.

- Distributed circuits
- Transmission lines
- Interconnection lines

Applications. Passive devices are used in electronic circuits for various applications such as given below.

i. In bias and matching network
ii. In lumped element circuits
iii. For resonator or decoupling purposes.

14.2 Introduction to Resistors

A resistor is a passive component used to introduce resistance against flow of current in a circuit. Resistance to flow of current in electronic circuits is required for various purposes such as given below.

- Voltage division
- Setting biases
- Controlling the gain
- Heat generation
- Matching and loading circuits, etc.

Resistance in electronic circuits for all above purposes is provided by 'resistors'.

Resistance is a basic property being possessed by a conducting material. It is given by $R = \rho\ (l/a)$. From this relation, it is evident that the resistance of a resistor depends upon its physical dimensions 'l' and 'a', and specific resistance 'ρ' of the material of which it is made. Details in this regard have already been discussed in chapter 4 and can be referred there.

14.2.1 Characteristics of Resistors

All resistors possess the following characteristics. These characteristics can be varied by proper design of resistors and choice of suitable materials for them.

Temperature coefficient of resistance: We have described earlier in chapter 4 that the resistance of materials changes with change in their temperature. Temperature coefficient of resistance α_T of a material is defined as the percentage change in resistance per unit change in temperature. From Eq. 4.19, we know that

$$R_{T2} = R_{T1}\ (1 \pm \alpha_T dT) \quad \text{where } dT = T_2 - T_1$$

$$R_{T2} = R_{T1} \pm R_{T1}\alpha_T dT$$

or

$$R_{T2} - R_{T1} = \pm R_{T1}\alpha_T dT$$

$\therefore$

$$a_T = \pm \frac{R_{T2} - R_{T1}}{R_{T1}(T_2 - T_1)} \tag{14.1a}$$

Here α +ve indicates increase and α –ve indicates decrease in the value of α_T.

If R_{T1} and R_{T2} are in 0°C, then α_T is expressed as per °C. The value of α_T may also be expressed as below.

$$a_T = \pm \frac{R_{T2} - R_{T1}}{R_{T1}(T_2 - T_1)} \times 10^8 \text{ ppm/° C} \tag{14.1b}$$

Voltage coefficient of resistance α_V. The value of resistance of materials also changes with change in the voltage applied on it. For example, if R_{V1} and R_{V2} are the values of resistance at V_1 and V_2 voltages respectively ($V_2 > V_1$), then we can write

$$R_{V2} = R_{V1}\ (1 \pm \alpha_V dV) \tag{14.2}$$

$$a_V = \pm \frac{R_{V2} - R_{V1}}{R_{V1}(V_2 - V_1)}$$

Here α_v is voltage coefficient of resistance and is defined as the percentage change in resistance per unit applied voltage. It is expressed in the unit of per volt.

Maximum voltage rating. It is the maximum voltage that can be applied upon a resistor without causing any damage to it. Mathematically, it is expressed as

$$V_{max} = \sqrt{P_{max}R} \tag{14.3}$$

where P_{max} is maximum power rating and R is the value of resistance.

Power dissipation P_d. It is defined as the maximum amount of heat dissipated by a resistor at the maximum specified voltage, without damaging the resistor. It is expressed in watt.

Power dissipated in an ideal resistor is specified by

$$P_d = VI = I^2R = V^2/R \text{ Watt} \tag{14.4}$$

Power dissipated is restricted by the power rating P_R. At higher temperatures, the power rating of resistor decreases.

Stability. It is the percentage change in resistance at rated power and hot spot temperature (normally 70°C) after a specified time. It is expressed as

$$S = (\Delta R/R)_t \tag{14.5}$$

where t is any specified time which may be 1000 hours or more, hotspot temperature is 375°C or less, and ΔR is normally more than 5% of R.

Frequency range. A resistor may be purely ohmic ($V/I=R$) i.e. pure resistive at low frequency, but may not be ohmic at high frequencies. At those frequencies, it may have inductive impedance X_L, or capacitive impedance X_C. The frequency upto which it is ohmic (i.e. resistive), is called frequency range.

Residual capacitance. This is shunt capacitance which increases mainly with increased pulse rating. Non-inductive elements have low residual capacitance. Typically a $2W$ metal oxide film resistor has 0.5pF residual

Maximum operating temperature. It is the maximum temperature upto which the resistor can work without failure or without any appreciable change in its dimensions.

Reliability. It is the percentage change in resistance at rated power and hot spot temperature after a specified time. It is dependent on the resistance value, power dissipation and ambient temperature.

Shelf life. It is defined as the change in the value of resistor during storage. Usually, it is quoted for one year and expressed in percentage.

14.3 Manufacturing Method of a Resistor

A typical resistor is shown in Fig. 14.1. It mainly consists of a rod, an insulating coating, end connectors, and wire lead. It can be manufactured by performing the following processes.

Preparation of base (or substrate). The base is made of an insulating material, over which the resistive element is either mounted or applied upon. The resistive element may be in the form of wire winding or coating. Glass, ceramics and plastics are generally used as base material. The base material is required to possess the following properties.

i. Good mechanical strength.
ii. High insulation resistance for the anticipated range of temperature.
iii. Moisture resistant.
iv. Good ability to withstand the thermal shock.
v. Thermal coefficient of expansion should be almost the same as that of the resistive element to avoid warping or cracks.

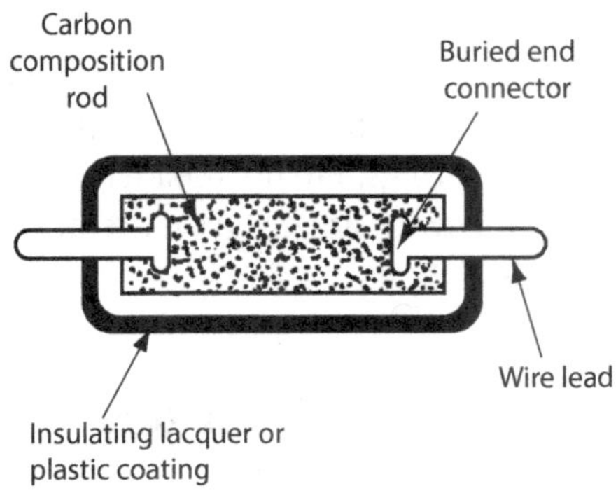

Figure 14.1 Constructional details of a typical resistor. We shall study later that this is a carbon composition type resistor

Following materials are more common to make the base.

- Glass
- Ceramics
- Plastics

Preparation of resistive element. The resistive elements are generally used in the forms of wires, film and slug. They are made of a material having lower temperature coefficient. Following materials are commonly used for it.

- Ni-Cu wire
- 80-20 Ni-Cr wire
- Ni-Cr-Al wire
- Carbon with binders
- Graphite with binders

Preparation of end-terminals (or connections). End-terminals are used to make electrical connection of resistive element in the circuit. This is accomplished by means of wire leads, metallic ferrules and metallic lugs. The wire lead may be of axial or radial configurations. The end-terminals are connected to end caps by mean of soldering, welding or press-fitting.

Provision of protective coating. The protective coating is provided on the resistance elements to protect them against external mechanical force, corrosive and environmental effects. For that the coating material should be such that the resistor may sustain higher temperatures. Generally the following materials are preferred for protective coating.

- Vitreous enamel
- Silicon
- Binder mixed cement

Ageing, testing and identification. Ageing of resistors is done at elevated temperatures in order to stabilize their characteristics. They are tested for their designated values and ratings. Finally the identification marking such as color coding and tolerances etc. are given on them. Details of identification is given later in art. 14.7.2.

14.4 Basic Classification of Resistors

A wide variety of resistors are available for various applications. Primarily, these are classified as shown in Fig. 14.2.

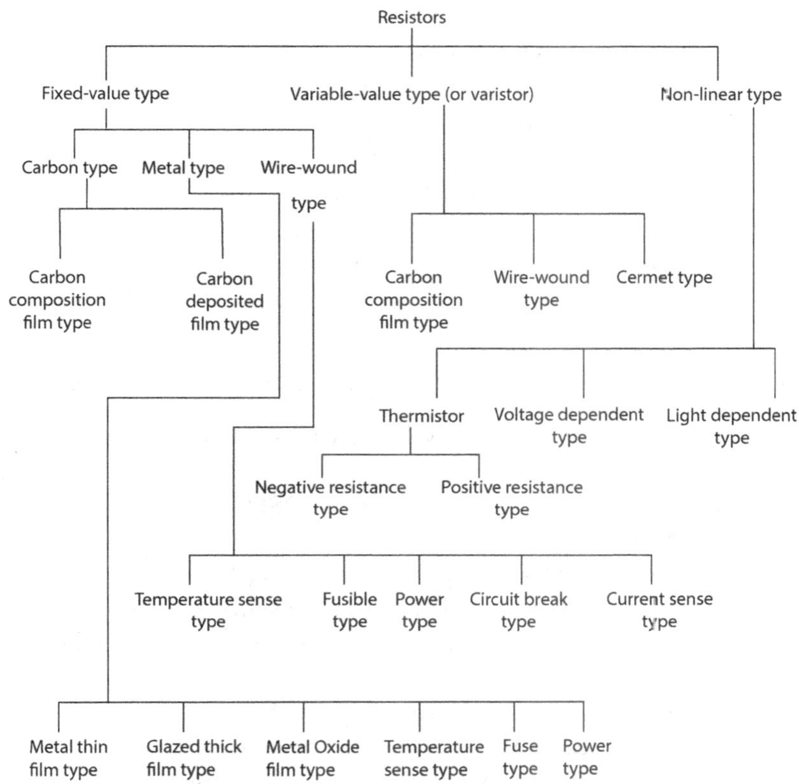

Figure 14.2 Basic classification of resistors.

14.4.1 Specific Types of Resistors

On the basis of different parameters, the resistors may be of following types.

1. On the Basis of Value of Resistance
    - Low resistance resistor
    - Medium resistance resistor
    - High resistance resistor
2. On the Basis of Temperature Rating
    - Low temperature resistor
    - High temperature resistor
3. On the Basis of Lead Configuration
    - Axial lead type
    - Radial lead type

4. On the Basis of Intended Applications
   - Power resistors of ≥ 1 W
   - Fusing resistor
   - Circuit breaking resistor
   - Temperature sensing resistor
   - Current sensing resistor
5. On the Basis of Film Thickness
   - Thin film resistor
   - Thick film resistor

14.4.2 Different Types of Resistors: Materials and Applications

Applications. Resistors are extensively used in following main applications.

1. Series R-C circuits for diode, MOSFET and thyristor snubbers.
2. Turn-off snubbers for bipolar transistors (BPT).
3. Current sharing for parallel connected semiconductors.
4. Voltage sharing for series connected semiconductors and capacitors.
5. In damping and voltage dropping circuits.
6. Resistor divider for proportional voltage sensing.
7. As pure dissipative element in power electronic circuits.
8. Current limiting path for charging and discharging currents.

Materials. Different materials of resistors and their properties are given in Table 14.1.

14.5 Constructional Details of Different Kinds of Fixed Resistors

14.5.1 Construction of Carbon Composition Resistor

Constructional details of a carbon composition resistor are shown in Fig. 14.3. In its construction, the carbon composition of finely ground carbon, resin binder and insulating filler material is made first. The resulting mixture is pressed into the shape of the resistor to be made, and then fired in a kiln or oven. In it the ratio of carbon content to the insulating filler determines the final value of resistance. The end connection is obtained by means of spraying metal at the ends, and then soldering the tinned copper

Table 14.1 Properties of resistor materials

Material	Sheet resistance (Ω square^{-1})	[α_T] (pmK^{-1})
Thin film resistor materials		
Ni-Cr	10–500	< 20–50
Oxidized Cr	100–1000	25–100
Cr-Si monoxide	10–1000	< 20–200
TaN or Ta_2N	10–200	50–200
Thick film resistor materials		
W-Ru	25–300	120–300
RuO_2	100–100000	150–300
IrO_2	1–100000	150–300
Polymer pastes	1–1000	50–250

Figure 14.3 Carbon composition resistor.

wire around it. It is also made by moulding the connecting wires directly and pressing a metal cap over metal sprayed end.

Finally, the whole resistor is either moulded in plastic or given several coats of insulating lacquer to provide electrical insulation and protection from moisture.

14.5.2 Construction of Carbon Film Resistor

Carbon film resistors are manufactured by depositing an even film of resistive material (pure carbon) on high grade insulator cores as shown in Fig. 14.4.

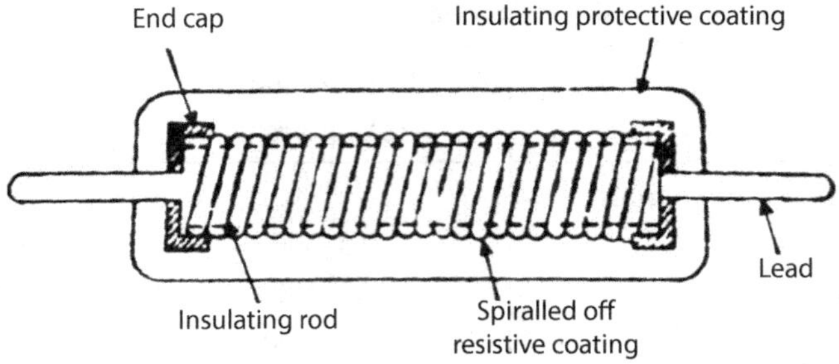

Figure 14.4 Carbon film resistor.

In its construction, the high grade non-porous electrical insulator cores with fine surface structure are heated in specially constructed furnaces with accurate temperature and vacuum controls. The hydrocarbon is then passed through the furnace. By pyrolytic action, the hydrocarbon cracks at high temperature inside the furnace and forms a very hard crystalline carbon coating on the insulator core. The coated rods are press-fitted with brass caps which are pre-welded with tinned electrolytic copper wire. These are then helically grooved with a high degree of precision so as to achieve the desired value and tolerance. The film is protected by an undercoat of silicon varnish and a special epoxy paint, which ensures good protection against tropical and climatic exposure and humidity.

Applications. In view of remarkable long term stability of the resistors under severe environmental conditions and long service life, these resistors are most suitable for all types of

- precision equipment
- defence communications
- industrial controls, and
- computers.

14.5.3 Construction of Metal Film Resistors

Metal film resistors are manufactured by depositing a film of metal, generally nickel and chromium, on high grade ceramic rod shown in Fig. 14.5.

These resistors have a non-porous high quality ceramic rods/cores, which are coated with a special metallic film of nickel-chromium. Its composition and thickness can be controlled by vacuum evaporation process. End cap are press fitted to the coated ceramic substrate along with specially coated axial leads, which are welded to ensure good mechanical and

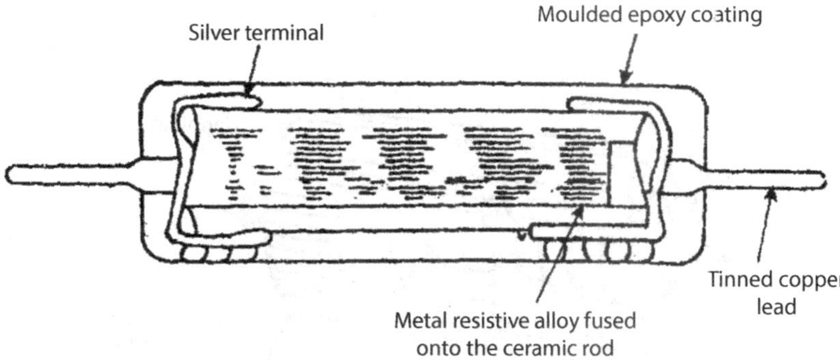

Figure 14.5 Metal film resistor

electrical contact. These cores are then helically grooved with high degree of precision, so as to achieve the desired value of resistance.

To withstand the severe environmental conditions, they are treated with a coat of special grade varnish and compounded resin of high thermal stability. The resistors are moulded with epoxy powder to assure perfect electrical insulation and moisture protection.

Applications. Since metal film resistors are most stable resistors among all film type resistors, they are widely used in all professional measuring and calibrating equipment such as given below.

- Oscilloscope
- Decade resistance boxes
- Measuring bridges
- Oscillators
- Precision voltage dividers
- High frequency communication systems.

14.5.4 Construction of Wire-Wound Resistor

The sectional view of an aluminium-housed power wire-wound resistor is shown Fig. 14.6. It can dissipate up to 300W with a suitable heatsink in air or up to 900 W when water cooled.

The former is a high purity, high thermal conductivity ceramic of either steatite or alumina tube, depending on the size. The matching resistive element is iron-free, 80: 20 nickel-chromium for high resistance values or copper-nickel alloy for low resistance. These alloys result in a wire or tape which has a high tensile strength and low temperature coefficient. The tape or wire

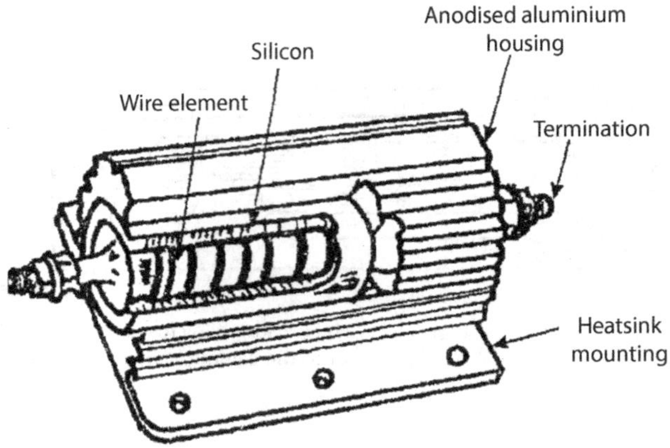

Figure 14.6 Wire wound aluminium clad resistor construction.

is evenly wound on to the 'tube former' with an even tension throughout. This construction gives a resistor repeated heat cycling without damage.

The assembled and wound rod is encapsulated in a high temperature thermal conducting silicone moulding material, and then cladded in an extruded, hard, and anodised aluminium housing. It ensures the electrical stability and reliability. Alternatives to the aluminium clad resistor, the encapsulation of the wound rod can be done in a vitreous enamel or a fireproof ceramic housing.

Wire-wound power resistors with a low temperature coefficient of less than $\pm 20 \times 10^6$/K, use a resistive element made of constantan or nichrome. The constantan is used for lower resistance up to several kilo-ohms, while the nichrome is applicable up to several hundred kilo-ohms. The resistance range depends on the ceramic core dimensions, hence power rating. The element is wound under negligible mechanical tension, resulting in a reliable, low temperature coefficient resistor which at rated power can safely attain the surface temperatures of over 350°C. Because these resistors can be used at very high temperatures, the thermally generated emf developed at the interface between resistive element and copper termination can be significant, particularly in the case of constantan which produces -40 μV/K. Nichrome has a coefficient of only $+1$ μV/K.

14.5.5 Construction of Circuit Breaker Resistors

The construction of a wire-wound circuit breaker resistor is shown in Fig. 14.7. Under overload conditions, the solder joint melts, producing an

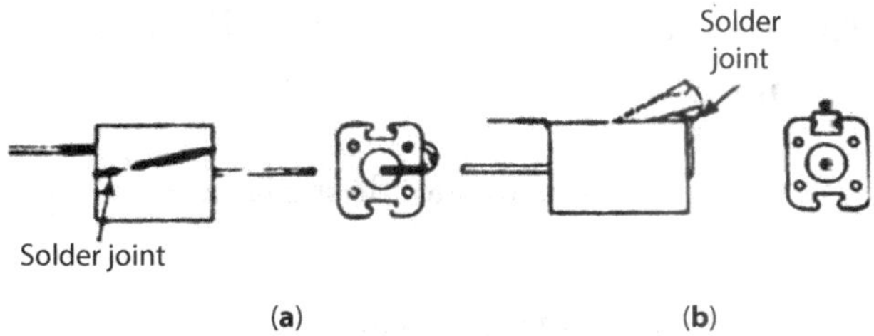

Figure 14.7 Construction of circuit breaker resistors of (a) Type 1, and (b) Type 2.

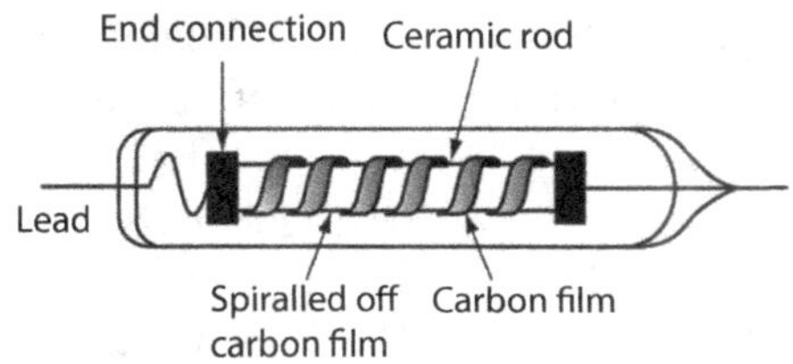

Figure 14.8 High value resistor

open circuit. After fusing, the solder joint can be resoldered with normal 60:40 Sn-Pb solder.

The joint melts at a specified temperature. To ensure reliable operation, the solder joint should not normally exceed 150°C. This characteristic is similar to that of fusible resistors. A typical power range of circuit breaker resister is 1–6 W at 70°C, with a resistance range of 75 mΩ to 82 kΩ, and temperature coefficient of -80 to $+500 \times 10^{-6}$/K depending on the resistance values.

14.5.6 Construction of High Value Resistors

These resistors (Fig. 14.8) are of carbon composition type. Resistive element consists of carbon film on insulating ceramic rod, which is sealed in evacuated glass tube. Mild connections are made of wires of copper clad nickel-iron alloy. The moisture effects are minimized by coating the outside of glass tube with silicon lacquer. The element can be spiralled for control of fine resistance. These resistors are used in measurement work for atomic energy, where it is often necessary to detect very small currents such as for photon counting, radiation monitoring etc. The high value resistors range upto ten million mega ohm (10^7 to $10^{13}\Omega$).

14.6 Comparison Among Different Types of Fixed Resistors

Various resistors have been compared for their characteristics in Table 14.2.

14.7 Specifications of Resistors

The resistors possess several electrical properties which vary significantly with their physical size and resistance value. Accordingly, they are specified by the following parameters/properties. Usual values of commercially available resistors are also given for a ready reference.

- Resistance range 0.01 Ω to 1 GΩ
- Power range at 70°C 1 W to > 300 W
- Working voltage 500 V to 100 kV
- Voltage coefficient (0.1 to 200) × 10^{-6}/volt
- Residual capacitance 0.2 to 0.5 pF
- Maximum temperature 100°C to 300°C

Table 14.2 Comparison of different fixed resistors

Description	Carbon composition type	Carbon film type	Metal oxide type	Metal metal glaze type	General Purpose wire-wound type
• Operating range	10Ω to 22Ω	10Ω to 2MΩ	10Ω to 1MΩ	10Ω to 100MΩ	0.25Ω to 10kΩ
• Power rating	250 mW	250 mW	500 mW	500 mW	2.5 W
• Maximum voltage	150 V	200 V	350 V	250 V	200 V
• Ambient temperature range	−40°C to +105°C	−40°C to +125°C	−55°C to +150°C	−55°C to +150°C	−55°C to +185°C
• Soldering effect	2%	0.5%	0.15%	0.15%	0.05%
• Shelf life 1 year	5% maximum	2% maximum	0.1% maximum	0.1% maximum	0.1% maximum

- Temperature coefficient (-3000 to + 5500) × 10^{-6}/K
- Thermal resistance 0.3 to 90 K/W
- Reliability (1 to 300) × 10^{-9}/h
- Stability (ΔR/R) % 0.1 to 5

14.7.1 Symbolic Representation of Resistors

It is customary to represent the electrical/electronical components by a specified symbol in circuits. Resistors of different types are also represented by specific symbols. These are shown in Fig. 14.9.

14.7.2 Identification of Resistors

To distinguish between the resistors of different types and for their easier identity, the resistors are provided with certain 'identification marks' on them. Following two kinds of identification coding is provided in them.

Colour coding. The values of resistances are coded on resistors using different color bands as shown in Fig. 14.10. Three bands are limited to only two significant figures and cannot identify the code value better than 5%. Color significance is shown below in Table 14.3.

Color coding on resistors is not a very good system. It is prone to damage from high dissipation temperatures. Also, it does not adhere well to

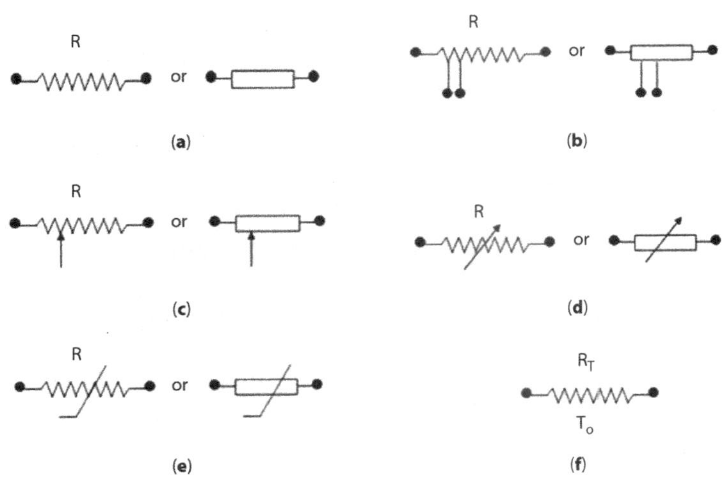

Figure 14.9 Symbolic representation of different types of resistors (a) fixed value resistor, (b) tapped resistor, (c) (d) rheostats, (e) varistor, and (f) thermistor

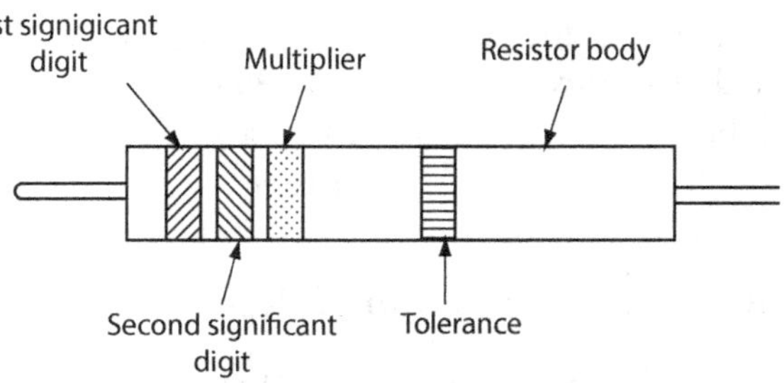

Figure 14.10 Color coding of resistors

Table 14.3 Color code for resistors

Color	Number	Multiplier	Tolerance
• Black	0	10^0	±20%
• Brown	1	10^1	
• Red	2	10^2	± 1 %
• Orange	3	10^3	
• Yellow	4	10^4	±2%
• Green	5	10^5	
• Blue	6	10^6	
• Violet	7	10^7	
• Grey	8	10^8	
• White	9	10^9	
• Gold	—	10^{-1}	±5%
• Silver	—	10^{-2}	+10%

some encasing materials. It generally finds greater utility in low wattage general purpose resistors having axial leads.

Alphanumeric coding. In this system, the value and the tolerance of resistors are printed on the resistor body instead of providing color bands. After the resistance value code, a letter is added to indicate the tolerance. For example,

- *5 k8F* alphanumeric code indicates a resistor of 5800 Ω and the letter *F* indicates a tolerance of ±1%.
- *R33M* alphanumeric code indicates a resistor of 0.33 Ω and the letter *M* indicates a tolerance of ±20%

14.8 Variable Resistors (i.e. Varistors)

As the name implies itself, the variable resistors do not have a fixed value. Their resistance can be varied between zero to a maximum specified value. The adjustment of the value of resistance is accomplished by means of a movable wiper that makes changing contact with resistance material.

Types. Varistor are classified on the basis of their applications and the materials of their construction. Accordingly, they may be of following types.

i. General purpose type
ii. Precision purpose type
iii. Carbon composition type
iv. Wire bound type
v. Cermets type

14.8.1 Carbon Composition Variable Resistor

A carbon composition variable resistor is a low power resistor made from carbon composition in solid track form or of carbon film coding. Both of these types consist of an annular ring (Fig. 14.11a-b) of substrate material (ceramic or plastic). A resistive carbon layer is grown over it on which the moveable contact slides. The carbon track can be made film type or moulded type.

In *moulded type* (Fig. 14.11a), the resistive material and base plate are moulded together along with stamping, terminal lugs and bushing. In *film type* (Fig. 14.11b), the carbon resistive paste is sprayed over the base annular ring. Different variations can be obtained by either controlling the

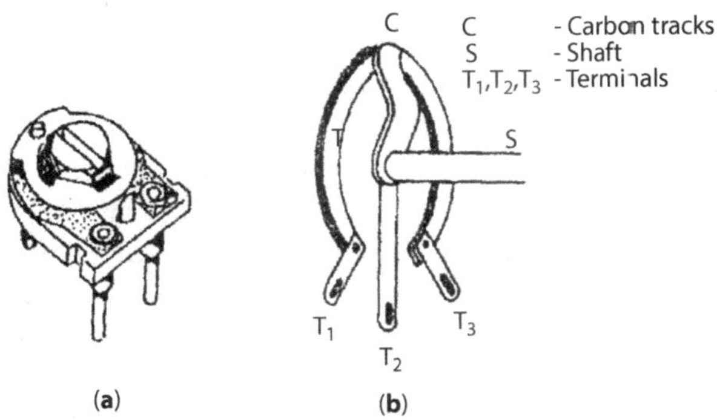

Figure 14.11 Carbon composition variable resistors (a) moulded type, and (b) film type.

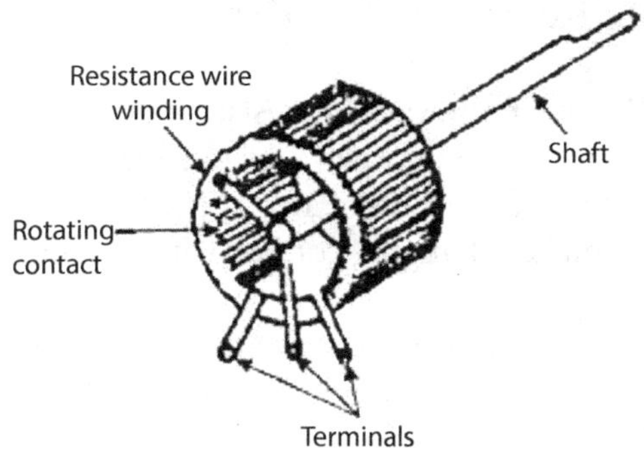

Figure 14.12 Wire wound variable resistor

width of track or composition of material. These resistors are made to follow linear and logarithmic variations. Carbon composition type variable resistors are available in following construction:

i. Single turn type, and
ii. Slide type

14.8.2 Wire Wound Variable Resistor

Wire wound variable resistors consist of a nichrome or other resistance wire, wound over insulating former, which is in the form of an arc as shown in Fig. 14.12. The sliding contact moves on the edge of the element. Wire wound resistors can be made *(i)* single-turn type, *(ii)* multi-turn type, or *(iii)* ganged type.

14.8.3 Cermets Type Variable Resistor

In cermets type variable resistor, the cermets compositions are screen printed on ceramic substrate and fired in a tunnel kiln at temperature ranging between 750°C to 850°C. Prior to firing, the conductor pads are screened on to the substrates and fired at about same temperature. The process of firing results in a resistance element which is permanently fused to its substrate. The variation in resistance is obtained by using the threaded screw or wiper assembly, depending upon the configuration used. The cermets resistors are made as single-turn type or multi-turn type.

14.8.4 Characteristics of Variable Resistors

In addition to the specification used for the fixed resistors such as tolerance, temperature coefficient, stability and power rating; the variable resistors are also expected to have the following characteristics.

i. **Absolute minimum resistance.** It is the resistance measured between the wiper terminal and each end terminal, with the wiper positioned to give a minimum value.
ii. **Contact resistance variation.** This is apparent resistance observed between the wiper and a resistive element, when the wiper is energized with a specified current. The output variations are measured over a specified frequency bandwidth. It is expressed in % of test resistance or in ohms.
iii. **Dielectric strength.** It is the ability to withstand the application of a specified potential of a given characteristics, between the terminals and all other external conducting members such as shaft housing, without exceeding the specified leakage current.
iv. **Equivalent noise resistance.** It refers to the variation in contact resistance of potentiometer along the length of resistive track.
v. **Linearity.** It is the amount by which the actual resistance at any point of contact varies from the expected value of their resistance for same angular resistance. It is expressed in percentage.
vi. **Mechanical angle of rotation.** It is the maximum angle traversed by the moving element while the resistance is changed from minimum to maximum. It is 360° for single-turn resistors and more than 360° for multi-turn resistors depending upon the number of turns.
vii. **Resistance law.** It is the relation between the change in resistance and movement of wiper. It may be
   - linear law
   - log law
   - sine law
   - cosine law
viii. **Rotational life.** The number of cycles obtainable under specified operating conditions, while remaining within the specified allowable degradation. A cycle is defined as the complete traversal of the wiper over resistive element in both directions.

14.9 Non-Linear Resistors

Non-linear resistors can be classified into following three categories.

1. Thermistor
2. Voltage dependent resistor (VDR), and
3. Light dependent resistor (LDR)

14.10 Thermistors

The term thermistor represents the thermal resistor. It is used for devices having high temperature coefficient. Based on the nature of temperature coefficient of resistance, there are two types of thermistors. These are

i. Negative temperature coefficient (NTC) thermistors whose resistance decreases with increase in temperature.
ii. Positive temperature coefficient (PTC) thermistors whose resistance increases with rise in temperature.

Since the carriers are created by thermal excitation, the resistivity of a semiconductor changes with temperature. Semiconductors used in temperature measuring devices are called thermistors. For this purpose, polycrystalline oxide ceramics such as MnO containing dissolved lithium ions, magnetite (Fe_3O_4), or extrinsic single crystal semiconductor (Ge) are useful. Thermistors capable of measuring temperatures up to about 450°C and others suitable down to 1° K are commercially available. Such materials have sensitivities of millidegrees; hence precision auxiliary equipment is not necessary if the measurements require only 0.1°K accuracy. Thermistors are also employed to advantage as temperature compensators in electronic circuitry.

14.10.1 NTC Thermistors

NTC thermistors are two terminal devices which are manufactured in three basic shapes as shown in Fig. 14.13a-c. For disc and rod shapes, the constituent oxides are mixed in suitable proportion, then a binder is added and mixture is pressed (or extruded) into required shape. The ceramic body is then heat treated at temperatures around 1200°C. By doing so, the binder material is burnt out and the constituent oxide reacts to form the desired composition. Thus, the material sinters to form a hard ceramic body. Silver paste is either painted or sprayed onto the opposite faces of the disc at the ends of the rod and then fired into ceramic to form ohmic contact. The leads are attached to this ohmic contact by soldering. Finally, the device is coated with insulating resin layer.

Bead thermistors are manufactured by forming a small blob of semiconducting oxide mixture, between two parallel platinum or platinum alloy wires. The wires are heat treated to about 1200°C so that the thermistor material shrinks on wire, forming a sintered bead. The platinum wires are then welded/soldered to thicker supporting leads. The bead is finally encapsulated in gas filled or evacuated glass envelope.

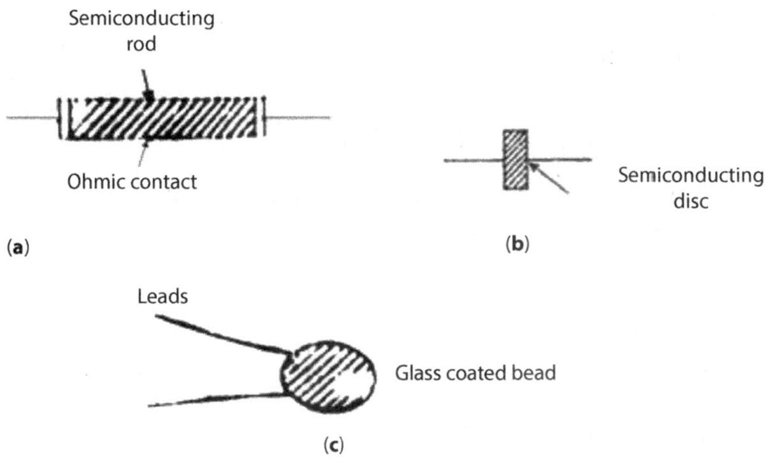

Figure 14.13 Thermistor types *(a)* rod form, *(b)* disc form, and *(c)* bead thermistor

14.10.2 PTC Thermistors

PTC thermistors are produced by using either *Ge* or *Si* on semiconducting barium titanate. The latter type thermistors are also known as switching PTC thermistors. The former type thermistors are prepared by forming an ohmic contact on opposite faces of rectangular block of suitably doped silicon. Axial lead wires are attached by soldering to the contact areas, and the whole unit is then encapsulated either in glass or moulded in epoxy resin. For producing switching PTC thermistors, care is taken to use high purity materials since the unwanted impurities can cause large variations in final properties of device. Oxides, carbonates of required elements are mixed together in binder and pressed into the desired shape. They are then sintered at about 1350°C. Leads are attached to the opposite faces of disc or ends of rod.

14.10.3 Applications of Thermistors

Thermistors are used in following main applications.

i. Temperature measurement. Its large temperature coefficient provides good accuracy and resolution.
ii. For temperature compensation over a wide range of temperature.
iii. Measurement of power at high frequency.
iv. Measurement of thermal conductivity.

v. Thermal relay.
vi. In thermally controlled devices.

14.10.4 Specifications of Thermistors

Specifications of thermistors are the following. For any thermistor, the undermentioned values should be quoted for it.

- Resistance at 25°C
- Thermal time constant
- Switching temperature
- Tolerance
- Maximum voltage at 55°C
- Temperature range
- Temperature coefficient
- Dissipation factor

14.11 Solved Examples

Example 14.1 Give a brief account of (a) chip resistors, and *(b)* network resistors?

Solution. *(a) Chip and network resistors* are made by thick film process that consists of firing the Ruthenium oxide (RuO) and glass paste, which have been screened on to a ceramic or glass substrate. Termination pads are added. The specific resistance value is obtained by laser trimming. *(b)* A number of metal film resistors packaged in single package constitute the resistor network. Mounting of resistor network instead of individual chip resistors results in reduced space requirement. It also reduces the mounting cost.

The major applications of these resistors are in

- Pagers,
- Disk drives
- Lap top computers
- Hand held phones,
- Heart pace maker, and

Example 14.2 What are alloy resistors?

Solution. Alloy resistors are the nickel-chromium alloy made resistors, which is made by evaporating on to a glass or ceramic substrate. The evaporated film makes a good adhesion to the base, hence a stable resistor. A resistive film of 50 Å thickness provides a very good stability. These resistors are used in precision applications.

Example 14.3 Write brief notes on (a) thermistors, *(b)* transducer, (c) metallization, and *(d)* metal-film resistors.

Solution.

a. **Thermistors** are temperature measuring devices, usually made of semiconducting materials. They have a negative temperature coefficient of resistance. That is why their resistance decreases with increase in temperature. Due to their high sensitivity of temperature changes, thermistors are used in temperature measurement, control and compensation of precision systems. Thermistors can operate in a temperature range of -100°C to 300°C. Commercial thermistors are made from sintered mixture of Mn_2O_3, NiO_2 and $CO_2 0_3$.
b. **Transducer** is a device that converts energy from one form to another. For example, a loudspeaker converts electrical energy into sound energy while an electric heater converts electrical energy into heat energy. Components of transducers are generally made of electrical, electronics and optical materials.
c. **Metallization.** This is a process of IC-fabrication in which the interconnections of components are formed on the chip. To accomplish this process, aluminium is vaporized so as to get a thin layer on the slice. Then a layer of photoresist is spread over it, to etch everything else except aluminium.
d. **Metal-film resistors** are manufactured by depositing a film of metal (generally *Cr, Ni, Cr-Ni*) on high grade ceramic rod. The end caps are force-fitted on the ceramic rod along with the specially coated axial leads. The 'leads' are welded to ensure good mechanical and electrical contact. The ceramic rod is then helically grooved to achieve the desired value of resistance. Such resistors are generally used in oscilloscopes, oscillators, and measuring bridges etc.

Example 14.4 What do you know about variable inductors?

Solution. Variable inductors are electrical devices that are used in tuning, timing, and calibration circuits for varying the value of inductance. They use a variable magnetic core of adjustable inductance.

Review Questions

1. Define passive components. How are they different from active components? State the examples of passive components and write their applications.
2. Discuss any five characteristics of a resistor.
3. Describe the various characteristics of resistors along with suitable mathematical expression.
4. Classify the resistors. How do the fixed value resistors differ from the variable value resistors?
5. Enlist major applications of resistors and state the main materials used in them.
6. Enumerate different types of fixed resistors, and describe the construction and working of any one of them.
7. Sketch and explain the construction and working of a carbon composition type resistor.
8. Sketch and explain the construction and working of a carbon film type resistor.
9. Sketch and explain the construction and working of a metal film type resistor.
10. Sketch and explain the construction and working of a wire-wound type resistor.
11. Sketch and explain the construction and working of a circuit breaker type resistor.
12. Sketch and explain the construction and working of a high value type resistor.
13. Distinguish between the following.
    a. Shelf life and load life of resistors
    b. Temperature coefficient and voltage coefficient of resistance in respect of resistors
    c. Thick film and thin film resistors
14. Compare the merits and demerits of different fixed type resistors.
15. Write the specifications of resistors. Also show their symbolic representation and discuss their identification.
16. What do you mean by varistor? State their different types and describe the constructional details of any one of them. Also write their applications.
17. Describe a carbon composition variable resistor with the help of a suitable diagram.

18. Describe a wire-wound variable resistor with the help of a suitable diagram.
19. Describe a cermets type variable resistor with the help of a suitable diagram.
20. Discuss the various characteristics of variable resistors. How are these different from the characteristics of fixed valve resistors?
21. What do you mean by non-linear resistors? What are their different types?
22. What is thermistor? What are its properties? With the help of a neat labeled diagram, describe the constructional details of a thermistor. Also write its applications.
23. Discuss the characteristics and properties of a thermistor. Briefly state their typical applications.
24. Sketch the various symbols of fixed value resistors and variable value resistors. Explain as to how these resistors are identified.
25. Bring out a detailed classification of different types of resistors and write their specifications.
26. Explain the functions and applications of various passive components other than resistors and capacitors.

15

Passive Components (Capacitors)

15.1 Capacitor: an Introduction

Capacitors are the devices which store electricity when put in an electrical/ electronic circuit, and release it as and when required. They are made of dielectric (i.e. insulating) materials. Several kinds of capacitors are in use viz. parallel plate capacitor, multi-plate type, cylindrical type etc. Although we have described dielectric properties and behaviour of dielectric materials in details in chapters 9 and 10; yet some more will be described here to meet the requirement of the subject.

Basic representation. A capacitor consists of two metal plates separated by a dielectric material as shown in Fig. 15.1.

When a voltage is applied across the plates, the capacitor gets charged. The capacitance of a capacitor is defined as the ratio of charge acquired to the voltage applied. Thus

$$C = \frac{Q}{V} \tag{15.1a}$$

where C = capacitance in farads, Q = charge in coulombs, and V voltage in volts.

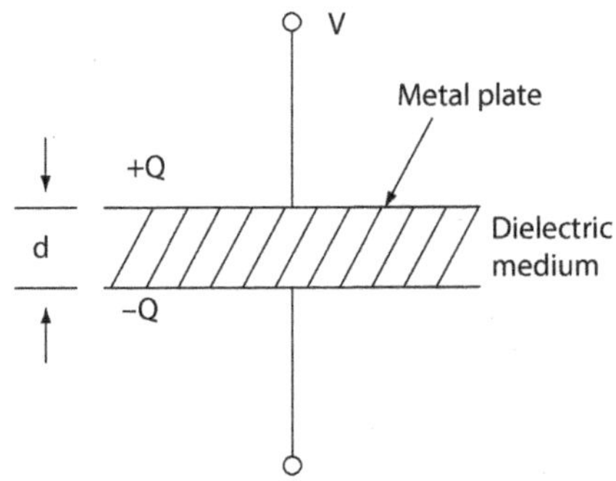

Figure 15.1 Basic representation of a capacitor

A capacitor is one of the most used components in electronic circuits. Recall Eqn. 9.8 and 9.9 that, for a parallel plate capacitor

$$C = \varepsilon \frac{A}{d} = \varepsilon_0 \varepsilon_r \frac{A}{d} \tag{15.1b}$$

where d = distance between plates, A = area of one plate, $\varepsilon = \varepsilon_o \varepsilon_r$ = absolute permittivity of dielectric, ε_o = permittivity of free space, and ε_r = dielectric constant. Above equation implies that to achieve the reasonable value of capacitance; the area of the plates must be large, ε_r should be high, and dielectric thickness should be small. Dielectric constants of different materials has already been listed in chapter 9.

Energy stored in a capacitor. When a capacitor is charged, the electric field within the dielectric is displaced, resulting in polarization of bound charges (See chapter 9) and difference in number of electrons on the two plates. Thus, the stored energy is in the form of electrostatic energy in dielectric and is mathematically expressed as

$$E = \frac{1}{2} CV^2 \tag{15.1c}$$

15.1.1 Equivalent Circuit

An equivalent circuit of a capacitor is shown in Fig. 15.2. For it, the impedances Z is given by

$$Z = \sqrt[2]{\left[R_s^2 + (X_L - X_C)^2\right]} \tag{15.2}$$

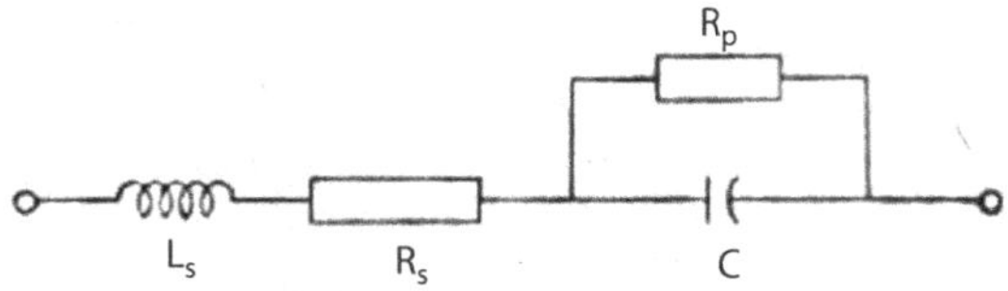

Figure 15.2 Equivalent circuit of a capacitor

where R_s is series resistance, L_s is series inductance, R_p is plate resistance, C is capacitance, $X_L = \omega L_s$ and $X_C = 1/\omega c$ are the inductive and capacitive resistances respectively.

From the phaser diagram and mathematical analysis, it can be shown that (for details, refer to a text on Electrical Engineering), below resonance the impedance is capacitive, at resonance it is resistive, and above resonance it becomes inductive. It is essential to keep the inductance low in order to increase the range of working frequency.

All parameters shown in equivalent circuit of the capacitor play vital role in applications of a capacitor. The inductance and resistive losses control the utility of the capacitor at high frequency, while the insulation resistance affects performance in timing and coupling circuits. Series resistance is a measure of heat dissipation in the capacitor.

15.1.2 Major Applications of Capacitors

The applications of capacitors are too wide. Many of them are already given in chapters 9 and 10, yet many more are given below.

- Filtering of electrical signals
- Tuning of the system
- Coupling and decoupling
- Bypassing
- Temperature compensation
- Buffer
- Energy storage
- Turn-off snubbering
- Transient voltage sharing of series connected devices
- Motors for single-phase supplies
- Voltage multipliers

15.1.3 Important Materials Used for Capacitors

Although a very detailed information on materials is given in chapters 9 and 10, yet some more common materials are listed below for remembrance.

1. Mica : ruby (or muscovita) and amber (or phlogopite) types
2. Porcelain : high voltage and low voltage types
3. Paper : capacitor grade paper
4. Plastic film : polyester, teflon, polypropylene etc.
5. Glass : high silica type
6. Tantalum oxide
7. Aluminium oxide
8. Other ceramics e.g. TiO_2, BaO_2, $BaTiO_3$
9. Air

Some other commonly used capacitor dielectrics are given below along with their dielectric constants.

Thin film dielectric materials	ε_r
SiO	6–8
SiO_2	4–4
Al_2O_3	7–10
Ta_2O_5	25
Si_3N_4	6.5–8
BCB (benzocyclobutene)	2.7
Polyimide	3.5

Thick film dielectric materials	ε_r
$BaTiO_3$	20–2000
Dupont 8229	300–800
ESL 4510	700–1300

15.2 Characteristics of Capacitors

The following characteristics are specified for a capacitor.

1. Capacitance
2. Temperature coefficient
3. Voltage coefficient
4. Working voltage
5. Insulation resistance
6. Dissipation factor
7. Shelf life

8. Tolerance
9. Load life stability
10. Equivalent series resistance
11. Capacitance-voltage (CV) product

Capacitance. The value of capacitance depends upon the area of conducting plates, thickness of dielectric material and its permittivity.

Working voltage. It is defined as the maximum voltage at which the capacitor can operate without failure. The working voltage is usually rated as dc value, which decreases with rise in frequency of operation.

Dissipation factor. This factor is a measure of losses in dielectric of the capacitor. The loss factor described in chapter 9 is the same as dissipation factor. Hence dissipation factor = tan δ.

Equivalent series resistance. This parameter reflects losses inside the capacitor. From equivalent circuit, the equivalent series resistance of the capacitor expresses the losses at a given frequency, and is given by:

$$R_{equ\,series} = \frac{Losses\ inside\ the\ capacitor}{I^2} \tag{15.3}$$

where I is the current flowing through the circuit.

Leakage current (or insulation resistance). A certain amount of current flows through the capacitor material. This is known as leakage current. Insulation resistance is the resistance measured across the terminals of a capacitor at specified voltage and temperature. This is expressed in Mega ohms.

Temperature coefficient. It indicates the change in capacitance due to change in temperature. Its value is generally positive.

Voltage coefficient. Voltage coefficient indicates a change in the value of capacitance with voltage. It is expressed in percentage.

Tolerance. This is the accuracy to which the value of capacitor can be made or selected.

Shelf life. Shelf life for a capacitor is defined as the change in the value of capacitance during storage. It is usually quoted for one year and expressed in percentage.

CV product. CV product (capacitance x voltage) is the measure of the efficiency of a capacitor and of the total charge which can be stored by the capacitor. Electrolytic capacitors have the highest CV ratio available.

Load life stability. It is expressed as change in the value of capacitance that alters the stated operating time at full load and 70°C. The operating time is usually taken as 1000 hours.

15.3 Classification of Capacitors

Capacitors can be classified on vivid basis such as its value, type of dielectric material used for its construction, and the nature of its polarization (See chapter 9 for details on polarization). These are given in Fig. 15.3.

Besides above, few more types of capacitors as given below are also used.

- Vitreous enameled capacitor

Fixed value capacitors are those, whose capacitance value cannot be changed and hence remains constant, whereas the variable capacitors are those for which the value of capacitance can be changed. This change is done by adjusting some parameters.

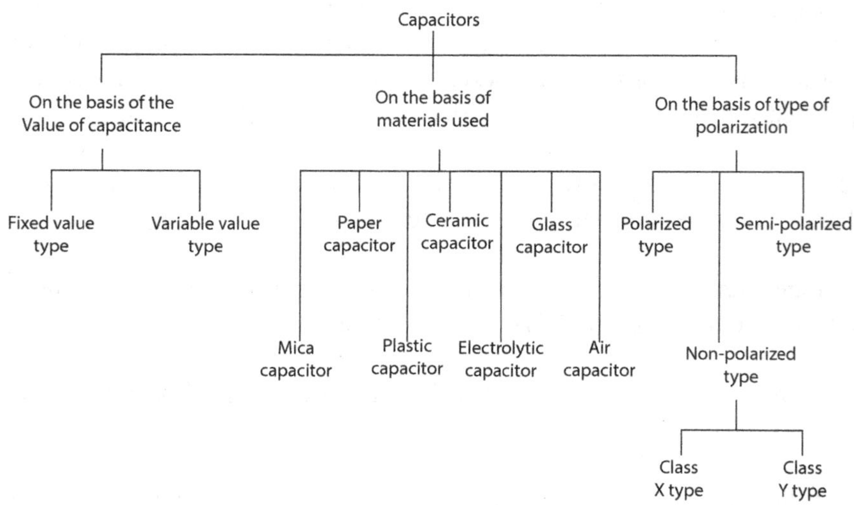

Figure 15.3 Different types of capacitors.

15.4 Forms and Materials of Common Types of Capacitors

Following are the forms and materials of commonly available capacitors.
Mica capacitor. These generally consist of thin sheets of mica, about 2.5 mm thick, coated on both sides with silver (Fig. 15.4a); an alternative form is of mica sheets sandwiched between the sheets of lead or aluminium foil. Mica capacitors are low loss type, good for radio frequencies, have good temperature stability and have values up to about 1000 pF. They do, however, tend to be bulky.

Paper capacitors. These consist of layers of waxed paper sandwiched between the layers of metal foil, the whole being wound into a roll (Fig. 15.4b). Such capacitors tend to be used where losses are not too important, and have a capacitance which changes significantly with temperature, a working life which is shorter than other capacitors and values between about 500 pF and 10 μF. Working voltages can be up to 150 kV.

Plastic capacitors. These have a similar construction as of paper capacitors, consisting of layers of a plastic film e.g. polystyrene or PTFE, between the layers of metal foil. They are very reliable capacitors, maintaining their capacitance values at high temperatures and over long periods of time.

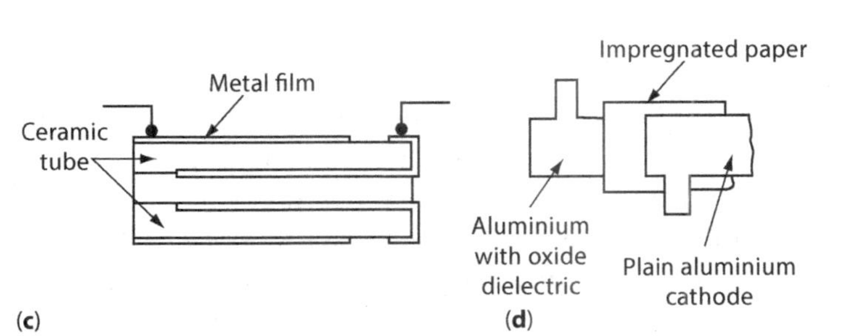

Figure 15.4 Different types of capacitors (a) mica capacitor, *(b)* paper capacitor (c) ceramic tubular capacitor, and (d) electrolytic capacitor

Ceramic capacitors. These can be in tube (Fig. 15.4c) or plate forms, the form of construction depending upon the capacitance value required. Ceramic materials have high relative permittivity and so the capacitors can be very compact. They have high working voltages, can be used at high frequencies, and have capacitances between about 1 pF and 0.1 μF.

Electrolytic capacitors. These consist of aluminium foil plates separated by a thick absorbent material e.g. paper, impregnated with an electrolyte such as ammonium borate (Fig. 15.4d). Electrolytic action occurs when a potential difference is connected between the plates and results in a thin layer of aluminium oxide being formed on the positive plate. This very thin layer forms the dielectric.

Another form of electrolytic capacitor uses tantalum instead of aluminium, with tantalum oxide forming the dielectric. The electrolytic capacitor must always be used with a D.C. supply and must always be connected with the correct polarity. Because of thinness of the dielectric, such capacitors have very high capacitances with the values tending to be in microfarads. Working voltages tend to be between ≈ 6 V and 600 V.

15.5 Constructional Details of Fixed Value Capacitors

15.5.1 Mica Dielectric Capacitor

The dielectric mica is a mineral which has a plane of easy cleavage enabling the large sheets of single crystal to be split into thin plates, typically 50 μm thick. Stacks of mica plates are interleaved with silver metal foils as shown in Fig. 15.5. The metal foils to which the leads are spot-welded, are made of silver, copper, brass, tin or lead. The stack is held together either by the encapsulation or a metal crimp. The assembled unit is encapsulated either by dipping it into a high melting temperature microcrystalline wax or coating it with epoxy resin.

Properties and applications. Mica capacitors are non-polar, low loss and stable upto about 30 MHz. Because of their relatively high cost of manufacture, the monolithic multi-layer type is favoured.

Maximum ratings are a few nanofarads at 5000 V, with dissipation factors of 0.1% at 1 kHz. For capacitance less than 1 nF, a 0.1% dissipation factor is obtainable at 1 MHz. An insulation resistance of 10^5 MΩ at 20°C down to 10^4 MΩ at 125°C is common for capacitance to 10 nF, after which the resistance falls off. Typical operating temperature range

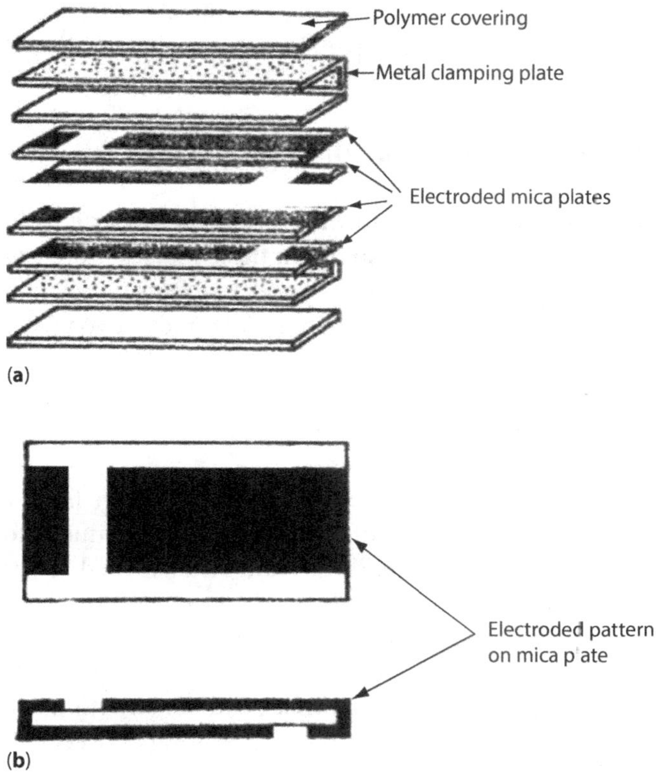

Figure 15.5 Silver mica capacitor *(a)* exploded construction view, and (b) electrode pattern of a silvered mica plate.

is from –55°C to 125°C, with a capacitance temperature coefficient of 0 to +70 ppm/K.

The maximum current depends on the edge connections and electrodes. Mica dielectric capacitors are sensitive to pressure.

15.5.2 Paper Capacitor

These capacitors employ paper as dielectric material. A paper capacitor is already shown in Fig 15.4b. The paper sheets used for this purpose are impregnated with oils or waxes to prevent absorption of moisture. Impregnation also increases the dielectric strength. These paper sheets are wound with thin aluminium foils. The contact to metal is made by welding the leads to it. Finally the capacitor is encapsulated in a 'metal can' or in resin. These capacitors are larger in size due to their foil size. Paper capacitors are used in single phase motors and power factor correction devices.

The size of paper capacitors can be reduced by metallization of aluminium by vacuum deposition on to the paper sheets.

15.6 Plastic Film Capacitors

Since polymeric materials can be formed in thin, highly uniform and non-porous films with high dielectric strength, these materials are preferred as the dielectric materials for capacitors. These capacitors use different plastic films as dielectric. These capacitors are manufactured in two ways viz, using

i. Metal foil, and
ii. Metalized film capacitor.

In *metal foil type* plastic capacitor, a number of thin films of plastic material are interleaved with aluminium foils, and is wound into a cylinder. The coil is then filled with end caps and encapsulated either in resin or insulating lacquer. In *metalized film type* capacitors, the electrodes are vacuum deposited on plastic films. Since the thickness of deposited film is small, a higher capacitance can be realized. This is encapsulated as earlier.

Depending upon the dielectric material and construction, the capacitance tolerances between 1 to 20% and voltage rating between 50 V to 400 V are usual.

15.6.1 Metalized Plastic Film Dielectric Capacitors

The dielectric of these capacitors consists of plastic film on to which the metal layers of approximately 0.02-0.1 μm are vacuum deposited. A margin of non-coated film is left as shown in Fig. 15.6a. The metalized films are either wound in a rolled cylinder or flattened to form a stacked black construction. In this construction, the metalized films are displaced so that one extends out at one end of the roll and the next layer extends out the other end as shown in Fig. 15.6a. This displaced layer construction is termed as *extended metallization* and facilitates electrical contact with the electrodes. A hot metal spray technique, called *Schooping*, is used for making the electrical contact to the extended edges of the metalized plastic winding. This large area contact method ensures a good ohmic contact, hence low loss and low impedance capacitor characteristics result in.

Common materials. The most common metallised plastic film capacitors are those employing the polyester, metallized polyester, polypropylene

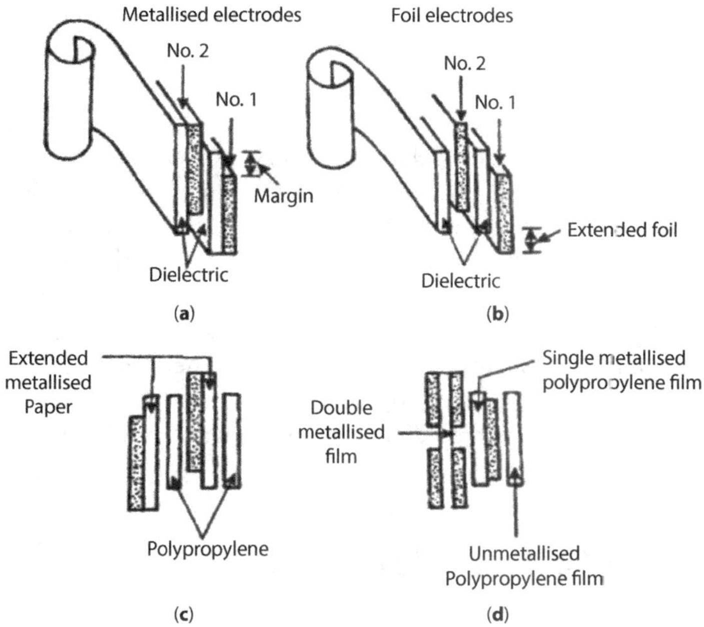

Figure 15.6 Construction of plastic capacitor (a) extended single metallization, (b) extended foil, (c) mixed dielectric, and (d) mixed dielectric, double metallization.

and metallized polypropylene. Polyester has a higher dielectric constant than polypropylene, and because of its stronger physical characteristics it is available in thinner gauges than that of the polypropylene. Very high capacitance values result in the smallest possible space in this capacitor. But polypropylene has a higher dielectric strength and lower dielectric losses, hence is favoured at higher ac voltages.

15.6.2 Foil and Plastic Film Capacitors

Foil capacitors normally use a plastic film dielectric which is a flexible biaxially aligned electro-insulator, such as polyester. Aluminium foils and/or tin foils are used as the electrodes. The thin strips are wound to form the capacitor as shown in Fig. 15.6b. An *extended foil* technique similar to the extended metallization method is used to enable the contact made to the extended foil electrodes.

15.6.3 Mixed Dielectric Capacitors

To further improve the electrical stress capabilities of a capacitor, combinations of different dielectrics are commonly used. Such capacitor of metalized plastics, metalized paper, discrete foils and dielectrics, and oil impregnation.

Fig 15.6c shows the layers of a mixed dielectric paper and polypropylene capacitor. A thin gauge of polypropylene dielectric is combined with textured metalized paper electrodes. The coarse porous nature of the paper allows for improved fluid impregnation of the dielectric material, which counters the occurrence of gas air bubbles in the dielectric. This construction has the electrical advantages of high dielectric strength, low losses and a self-healing mechanism, all at high voltages.

Two plastic dielectrics can be combined, as shown in Fig. 15.6d, to form a *mixed layer* capacitor. It involves a double metalized polyethylene-terephthalate ate film and polypropylene films. These dielectric combinations give low inductance, high dielectric strength and low losses with high ac voltage capacity.

15.7 Ceramic Dielectric Capacitors

Ceramic capacitors as a group have in common an oxide ceramic dielectric. The dielectric is an inorganic, non-metal polycrystalline structure formed into a solid body by high temperature sintering at 1000–1300°C. The resultant crystals are usually between 1 and 100 μm in diameter.

The basic oxide material for ceramic capacitors is titanium dioxide (TiO_2) which has a relative permittivity of about 100. This oxide together with barium oxide (BaO_2) forms barium titanate ($BaTiO_3$) which is a ferroelectric material with a high permittivity, typically 10^4. Alternatively, the strontium titanate may be utilized. The same materials are used to make the positive temperature coefficient resistors - thermistors, where dopants are added to allow conduction.

Metal plates of silver or nickel (with minimal palladium and platinum) are used to form the capacitor. Single plate, or a disc construction, is common as is a multi-layer monolithic type construction.

15.7.1 Classes of Ceramic Dielectrics

The ceramic dielectrics are split into two classes, as given below.

Class I Dielectrics. This class of dielectric consists mainly of TiO_2 and additions of BaO, La_2O_3 or Nd_2O_5, which provides a virtually linear, approximately constant and low temperature coefficient.

COG capacitors belong to the class I dielectrics and have a low temperature coefficient over a wide temperature range. They provide stability and minimum dissipation properties. In attaining these properties, a low dielectric constant results -in and these capacitors are, therefore, termed as low K. Because of the low dielectric constant, value of the capacitance is limited.

Class II Dielectrics. Ceramic capacitors in this class are usually based on a high permittivity ferro-electric dielectric $BaTiO_3$, hence they are termed as *high K capacitors.* Large capacitance in a small volume can be attained, but only by sacrificing the temperature, frequency and voltage properties, all of which are nonlinear. Their characteristics are less stable, non-linear, have higher losses than the class I ceramic.

A comparison between class I and class II dielectrics is shown in Table 15.1.

15.7.2 Applications

Flat circular disc ceramic capacitors have a 2000 V dc, 550 V ac rating with capacitances of up to 47 nF. An exploitable drawback of such a mimic capacitor is that its permittivity decreases with increased voltage, i.e. the capacitance decreases with increased voltage. Such a capacitor can be used in turn-off snubber for thyristors and transistors.

Table 15.1 Comparison between the characteristics of class I and class II type dielectrics

S.No.	Class I	Class II
1.	Almost linear capacitance/ temperature function	Non-linear capacitance temperature function
2.	No voltage dependency of capacitance and loss angle	—
3.	No ageing	Slight ageing of capacitance
4.	High insulation resistance	High insulation resistance, extremely high capacitance value per unit volume
5.	Very small dielectric loss	—
6.	High dielectric strength	—
7.	Normal capacitance tolerance 1% to ±10%	Normal capacitance tolerance ±5% to ±20

15.8 Electrolytic Capacitors

The capacitor dielectric layer consists of aluminium oxide Al_2O_3 or tantalum oxide Ta_2O_3 which is formed by an electrochemical oxidizing process of aluminium foil or sintered tantalum powder. These starting metals form the capacitor anode. The oxide layer withstands very high shuttle field strengths, typically 8×10^8 V/m for Al_2O_3. This field strength is maintained during the oxidizing process, so that the oxide thickness is dependent and practically proportional to the *forming voltage* V_F. To avoid changing the oxide thickness during normal use, the component *operated rated voltage* V_R should always be lower than the forming voltage. The difference $V_F - V_R$ is the *overoxidisation voltage* and substantially determines the capacitor operational reliability. For general-purpose electrolytic capacitors, the value of V_F/V_R is about 0.8, while the solid capacitors are rated at 0.25.

The oxide dielectric constant ε_r is approximately 10 for Al_2O_3 and 25 for Ta_2O_3, while the paper-based dielectrics have a value of approximately 5. An oxide thickness of 0.7 μm is sufficient for high voltage capacitors (≥ 160 V) as compared with the minimum thickness of about 6 μm for paper dielectric.

Formation of the capacitor. The capacitor is formed by the placement of cathode on to the oxide layer. In case of the electrolytic capacitor, a highly conductive organic acid electrolytic (based on dimethylacetamide) impregnated into porous paper, forms the capacitor cathode. The electrolyte largely determines the equivalent series resistance (ESR), hence it must have a low resistivity over a wide temperature range. It must also have a breakdown voltage well above the capacitor rated voltage at maximum operating temperature. For long life, the electrolytes with water content must be avoided. For this purpose, the Teflon spacers are sometimes used rather than the paper. In the case of solid capacitors, a high conductive cathode is formed by a solid semiconductor metal oxide, such as manganese dioxide. The electrical contact to the cathode is by means of a layer of etched aluminium, which has a thin oxide layer. In solid oxide capacitor, the manganese dioxide is dipped into graphite which is coated with silver epoxy for soldering.

The four possibilities of construction of metal oxide capacitors are shown in Fig. 15.7. A porous paper of glass fibre is used as a space keeping agent in order to avoid the short circuits and direct mechanical contact. Long strips of the cross-sections are wound into the cylindrical bodies and encased as shown in Fig. 15.7.

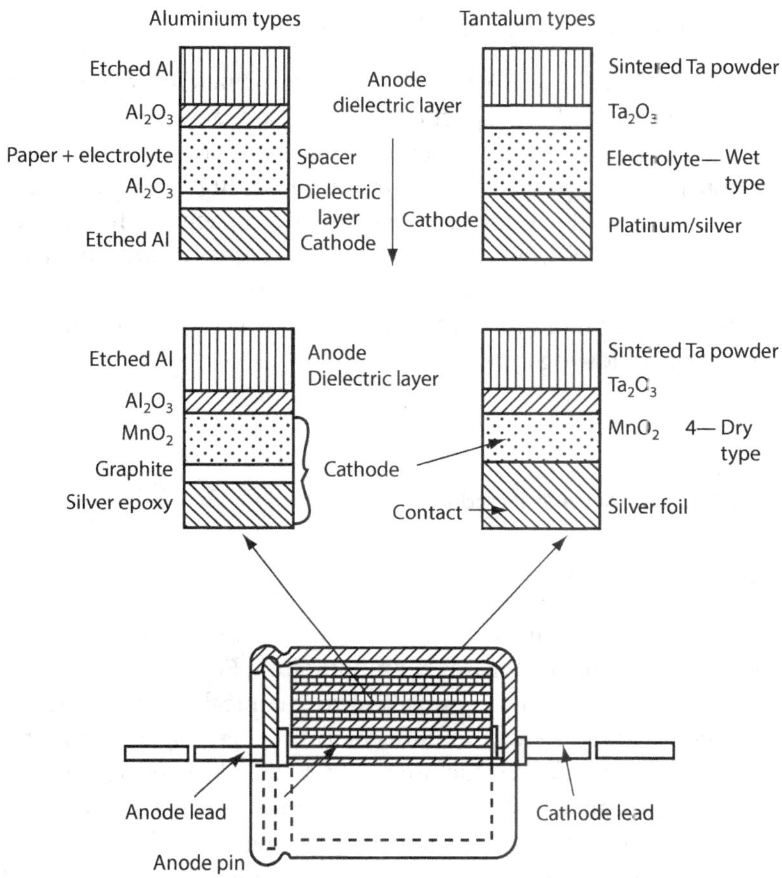

Figure 15.7 Construction of metal oxide capacitors

15.8.1 Aluminium Electrolytic Capacitor

In these capacitors, a highly etched aluminium foil is used as anode, Fig. 15.7. Aluminium oxide film is grown over it which acts a dielectric film. The electrolyte in liquid form is held in contact with dielectric film, with the help of a separate paper. The cathode is made of an etched foil of aluminium. In order to reduce the evaporation of electrolyte solution, these are sealed in aluminium case. These capacitors suffer from the disadvantage of solidification of electrolyte, making them unsuitable for operation at low temperatures.

15.9 Tantalum Electrolytic Capacitor

Tantalum capacitors have a very high volumetric efficiency as compared to the aluminium electrolytic capacitors. These also have improved electrical and reliability characteristics. Tantalum capacitors have lower leakage and dissipation factor, improved frequency response, high stability, and long life as compared to the aluminium electrolytic capacitors. In addition to these, they do not have limited shelf life as is the case of aluminium electrolytic capacitors. Their life depends upon the purity of oxide layer formed in the capacitor. The tantalum electrolytic capacitors are further subdivided as being of the following types.

i. Foil type
ii. Solid type
iii. Wet type with sintered anode.

15.9.1 Tantalum Foil capacitors

These are made in the same way as the electrolytic capacitors. In this case, the Tantalum foils are used in place of aluminium foils, Fig. 15.8a. The Tantalum oxide dielectric film is grown using electro-chemical reaction. Electrolyte used is either sulphuric acid or lithium chloride depending upon the characteristics desired. The whole structure is encapsulated in hermetically sealed package. These capacitors offer higher capacitance density because of higher dielectric constant of tantalum oxide. They can be used at higher temperature (about 200°C) because of the higher boiling point of electrolyte.

15.9.2 Tantalum Solid Capacitor

In this construction, the anode consists of the tantalum powder mixed with an organic binder, and pressed into pellet form. The constructional details of such capacitors are shown in Fig. 15.8b. The pellets are sintered to decompose and evaporate the binder. This results in the formation of a pellet of high porosity and high surface area. These pellets are anodized to get a layer of tantalum pentaoxide on the surface. They are then impregnated with a solution of manganese nitrate. This is then heated to 300°C, which gives a semiconductor layer of manganese dioxide (MnO_2). Further, it is coated with a mixture of graphite and silver. After the anode and cathode leads are attached, the capacitor assembly is encapsulated in a resin or a metal case.

These capacitors have high volumetric efficiency, and good stability with time and temperature. The limitations of solid tantalum capacitors are the following.

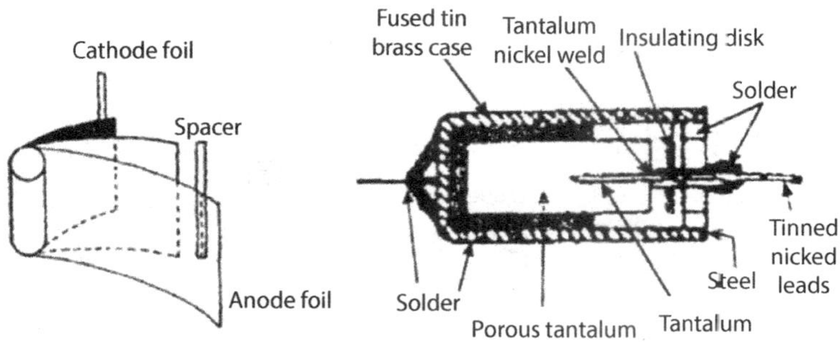

Figure 15.8 Tantalum capacitors (a) foil type, and (b) solid type

i. High leakage current, and
ii. Limited voltage range.

These capacitors are used in those applications where low frequency components are to be by-passed or filtered out, where space limitation is a consideration; and where stability, size, weight and shelf life are important factors.

15.10 Air Capacitor

Air dielectric capacitors are mainly used as laboratory standards of capacitance for measurement purpose. These capacitors are very accurate (0.01 to 0.02%), but their size is large. Air capacitors are available in working voltage upto 45 kV, and the range of capacitance available is 0.025 pF to 6 pF. The insulating material used in these capacitors is fused silica. The whole assembly is shielded in a metal case.

15.11 Polarized and Non-Polarized Capacitors

Aluminium oxide or tantalum oxide film grown over the anode has a very low resistance in one direction and high resistance in another direction. Because of this reason, the reversal of polarity of applied voltage either causes dielectric breakdown or the formation of oxide layer on cathode, with a resultant decrease in capacitance. On the basis of polarizing nature, the electrolytic capacitors have three basic configurations as given below.

1. Polarized capacitor
2. Semi-polarized capacitor
3. Non-polarized capacitor

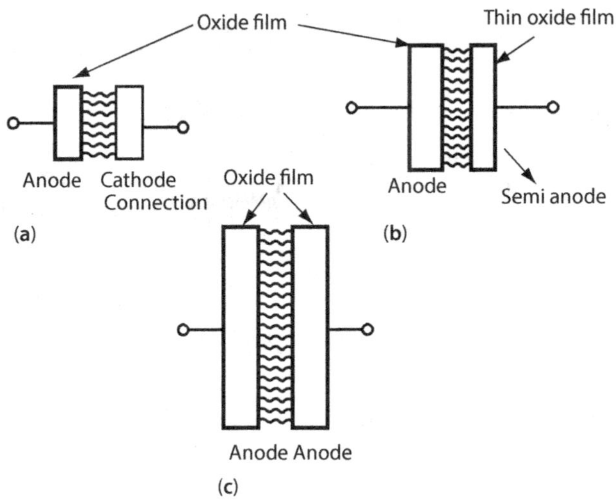

Figure 15.9 (a) Polarized capacitors, (b) semipolarized capacitors, and (c) non-polarized capacitors

In case of polarized capacitors, the oxide layer is grown over the anode foil. Semi-polarized electrolytic capacitor has the same construction as that of a polarized capacitor except that a thin oxide layer is grown over cathode. This minimizes the effects of voltage reversal. Applications to energy storage often create such conditions. In applications where the polarity reversal is often observed, a non-polarized structure is preferable. In this structure, the oxide film is grown over the anode and cathode electrodes to same thickness. These configurations have been depicted in Figs. 15.9a-b-c, which also indicates the relative size of capacitor for same capacitance value.

15.11.1 Non-Polarized Capacitors

Non-polarized capacitors are used in rfi filters for electrical appliances and equipment. The capacitors used between the supply line and the neutral are termed as class X, while those used to 'earth' are termed as class Y.

Class X capacitors. Class X capacitors are suitable for use in situations where failure of the capacitor would not lead to danger of electric shock. X capacitors are divided into two subclasses according to the ac power line voltage applied. The X1 subclass supports a peak voltage in excess of 1.2 kV in service, while X2 capacitors have peak service voltage capabilities of less than 1.2 kV.

In order to obtain the peak voltage requirement of X1 capacitors, a construction comprising impregnated paper dielectric and metal foil electrodes is essential. The common capacitance range is 10 nF to 0.2 μF.

Class Y capacitors. Class Y capacitors are suitable for use in those situations where the failure of the capacitor could lead to danger of electric shock. These capacitors have high electrical and mechanical safety margins so as to increase the reliability and prevent the short circuit. They are limited in capacitance so as to restrict any ac current flowing through the capacitor, hence decreasing the stored energy to a non-dangerous level.

An impregnated paper dielectric with metal foil electrodes is a common construction and values between 2.5 nF and 35 nF are extensively used. Capacitance as low as 0.5 nF is not uncommon.

15.12 Variable Capacitors

Variable capacitors are invaluable in design of electronic equipment. They are commonly used in applications where the exact value of capacitance cannot be calculated using normal design procedure. In this case a variable capacitor is employed to provide the needed capacitance range.

As can be seen from the expression for capacitance, the capacitance can be varied by changing three parameters viz. the distance between the plates, dielectric constant and area of the plates. In most of the variable capacitors, the area of plates is changed. In some cases, thickness of the dielectric medium i.e. distance between the plates is varied. Dielectric materials used in most of the cases of variable capacitors are air, ceramic, glass, plastic and mica.

15.12.1 Air Variable Capacitor

Air variable capacitor is fore-runner of all variable capacitors and is used as a tuning element. These are available in concentric and rotary type configurations. The rotary configuration consists of two basic parts: the rotor and the stator. The rotor is the rotating part and is usually constructed of several semi-circular disks or plates, affixed to the shaft. The stator is the fixed portion and consists of similar shaped plates as shown in Fig. 15.10. These are mounted in such a way that the rotor plates can be intermeshed with the stator plates by rotating the rotor shaft. Since the dielectric constant of air is 1.0, these units are relatively larger in size.

Concentric type variable capacitors consist of two cups of aluminium, one moving inside the other. The movement is actuated by a threaded screw.

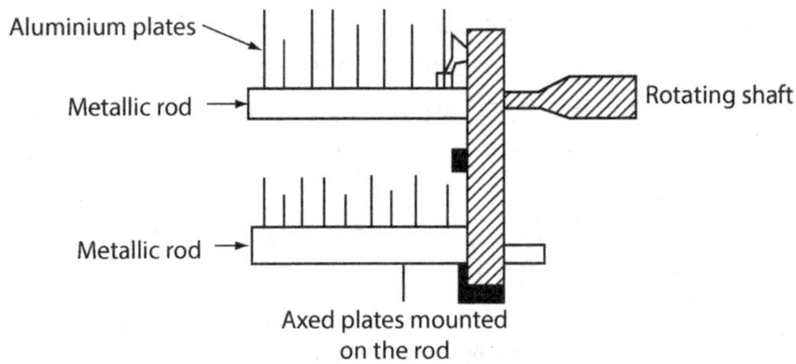

Figure 15.10 Basic construction of an air capacitor

These capacitors are most temperature stable. Change in capacitance may follow one of the following laws, depending upon the shape of the plates.

i. Linear
ii. Inverse square law
iii. Square law
iv. Logarithmic law

15.12.2 Ceramic Variable Capacitor

These are constructed in two forms viz. rotary and concentric type, Fig. 15.11a-b.

Rotary type ceramic capacitor contains three parts, viz. rotor, stator and dielectric. The *rotor* carries semicircular pattern. This pattern consists of metallization of rotor to provide an electrode. The rotor pattern is usually semicircular so as to utilize the maximum resolution capability during tuning. The *stator* consists of the metalized pattern deposited on to a low loss steatite. The dielectric, usually titanate ceramic, is placed between the rotor and stator electrodes. To assure stability, the interfaces of all bearing surfaces are kept in intimate touch with each other. It gives straight line capacitance law. These capacitors are available up to 120 pF.

In case of concentric type ceramic capacitor, a long metallic screw runs inside a ceramic tube, which is silvered on outside. The capacitance variation follows linearity. These are available upto 10 pF.

15.12.3 Glass Capacitors

These variable capacitors also change the capacitance by varying the effective plate area. In its construction, a glass tube is coated on outside

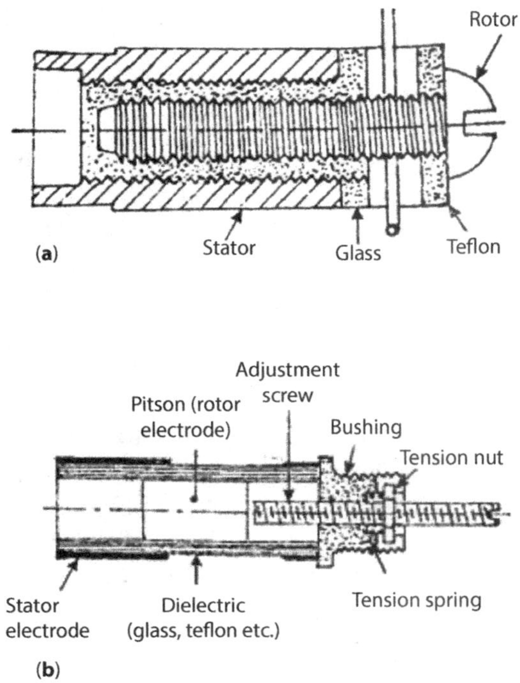

Figure 15.11 Construction of ceramic variable capacitor (a) rotary type, and *(b)* concentric type.

with a conducting material. The metallization is deposited from one end of the tube towards another end. This constitutes the stator. The rotor *is* a piston usually mounted on a threaded rod and inserted from other end of the tube. This piston, when moved in and out of the metalized section of the tube, varies the capacitance by changing the effective area of the plates. The connection to device is made by means of straps and clamps embedded in the metallization on outside of the tube and by brushes in intimate contact with the threaded rod on piston. These units are usually low in capacitance and range from 5 to 30 pF. These have temperature coefficient of 100 ppm/^{0}C, and are used in applications where precise variation is required.

15.13 Specifications of Capacitors

The capacitors possess several characteristics which vary significantly with their physical size and capacitance values. Accordingly, the capacitors are

specified by the parameters given below. Their numerical values over a range are given for a ready reference.

• Material of capacitor	Mica, paper, plastic, ceramic, electrolytic: Al foil or Ta foil
• Range of capacitor	5 pF to 100000 μF
• Dissipation factor (= tanδ)	0.0005 to 0.08
• Service lifetime	24000 to 28000 hours or more than 3.5 years
• Temperature range	–55°C to 125°C
• Tolerance	±0 5% to ±20%
• Workable A.C. voltage	150 V to 600 V
• Workable D.C. voltage	1 V to 10 kV
• Temperature coefficient	±30 to 1500 ppm/°C
• Leakage resistance	10^8 to 10^{12} Ω
• Dielectric constant of material	1 for air, 10–25 for Ta_2O_3

(See more values in chapters 9 and 10)

15.13.1 Symbolic Representation of Capacitors

Electrical/electronic components are represented by certain specified symbols in circuits. Capacitors of different types are also represented by specific symbols. These are shown in Fig.15.12a-h.

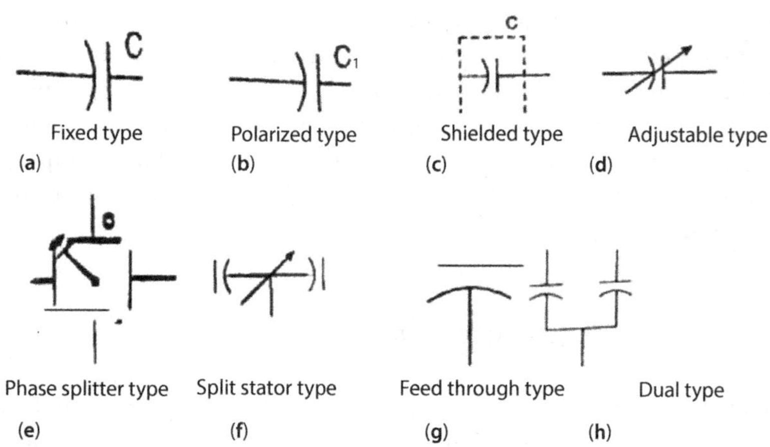

Figure 15.12 a-h Symbolic representation of different types of capacitors.

15.14 Identification of Capacitors

To distinguish among different types of capacitors and for their easier identification, the capacitors are provided with certain 'identification marks' on them. The identification of capacitors generally provided, are the following.

Colour coding. The information in colour code identification includes some or all of the parameters such as capacitance, tolerance, temperature coefficient, breakdown voltage, capacitance etc. Colour code schemes are different for different types of capacitors as given below, Fig. 15.13a-b.

15.14.1 Colour Codes for Tubular Ceramic Capacitors

Tubular ceramic capacitors use coloured dots or coloured bands in five or six-colour identifications. It is read from left to right, with the temperature coefficient band being the largest in series appearing on the left, Fig. 15.13a-b. Other related details are shown in Table 15.2.

15.14.2 Colour Code for Mica Capacitors

Mica capacitors employ a 6-dot and 9-dot colour arrangement to indicate the capacitance as shown in Table 15.3.

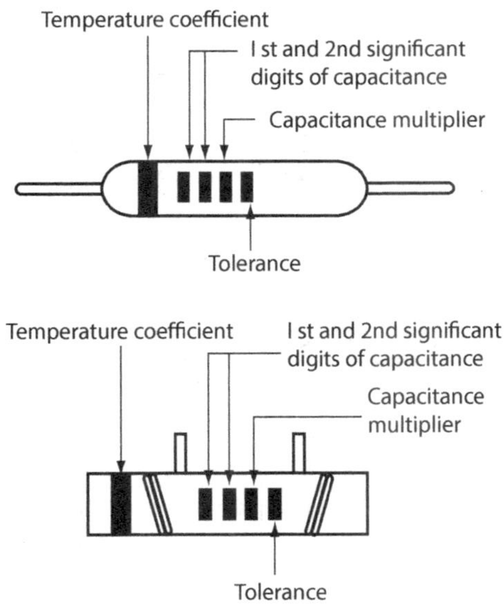

Figure 15.13 Colour code for five-dot ceramic tubular capacitor having (a) axial load, and *(b)* radial load.

Table 15.2 Colour codes for tubular ceramic capacitors

Colour of dot or band	Capacitance in pF			Tolerance	
	Significant digit's		Capacitance multiplier		
	1st	2nd		≤ 10 pF	> 10 pF
Black	0	0	1	±2 pF	±20%
Brown	1	1	10	±0.1% pF	±1%
Red	2	2	100	-	±2%
Orange	3	3	1000	-	±3%
Yellow	4	4	-	-	-
Green	5	5	-	±0.5% pF	±5%
Blue	6	6	-	-	-
Violet	7	7	-	-	-
Gray	8	8	0.01	±0.25 pF	-
White	9	9	0.1	±1 pF	±10%
Silver	-	-	-	-	-
Gold	-	-	-	-	-

Table 15.3 Colour code for mica capacitors

Colour	Characteristic	Digits		Multiplier	Tolerance
		1st	2nd		
Black	-	0	0	1	±20%
Brown	B	1	1	10	±1%
Red	C	2	2	100	±2%
Orange	D	3	3	1000	-
Yellow	E	4	4	10000	-
Green	F	5	5	-	±5%
Blue	-	6	6	-	-
Violet	-	7	7	-	-
Gray	-	8	8	-	-
White	-	9	9	-	-
Silver	-	-	-	0.1	±0.5%
Gold	-	-	-	0.01	±10%

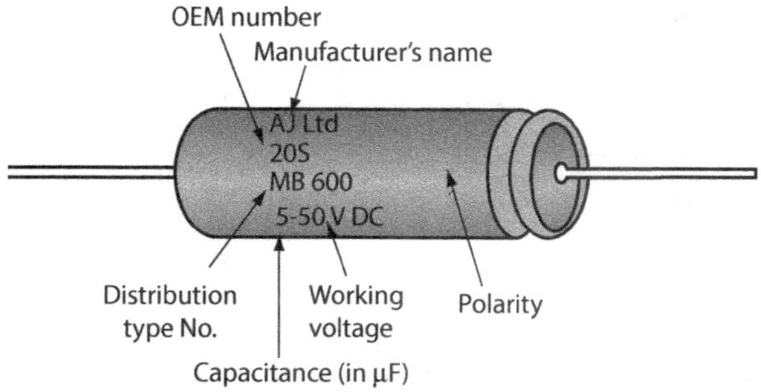

Figure 15.14 Marking on an electrolytic capacitor.

15.14.3 Marking of Capacitors

Figure 15.14 shows an example of marking of an electrolytic capacitor. The manufacturer's name, original equipment manufacturer (OEM), distribution type number, capacitance in microfarads, and dc working voltage are stamped on the outside of the capacitor.

15.15 Solved Examples

Example 15.1 A 470 Ω resistor is constructed from a film of carbon having a resistivity of 3.5×10^{-5} Ω m, deposited on a non-conducting ceramic bar of 3 mm diameter and 6 mm length. Calculate the thickness of the film required, ignoring the end connection effects.

Solution. Let the thickness of the film be t meter, then

Cross-sectional area $a = \pi \times 3 \times 10^{-3} \times t$

As $R = \rho l/A$,

$$\therefore \quad 470 = 3.5 \times 10^{-5} \times (6 \times 10^{-3}/\pi) \times 3 \times 10^{-3} \times t$$

$$t = \mathbf{0.0474\ \mu m}$$

Example 15.2 Discuss the different causes of failure of capacitors.

Solution. Generally the capacitors are metal foils or electrodes, separated by a dielectric. The possible failures in them are the following.

i. Gradual fall in insulating resistance or increase in leakage current
ii. Rise in series resistance i.e. increase in dissipation factor

iii. Manufacturing defects
iv. Open circuit, due to the failure of end connection
v. Environmental factors like shock and vibration, temperature variation, pressure variation
vi. Short circuit due to dielectric breakdown
vii. Misuse

Example 15.3 What are the operational limitations of a capacitor?

Solution. The effectiveness of the operation of capacitors is limited by the following factors.

i. Temperature effect
ii. Ripple current limitation
iii. Dielectric absorption
iv. Dielectric losses
v. Inductive effect or frequency limitation
vi. Current or VA limitation.

Example 15.4 Name different polymers that are used to make plastic capacitors.

Solution. These are given as follows.

Polyester capacitors. Polyester is the lowest cost material with an upper temperature limit of about 125°C and maximum capacitance of about 30 μF.

Polycarbonate capacitors. These capacitors employ polycarbonate film as dielectric. These have an upper temperature limit of about 100°C with capacitance variation of 2%. Maximum capacitance is about 30 μF.

Polyethylene capacitors. These capacitors have high frequency stability. Most of the polyethylenes used are low density type as well as high density film type.

Polystyrene capacitors. Polystyrene is a thermoplastic material. In capacitors, it is used in the form of film foil. These capacitors have a maximum temperature rating of about 85°C. They have low dielectric absorption and low linear temperature coefficient. These are suited for computer and tuning applications.

Polypropylene capacitors. These have excellent stability under extreme environmental conditions. They have high insulation resistance and low dielectric absorption. Their temperature coefficient is higher.

PTFE (Teflon) capacitors. These capacitors use teflon film as dielectric. They have good stability under extreme environmental conditions. These have very high insulation resistance and low dielectric losses.

Example 15.5 Distinguish between the low-loss low-permittivity ceramic capacitor and high-permittivity ceramic capacitor.

Solution. Low loss permittivity type capacitors are usually made of steatite which is naturally occurring substance. It is finely ground, compressed and heated to 900°C to remove the impurities. It is then reground and reformed at about 1300°C. A thin plate or film is metalized on both sides, and the leads are soldered to the metalized layers. These capacitors are made in disc, tubular and rectangular forms. These capacitors have low temperature coefficient (±30 ppm/°C) and are used in temperature compensating applications and bypassing.

High permittivity type capacitors employ high permittivity ceramic material as dielectric. The material used is barium titanate ($BaTiO_3$). These are manufactured in same way as the low loss permittivity capacitors. These capacitors have relatively large capacitance in small volume. They are used for general purpose coupling and decoupling; where wide variation in capacitance value due to temperature, frequency, voltage and time can be tolerated.

Example 15.6 Write a note on vitreous enamel capacitor.

Solution. Vitreous enamel capacitors are manufactured by spraying the vitreous enamel on the metal plates that are stacked and fired at a temperature high enough to vitrify the glaze. These capacitors have a superior high frequency characteristic and can be used at high temperatures (150–200°C). They can also withstand high humid conditions.

Review Questions

1. What are capacitors? What are their functions? Draw its equivalent circuit and explain.
2. Write the important applications and main materials used for the construction of different types of capacitors.

3. Enlist various characteristics of a capacitor and briefly explain them.
4. Bring out a detailed classification of capacitors. How do the fixed value capacitors differ from the variable value capacitors?
5. Describe different forms of capacitors and the materials used for them.
6. Describe the construction of the fixed value mica dielectric capacitor with the help of suitable diagrams. Also write its salient features.
7. Describe the construction and working of a fixed value paper capacitor with the help of suitable diagrams. Also write its salient features.
8. Describe the construction and working of a fixed value plastic film capacitor with the help of suitable diagrams. Also write its salient features.
9. Describe the construction and working of a fixed value metallised plastic film capacitor with the help of suitable diagrams. Also write its salient features.
10. Describe the construction and working of a fixed value foil and plastic film capacitor with the help of suitable diagrams. Also write its salient features.
11. Describe the construction and working of a fixed value mixed dielectric capacitor with the help of suitable diagrams. Also write its salient features.
12. Describe the construction and working of a fixed value ceramic capacitor with the help of suitable diagrams. Also write its salient features.
13. Describe the construction and working of an electrolytic capacitor with the help of suitable diagrams. Also write its salient features.
14. Describe the construction and working of an aluminium capacitor with the help of suitable diagrams. Also write its salient features.
15. Describe the construction and working of a tantalum capacitor with the help of suitable diagrams. Also write its salient features.
16. Discuss the construction and working of an electrolytic capacitor with the help of suitable diagrams.
17. What are polarized capacitors? How are they different from the semi-polarized and non-polarized capacitors?

18. What are different types of variable capacitors? Explain the construction and working of any one of them with the help of suitable diagrams.
19. Distinguish between the following.
    a. Class I and Class II type fixed value capacitors
    b. Class X and Class Y type non-polarized capacitors
20. Write notes on the following.
    a. Variable type air capacitor
    b. Variable type ceramic capacitor
    c. Variable type glass capacitor
21. How are the capacitors represented symbolically in a circuit? Express the colour code identification and marking for them. Also write the specifications of a capacitor.
22. What are different causes of failure of capacitors and what are their operational limitations?
23. What is the insulation resistance of a capacitor? Why are the ceramic capacitors preferred in high frequency circuits?
24. What are the important factors that affect the life of a capacitor? What will happen if we connect electrolytic (unipolar) capacitor in an ac circuit?
25. Explain with suitable diagrams the difference between film capacitors and metallised film capacitors. Why are the metalized film capacitors more preferred?
26. How are the tantalum capacitors different from aluminium capacitors. Discuss briefly different types of tantalum capacitors.
27. Explain as to why (a) the capacitors are preferred for tuning circuits, and (b) Al-electrolytic is not used in precision applications.
28. Write the specifications of capacitors. What are the symbolic representation of different types of capacitors?
29. How is the identification done for capacitors? Explain the arrangement of colour codes.

16

Printed Circuit Board (PCB) Fabrication

16.1 Printed Circuit Board

The interconnection of various electrical/electronic components and their assembly by means of wiring is a complex and cumbersome technique. Interconnection by wiring occupies more space, remains entangled, and chances of disconnection of wires are always there. That is why this method has become obsolete from electronic equipment. Also the miniaturization of electronic gadgets/devices/systems requires other advanced methods of interconnection, owing to these reasons. This style has therefore been replaced by a newer technique in which the inter-component connection is done by means of 'printed circuit'. Printed circuit is a metallic circuitry which does not require any wiring. It is formed photo-chemically upon an insulating substrate material or board. Since the printed circuit is formed on this board, it is known as *printed circuit board* (PCB).

The PCB provides sufficient mechanical support and necessary electrical connections for an electronic circuit. Part of a printed circuit board is shown in Fig.16.1. Its basic features are the following.

i. **Track.** This is a metallic conducting path that runs on through the substrate to make the inter-connection between

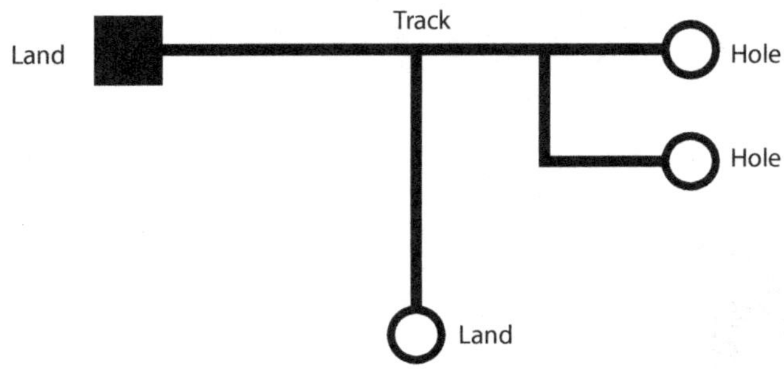

Figure 16.1 Part of a printed circuit board showing its basic features.

the components. Tracks are connected to the components by means of a larger conductor area, which is known as 'land'. The tracks are made of thin copper film.

ii. **Land (or pad).** It is a larger area of conductor on which the tracks meet the component. The electrical connection between the land and terminal of components is achieved by means of a solder joint. The lands are of different shapes (rectangular, square, circular etc.) and sizes depending upon the shape and size of the component to be connected. Holes are also drilled through the lands to accommodate the leads of the component or to make the connection with another track. Similar to the tracks, the lands are also made of thin copper film.

The thin copper film area for tracks and lands are created by either the *plating process* or by *etching process*. The etching process is more common. In this process, the entire copper skin laminated on substrate base material, is etched.

16.1.1 Advantages of PCBs

Printed circuit boards have the following advantages.

1. Circuit characteristics can be maintained without introducing any variation in inter-circuit capacitance.
2. The size of component assembly can be reduced along with corresponding decrease in weight.
3. Component wiring and assembly can be mechanized by wave soldering or vapour phase reflow soldering.

4. Inspection time is reduced as the probability of error is eliminated.
5. Mass production can be achieved at lower cost.

16.2 Types of PCBs

Broadly, the PCBs are of following four kinds.

1. Single sided PCB
2. Double sided PCB
3. Multi-layered PCB
4. Flexible PCB
   i. static flexible PCB, and
   ii. dynamic flexible PCB

In a *single sided PCB,* the copper tracks are formed on the side only of the board (Fig. 16.2a); while in *double sided PCB,* the tracks are formed on both sides of the substrate. Since the connections are required between both sides of the board, it is accomplished by plating through holes as shown in Fig. 16.2b.

In a *multi-layered PCB,* 2 or more printed circuit formed substrates are stacked one over the other and then bounded together (Fig. 16.2c). Electrical connections are established through drilled holes, which are plated with copper. Flexible PCBs are just the opposite of conventional (i.e. rigid) PCBs. The *static flexible* PCBs can be made in bent profiles of particular configuration (Fig. 16.2d), while the *dynamic flexible* PCBs can be deformed continuously during operation.

Each type of PCB has its own advantages and disadvantages. A comparison between them is given below in the chart along with their broader applications.

Description	Single sided PCB	Double sided PCB	Multi-layered PCB	Flexible PCB
• Level of complexity of printed circuit	Simplest	Complex	More complex	Complex
• Manufacturing complexity	Simplest	Complex	More complex	Saves space

Description	Single sided PCB	Double sided PCB	Multi-layered PCB	Flexible PCB
• Cost • Applications	Cheapest Digital circuits, high frequency uses, analog circuits	Costlier Fast pulse uses, power electronic use, microwave uses	Costlier Military, aerospace	Cheaper Medical, aerospace, computers, industrial, automotives

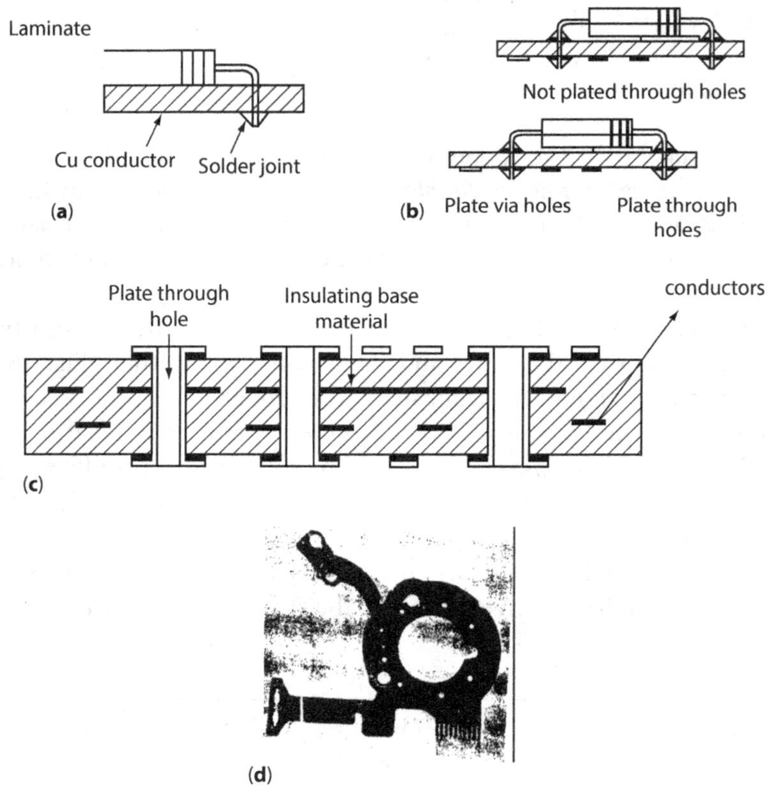

Figure 16.2 Different types of PCBs (a) single sided, *(b)* double sided, (c) multi-layered, and *(d)* static flexible type.

16.2.1 Measures of Level of Complexity in PCB

Whether a PCB is simple, complex, or much complex; depends on the complexity of interconnection. In this regard, the indication of level of the complexity can be determined by considering the packing density of PCB, whose measure may be the quantities as given below.

i. Number of holes/unit area of usable surface of the board.
ii. Number of centimeters (or millimeters) of conductor contained within one cm^2 (or mm^2) area of useable board.

The typical values of these measures are the following.

- For single-sided PCB 1 to 2 holes/cm^2
- For double-sided PCB 2 to 4 holes/cm^2
- For multilayered PCB 4 or more holes/cm^2

16.3 Types of PCB Substrates (or Laminates)

Printed circuit boards are built on a variety of substrates. A particular substrate/laminate used for a particular application is chosen on the basis of function to be performed, operating environment and cost. Functional considerations include the mechanical, electrical, and flammability factors, while the environmental considerations include temperature, humidity, vibration and thermal stresses. Since it is desirable to use less costly material, the cost must also be an important factor. Different types of substrates/laminates in common use are the following.

1. **Epoxy laminates.** Majority of the printed circuit boards used in consumer electronics are manufactured on epoxy laminate. Epoxy laminates are tough, have outstanding adhesion, high mechanical properties, good chemical resistance and electrical properties. They have low water absorption also.
2. **Phenolic laminates.** Phenolic resins are reinforced with paper fillers for making the copper clad laminate. They are stable enough under a variety of conditions. They are easy to punch and fabricate. Due to low cost, such laminates are mostly used for commercial applications.
3. **Polyimide laminates.** This material has excellent electrical and mechanical properties, which are maintained at elevated operating temperatures also. Polyimide is very hard and requires frequent changes of drill bit during drilling. Reinforced with woven glass fabric, the laminates are used in military and aerospace applications and multilayer circuits.

These laminates have higher copper bond strength at soldering temperature as compared to epoxy laminates.

4. **Polyester laminates.** Polyesters are unsaturated resin in copolymerisable monomers such as styrene etc. These are used with glass fibre reinforcement. The resins themselves have good arc and track resistance, which can be improved by addition of fillers and additives. Their dimensional stability and water resistance are also good. They are not used extensively in electronic industry as epoxies have better dielectric properties.
5. **Silicon laminates.** Silicon laminates are made of silicon resins with glass reinforcements. Silicon resins contain silicon, carbon, oxygen and. hydrogen and are outstanding in their heat resistance. The glass based laminates have excellent arc resistance as well as good electrical properties up to 250°C for larger duration and for short periods of time at higher temperature.
6. **Melamine laminates.** Melamine resins can be combined with a variety of reinforcing fillers, but the best properties are achieved with glass fabric. These laminates have high arc resistance. Melamine glass laminate has poor dimensional stability, particularly when exposed to alternating cycles of high and low humidity. Higher mechanical strength cannot be achieved by using this resin system.
7. **Teflon (PMMA) laminates.** Teflon is a thermoplastic which when reinforced with glass, results in laminate that has low dielectric constant and low dissipation factor over the operating range of humidity, temperature and frequency. These laminates have very good electrical insulation under harsh environmental conditions. Teflon laminates are used for VHF and microwave applications.
8. **Diallyl phthalate (DAP) laminates.** These are glass reinforced and used extensively in electronics as insulation. Their low loss characteristics are attractive over a wide range of frequencies and up to a maximum temperature range of about 165°C.

16.3.1 Merits and Demerits of Different PCB Laminates

Each type of laminate has its own merits and demerits. Accordingly, they are selected for different applications. A comparison among different types of laminates is given below in this regard.

Description	Epoxy laminate	Phenolic laminate	Melamine laminate	Polyester laminate	Silicon laminate
• Effect of acids	Least	No	-	A little	No
• Effect of alkalis	Least	Yes	-	Slight	Very little -
• Effect of organic solvent	Least	No	-	-	
• Arc resistance	Fair	Excellent	Excellent	Good	Excellent
• Water resistance	Excellent	Very good	Fair	Very good	Good
• Heat resistance	Fair	Excellent	Excellent	Good	Excellent
• Mechanical properties	Very good	Very good	Very good	Very good	Fair
• Adhesion quality of bond with copper foil	Very good	-	Very good	-	Different
• Dimensional stability	Good	Good	-	Fair	-
• Electrical properties	Excellent	Good	Excellent	Excellent	Excellent
• Flammability	Burns slowly	Self-extingui-shing	Self-extingui-shing	Burns slowly	-
• Limitations	Poor mould release	Color	High cost	Large care for shrinkage	High cost
• Main advantage	Low shrinkage	Good general properties	High arc resistance	Fluid before cure	High heat resistance
• Shrinkage	2%	2%	-	4%	-
• Normal usable temperature	90-120°C	250-300°C	-	80-180°C	250-300°C
• Cost	Low	High	High	Medium	-

16.3.2 Fabrication of laminate and Reinforcing Materials

Different kinds of PCB laminates described above are basically the polymeric composite materials. They are reinforced by glass fibres and paper. The glass fibre reinforcement is available in different configurations such as glass fabric, glass mat, glass roving, glass chopped strands etc. These are shown in Fig. 16.3.

Depending upon the type of reinforcement used, the PCB laminates may also he named as follows.

1. Glass fabric-polymer based PCB
2. Paper-polymer based PCB

Their details are described throughout this chapter.

16.3.3 Properties of Different PCB Laminates and Commercial Grade Copper Cladded Laminates

Some salient properties of different PCB laminates are displayed in Table 16.1 for a ready reference. In it, the desired properties are the following. The given values are averaged and approximate.

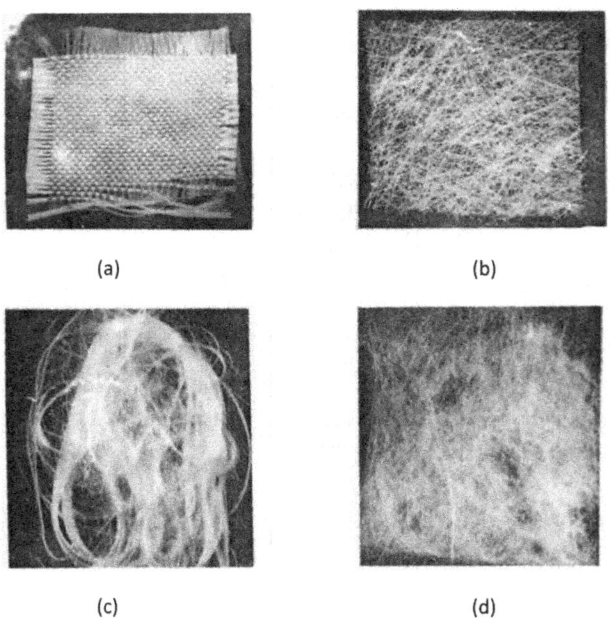

Figure 16.3 Different forms of glass reinforcement used in the construction of a PCB laminate, showing (a) glass fabric, *(b)* chopped strand mat, *(c)* fibreglass roving, and *(d)* fibrewool mat.

Table 16.1 Salient properties of different PCB laminates

Property	Epoxy laminate		Polyester laminate		PTFE laminate	Phenolic laminate
	Glass fabric based	**Paper based**	**Glass fabric based**	**Paper based**	**Glass fabric based**	**Paper based**
Insulation resistance (MΩ)	5×10^6	0.5×10^6	0.1×10^6	–	0.2×10^6	35×10^6
Volume resistivity (MΩ-cm)	10^9	4.5×10^4	-	10^7	-	10^4–10^7
Dielectric strength (kV/mm)	11–24	12–22	8–24	15–24	10–28	10–32
Arc resistance (s)	15–180	30–120	80–140	25–75	180	4–75
Tensile strength (MPa)	230–580	70–130	120–440	40–100	80–135	55–130
Water Absorption (%)	0.05–0.3	0.15–0.5	0.15–2.5	0.1–5	0.02	0.2–6
Specific gravity	1.7–2	1.4–1.5	1.5–2.1	1.2–1.5	2.2	1.25–1.4
Resistance to electrical heating (°C)	105–200	90–110	120–200	120–150	290	105–140
Dielectric constant at 10^6 Hz	4.5–5.3	4–5	3–4	3–4.2	2.4–2.7	3.5–6.0

16.4 Manufacturing Process of Copper Cladded Laminate

Manufacturing of copper cladded laminate for making PCB employs the following two materials.

- Base material
- Foils as conductor material

The quality of base material decides the reliability of PCB. The foils are bonded to or deposited on the base material by some process. These bonding or depositions form the surface of the cladded laminate.

Base materials. These are generally made of plastics or polymeric composites such as given below.

i. Glass based epoxy for
   - higher mechanical strength
   - low dimensional change, and
   - good fungus resistance

ii. Paper based phenolic resin for
   - less expensive boards

iii. Glass based melamine for
   - high abrasion resistance

iv. Glass base silicon for
   - high temperature (usually upto 180°C) uses

v. Nylon with phenolic resin for
   - high insulation resistance
   - low moisture absorption

vi. Teflon for
   - microwave application

Conductor materials. These are generally made of the following metals in foil form, however the copper foils are most widely used.

i. High purity electrolytic copper foil
ii. High purity silver
iii. High purity aluminium
iv. High purity gold
v. Brass

The foils are available in various thicknesses viz. 0.01 mm to 0.5 mm.

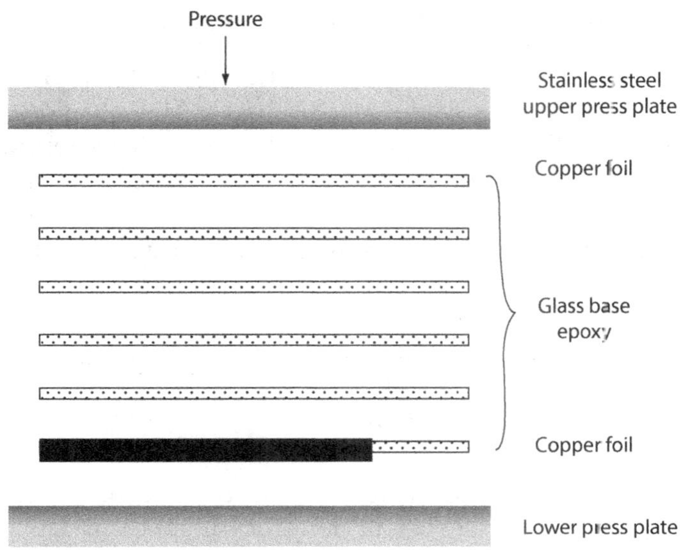

Figure 16.4 Manufacturing of copper-cladded laminate in a hydraulic press

16.4.1 Processing of a Copper Cladded Laminate

Copper cladded laminate manufacturing process is performed in a heated press which has a highly polished stainless steel press plates (Fig. 16.4). The manufacturing starts by placing a sheet of copper foil on to the lower press plate. The copper sheet forms the surface of copper clad, board and a base on which the board structure is built. The copper sheet is produced by either rolling a piece of copper or by an electroplating process. Out of these, the electroplating is preferred as it produces a more uniform and controllable film thickness.

Layers of the base material are built up on top of the copper foil, until the required board thickness is achieved. Once the layers of resin impregnated base material are built up, a final layer is added. The final layer may be a second sheet of copper foil for double-sided boards or may be a material which acts as a release film for single-sided boards.

The entire assembly thus produced is placed in a heated hydraulic press which bonds the different layers into a complete board. The pressing operation proceeds by applying an appropriate pressure, and heating the plates to the curing temperature of the resin. After appropriate curing time, the plates are cooled to room temperature and the boards are removed from the press. The boards are now trimmed to remove any extruded resin.

16.4.2 Desired Properties of Copper-Cladded Laminates

A copper-cladded laminate possesses electrical properties such as dielectric strength, dielectric constant, dielectric loss etc. They also possess other properties such as flexural strength, water absorption, flame resistance etc. However, the following properties are very much desired in them.

Insulation resistance. Insulation resistance is defined as the ratio of

$$R_{insulation} = \frac{Voltage\,applied\,onto\,the\,base\,laminate}{Current\,flowing\,through\,the\,base\,laminate} \tag{16.1}$$

A high insulation resistance is desired.

Surface resistivity. This property refers to the resistance of copper cladded laminate against the leakage current along the surface of insulating board/ laminate.

Volume resistivity. It refers to the resistance against leakage current through the body of the insulating laminate. A high volume resistivity is desired. It decreases with increasing temperature and is also affected by moisture absorption.

Warp and twist. The thermal expansion and contraction of copper foil and plastic base board are different. Their moisture absorption ability are also different. On account of these factors, the copper cladded laminate exhibits warp and twist under conditions of heat and moisture.

A good copper cladded laminate should have least warp and twist.

Flame resistance. A good copper-cladded laminate should have high flame resistance. It can be made effective flame resistant by adding the chemical flame retardants along with the glass cloth reinforcement.

Punchability. Punching of laminate sheets is an essential process to make them workable. Therefore, the copper cladded laminate should have good punchability, which means

i. no cracking in the vicinity of the punch,
ii. no undulations around the punched hole,
iii. smooth inner surface of the hole, and
iv. least shrinkage when punching is done at high temperatures.

Peel strength. Peel strength refers to the pulling force required to peel off the copper foil from the laminate. High peel strength is desired so that the copper foil does not debond.

16.5 Layout and Design of a Printed Circuit Board

Layout of a PCB incorporates all details of the circuit and of equipment. The layout should be viewed from component side (i.e. top side) and should be drawn after getting the absolutely clear circuit diagram with the list of components. The circuit should be divided into functional sub-units so that the PCB has a functional reliability as well as faster testing and servicing.

A proper PCB layout means to design the interconnections for components, and to minimize the magnitude and influence of parasitic effects. This effect may be caused by resistance or inductance of a conductor, or capacitance. They may generate heat also. Hence all these factors must be carefully considered in the layout.

16.5.1 Planning of the PCB Layout

Various considerations that are made in planning the layout of a PCB are the following.

1. The artwork should be prepared to some scale viz. 1: 1 (full size), 2 : 1 or more.
2. Use of grid system should be incorporated in the layout, as the components fit well into grid intersections with their leads.
3. Numerically controlled programmable drilling should be incorporated for higher accuracy.
4. As far as possible, the single -sided PCBs should be used to cut-done the cost of fabrication.
5. Layout should be made in accordance with certain standards e.g. DIN, IS, BS, NEMA etc.
6. The component layout should be completed first, followed by interconnections; or alternatively both can be done simultaneously.
7. The larger components should be placed first and the space in between should be filled with smaller components.

A typical PCB layout is shown in Fig. 16.5 for a certain circuit.

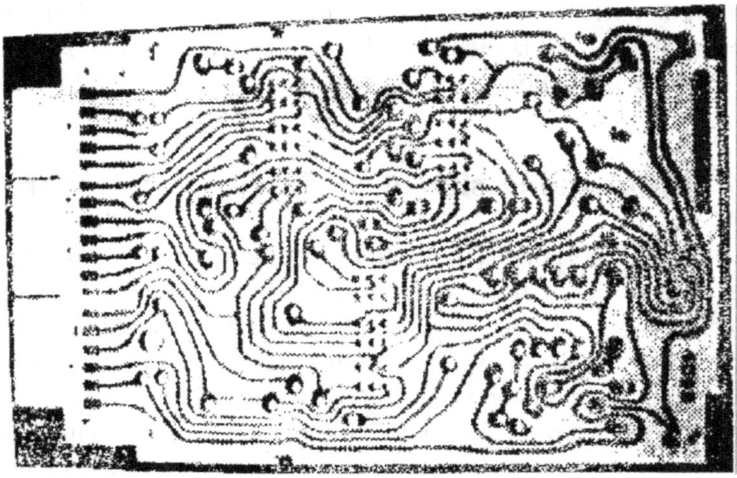

Figure 16.5 Typical layout of a circuit on PCB.

16.5.2 Design Considerations in Making PCBs

A good PCB is the outcome of a well-thought design consideration. If not designed properly, the problems such as given below can arise during operation.

- Interference between neighboring signal lines resulting in cross-talks.
- Electromagnetic interference from pulse-type electromagnetic fields causing high-noise situation.
- Reflections causing signal delays and double pulsing. It results in conversion of one pulse into two or more pulses.
- Noise in supply -line and ground.

Following considerations are usually made in the design of PCBs.

1. The grid system should neither be too coarse nor too fine. A grid of 2.5 mm x 2.5 mm is coarse, but 0.5 mm x 0.5 mm grid is acceptable.
2. PCBs should have sufficient mechanical strength. They should neither be too thick nor too thin. A thickness of 1.6 mm is quite reasonable.
3. PCB can be manufactured in different sizes. Size less than 100 mm x 150 mm is for small PCBs. A standard size board should be of 150 mm x 250 mm dimension.

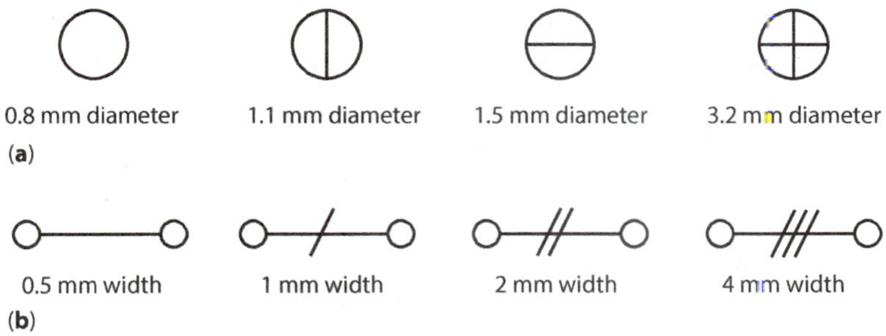

Figure 16.6 Depiction of (a) standard size holes, and (b) standard width of conductors; on a PCB

4. Heavy parts should be mounted near supporting devices such as stiffener or connector.
5. Too large boards and longer interconnections must be avoided.
6. About 95% area of the board should be utilized, leaving at least 5% area for circuit modification, later.
7. Standard size of holes on PCBs should be 0.8 mm. Other hole diameters should be 1.1 mm, 1.5 mm and 3.2 mm. These are shown in Fig. 16.6.
8. Standard width of conductors is 0.5 mm. Other acceptable widths are 1 mm, 2 mm and 4 mm.

16.6 Manufacturing Processes For PCB

A number of processes are adopted to manufacture a PCB from a copper cladded base material. These processes are the following.

1. Preprocessing
2. Photolithography
3. Etching
4. Drilling
5. Solder masking
6. Electroplating
7. Board testing
8. Board finishing

Actual process used for manufacturing a PCB, depends on the board complexity and the application for which the PCB is intended. These processes are described as follows.

16.6.1 Preprocessing

This is initial process of preparing a copper clad laminate ready for subsequent processes. Copper clad laminates are available in a range of sizes.

The oversized laminates are cut to the required PCB size. Once a piece of copper clad laminate is selected and cut to size, the next stage is to drill the holes and to form the special features in the board. Additional holes may be required to provide mechanical support in order to prevent the warping during soldering. Certain features such as tabs may be needed to form the edges of the boards for use with Automatic assembly equipment.

The copper surface of cladded laminate needs to be thoroughly cleaned before actual processing. Cleaning is performed by passing the board through the rollers in an abrasive slurry. When very fine tracks are required on PCB, it is very important to ensure that the board and processing environments are clean.

16.6.2 Photolithography

Copper track and land patterns are defined on a copper cladded board by means of photolithographic process. This process for PCBs involves the exposure of a photoresist material to the light through a mask. Both positive and negative photoresist materials can be used depending upon the actual process. Use of a negative photoresist which becomes insoluble on exposure, is almost universal. The exposure uses UV light through a mask. A preprocessed board is laminated with a photosensitive resist material. The photoresist can be applied either in liquid form or as a dry film. Liquid resists are applied by dip coating, spray coating, or a roller application process. Dry film resist materials take the form of a photosensitive polymer, sandwiched between the polymer films. The dry film is applied to the board by removing a cover film and allowing the photosensitive polymer to adhere directly to the board's surface. The second cover of the film can be removed after application onto the board or at later stage of the processing.

The exposure of photolithography process involves the exposure of 'film coated copper clad board' to UV light through mask. The masks can take the form of either a photographic positive image or a photographic negative of the required track layout. The image is correctly aligned with the board's geometry, especially in case of double sided and multilayered boards.

Development of image. After exposure, the image needs to be developed. The developmental stage involves the removal of less soluble areas of resist material. This corresponds to the unexposed areas of resist material with negatively acting photoresist material. In case of dry film resists, the developing agent is usually either a solvent such as trichloroethylene or a sodium carbonate solution, depending on the actual material used.

Once the selective processing which requires the photolithographic process has been performed; it is necessary to remove the remaining resist material from the board's surface. In case of dry film resists and depending on the actual materials used in the resist, this removal process uses either Methylene Chloride solvent at a temperature of about 20°C or a solution containing sodium hydroxide at about 60°C.

16.6.3 Etching

This process is used for removal of copper surface of a PCB which is not protected by an etch resist material. Thus, the final copper pattern is formed after etching. The etching process is performed by exposing the surface of the board to an etchant solution which dissolves the exposed copper areas, thereby leaving the desired conductor pattern on the board. Different types of etchant solutions are in use. Main among them are the following.

i. Ferric chloride
ii. Cupric chloride
iii. Chromatic acid
iv. Alkaline ammonia

The choice of suitable etchant for a PCB production, depends on the factors like compatibility with etch resist, etching speed, copper dissolving capacity, etchant price and pollution characteristic. The etching process can be performed by simply immersing the board in an etchant solution and then agitating. This is however, difficult to control the process and can result in non-uniform etching.

Spray etching. Commercially, the etching of PCB is done in spray type etching machines. In spray etching, the etchant is pumped under pressure from the sump via a pipe network to the nozzles, and from there it is splashed on to the boards. Spray etching machines offer high etching uniformity and fast etching rate. The high etching uniformity is achieved by spraying through a number of equally distributed nozzles in ring supply. High etching rate results from the fresh flow of the etchant over the boards. However, it is important to control the etching process. Insufficient etching will result in excess copper being deposited on the board and can cause short circuiting between the tracks. Over-etching results in etchant caused under-cutting of the track area defined by the etch resist layer. It can result in broken tracks or thin tracks. Figure 16.7 shows different etching profiles.

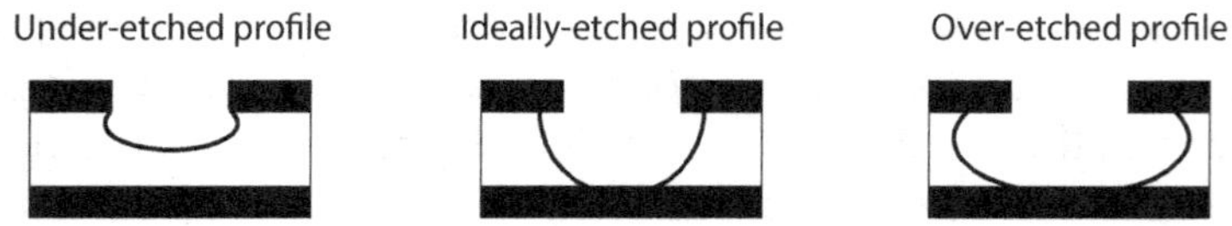

Figure 16.7 Different etched profiles.

The use of etchant solution should be carefully controlled as it becomes contaminated with copper (dissolved), thereby the performance is adversely affected. The copper contained in the exhausted etchant solution can be regenerated by making use of electrochemical extraction methods. After completion of the etching process, the etch resist material is removed chemically. This leaves the PCB with desired copper pattern.

16.6.4 Drilling

Drilling of holes into PCB for component mounting is an important mechanical operation in PCB production processes. Drilling is used to create the component lead holes and 'through holes' in a PCB. These holes pass through the land areas and are positioned properly. In addition to the component holes and via holes, the holes are also required for providing mechanical connections. Holes may also be required for heat sink assembly or connectors connection to the board. The drilling can be done before or after the track areas have been defined, depending on the manufacturing sequence.

Drilling machines for PCB applications are available in wide range of designs. They have high speed range which is required for economic and efficient drilling of various base materials. The drilling process can be performed by using manually operated drilling machines or by using CNC drilling machines. The later is generally used commercially. For CNC drilling machine, the drilling programme can be generated from the computer aided design system's drill template output. For a particular PCB, a wide range of drill holes may be required e.g. smaller holes are required for component leads, whereas the larger diameter holes are required for bottling heat sinks, sockets etc.

Generally many boards are drilled simultaneously. Using suitable jig, the boards of same size are stacked and mounted on drilling machine. Thus, many boards can be drilled in a single drilling operation. The boards to be drilled are sandwiched between two sheets of material, in order to reduce the drill burrs on actual boards. Drill bit, due to its smaller diameter requires careful monitoring. Boards are deburred after drilling operation by passing through the abrasive rollers.

16.6.5 Solder Masking

Solder masking is the process of applying organic coatings selectively to those areas where no solder wetting is needed. Thus in a PCB with solder mask, only such areas are exposed to which the components are to be soldered. The solder masks are one or two component system and need to be cured. Curing can be done thermally or by exposing to UV light. The solder mask is applied by screen printing using the artwork; which enables the entire board to be covered except holes, pads and contact fingers. The use of solder mask offers the following advantages.

i. It avoids solder bridging during soldering.
ii. It reduces contamination of the solder alloy during soldering.
iii. It improves mechanical properties of the PCB.
iv. It acts as a barrier between atmosphere and the board, thus eliminates the danger of corrosion and leakage currents.

The surfaces under the solder mask should be clean; otherwise it may come out or peel off. Solder mask coatings are deep colored but are still sufficient translucent to show the substrate. This is required for observing the defects in circuit pattern.

16.6.6 Electroplating

Printed circuit boards are collected together before being taken for final assembly of components. The PCB is required to retain its solderability for long time so that the reliable solder joint can be produced during assembly. The plating is done for the following reasons.

- It ensures protection of the copper tracks.
- It establishes connection between different layers of double-sided and multilayer boards.
- It increases wear resistance and decreases contact resistance.

The plating can be of three different types as given below.

i. **Immersion plating.** It is the deposition of metallic coating on a substrate from the solution of salt of the coating metal. This process is used for plating the tin and its alloys, and gold. The PCB should be thoroughly rinsed after immersion in order to eliminate all residue chemicals to ensure acceptable shelf life.

ii. **Electroless copper plating.** The manufacturing of double-sided and multilayer PCB requires electrical connections through the board in order to link the interconnection layers. This is usually done by making the inside surfaces of the component electrically conductive in a plating through hole. Electroless plating deposits sufficient amount of copper on the inside of holes to make the entire board surface conductive. After this the copper layer can be built up in desired thickness using electroplating process.

iii. **Electroplating.** Electroplating is done to increase the thickness of the copper applied upon by the electroless plating process. A part of it which has been covered by the layer of copper deposited by electroless plating, acts as cathode.

There are two electroplating processes in common use depending upon the chemical composition of electroplating solution. These are referred to as 'acid copper' and `pyro copper'. The acid copper process uses an acidic copper sulphate solution, while the pyro process makes use of copper pyro-sulphate solution. This process is more efficient at copper depositing but is affected by change in temperature and pH value.

16.6.7 Board Testing

Each board needs to ensure that there are no short circuits in the required connections, and that the drill holes are properly placed. The testing usually consists of visual inspection and continuity testing. Complex boards require both. In these testings, the

i. Visual inspection is carried out to ensure proper track width and proper conductor pattern

ii. Continuity testing ensures proper connections between different points. Open circuits may result due to faulty artwork or over- etching etc.

16.6.8 Board Finishing

Finally, the solder mask layer is applied as discussed earlier. The legend layer consists of the components positional information in the form of the diagram of component's outlines and identification. For example, the codes such as C1 are for the first capacitor. This legend layer is applied upon by silk printing, after the board has been degreased in a solvent bath.

16.7 Manufacturing of Single Sided PCBs

These are the simplest types of boards. They consist of an insulating base material bearing the copper tracks and lands only on one side.

i. Manufacturing starts from a single-sided copper clad sheet/board.
ii. These boards undergo preprocessing operations which include cutting the board to correct size, creating tooling holes, and cleaning of the board surface.
iii. After cleaning, the board is subjected to photo processing. Then a layer of photoresist material is applied on to the board surface and the board is exposed to ultraviolet light through required artwork mask.
iv. After development of resist image, the remaining photoresist material covers the areas with required copper pattern.
v. Etching is done to remove the exposed copper areas.
vi. The remaining photoresist material is removed after etching.
vii. Now the board is placed on CNC machine to drill the component lead holes and any fixing holes.

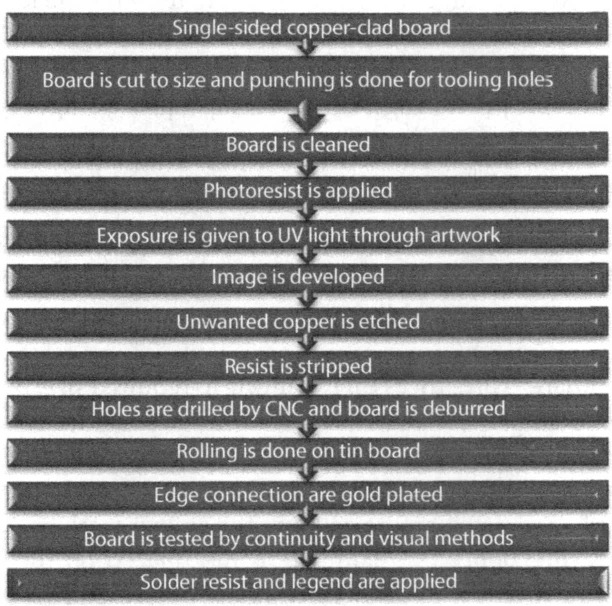

Figure 16.8 Manufacturing of single sided PCBs

viii. The drilling by CNC is done according to the program generated from the artwork.
ix. Boards are deburred after drilling, to remove the burrs.
x. Copper track and land areas of the board are solder plated by passing the copper-side of the board over rollers, which are running in a molten solder bath.
xi. Boards with edge connector are required to undergo the gold plating process on the connector track areas.
xii. Finally, the boards are tested, and solder resist and legend layer are applied.

16.8 Manufacturing of Double-Sided PCBs

Double-sided boards require additional manufacturing process as they have copper tracks on both sides of the board that need to be interconnected. Thus, additional holes are required along with the normal holes for component insertion. The additional holes are used to connect copper tracks on opposite sides of the board and are known as via holes. The electrical connection between two sides of the board, through via holes and component lead holes, is achieved by the plating process. The PCBs with plated holes are referred to as having plated through holes (PTH). The following two manufacturing processes are used for the fabrication of double-sided boards with plated through holes:

1. Logic layer processing 2. Outer layer processing

16.8.1 Logic Layer Processing

This process is used for manufacturing the intermediate logic layers of the multilayer PCBs. The manufacturing sequence for double-sided PCBs with logic layer processing are as follows.

1. The board is preprocessed by cutting to required size, punching tooling holes and identification marks. The cleaning process is performed similar to as has been done for single-sided boards. In logic layer processing, the 'plated through holes' are created before defining the copper track patterns.
2. Thus the copper clad laminate is first placed on the CNC drilling machine to drill the component lead and via holes. The CNC drilling program is generated from the drill template of the design system.

3. After drilling process, it is required to plate the holes to provide an electrical connection between two sides of the board. The plating process is performed in two stages, viz. electroless plating and electroplating as discussed earlier. Electroless plating applies a thin layer of copper to entire board surface, whereas the electroplating thickens the copper layer to the required thickness.
4. After plating the 'via holes' and 'component lead holes', the track pattern is created on both sides of the board by using photo processes. Photoresist layer is applied on both sides of the board and exposed to UV light through the artwork masks. The remaining photoresist covers the areas with desired track pattern. Through holes are also covered by the

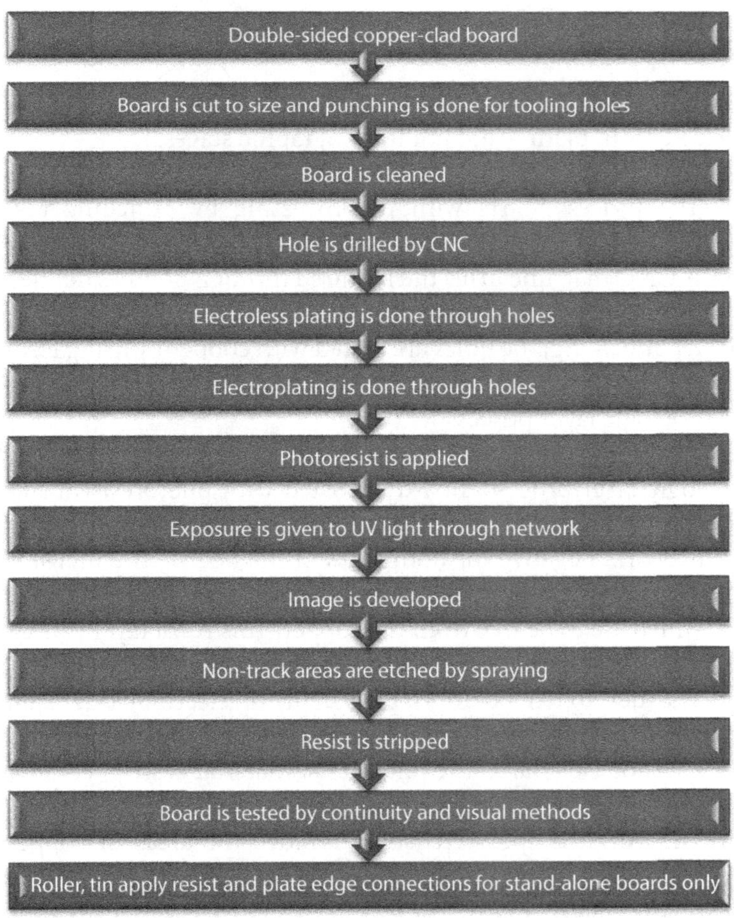

Figure 16.9 Fabrication of double-sided boards with logic layer processing

photoresist to ensure that the plating is not removed by later etching process.

5. Developed board is passed through an etchant spray process to remove the undesirable surface copper from the non-track areas. After etching, the photoresist layer is stripped from the track areas. Additional holes, if required for later processes are punched at this stage.
6. When the logic layer board is to become the component part of a multilayer board, a detailed inspection of the track and land areas is required. The inspection includes visual inspection and continuity test.

The double-sided PCBs undergo the roller tinning, solder resist layer and legend layer application as for the single-sided PCBs.

16.8.2 Outer Layer Processing

This process differs from the earlier process used for manufacturing double-sided PCBs in plating and photolithographic stages.

i. The process starts with preprocessing by punching and cleaning of a double-sided copper clad board.
ii. A CNC machine drills the required through holes as defined by drill template artwork.
iii. After drilling, the holes are plated with copper by using electroless process, followed by electroplating process similar to the logic layer process.
iv. The photolithography operates in reverse to corresponding logic layer process. Board is coated with photoresist material and exposed to UV light through the track definition artwork, negative for both sides of the board. The image is developed to photoresist material on the areas of the board which are to be free from the copper. This resist layer is used as plating mask for pattern plating process.
v. The developed board is placed in an electroplating bath where solder is plated on to the track, pad and 'through hole' areas of the board. The photoresist material is stripped after plating.
vi. The board is spray-etched to remove the areas of copper which have not been solder-plated. After etching process the board undergoes a reflow process which involves heating the

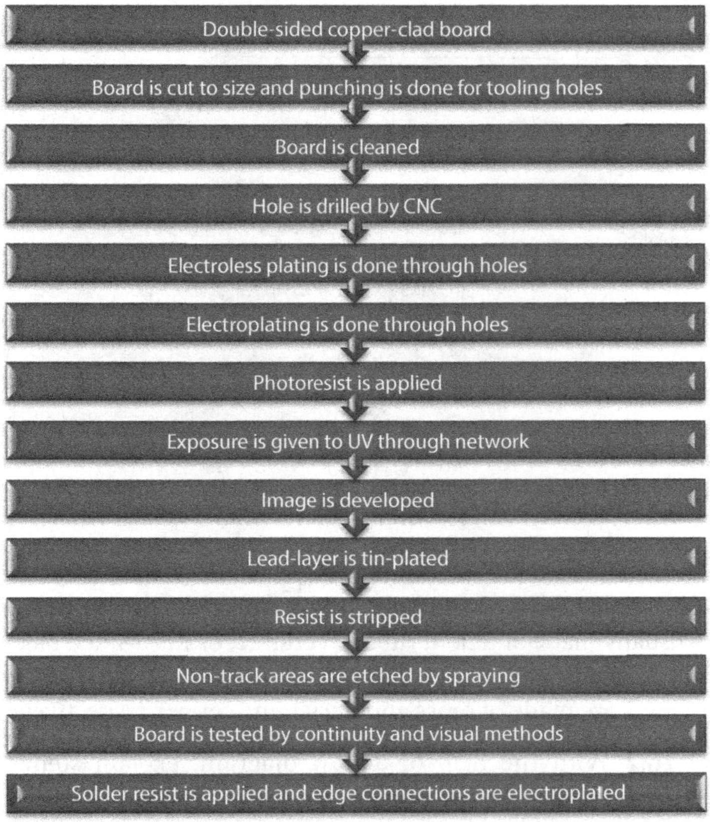

Figure 16.10 Fabrication of double-sided boards without layer processing

board and passing it through flux impregnated rollers. The purpose of this process is to melt the solder layer to remove any oxidation and to produce a smooth surface finish.

vii. After reflowing, the board is cleaned in a solvent to remove excessive flux. Solder resist and legend layer are applied by screen printing.

16.9 Solved Examples

Example 16.1 Briefly describe any connector which is used on a PCB.

Solution. Coaxial connectors are used to provide interconnection between various components of an electronic equipment/system or a connection with printed circuit board (PCB). Several configurations

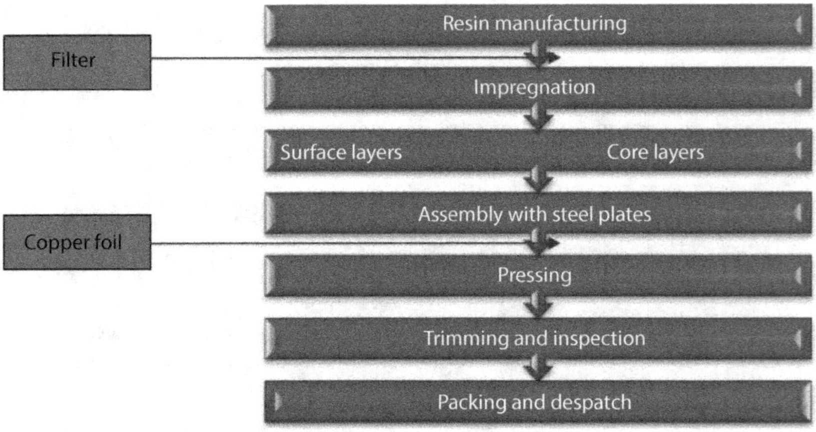

Figure 16.11 Flow diagram showing the manufacturing steps involved in making a copper-clad laminate

of connectors are employed, among which coaxial is the main type. In it, the braid shielding is crimped to the connector. The centre insulation is usually made of teflon, rubber, ceramic, or mica-filled bakelite; and the connector bodies of copper, brass or zinc. The male and female pin contacts are made of brass and beryllium-copper respectively.

Example 16.2 With the help of a flow diagram, explain stepwise the sequence involved in manufacturing of a copper-cladded laminate.

Solution. It is as given in Fig. 16.11.

Example 16.3 Which devices can be mounted on PCBs without drilling holes in them? What are their advantages?

Solution. Surface mounted devices (SMDs) can be used without drilling holes on PCBs because, they do not have lead. The uses of SMDs offer a number of benefits over leaded components. Main among them is given below.

i. The SMDs are small in size and can be mounted on both sides of PCB. This results in high density of components leading to compact electronic products.
ii. The number of through holes required for PCB is very low. This reduces the drilling cost, considerably.
iii. Most of the components in SMD form are in rectangular shape. Therefore, the assembly is simpler and faster. This results in reduced assembly costs.

iv. Due to small or no lead on SMDs, the parasitic inductances and capacitances are lower. Thus, it is simpler to manufacture the SMDs for high frequencies.
v. The electronic products made from SMDs are cheaper, portable due to their smaller size, and more reliable due to higher level of automation.
vi. The lack of lead wires simplifies the handling of SMDs. Their components are less liable to damage as no lead preparations are required.

Example 16.4 What is photolithography?

Solution. Photolithography is a process of planar technology to fabricate an IC in which, a desired pattern of doping is transferred onto the silicon wafer. The entire process involves (i) coating of SiO_2 layer with a film of photosensitive emulsion called *photoresist* (ii) blackening the portions to be doped on a transparent sheet, placing this sheet on photoresist layer and then exposing to ultraviolet light (iii) removing the mask and dipping the wafer in trichloroethylene, and (iv) *etching* the SiO_2 with hydrofluoric (HF) solution.

Review Questions

1. What is PCB? State its advantageous features over wired interconnecting of components. What are the basic features of a PCB?
2. What are different types of PCBs? Discuss each of them briefly.
3. What are different types of substrates or laminates used in PCBs? Describe them briefly and compare their merits and demerits.
4. Describe as to how the copper-cladded laminates are manufactured. Which base and conductor materials are used in them?
5. Explain the various electrical and mechanical properties of a copper-clad laminates.
6. What are main characteristics of a copper-clad laminate? Discuss each of them briefly.
7. Explain the following processes used for PCB fabrication.
   i. Preprocessing
   ii. Photolithography

   iii. Electroplating
   iv. Etching
8. What do you mean by PCB layout? What considerations are required in making the layout for a PCB?
9. Describe the salient features to be included in the design of a PCB.
10. Describe any one technique used for manufacturing the printed circuit board (single sided).
11. With the help of a flow chart, discuss the manufacturing of a single-sided PCB.
12. Explain as to how a double-sided board is fabricated.
13. Describe any one technique used for manufacturing a double- sided printed circuit board.
14. Compare the following PCBs and processes.
    a. Single-sided and double-sided PCB
    b. Logic layer processing and outer layer processing
15. Explain where the following types of printed circuit board are used.
    i. Multi-layer PCBs
    ii. Flexible circuit PCB
    iii. Single-sided PCBs
    iv. Double-sided PCBs
16. Explain the following characteristics for copper clad laminate:
    i. Single resistivity
    ii. Punchability
    iii. Volume resistivity
    iv. Insulation resistance
    v. Warp and twist
    vi. Flame resistance
    vii. Peel strength
    viii. Dissipation factor

17

Optical Properties of Materials, and Materials for Opto-Electronic Devices

17.1 Introduction

Meaning. By optical property we mean the response of a material against exposure to electromagnetic radiations, particularly to visible light. When the light falls on a material, several processes such as reflection, refraction, absorption, scattering etc. take place due to interaction between the light and material. The e.m.f. of the light radiation interacts with the changes contained in a material to exhibit its different optical properties. The nature and distribution of charges in a material play an important role in determining the optical properties.

Optical requirements of materials. Lenses, optical fibres, optical glasses, diamond etc. have to meet the following requirements.

1. Optical lenses refract the light along a desired optical path. In eyesight purpose lenses, the refraction is controlled by grinding the surface to a suitable curvature.

2. The optical fibres for long-distance communication need to have almost zero-light absorption.
3. The optical-glasses meant for producing mirrors, lenses, and prisms have to follow stringent requirements of composition and homogeneity to ensure clarity and precise optical paths for high resolution.
4. The brilliance of diamond is attributed to its high refractive index and dispersion.

17.1.1 Important Terminologies Related to Optical Materials

Electro-optic effect. It is the behaviour of a material in which its optical isotropic nature changes to anisotropic nature on application of an electric field. This effect is seen in $LiNbO_3$, $LiTiO_3$ etc.

Photoelectric effect. It refers to a phenomenon in which the ejection of electrons from a metal surface takes place, when the metal surface is illuminated by light or any other radiation of suitable frequency (or wavelength). Several devices such as phototube, solar cell, fire alarm etc. work on this effect (principle).

Photoemissivity. It refers to the phenomenon of emission of electrons from a metal cathode, when exposed to light or any other radiations.

Brightness. It is the power emitted by a source per unit area per unit solid angle.

Photons are discrete radiation emitted from the materials in the form of small energy packets.These packets are also called as '*quanta*'. They possess energy, mass and momentum.

Optical absorption means absorption of electromagnetic radiations of light by the semiconducting material. Whether the optical absorption will be less or more, depends on the property of material.

Luminescence is the ability of a material by virtue of which it remits visible light on absorption of energy.

Photoconductivity refers to the phenomenon of increase in conductivity of a semiconductor on account of excess carriers arisen from optical luminescence.

17.2 Optical Phenomena

When the light falls on a material, one or more of the following phenomena may occur:

1. Reflection
2. Refraction
3. Transmission
4. Scattering
5. Absorption
6. Luminescence

All these phenomena can be understood in terms of the interaction of incident light (electromagnetic radiation) with electrons of the material.

Spectrum of electromagnetic radiations is displayed in Fig. 17.1 which is helpful in designing of optical/optoelectronic devices.

17.2.1 Interaction of Light with Solids

When a beam of electromagnetic (EM) radiation falls on a solid medium, the light interacts in different ways as given below.

1. Some light is reflected from the surface of the medium
2. Some light is absorbed within the medium, and
3. Some light is transmitted through the medium.

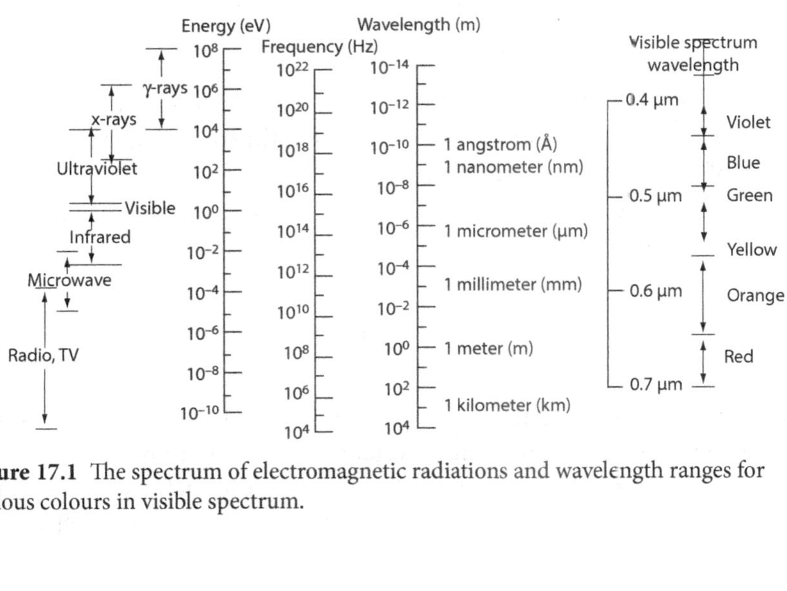

Figure 17.1 The spectrum of electromagnetic radiations and wavelength ranges for various colours in visible spectrum.

If I_R, I_A and I_T are the component intensities of reflected, absorbed and transmitted EM beam of total intensity I_0 incident on the surface of the solid medium, then

$$I_0 = I_R + I_A + I_T \tag{17.1}$$

or

$$\frac{I_R}{I_0} + \frac{I_A}{I_0} + \frac{I_T}{I_0} = 1$$

where $I_R / I_0 = \alpha_R$, $I_A / I_0 = \alpha_A$, $I_T / I_0 = \alpha_T$ are known as *reflectivity, absorptivity* and *transmittivity.*

17.2.2 Types of Optical Materials

Based on the reflection, absorption and transmission behaviour of materials, the optical materials are classified into following main categories.

1. **Transparent materials** which are capable of transmitting the light with relatively low reflection and absorption. They are such materials through which one can see across. Some examples of this category of materials are:
   - Optical crown glass
   - Optical flint glass
2. **Translucent materials** which are capable of transmitting the light partially, and having more scattering effect within the solid. Examples of this category of materials are:
   - Thin plastic films
   - Diamond
3. **Opaque materials** which do not allow light to be transmitted through the solid. One cannot see through it. Examples of such materials are the following.
   - Metals
   - Asbestos

17.3 Reflection

Most of the absorbed radiation by the metal surface is remitted in the form of visible light of nearly the same wavelength. This phenomenon is known as *reflection of light.* The light rays reflected from the metal surface exhibits a bright silvery appearance. The reflectivity $\alpha_R = I_R/I_0$, for most metals lies between 0.90 to 0.95. The reflectivity and refractive index n are interrelated as follows.

i. If the incident light is normal (or perpendicular) to the interface,

$$a_R = \left(\frac{n_2 - n_1}{n_2 + n_1}\right)^2 \tag{17.2a}$$

ii. and if the incident light is not normal to the interface then,

$$a_R = \left(\frac{n_2 - 1}{n_3 + 1}\right)^2 \tag{17.2b}$$

where n_1 and n_2 are refractive indices of the two media, n_3 is refractive index when light transmits from air to solid.

17.4 Refraction

Not all the radiation that is incident on a material gets scattered out of the material. The newly generated radiation due to the oscillating dipoles, interferes with the incident radiation and propagates into the material with a phase retardation. The speed of the wave that propagates inside the material is lower than that of the incident radiation. The speed depends on the number of electrons that are encountered per unit length and their polarizability (because the oscillating electric field of radiation generates polarization in the material). The speed is said to depend on the *optical density* of the material. At optical frequencies, the polarization that is effective, is only the electronic polarization.

17.4.1 Refractive Index

A characteristic parameter called the *refractive index* is a measure of change in speed of the wave. Refractive index n is defined as the ratio of the speed of electromagnetic wave in vacuum (or air) v_{vacuum}, to its speed in material v_{material}. Thus

$$n = \frac{v_{\text{vacuum}}}{v_{\text{material}}} \tag{17.3a}$$

The electromagnetic radiation of wavelength λ and frequency ν pass through the vacuum at same velocity as that of the light c ($c = 3 \times 10^8$ m/sec), hence

$$c = \lambda v \tag{17.3b}$$

The refractive index n defined above is the absolute refractive index because it is defined with respect to vacuum. If the radiation travels from medium 1 of absolute refractive index n_1 into another medium 2 of absolute refractive index n_2, then

$$\frac{n_2}{n_1} = \frac{v_1}{v_2} = n_{21} \tag{17.4a}$$

where n_{21} is called the *relative refractive index* (of medium 2 with respect to medium 1). n_{21} will be greater than one ($n_{21} > 1$) if medium 2 is optically denser than medium 1, and it will be less than one ($n_{21} < 1$) if medium 1 is optically denser than medium 2.

Alternatively, when the light passes from one material to another, the ray is refracted according to Snell's law. This is shown in Fig. 17.2 and expressed as follows.

$$\frac{n_1}{n_2} = \frac{v_2}{v_1} = \frac{\sin \phi_2}{\sin \phi_1} = n_{12} \tag{17.4b}$$

Values of refractive index of some materials is displayed in Table 17.1.

The refracted ray as it passes through the material, may get partially *absorbed* by the material and the rest may get *transmitted*.

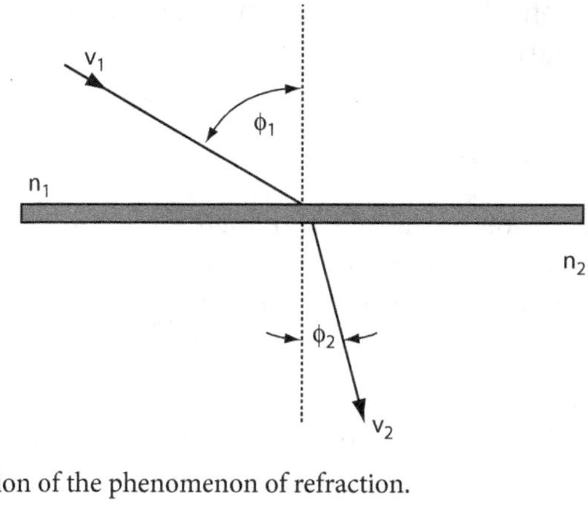

Figure 17.2 Depiction of the phenomenon of refraction.

Table 17.1 Refractive Index of Some Materials

Material	Refractive index n	Material	Refractive index n
Air	1.00028	LDPE	1.51
Water	1.333	HDPE	1.54
PTFE	1.4	Quartz	1.544
Polypropylene	1.47	Polystyrene	1.60
PMMA	1.49	Calcite	1.658

LDPE = Low density polyethylene, HDPE = High density polyethylene

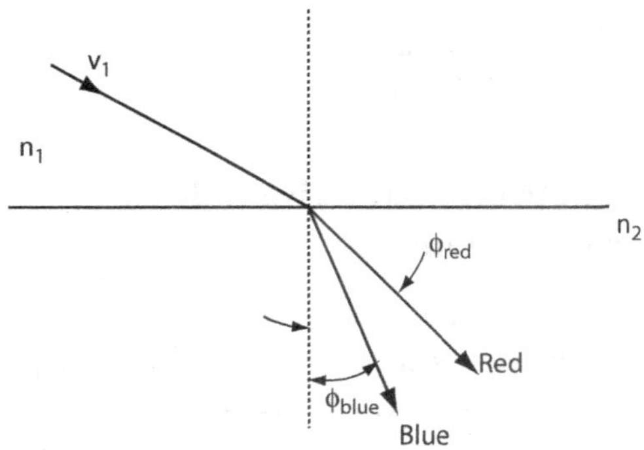

Figure 17.3 Depiction of the meaning of double refraction (i.e. birefringence).

17.4.2 Bi-Refringence

Definition. Birefringence is a phenomenon in which the material displays double refraction. It so happens when the material/fibre behaves as a birefringent medium due to difference in effective refractive indices and hence phase velocities. Birefringence also occurs due to different crystal orientations and anisotropy of fibre cross-section.

In materials having non-cubic structures, the electronic polarization changes with the direction of light vibration. It means that the refractive index is of anisotropic nature. For example in calcite ($CaCO_3$), the electronic polarization is 24% greater for the light vibrating in a plane perpendicular to vertical axis, *parallel* to (0001) plane than the light vibrating in a plane parallel to vertical axis i.e *perpendicular* to (0001) plane. This produces two refractive indices valued as 1.658 and 1.486. This difference appears as double refraction (Fig. 17.3) and is called *birefringence.* Examples of some birefringence materials and their refractive indices are displayed in Table 17.2.

Table 17.2 Double Refraction (i.e. Birefringence) Causes Dual Refractive Indices

Material	Refractive indices		Birefringence $(\mu_1 - \mu_2)$
	μ_1	μ_2	
Fused silica	1.458	1.457	0.001
Optical (crown) glass	1.517	1.514	0.003
Optical (heavy flint) glass	1.650	1.644	0.006
Diamond	2.438	2.426	0.012

μ_1 is at lower wavelength $\lambda = 589.3$ nm and μ_2 is at higher wavelength $\lambda = 656.3$ nm

Dispersion. The spread in refractive index between the red and blue as shown in Fig. 17.3 is called *dispersion.*

17.4.3 Relation between Refractive Index and Dielectric Constant

The refractive index of a material is related to its polarization. Since only electronic polarization occurs in response to light frequencies of $\approx 10^{15}$ Hz, therefore the relationship exists between n and electronic dielectric constant ε_{re} only. It is

$$n = \sqrt{\varepsilon_{re}} \tag{17.5}$$

The value of ε_{re} depends on the number of electrons encountered per unit length and their polarizability. Thus a denser material has larger n. For example, the quartz of specific gravity $\rho = 2.65$ has the value of $n = 1.544$ as compared to silica of specific gravity $\rho = 2.2$ which has the value of $n = 1.458$, although both are silica.

17.5 Transmittivity

The part of the incident radiation that is not reflected, will get partly transmitted and partly absorbed. The *transmittivity* is defined as the ratio of the intensity of light transmitted through the material to the intensity of light that is incident upon the material. Thus

$$a_T = \frac{I_T}{I_0} \tag{17.6a}$$

If there is no absorption, then

$$\alpha_R + \alpha_T = 1 \tag{17.6b}$$

The optical requirement of transmission without distortion is very essential in applications such as automobile windows and windows for architectural values/sense. Distortion-free transmission desires a flat and parallel surface without internal flaws.

17.6 Scattering

When the electromagnetic radiation is incident on a material, the oscillating electric field of the incident radiation forces the electrons in the material to oscillate at the frequency of the incident radiation. An oscillating charge acts as a source of electromagnetic radiation. As the electrons oscillate, they produce radiation of same frequency as that of the incident radiation. The radiation emanating from the oscillating electrons, which travels in all directions, is called *scattered radiation.* Almost all objects scatter radiation and this makes the object visible to us.

17.7 Optical Absorption

By optical absorption we mean the absorption of electromagnetic radiations of light by the materials. Whether the optical absorption will be less or more, depends on the property of material. An important use of this phenomenon is made in

- measuring the energy of band gap.

Measurement of band gap energy of semiconductors is often required for the analysis and proper design of semiconductor devices. It is generally accomplished by a technique involving absorption of incident photons by the material.

All incident photons on a semiconductor are not absorbed. While some of them are absorbed, the rest are transmitted. The absorption or transmission depends on the following condition.

1. Photons are *absorbed* if $E_{\text{Photon}} > E_g$, and
2. Photons are *transmitted* if $E_{\text{Photon}} < E_g$.

Here, $E_{photon} = hf$ is the energy of photon where h is Planck's constant and f is frequency of radiation. The absorption at energies greater than band gap energy is due to the fact stated below.

> "While the valence band contains several electrons, the conduction band has many empty states. Therefore, the electrons can be easily excited from valence band and hence the probability of photon absorption is high. However, a photon having energy less than E_g cannot excite electrons from the valence band to the conduction band. That is why the absorption of photons is negligible in pure (intrinsic) semiconductors like Si and Ge".

17.7.1 Capability of a Material to Absorb Light

The capability of a material to absorb light depends on the type of charges contained in it. The absorption takes place due to the interaction of electromagnetic radiation of light and the charges in the material. In general, following three types of charge distribution exist in materials.

i. Tightly bound charges such as the inner shell electrons of the lattice ions or the lattice ions themselves.
ii. Loosely bound charges such as outer shell electrons of the lattice ions.
iii. Free charges like conduction electrons in metals and semiconductors.

Tightly bound charges provide the phenomenon of optical dispersion. In such materials, the excitation of light near the frequency of resonance leads to a pronounced absorption. The presence of both : the bound electrons and ions, thus lead to absorption in the UV and infrared regions of the spectrum.

The loosely bound charges can be best described by band theory of electrons. The absorption in such charges can result from interband excitation. The absorption will occur only if the photon energy is more than the band gap energy.

The free electron materials are very much polarizable and interact readily with an incident electromagnetic wave. They do not move independently of the lattice structure, but during their motion may interact with a photon. A loss of energy may result before the charge re-radiates the light and hence absorption occurs. In such materials, the absorption increases with the number of electrons, and the absorption coefficient also varies proportionately to the square of the wavelength.

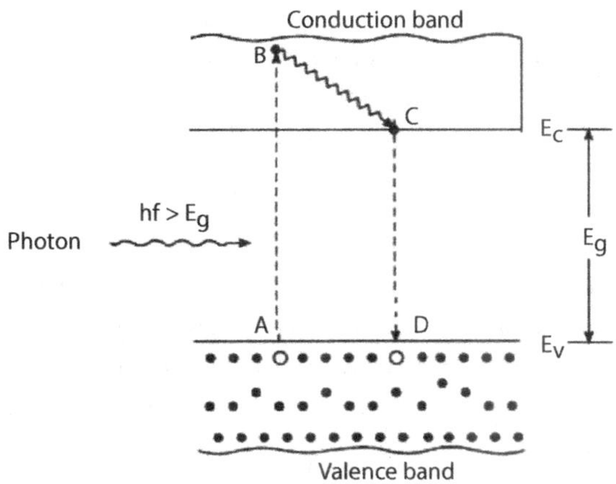

Figure 17.4 Depiction of mechanism of optical absorption.

17.7.2 Mechanism of Optical Absorption

Optical absorption of a photon is shown in Fig. 17.4. It shows that an electron excited from location *A* in valence band to location *B* in conduction band by optical absorption may have energy more than E_c.

The excited electron leaves a hole at *A*. It is common fact that almost all electrons have energy equal to or near to E_c, if the semiconductor sample is not heavily doped. Therefore in case of scattering, the excited electron loses energy to the lattice, until its velocity reaches the velocity of other conduction band electrons at thermal equilibrium. The electron and hole created by the above absorption process are referred to as 'excess carriers'. Although these carriers exist in their respective bands, but they freely contribute to the conductivity of semiconductor. The phenomenon of optical absorption of photon shown in Fig. 17.4, can be illustrated as follows.

i. Step *AB* shows creation of an electron-hole pair.
ii. Step *BC* shows giving-up of energy by excited electron to the lattice due to scattering.
iii. Step *CD* shows recombination of electron with the hole in valence band.

17.7.3 Absorption Coefficient

When a beam of photons falls on a semiconductor; a part of it is reflected, some fraction is transmitted and the rest is absorbed. Therefore for a given wavelength; if the absorbed, reflected and transmitted fractions are expressed by α_A, α_R, and α_T respectively; then we have

$$a_A + a_R + a_T = 1 \tag{17.7}$$

The fraction of incident light absorbed, depends on the wavelength of photon and the sample thickness on which it falls.

In order to determine the absorptivity, we assume a beam of photon of intensity I_0 (photons/cm^2- s) and wavelength λ that falls on a sample of thickness t. This intensity decreases with increase in distance x along its thickness as shown in Fig. 17.5. Since the degradation of intensity is $\frac{dI}{dx}$ proportional to remaining intensity at distance x, therefore we can write

$$-\frac{dI}{dx} \propto I \tag{17.8a}$$

or

$$-\frac{dI}{dx} = aI \tag{17.8b}$$

where α is proportionality constant and is known as *absorption coefficient.*

Solution of this first order differential equation may be obtained as

$$I = I_0 e^{-ax} \tag{17.9}$$

It depicts an exponential decrease as shown in Fig. 17.5. Its unit is cm^{-1}. This solution can be used to calculate the intensity of transmitted light I_t through the sample of thickness t. It is given by

$$I_t = I_0 e^{-ax} \tag{17.10}$$

17.7.4 Factors Affecting the Absorption Coefficient

The absorption coefficient mainly depends on the following two factors.

1. Photon wavelength λ, and
2. Type of semiconducting material.

The effect of wavelength on optical absorption is shown in Fig. 17.6. It shows too little absorption for longer wavelengths (*i.e.* smaller *hf*) and substantial absorption for smaller wavelengths (*i.e.* larger *hf*). Since the photon energy E_{photon} and wavelength are related as $E_{photon} = hc/\lambda$, where c is

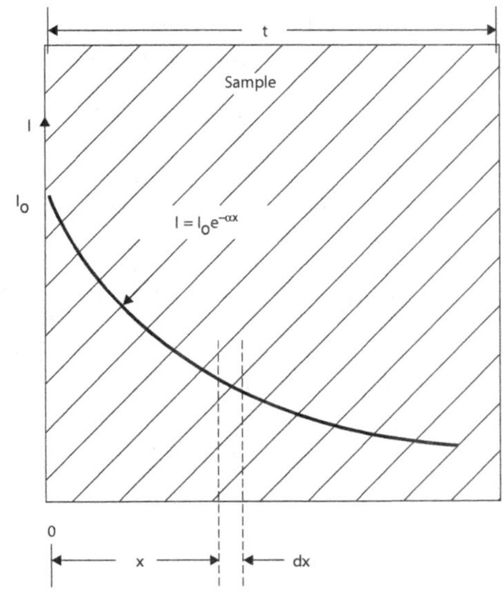

Figure 17.5 A decrease in intensity of photon along the thickness of specimen is due to absorption.

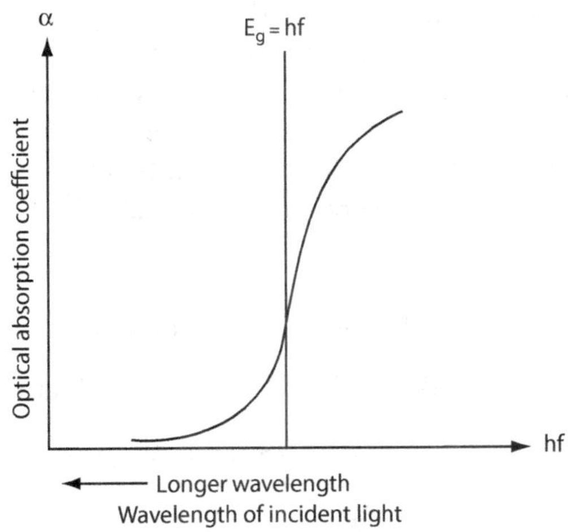

Figure 17.6 Dependence of optical absorption coefficient α for a semiconductor as a function of wavelength of incident light.

velocity of the light; hence on substituting $h = 6.626 \times 10^{-34}$ Js and $c = 2.998 \times 10^8$ m/s, we get

$$E_{\text{photon}} = \frac{hc}{\lambda} = \frac{1.24}{\lambda} \text{ electron-volt} \tag{17.11}$$

in which λ is in μm.

Example 17.1 Calculate the energy of a photon of sodium light having wavelength of 5.893×10^{-7} m (a) in joule, and (b) in electron-volt. Take $h = 6.62 \times 10^{-34}$ Js and velocity of light $c = 3 \times 10^8$ m/s.

Solution. Given are: $\lambda = 5.893 \times 10^{-7}$ m

(a) $\because$ $$E_{\text{photon}} = hf = \frac{hc}{\lambda}$$

$\therefore$ $$= \frac{(6.62\times10^{-34})\times(3\times10^{8})}{5.893\times10^{-7}}\text{Joule}$$

$$= 3.375 \times 10^{-19} \text{ J}$$

(b) $\because$ $$1 \text{ eV} = 1.6 \times 10^{-19} \text{ J}$$

$\therefore$ $$E_{\text{photon}} = \frac{3.375\times10^{-19}}{1.6\times10^{-19}}$$

$$= 2.11 \text{ eV}$$

Example 17.2 A 0.45 μm thick sample of GaAs is illuminated by a monochromatic light of photon energy 3 eV. The incident power on the sample is 15 mW. If the absorption coefficient is 50 000/cm, determine (*a*) the total energy absorbed by the sample per second, (*b*) the rate of excess thermal energy given up by the electrons to the lattice before recombination, and (*c*) the number of photons given off from recombination. Assume perfect quantum efficiency. The outgoing radiation is of 2.35 eV.

Solution. Given are:

$$t = 0.45 \text{ μm} = 0.45 \times 10^{-6} \text{ m} = 0.45 \times 10^{-4} \text{ cm}$$

$$E_{\text{photon}} = 3 \text{ eV}, \alpha = 50\ 000/\text{cm} = 5 \times 10^4 \text{ cm}^{-1},$$

$$I_0 = 15 \text{ mW} = 15 \times 10^{-3} \text{ W}$$

(a) Using Eqn. 17.10, we write

$$I_t = I_0 e^{-at} = 15\times10^{-3} e^{(-5\times10^4\times0.45\times10^{-4})}$$

$$= 0.015 \times 0.105 = 1.575 \times 10^{-3} \text{ W}$$

$$= 1.575 \text{ mW}$$

Thus the total power absorbed is

$$I_{\text{absorbed}} = I_0 - I_t = 15 - 1.575$$

$$= 13.425 \text{ mW} = 13.425 \times 10^{-3} \text{ J/s}$$

(*b*) $$\begin{pmatrix}\text{The rate of excess thermal} \\ \text{energy given by the} \\ \text{electrons to the lattice} \\ \text{before recombination}\end{pmatrix} = \begin{pmatrix}\text{The amount of} \\ \text{energy } E \text{ converted} \\ \text{to heat per second}\end{pmatrix}$$

As the fraction of each photon energy unit which is converted to heat is

$$\frac{hf_1 - hf_2}{hf_1} = \frac{3 - 2.35}{3} = 0.216$$

∴ Total amount of energy converted to heat per second is

$$= 0.216 \times (13.425 \times 10^{-3})$$

$$= 2.899 \times 10^{-3}\ \text{J/s}$$

(c) As the quantum efficiency is perfect, therefore there is a single (*i.e.* one) emitted photon for each absorbed photon. Hence the number of photons given off from recombination per second is

$$n_{\text{photon}} = \frac{I_{\text{absorbed}}}{e \times E_{\text{photon}}} = \frac{13.425 \times 10^{-3}\ \text{J/s}}{1.6 \times 10^{-19} \dfrac{\text{J}}{\text{eV}} \times \dfrac{3\text{eV}}{\text{photon}}}$$

$$= 8.39 \times 10^{6}\ \text{photons / s}$$

17.7.5 The Absorption Spectra of Materials

Study of optical absorption of materials gives information about the band gaps and the energies corresponding to the inter-band and intra-band transitions. Obviously, due to the quasi-continuous energy levels within a band, the optical spectra are not sharp. The absorption bands have widths spread over a range of wavelengths.

17.8 Optical Properties of Non-Metals

Non-metallic materials may be transparent, translucent, or opaque. Therefore, they exhibit different optical properties such as reflection, refraction, absorption and transmission. The phenomenon of refraction is more dominant in them. Values of refractive indices for some polymers, ceramics and other non-metals are already shown in Tables 17.1 and 17.2.

The non-metals which are transparent, are generally coloured due to light absorption and re- emission in the visible region by them. Absorption of light occurs due to

i. electronic polarization,
ii. excitation of electrons from filled valence band to empty state within conduction band, and
iii. wide band gaps in dielectric materials.

The non-metallic transparent materials transmit light due to net energy formed by absorption and reflection processes.

17.9 Optical Properties of Metals

Reasons of the metals being opaque and lustrous in appearance. In metals, the valence band is partially filled and so there are large number of quasi continuous vacant energy levels available within the valence band. When the light is incident upon the metals, the valence electrons absorb all frequencies of visible light and get excited to vacant states within the valence band (intra-band transitions). This accounts for the opacity of metals. The total absorption of light by the metal surface is within a very thin outer layer of less than 0.1 μm. The excited electrons revert back to lower energy states thereby causing emission of radiation from the surface of the metal in the form of visible light of the same wavelength. This emitted light which appears as the reflected light is the cause of the lustrous appearance of metals.

17.9.1 Reasons of Copper Appearing Reddish-orange, Silver and Aluminium White, and Gold Yellow

In addition to intra-band transitions, inter-band transitions also occur in metals from the valence band to the conduction band.

- In copper, inter-band transitions occur for energies greater than 2.2 eV. i.e. the photons of energy greater than 2.2 eV are strongly absorbed. This energy corresponds to wavelength below 5625 Å. This means that the radiation in the blue-violet range is absorbed. This accounts for the reddish-orange colour of copper.

- In silver and aluminum, there is no absorption in the full range of visible radiation. So, the re-emission occurs over the entire wavelength range of the visible spectrum which accounts for the white colour of these metals.
- Gold appears yellow because there is absorption in green portion and reflection in yellow and red region.

17.10 Optical Properties of Semiconductors

Intrinsic semiconductors at low temperatures have a completely filled valence band and an empty conduction band. Therefore, no intra-band transitions can occur in semiconductors. Radiation of low frequencies, i.e. infrared radiation are not absorbed and so semiconductors are transparent to infra-red radiation. The energy gaps in semiconductors are in the range of 0.5 - 3 eV. So inter-band absorption occurs for radiation in this range which corresponds to near infra-red and visible range, Fig. 17.7. This accounts for the opacity of semiconductors.

17.10.1 Visibility Range of Light Spectrum

The effect of different types of semiconductor on visibility range of light is shown in Fig. 17.7. It depicts a spectrum of light for different band gap energies E_g. The salient observations of it are given as follows.

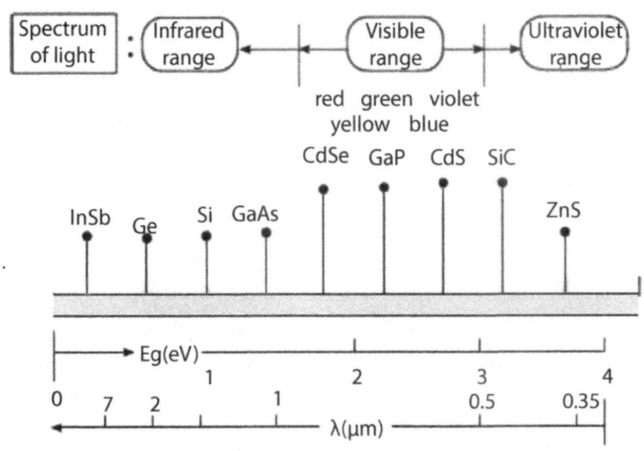

Figure 17.7 Some semiconductors under different visibility range of light spectrum. Values of E_g and λ are not to the scale.

i. Semiconductors lying in infrared region are
   - InSb • Ge • Si • GaAs
ii. Semiconductors lying in visible region are
   - CdSe (with red visibility)
   - GaP (with yellow visibility)
   - CdS (with green visibility)
   - SiC (with blue visibility)
iii. Semiconductors lying in ultraviolet region are
   - ZnS

17.11 Optical Properties of Insulators

Insulators have completely filled valence band and therefore as in semiconductors, no intra-band transitions can occur. The energy gap in insulators are greater than 5 eV and so no inter-band transition can occur in the visible range of radiation. Absorption occurs only for the ultraviolet radiation. Insulators are transparent from infra-red up to the ultra-violet radiation. Examples of transparent insulators are

- Perfect diamond crystal
- Fused quartz
- Window glass

Not all insulators or dielectrics are transparent. For example,

- Enamels,
- Porcelains,
- Opal glass etc.

are opaque. This is because the incident radiation gets scattered in all directions by the small particles present in these materials. Due to this, there cannot be perfect transmission. Part of the radiation is diffusely transmitted and part is diffusely reflected. This makes the materials appear opaque. If the particle size is of the order of the wavelength of visible radiation, there will be maximum scattering. For some applications, such particles are deliberately introduced in dielectrics to make them opaque.

17.11.1 Optical Absorption in Ionic Crystals

Ionic crystals are insulators. The energy gap in these crystals are in the range of 5-8 eV. The electrons cannot absorb photons in the visible radiation and get excited to the conduction band. So the complete range of visible radiation is transmitted by ionic crystals and they are transparent.

The absorption properties of ionic crystals change drastically if point defects such as lattice vacancy or Schottky defects are present in them. The F centres and other electronic centres are called ***colour centres*** because their presence results in imparting the characteristic colour of the crystals.

Another method by which the optical absorption in ionic crystals can be changed, is by adding impurities. The addition of specific impurities cause energy levels to be introduced in the energy gap, causing absorption in the visible range. For example, pure Al_2O_3.

17.12 Luminescence

Luminescence is the property by virtue of which a material emits the light. In semiconducting materials, the light is emitted under certain conditions such as given below.

i. When electron-hole pairs (EHP) are generated, or
ii. When the carriers fall to their equilibrium state after being excited to higher impurity levels.

Several semiconductors exhibit the property of luminescence, specially the compound semiconductors with direct band gaps.

17.12.1 Different Types of Luminescence

Depending upon the excitation mechanism involved, the luminescence can be classified as follows.

1. Photo-luminescence in which the carriers are excited by photon absorption.
2. Cathode-luminescence in which the carriers are created by high-energy bombardment of electrons on the material.
3. Electro-luminescence in which the carriers are excited by introducing current in the material.
4. Chemi-luminescence in which the luminescence is created by chemical action.

Amongst above types, the first three types of excitation are most important mechanisms for applications in vivid electronic devices.

17.12.2 Photo-Luminescence

It is the phenomenon of emission of light from a semiconductor on account of recombination of excited electron-hole pair (EHP). The recombination may occur directly or indirectly. If the excitation of EHP occurs under steady state condition, the rate of recombination remains the same as the rate of generation. In this case, one photon is emitted for each photon absorbed. Recombination in semiconductors takes place at varying rates: fast and slow. Accordingly, the photo-luminescence may be of following two types.

1. **Fluorescence.** It is a fast process property of material in which the emission of photon stops in about 10^{-8} s after the excitation is removed. The materials belonging to this category are the following.
   - Glass surface coated with tungstates or silicates such as in fluorescent lamps.
   - Television screen coated with sulphides, oxides, tungstates etc.
2. **Phosphorescence.** It is a slow process property of material in which the emission of photon continues for a longer duration, lasting for seconds and minutes after removal of excitation. Materials falling in this category are termed as **phosphors**, and their examples are the following.
   - ZnS coated with Cu as impurity
   - CdS coated with Ag as impurity
   - KCl coated with Tl as impurity
   - NaI coated with Tl as impurity.

Example of photoluminescence. The 'fluorescent lamp' is a common example of photoluminescence. It comprises of a glass tube filled with a gas, which is generally the mixture of *mercury vapours* and *argon*. The inside of the tube has a fluorescent coating. When an electric discharge is induced between electrodes of the tube, the atoms of the gas are excited and emit photons. This emission is largely in the visible and U/V regions of the spectrum. The light thus produced is absorbed by the luminescent coating and hence the visible photons are emitted.

17.12.3 Electro-Luminescence

We have studied earlier that the electro-luminescenes effect can be created by introducing the electric current into a semiconductor. This electrical current can be used in different ways to generate the photon emission from semiconductors. One such way is 'injection'. Therefore, the name of the process is *injection electro-luminescence* whose use is made in light-emitting diodes (LEDs). In them the minority carriers are injected by electric current, into the regions of a crystal where they can recombine with majority carriers. It results in emission of recombination radiation.

The effect of electro-luminiscence can be seen in devices incorporating the phosphor powder (such as of ZnS) in a plastic binder. The phosphor gives-off the light when an alternating current (a.c.) field is applied on it. Such device is known as 'electro-luminsecence cell', which can be used as *lighting panel.*

Destriau effect. The above explanation stating that the emission of photons in certain phosphors occurs when they are subjected to alternating electric field, was observed for the first time by Destriau. Hence this phenomenon is known as 'Destriau effect'.

17.13 Opto-Electronic Devices

Optoelectronic devices are those devices whose functions involve the interaction of photons with semiconductors. These devices are different from the electronic devices. They provide optical sources and optical detectors for different devices. In their working, some of the optoelectronic devices detect photons, while some others emit photons. The optoelectronic devices may broadly be grouped into following two categories.

1. Devices that convert optical energy into electrical energy, such as
   - Photodiodes
   - Solar cells
2. Devices that emit photons, such as
   - incoherent sources like light-emitting diodes (LEDs)
   - coherent sources such as lasers.

Details of optical materials used in these devices are discussed in subsequent sections. Solar cells, LEDs, and lasers are already discussed earlier.

17.14 Photoconductivity

On creation of excess electrons and holes in a semiconductor, the conductivity of material increases. This has been described by the equation $\sigma = n_e e_e \mu_e + n_h e_h \mu_h$ in chapter 6 also. Although the excess electrons and holes can be created by different means; but if these arise from optical luminescence, the increase in conductivity is called as 'photoconductivity'. This is an important phenomenon, which is utilised in the analysis and operation of semiconductor devices. Some of the practical applications are found in following devices.

- Diodes
- Transistors
- Photocells
- Lasers

17.14.1 Applications of Photoconductive Devices

There are numerous applications of photoconductive devices which change their resistance when exposed to light. Main among them are the following.

1. Light detectors such as used in
   i. automatic controlling of night lights at homes, streets, industries etc. These lights turn-on at dusk and turn-off at dawn, automatically.
   ii. exposure meters of cameras to measure the illumination level.
2. Moving object counters such as used in
   i. detecting a submarine, ship etc.
   ii. detecting an aircraft
   iii. low speed counting
3. Burglar alarms
4. Optical signalling systems
5. Television cameras
6. Photographic light meters
7. Infrared detectors
8. Automatic door opener
9. Fire alarms
10. Cinematography, etc.

17.14.2 Photoconductive Materials

A number of photoconductive materials are used in various semiconductor devices. Main among them are the following.

S.no.	Example	Symbol	E_g (eV)	Main application
1.	Germanium	Ge	0.67	In infrared portion of spectrum
2.	Telurium	Te	0.34	—
3.	Cadminum sulphide	CdS	2.42	Photoconductor in green light (visible range).
4.	Cadminum selluride	CdSe	1.74	Photoconductor in red light
5.	Cadmium telluride	CdTe	1.45	Photoconductor in infrared region
6.	Lead sulphide	PbS	≈0.38	Photoconductors
7.	Lead selluride	PbSe	≈0.32	Photoconductors
8.	Lead telluride	PbTe	≈0.33	Photoconductors
9.	Zinc sulphide	ZnS	3.82	Photo-luminescence
10.	Zinc oxide	ZnO	3.30	—
11.	Gallium selluride	GaSe	2.09	—
12.	Gallium arsenide	GaAs	1.37	Diode, laser
13.	Gallium antimonide	GaSb	0.82	Laser
14.	Indium antimonide	InSb	0.18	In infrared portion of spectrum
15.	Indium phosphide	InP	1.37	Photon detection
16.	Alloy	$CdS_x Se_{1-x}$	1.80–2.50	—
17.	Alloy	$GaAs_x P_{1-x}$	1.41–1.95	—

A semiconductor suitable for certain application, may not be suitable for other application. Similarly, all semiconductors are not suitable for making the same device. Their selection is based on different factors. These are described below in subsequent sections.

17.14.3 Factors Affecting the Selection of Semiconductor

Selection of an appropriate photoconducting material for a specific application depends on the following factors.

- Sensitive wavelength range
- Time response, and
- Optical sensitivity of the material

In this respect, the undermentioned characteristics form the basis for selection of semiconductors.

1. Semiconductors are most sensitive to those photons whose energies are equal to or a bit more than the band gap E_g. Therefore,
   i. less energetic photons are not absorbed.
   ii. highly energetic photons ($hf >> E_g$) are absorbed at the surface, but they contribute only a little towards bulk conductivity.
2. Some semiconductors are sensitive to photons of those energies which are lesser than that of the band gap energy (*i.e.* $hf < E_g$). It is because they respond to carrier excitations from those impurity levels which lie within the band gap.
3. Semiconductors used to make photoconductive cells have limited *time response* due to trapping of carriers, recombination time, and the time needed for carriers to drift through the device in an electric field. However, with proper choice of materials and device geometry, the time response properties can be improved.
4. Semiconductors meant to be used as photoconductors should have high *optical sensitivity*. For maximum photoconductive response, the material should have high mobilities and long lifetime. That is why, InSb is a good choice for making photoconductive devices. Its electron mobility is about 10^5 cm^2/Vs, and it is used as sensitive infrared detector.

17.15 Photoconductive Cell

Its operation is based on the principle that the resistivity of semiconductor materials like Se, CdS, PbS, and thalmium-sulphide (TlS) etc. is decreased when irradiated. In means that, such materials have 'dark' high resistance and irradiated low resistance.

Construction. The simplest form of such a cell using selenium is shown in Fig. 17.8*a-b*. It consists of a semiconductor material with two electrodes attached to it. The resistance of the cell, when unilluminated (dark) is always high so that the current through the circuit shown in Fig. 17.8*a* is low. When the cell is illuminated, the cell resistance is decreased and the circuit current becomes large. The shape of the semiconductor material is so made, as to obtain a large ratio of 'dark to light' resistance.

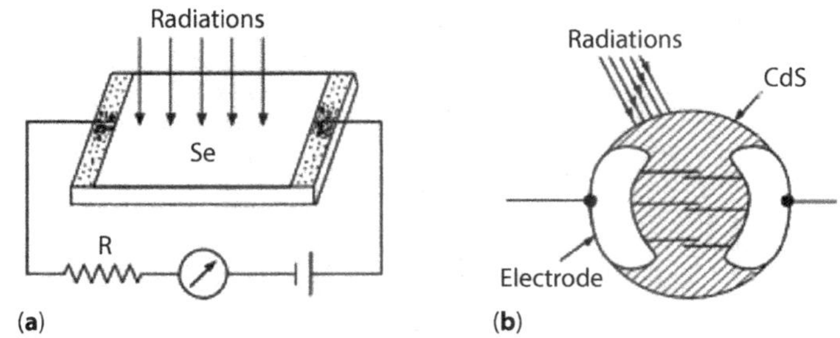

Figure 17.8 Photoconductive cell showing (a) Se in circuit, and (b) CdS.

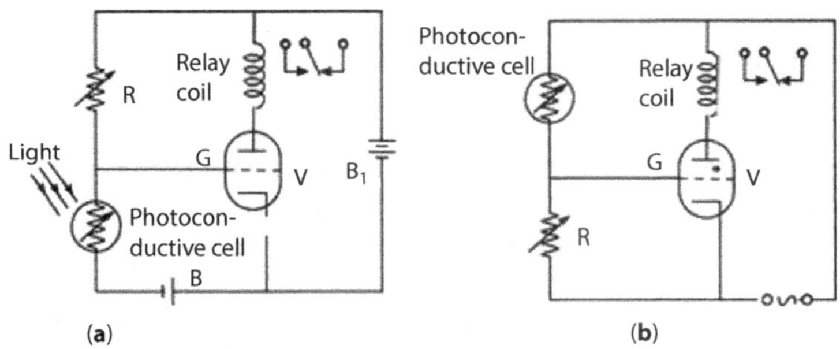

Figure 17.9 A fire alarm circuit incorporating a photoconductive cell with a (a) battery, and (b) an a.c. supply.

The commonly-used CdS cell is shown in Fig. 17.8*b*. It has a very high dark/light ratio and gives maximum response at about 5000 atomic unit. In order to increase the contact area with the sensitive material, the two electrodes are extended in an inter-digital pattern.

Uses. Photoconductive cells employing TIS and PbS are used for detecting the ships and aircrafts with the help of radiations given out by their exhausts or funnels, and for telephony by modulated infrared light.

A circuit employing such a cell which may be used for low-speed counting or as a fire or burglar alarm is shown in Fig. 17.9*a*. Figure 17.9*b* shows a similar circuit using a thyratron and an a.c. supply. In each case when sufficient light falls on the cell, the valve conducts and energises the relay. Resistor *R* can be changed to adjust the light level for relay operation.

17.15.1 Photo-multiplier Tube

In a phototube, the current is amplified by electron multiplication through secondary emission. Cross-section of one such multiplier tube is shown in Fig. 17.10. It consists of 9 electrodes (called dynodes) which are maintained

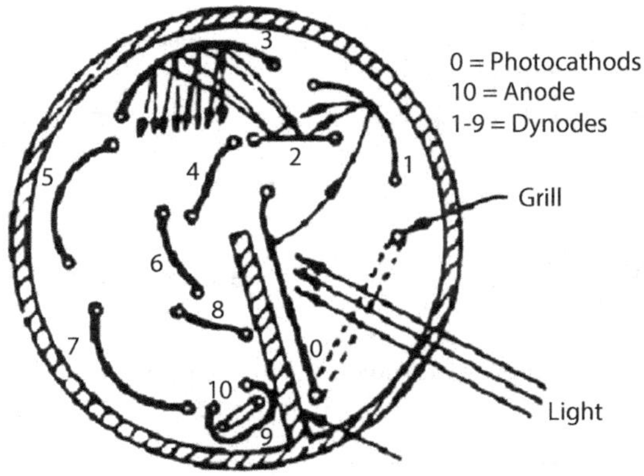

Figure 17.10 Cross-section of a photo-multiplier tube.

at increasing potentials in sequence from the photocathode 0 to the anode marked 10 in the diagram. When light falls on the cathode, photoelectrons are emitted which are accelerated towards dynode no. 1, since it is at a higher potential. On colliding with this dynode, they liberate secondary electrons which are attracted by dynode no. 2 since it is at a relatively still higher potential. These electrons release further secondary electrons from it which are attracted by dynode no. 3. This process of secondary emission from different dynodes continues till the electrons are finally collected by the anode. It is obvious that at each dynode, the number of secondary electrons keeps on multiplying.

17.16 Solved Examples

Example 17.3 Photon of energy 1.5×10^{-19} J is incident on a photodiode whose quantum efficiency is 60%. Determine (*a*) the wavelength at which the photodiode is operating, and (*b*) incident optical power to obtain a photon current of 3 μA.

Solution. Given are: $\eta_{quantum} = 60\% = 0.6$, $I_{opc} = 3\ \mu A = 3 \times 10^{-6}$ A,

$$E = 1.5 \times 10^{-19}\ J$$

a. $\because \quad \lambda = \frac{hc}{E}$

$$\therefore \quad \lambda = \frac{6.626 \times 10^{-34} \times 3 \times 10^{8}}{1.5 \times 10^{-19}} = 1.32\ \mu m$$

b. Using Eqn. (*i*) given below, where P_{iop} = incident optical power, I_{opc} = output photocurrent, and R = responsivity of photodiode,

$$\because \quad R = \eta_{\text{quantum}} . e / hf = \frac{0.6 \times (1.6 \times 10^{-19})}{1.5 \times 10^{-19}}$$

$$= 0.64 \text{ AW}^{-1}$$

and also
$$P_{iop} = \frac{I_{opc}}{R} \quad \text{(i)}$$

$$= \frac{3 \times 10^{-6}}{0.64} = 4.67 \ \mu\text{W}$$

Example 17.4 Energy band gap in a Ge crystal is 0.75 eV. Determine the wavelength at which the Ge starts to absorb light.

Solution. Given are: $E_g = 0.75 \text{ eV} = 0.75 \times 1.6 \times 10^{-19}$ J

$$\because \quad E_g = hf = \frac{hc}{\lambda}$$

$$\therefore \quad \lambda = \frac{hc}{E_g} = \frac{6.62 \times 10^{-34} \times 3 \times 10^{8}}{0.75 \times 1.6 \times 10^{-19}} = 16560 \text{ Å}$$

Example 17.5 What is the utility of phosphor in the fabrication of a colour T.V. screen?

Solution. The colour of the light emitted by phosphor depends mainly on the impurities present in it. It is so because many radiative transitions involve impurity levels within the band gap of certain materials such as ZnS. Hence, selection of proper phosphors is useful in fabrication of colour T.V. screen.

Example 17.6 Find the maximum energy of the emitted photoelectrons when the light of frequency 1.5×10^9 MHz falls on the surface of a material whose threshold frequency is 1.2×10^9 MHz.

Solution. As
$$E = hf \text{ where } f = 1.5 \times 10^{9} \text{ MHz} = 1.5 \times 10^{15}\text{Hz}$$

$$\therefore \quad E_{\max} = h(f - f_0)\text{joule}$$

$$= 6.62 \times 10^{-34} \times \left[\left(1.5 \times 10^{15}\right) - \left(1.2 \times 10^{15}\right)\right]$$

$$= 6.62 \times 10^{-34} \times (1.5 - 1.2) \times 10^{15} \text{joule}$$

$$= \frac{6.62 \times 10^{-34} \times 0.3 \times 10^{15}}{1.602 \times 10^{-19}} \text{eV} = 1.24 \text{ eV}$$

Example 17.7 Which material is most suitable for producing blue LEDs? Explain why? Write its applications also.

Solution. Blue LEDs are short wavelength emitters. GaN is most suitable material for it, as it gives a very high efficiency. Blue LEDs are used in outdoor displays and TV screens. They are combined with LEDs of other colours also to form very intense white light sources. Luminous efficiency of such light source exceeds over the conventional light bulbs.

Example 17.8 What is meant by optoelectronic pair? What is its importance?

Solution. An optoelectronic pair means a pair of a *light emitter* and a *photodiode*. It provides electrical isolation between the input and output. In an optoelectronic isolator, both these devices are mounted on a ceramic substrate and packed together to form a compact unit. This unit, while maintaining isolation, passes the information.

Example 17.9 What is meant by dark resistance of a photoconductor?

Solution. It is the resistance of a photoconducting cell when it is unilluminated. Dark resistance is always higher than the bright resistance (*i.e.* the resistance of the cell when illuminated).

Quick Revision Summary

Birefringence is a phenomenon in which the material displays double refraction. It so happens when the material/fibre behaves as a birefringent medium due to difference in effective refractive indices and hence phase velocities. Birefringence also occurs due to different crystal orientations and anisotropy of fibre cross-section.

Brightness. It is the power emitted by a source per unit area per unit solid angle.

Carrier lifetime is the duration until the excess carriers (electrons or holes) do not decay.

Diffusion is a flow process in which the carriers move from the regions of high concentration to the regions of low concentration.

Electro-optic effect. It is the behaviour of a material in which its optical isotropic nature changes to anisotropic nature on application of an electric field. This effect is seen in $LiNbO_3$, $LiTiO_3$ etc.

Excess carriers are those charge carriers which are in excess of the charges available at thermal equilibrium. The excess carriers can be created by optical excitation, electron bombardment, or other means.

Luminescence is the ability of a material by virtue of which it remits visible light on absorption of energy.

Optical absorption means absorption of electromagnetic (EM) radiations of light by the semiconducting material. Whether the optical absorption will be less or more, depends on the property of material.

Photoconductivity refers to the phenomenon of increase in conductivity of a semiconductor on account of excess carriers arisen from optical luminescence.

Photoelectric effect. It refers to a phenomenon in which the ejection of electrons from a metal surface takes place, when the metal surface is illuminated by light or any other radiation of suitable frequency (or wavelength). Several devices such as phototube, solar cell, fire alarm etc. work on this effect (principle).

Photoemissivity. It refers to the phenomenon of emission of electrons from a metal cathode, when exposed to light or any other radiations.

Photons are discrete radiation emitted from the materials in the form of small energy packets. These packets are also called as '*quanta*'. They possess energy, mass and momentum.

Recombination is a process of formation of an electron-hole pair (EHP), when an electron in conduction band makes a direct or indirect transition to an empty state (*i.e.* hole) in valence band.

Trap. It is an energy level that can capture either electrons or holes easily, but not the both.

Translucent materials are those through which the light is transmitted diffusely *i.e.* it scatters within the material. As a result of this behaviour, the objects are not clearly distinguishable when viewed through a translucent specimen.

Review Questions

1. What are the optical requirements of materials? State them in consideration of lenses and prisms, optical fibres, and diamond stone.
2. Define the following.
   a. Photon
   b. Electro-optic effect
   c. Photoelectric effect
   d. Photoemissivity
   e. Photoconductivity
   f. Dark resistance of photoconductor
3. Enlist various optical phenomena that are observed when light falls on a material? Briefly explain each of them. Explain: reflectivity, absorptivity and transmissivity.
4. Differentiate between transparent, translucent and opaque materials. Quote examples of each of them.
5. What is refraction? What is refractive index? Explain as to how the (*a*) reflectivity and refractive index are inter-related, and (*b*) refractive index and dielectric constant are inter-related.
6. Define birefringence. State the reasons of its occurrence. State the consequences of this phenomenon.
7. What do you mean by optical absorption ? State the conditions of photon absorption and discuss its mechanism.
8. What is absorption coefficient? Derive the equation to determine the intensity of transmitted light through a semiconductor sample.
9. Discuss the factors that influence the absorption coefficient of a semiconductor. Enlist semiconductors lying in different regions of spectrum of light.
10. Discuss the optical properties of non-metals, metals, semiconductors and insulators. Why does the gold appear yellow, but silver appears white?
11. Explain the phenomenon of luminescence. What are its different types? How does fluorescence differ from phosphorescence?
12. What is photo-luminescence? What are its different types? Explain the difference between fluorescence and phosphorescence.

13. Enumerate the various optoelectronic semiconductor devices and give a brief account of materials used in each of them.
14. What is photoconductivity? What are its various applications? Write the names of photoconductive materials and enlist their main applications.
15. Discuss the various factors that affect the selection of semiconducting materials for photoconductive applications.
16. Sketch and explain the construction and working of a photoconductive cell.
17. Sketch and explain the construction and working of a photomultiplier tube.

Numerical Problems

1. Calculate the energy content of ultraviolet light photons of wave length 3000 Å in (*a*) joule, and (*b*) electron-volt.
2. Determine the energy of photons of a spectral emission line whose wavelength is 210 mm.
3. Find the frequency and amount of energy associated with a quantum of electromagnetic radiation at each of the following wavelengths (*a*) 1 m, (*b*) 1 mm, (*c*) 5000 Å, and 1 Å.
4. A compound semiconductor sample 0.46 mm thick, is illuminated by a monochromatic light of photon energy = 2 eV. The power incident on the sample is 10 mW. If the absorption coefficient is 5×10^4 cm^{-1}, calculate (*a*) the total energy absorbed by the same per second, (*b*) the rate of excess thermal energy given up by the electrons to the lattice before recombination, and (*c*) the number of photons given off per second from recombination. The quantum efficiency is perfect.
5. The wavelength of light from a GaP laser is 5490A°. Compute the energy gap of it.

Objective Questions

1. Consider the following statements.
   A. Excess charge carriers can be created by electron bombardment also, instead of optical excitation.

B. Mean lifetime of electron-hole pair recombination is of the order of 10^{-8} s.
C. Drift current flows due to electric field.
D. CdS lies in infrared region of spectrum of light.

Of these, the correct answers are

a. *A* and *C* b. B and D
c. A, B and *C* d. *A*, *C* and *D*

2. Which of the following is *incorrect*?
   a. An application of photoconductivity is found in automatic door opener.
   b. ZnS coated with Ag is an example of phosphorescence material.
   c. CdSe lies in red visibility of spectrum of light.
   d. Photons are absorbed by a semiconductor, if the energy of photon is less than the band gap energy of semiconductor.

3. Consider the following statements.
   A. Birefringence occurs when the index of refraction is a function of crystal orientation.
   B. A good photo detector should possess short response time to obtain suitable bandwidth.
   C. Efficiency of Si solar cells increases at high temperatures.
   D. Output of semiconductor lasers can be modulated by controlling the junction current.

   Of these, the correct statements are

   a. *A*, *B* and *C* b. *B*, *C* and *D*
   c. *A*, *B* and *D* d. *A*, *B*, *C* and *D*

4. Which of the following pairs of devices and materials does *not* match?
   a. Solar cell : $CuInSe_2$ as absorber and CdS as collector
   b. LED : GaAs-AlGaAs
   c. Optical fibre: Vitreous high-silica glass doped with germania
   d. Semiconductor laser : Nd YAG

18

Specific Materials for Electrical, Electronics, Computers, Instruments, Robotics, and Other Applications

18.1 Recent Developments

- **Magnetic Ink.** Magnetic ink character recognition (MICR) is an electronic security system to check the authenticity of bank cheques coding, and desired information on these cheques are printed by means of a special ink made of magnetic material. When the pre-coded cheques are passed through special machines, the printed information is *magnetized* and read by them. Thus the genuineness of the cheque is established.
- **Electronic Blackboard.** It is a pressure-sensitive chalkboard that digitizes and displays the written content on it, on a monitor. Hence if the monitor is connected to a network, the classroom (chalkboard) contents can be seen/read from a remote location.

- **Electrosmog.** Electrosmog is a nonvisible radiation emitted by electronic devices. According to an specialist Olaf Shulz of Federal Radiation Defence Office (Germany), this radiation is very dangerous, induces disorder in body, and may cause cancer also. The degree of danger depends upon the frequency of emitted radiation. It is less in low-frequency emitting devices such as *cloth and utencils washing machines* and *vacuum cleaners,* but more in high-frequency emitting equipment such as *mobile phones,* computers and *microwave ovens.*

  The ill-effects on human body are caused due to electromagnetic field produced around these electrical equipment. Consequently the symptoms of headache, allergy, reduced immuniability, and lack of concentration are developed.
- **Bubble Memory**. It is a magnetic representation of data on the surface of semiconductor chips. Each bubble represents a binary 1, the absence of a bubble represents a binary 0.
- **OMR (Optical mark reader) and HB Pencil.** Several class of pencils are known to us. These are 2H, H, HB, B, 2B etc. Pencils are made by blending clay in graphite. The degree of hardness in them is decided by the clay content used and its nature. Amongst these, HB-pencil is used to create circle marks on answer sheets that are checked by OMR. It is because the HB pencil contains a large number of magnetic particles that can be easily sensed by OMR system. Thus, the checking of answer sheet becomes quick and foolproof. The OMR unit is attached to a computer for direct transfer of correct answers (HB marked circles) to a file.

18.2 Specific Materials for Electrical Applications

Machines and devices require an energy source for their operation. They may be mechanically operated, hydraulically, pneumatically, or electrically operated. Most of the present-day machines and devices are electrically operated. For that the electrical power has to be generated at power stations, transmitted to various locations, distributed, and utilized suitably. All these require appropriate materials for efficient and effective use of electricity. Some of these materials are given in Table 18.1 for a ready reference. Figure 18.1 illustrates its various details.

Table 18.1 Specific materials for electrical applications

Equipment/Component	**Materials**
Alternator	
• magnet	Alnico, Aclomax, W-steel
• armature	Annealed copper
• commutator	Mica paper, muscovita
• brushes	Carbon graphite, copper graphite
Motor	
• slot lining	Flexible micanite
• insulation	Flexible micanite, cotton tape, varnish, glass
Transformer	
• core	4% Si iron (i.e. Fe-Si alloy), permalloy, supermalloy
• winding wire/coil	Annealed copper
Conductor	
• at high voltage	ACSR, Teflon covered Cu
• at low temperature	OFHC copper
• for overhead transmission line	Hard drawn copper, Al
Cables	
• power cable	Annealed copper
• flexible wire	Annealed copper
• domestic wiring	Annealed copper, Al
Contact material	
• for precision devices	Ag, Pt, Au, Rh
• magnetos of automobiles	W, Mo, Pt
• for circuit breaker	Jr-Rh, Ag-Pa, WC-Ag
Filament	
• for incandescent lamp	C, Ta, W
• for arc lamp	Electrographite carbon
Electrode	Carbon, Zn, Mg
Resistor	Carbon, nichrome
Busbar	Hard drawn copper, Al
Rheostat	Constantan *i.e.* Eureka (60% Cu + 40% Ni)
Heating element for motor starter	Constantan, manganin (87% Cu + 13% Mn)
Eddy current brake	High energy hard magnetic materials

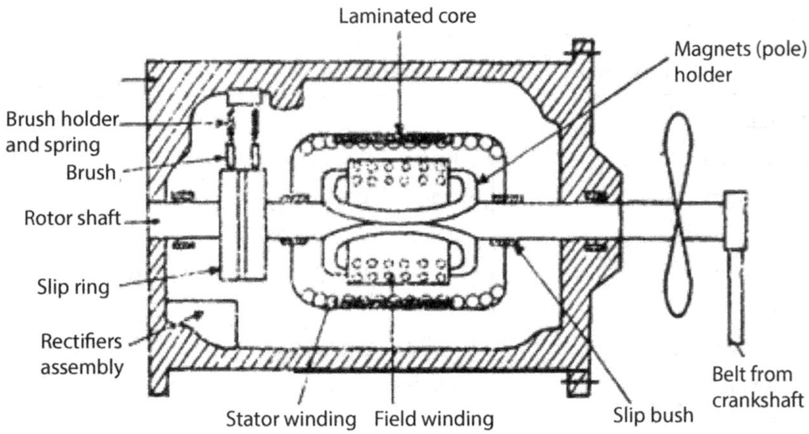

Figure 18.1 An alternator for which the electrical materials are described in Table 18.1

18.3 Specific Materials For A Typical Battery

An illustrious application of a typical battery is shown in Fig. 18.2, whose various components are made of one or the other type of electrical materials described in Table 18.2.

18.3.1 Separator and its Materials

It is used to prevent positive and negative plates from coming into contact with each other, and at the same time to allow necessary amount of electrolyte to be retained in inter-plate space. Separators are made of

i. Wood
ii. Hard rubber with microscopic pores
iii. Felted glass fibre
iv. Porous plastics
v. Perforated vinyl plastics
vi. Perforated and slotted ebonite

Among these, the hard rubber and PVC separators find wider use due to their greater reliability against short-circuits protection. Figure 18.3a-b show the container and positive plates, and Figure 18.3c-e show different kinds of separators.

Separators are also used in combination with each other to enhance the mechanical strength, chemical resistance, and also to cut-down the cost. A *perforated* separator in combination with a *microporous* separator, and a *felted glass fibre* separator in combination with the *wood* are such examples.

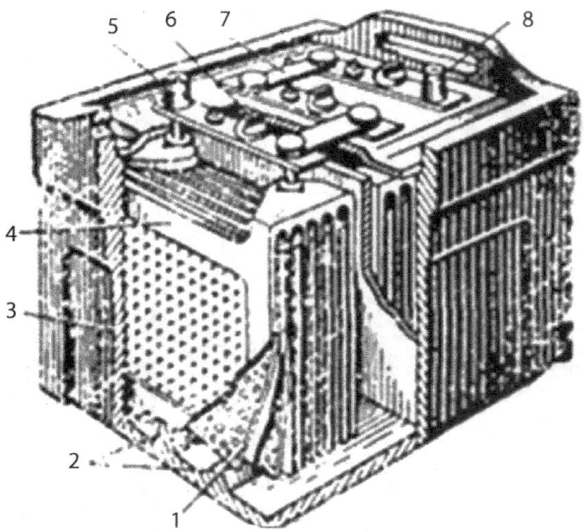

Figure 18.2 Depiction of various components of a battery

Table 18.2 Specific materials for a typical battery

Equipment/Components	Materials
Separator	Felted glass fibre, wood, perforated PVC, hard rubber, polyethylene with glass mat lining
Support ribs	Ebonite, moulded plastics, polypropylene
Container	Ebonite, polypropylene, ceramics, glass, wood lined with lead sheet
Negative plate group	Spongy lead
Positive plate group	Pb-Sb grid filled with PbO_2
Plug	Ebonite, moulded plastics
Inter-cell connector	Metal strap or bar
Terminal post	Cold forged inserts

18.3.2 Gridwork Construction of Plates

The plates of lead-acid cells are of diverse design, however, they all consist of some form of grid made of lead and the active material. Depending upon the constructional features of the gridwork, the plates can be one of the following types.

1. Faure (pasted) plate
2. Plante (surface type) plate

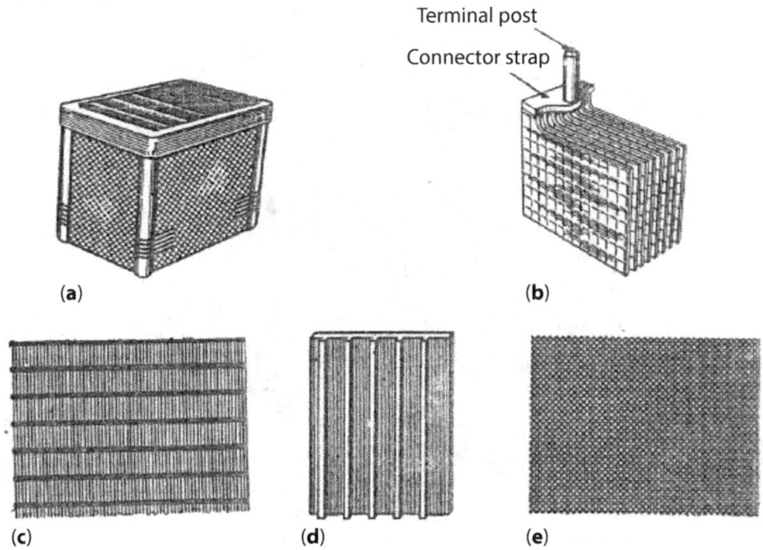

Figure 18.3 Components of a lead-acid battery shows (a) container, (b) grid form of positive plates; and separators of (c) perforated pvc, (d) hard rubber, and (e) micro-porous pvc.

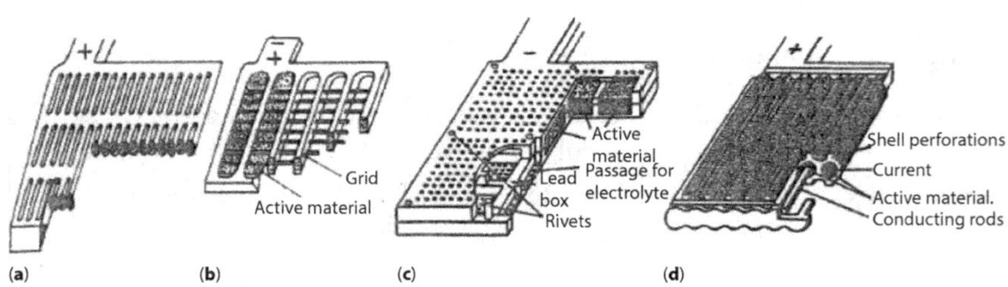

Figure 18.4 Constructional features of lead-acid cell plates (a) Faure plate, (b) Plante plate, (c) box plate, and (d) armoured type plate.

3. Box plate
4. Armoured type plate

The details of their construction are shown in Figs. 18.4a-d.

The active materials are most often prepared of lead powder and litharge (lead monoxide). Less frequently, use is also made of a mixture of lead monoxide and red lead (minimum). To decrease the possibility of shrinkage and hardening of the negative-plate active material during services, some expander is added. These expanders may be lamp black, oak flour, barium sulfate, cotton combings, humic acid, lignosulphonic acid and some tanning agent.

18.4 Specific Materials for Electronics Applications

Various electronic devices, components and equipment are shown in Figs. 18.5a-c. They are generally made of the materials given in Table 18.3.

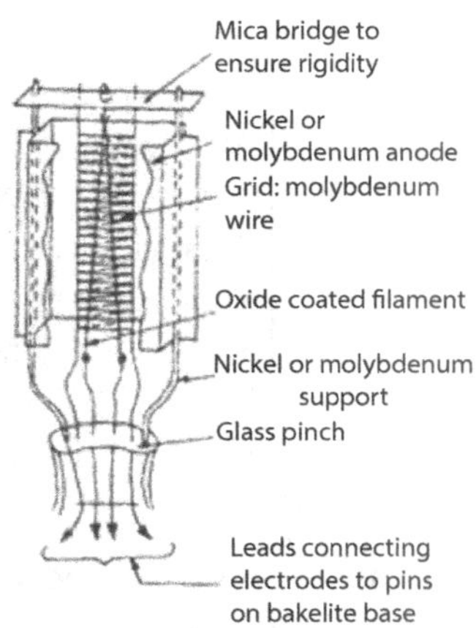

(**a**) Depiction of electronic materials in a triode

(**b**) Three IC circuit chips (**c**) Printed Circuit Board (NorthStar Horizon Z80 processor)

Figure 18.5 Various electronic devices, components and equipment

(b) [Source: http://en.wikipedia.org/wiki/File:Three_IC_circuit_chips.JPG]
(c)[Source:http://simple.wikipedia.org/wiki/File:NorthStar_Horizon_Z80_processor_board.jpg board]

Table 18.3 Specific materials for electronics applications

Equipment	Materials
Diode	Si, Ge
Transistor	Si, Ge
Valve anode	Carbon
Valve grid	Carbon
Light emitting diode (LED)	
• for visible green colour light	Ga phosphide
• for visible red colour light	Ga arsenide phosphide
• for invisible infrared light	Ga arsenide
• wire band	Au wire
• frame	Epoxy (transparent or coloured)
Electron emitters	W, thoriated W, Ni coated with BaO and strontium oxide
Resistors	Manganin (Cu 87% + Mn 13%), C, glass
Rectifiers	Se, copper oxide (CuO), Si
Electrodes	SnO
Microchip sealant	Expanding plastics
Anti-radar paints	Polyimides
Communication equipment	LCPs *e.g.* polyacrylate, polyamides
Gramophone records	PVC
Chemical sensors	Conducting electro active polymers
Cathode ray tube (CRT)	
• filters	Soft ferrite
• fluorescent screen	Phosphor coating of oxides and sulphides of Zn, Cd, Be
• cathode	BaO
• control grid	W, thoriated W, konel (Co + Ni + Ti + Fe)
Telephone receivers	Phenolics
Integrated circuits (ICs)	Single crystal silicon chip, hybrid chips of GaAs and Si
Photocells	Se, PbS, CdS
Thermistors	CuO, FeO, ZnO, CoO, MnO
Capacitors	Mica, paper, plastic film, air, mylar, tantalum oxide, glass

Table 18.3 (Cont.)

Equipment	Materials
Connectors	Teflon, rexolite, mica filled bakelite, rubber
Printed circuit boards (PCBs)	
• laminates	Phenolic, epoxy, polyimide, melamine, polyester
• conductor	Copper, silver
• microcircuits	Copper, silver
• joints	Solder material

18.5 Specific Materials for Computer Applications

Different hardware of a computer and various systems of its periphery are shown in Figs. 18.6a-b. Specific materials used in them and their components are listed in Table 18.4.

18.6 Specific Materials for Instruments and Control Applications

Various types of devices, instruments and controls are required in vivid engineering applications. Some of these are shown in Figs. 18.7a-c, and the specific materials used for them are listed in Table18.5.

18.7 Materials Used in Robots Construction

Robotics is an inter-disciplinary subject. The technologies in its construction and working involve the following engineering.

1. Mechanical engineering for design of manipulators, robot arms, joints, slides and bearings etc.
2. Electrical engineering in design of drives and controls.
3. Electronics engineering in design of circuits, controls, feedback etc.
4. Computer engineering in design, hardware and programming.

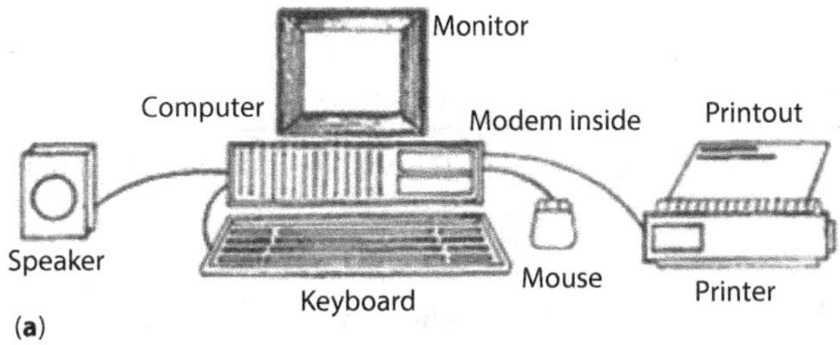

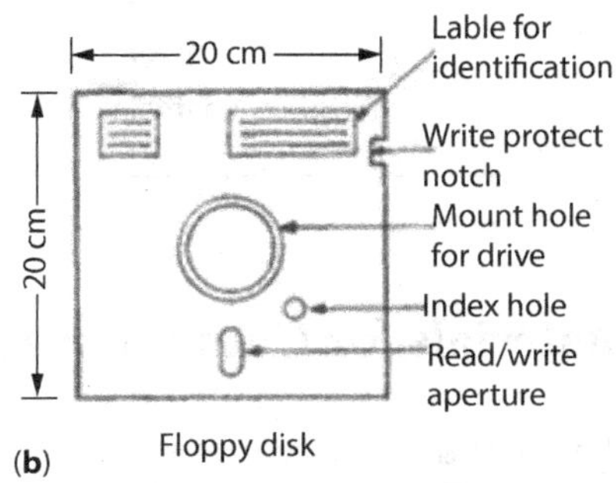

Figure 18.6 *Different hardware of a computer and various peripheral systems*

The choice of materials in robot construction is unlimited. A schematic robot is shown in Fig. 18.8.Important and commonly used materials are illustrated in Table 18.6.

18.8 Information Transmission from Cricket Field to Worldwide Televisions

Whether it is a Test match or a 1-day cricket match, every interested viewer sits in front of his/her television and enjoys the thrills of cricket. The thrill reaches to a climax when very close judgment for a run-out appeal, LBW (leg before wicket), stumping, or to catch for a bleak click sound is to be announced. The decision mainly comes through very minute vision by electronic systems. Here, we shall know as to how the complete telecasts of the game including these thriller moments are transmitted.

The information regarding the game of cricket is transmitted live through an electronic system shown in Fig. 18.9. It includes a miniature camera fitted inside the middle stump and its connection with various computers (installed outside the field) through the cables and control box.

Table 18.4 Specific materials for computer applications

Components	**Materials**
Computer body	Polymeric composite
• cabinet	Sheet steel
Central processing unit (CPU)	
• memory core	Ferrites of Cu-Mn, Mg-Mn, and Zn-Mn
• hard disk	Rigid metallic platter with magnetic particles
• microprocessor	AlGaAs chips
• motherboard	Epoxy, composites, sheet steel
• memory devices	Iron garnet, magnetic bubble, Mn-Mg ferrite
Magnetic head recorder	Metallic glass (Co-based glass alloy)
Computer storage	
• magnetic film and tape	$Gd_3Fe_5O_{12}$, CrO_2, 81.5% Ni + Co steel alloy
• compact disk	Magnetic mylar plastics, polyvinyl cinammnate
Stationary	Coated paper, photoglossy paper
Input devices	
• keyboard	Thermosets *e.g.* phenolics
• electrical contacts on keyboard	Gold
Electric and magnetic shielding cables	Copper braided coaxial cable
Output devices	
• floppy disk	Flexible mylar plastic
• display screen	Fluorescent coated plastics

The stump camera (or stump vision) is about 12.5 mm wide and 50 mm long, and is fitted 550 mm above the pitch surface. Weighing about 455 gm, this camera faces a prism at 60°. To safeguard against any damage, the internal parts of the stump are covered with a nylon and copper layer. A control box is located at about 1 m distance from the camera. It remains hidden under the wickets and covered below the astroturf. Its one end is connected to the camera while the other end is attached with broadcasting

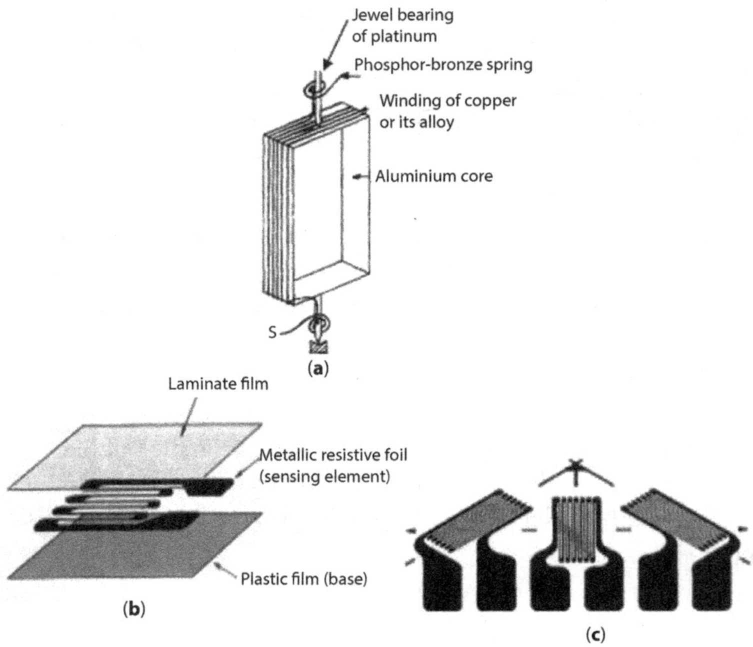

Figure 18.7 Specific materials for instruments and control applications
(a) moving coil instrument, (b) strain gauge, and (c) typical strain gauge rosettes[*Source: www.vishaymg.com*]

system, placed outside the boundary lines, through the underground wires/cables.

When the bowler delivers the ball, its image is formed on the prism through a window-type gate on it. This image is then reflected to the camera from where it reaches the control box. Signal of the image is then transferred to the broadcasting system (vehicle) for live telecast. The stump vision is highly accurate and can see the minutest actions occurring between the 'bat', 'pad', and 'ball'. Its construction is so robust that neither the wicket nor the camera breaks due to impact of ball.

18.9 Specific Materials for Networking Applications

Computer network is a mode of electronic communication. The network may be of a small group of computers, or a large number of computers connected with communication links at different locations. Depending upon the operational area of the network, the networking may be classified as under.

Table 18.5 Specific materials for instruments and control applications

Instruments & Components	**Materials**
Strain gauge	
• resistance type	Nichrome, constantan
• semiconductor type	Si, Ge
Thermocouple	Nichrome, constantan
Microphone	Ferroelectrics
Sonar device	Ferroelectrics
Miniature capacitors	Ferroelectrics
Foils	Catridge brass (70% Cu + 30% Zn)
Instruments	
• needle bearings	Silver bearing metal (95% Ag+ 4% Pb + 1% Sn), iridium
• springs	Elnivar (32% Ni + 68% Fe)
Electrically heated appliances	Constantan
Sprinkler system against fire and explosion prevention	Fusible alloys e.g. Wood's metal, Rose metal, Newton metal
Linear slide	Antifriction materials
Magnetic shielding	Gold, graphite, diamond
Audio transformers	Lithium ferrite
Television transformers	Ni-Zn ferrite
Hearing aids	High energy hard magneticmaterials
Underwater sound projectors	Magnetostrictive materialand sound detectors

(1) Local Area Network (LAN), and
(2) Wide Area Network (WAN).

A Local Area Network is a small group of computers (generally 3 to 50 terminals) interconnected together, and located inside a single building, within the buildings of same organization or institution, or within a limited area. All the terminals are connected with a main computer which is widely known as **server.** A device called Network Interface Unit (NIU) establishes the link between the networks. All these are connected through cables which are similar to 'Cable Television' connection. A typical construction of cable is shown in Fig. 18.10.

Wide Area Network (WAN) is used to connect a wider geographical area like different cities or even different countries. These communication

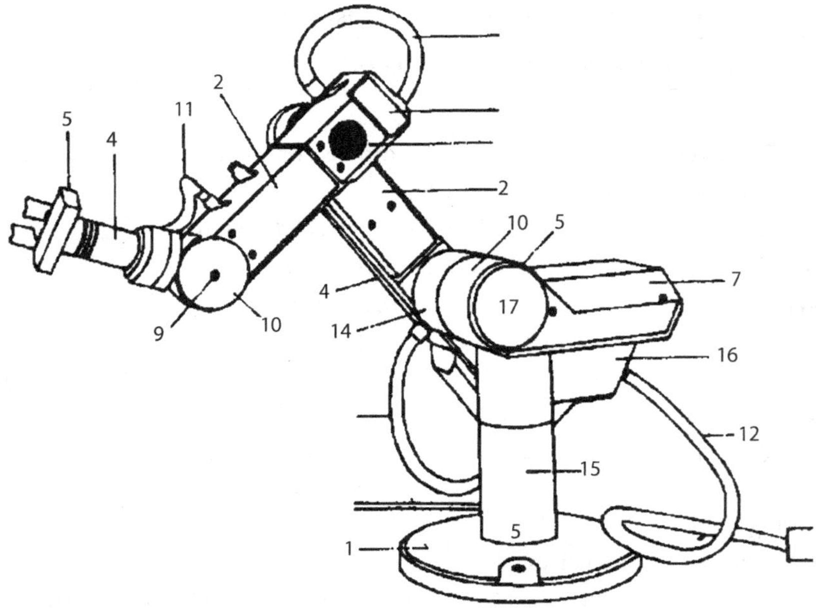

Figure 18.8 Diagram showing a Schematic Robot

links are established through telephone lines or microwave links via satellite. The Wide Area Network supports the global networking also, as can be seen in Fig. 18.11.

18.9.1 Networking of Networks and Connecting Devices

Interconnection between different networks is often required for extension in working range of business operations. Occasionally, the independent LANs are also connected to each other for the same purpose. To accomplish such connection; the host computer, terminal computer and other related systems are attached together by means of various connecting devices. Important among these devices are given below.

1. Bridge
2. Router
3. Gateway
4. Repeater
5. Hub
6. Brouter

Table 18.6 List of some commonly used materials employed in robot construction

System	Material	Specialty	Code number in Fig. 18.8
• Manipulator			
-structure and base	Aircraft grade Al alloys, Graphite fibre-reinforced plastics (GrRP)	σ_y = 500 MPa σ_u = 580 MPa	1
-robot arm	*Li-Al alloys, boron* carbide (BC), Lockalloy *(Be 62% + Al 38%)*	E = 190 GPa γ = 2100 kg/m^3	2
-sensor	Nichrome (Ni + Cr) thermocouple, Halleffect sensor (of semiconducting material)		3
-links	Ceramic		4
-bearings	Elastomers, rubber		5
-end of arm devices	Mg, Al, Ti		6
-guideways	Ball, roller and needle bearings of stainless steel		7
-damping devices	TiC, alloyed W		8
-joints	Brass bushings and nylon sleeves		9
-gears	Aluminium alloys	σ_y = 150 MPa σ_u = 165 MPa	10
-combination link	Incramute (Cu + Mn + Al) Nitinol (Ni + Ti + Al)	E = 95 GPa γ = 7630 kg/m^3 E = 80 GPa γ = 6600 kg/m^3	11
-cables and hoses	Optical fibre, flexible hose		12
• Control Unit			
-input/output	*npn and pnp switches*		*13*

(Continued)

Table 18.6 (Cont.)

System	Material	Specialty	Code number in Fig. 18.8
-safety	Collision protection (software and electrical drive)		13
-CPU	Microprocessor and micro-computer (semiconducting materials)		13
-coordinate system	XYZ and robot joints		14
-programming language	Real time robotic language*		not shown
• Drive System			
-actuator	D.C. servo motor,hydraulic, pneumatic, A.C. servo motor		15
-transmission	Harmonic drives		16
-feedback	Optical encoders		17
-grippers	Vacuum (pneumatic), D.C. (electrical)		—
- gripper connections	Optical fibres, metal conductors		—
• Accessories			
- pendant	Hand-held (direct control) computers using materials as in computers		18
- rotary table	Aluminium alloy		—
- conveyor	Asbestos, rubber, flax		—
- linear slide base	Antifriction materials		—
σ_y = yield strength	σ_u = ultimate strength E = Young's modulus * not a material	γ = density	

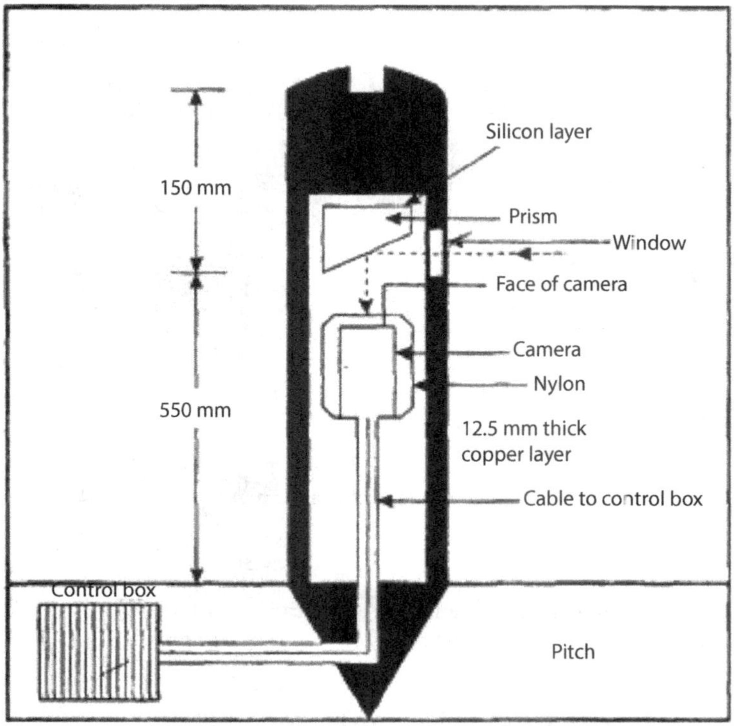

Figure 18.9 Schematic layout showing transmission of cricket game from a cricket field to worldwide TVs

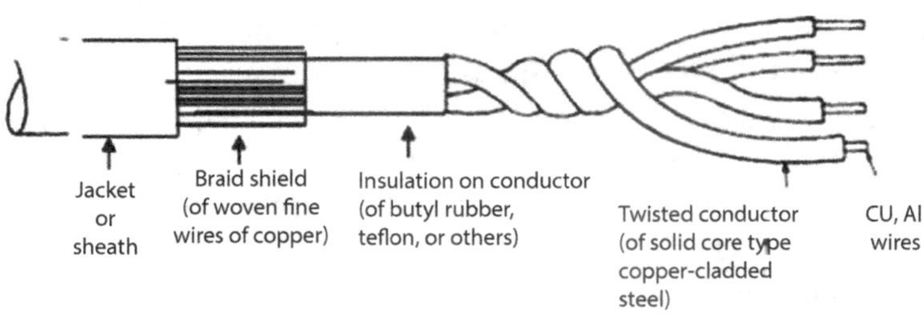

Figure 18.10 Depiction of different electrical materials in a multi-pair cable used for telephony and other circuits

LANs are connected by a 'bridge', a 'router' or a 'gateway'. A **bridge** is used to connect two similar LANs, a **router** is used to connect two or more similar LANs, but a **gateway** is used to connect two dissimilar LANs. A **repeater** is used to copy incoming signals and to amplify it for retransmission. A **hub** is used to receive the data packet and to distribute them to network connected computers. A **brouter** functions as a router and a bridge. Some of these devices are shown in Figs.18.12 a-e.

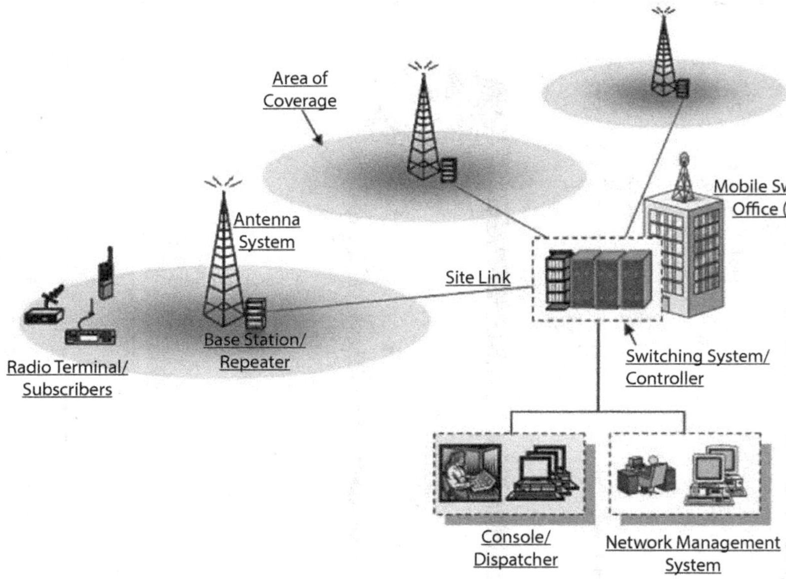

Figure 18.11 schematic arrangement of Wide Area Network (WAN) *[Source: http://www.about2wayradio.com/The_Component.htm]*

18.9.2 Hardware Used in Networking

Various hardware required for networking operations and information collection activities are the following.

1. A multimedia based computer and its components such as
   i. Central Processing Unit (CPU)
   ii. Data storage devices such as magnetic tape and disk, floppy disk, hard disk, optical disk etc.
   iii. Input devices such as keyboard, mouse, touchpad, joysticks etc.
   iv. Output devices such as monitors, audio etc.
   v. Voice recognition devices such as video digitizers.
2. Peripheral devices such as
   i. Printer
   ii. Scanner
   iii. Modem

In addition to above; the following systems, equipment, accessories and facilities are also essential.

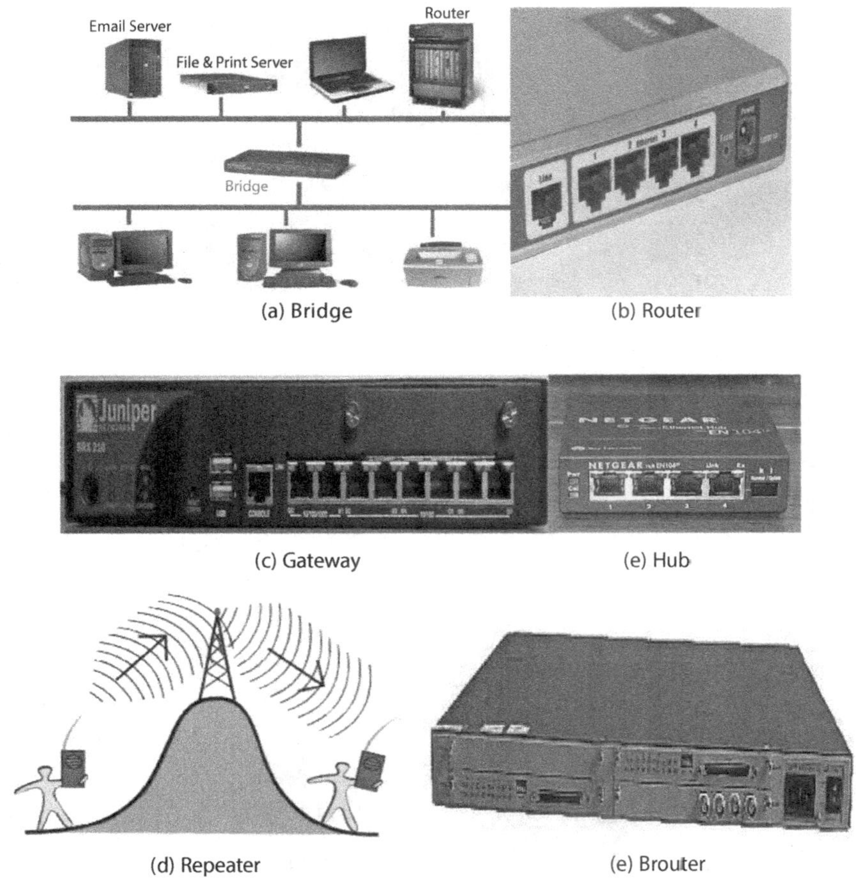

Figure 18.12 **(a-e)** Networking hardwares

1. Uninterrupted power supplier (UPS)
2. A two-way V-set for Internet uses
3. A telephone for Internet connectivity
4. Web camera
5. Camcorder for preparing videos and business purpose visiting cards

A system containing various hardware and other peripheral devices are shown in Figs. 18.13

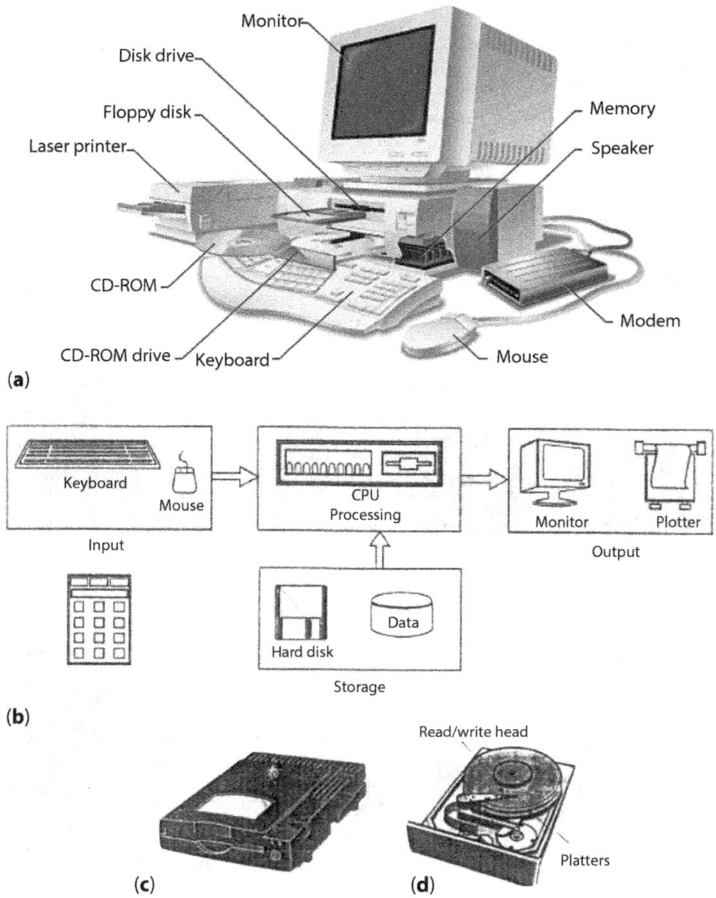

Figure 18.13 (a) Different types of computer hardware systems (b) Hardware organization of a computer system. (c) A hard disk of portable zip drive type that uses a 3.5" cartridge, and (d) showing several platters and read/write head (a) *[Source:http://www.warepin.com/classifications-of-computer-hardware-systems/]*

Multimedia Computer. A multimedia computer with a vast memory and a large hard disk is desired for business purposes. It is capable of displaying the pictures along-with the sound. A multimedia computer mainly consists of the following hardware items.

1. Keyboard
2. Mouse
3. CPU

4. Monitor
5. Magnetic disk
   i. Hard disk
   ii. Floppy disk
6. Disk drives
   i. Floppy drive ii. CD drive
   iii. CD writer iv. DVD
7. Storage medium
   i. CD-ROM ii. RAM
8. In-built Speakers and Microphone
9. Motherboard
10. Cards and Ports.

Hard Disk and its materials. Working of a hard disk is similar to that of a floppy disk, but it operates much faster and can store much more data. It is called *'hard'* disk because of its constructional feature (it is made of rigid *metallic platters* instead of flexible *plastic* as in floppy disk). Hard disks may be further sub-classified as

1. Permanently encased type (in disk drive), or
2. Removable cartridge type (encased in plastic cartridge).

Hard disk is storage device of a computer.

Disk Drive. This is a mechanical device used for rotating the disk. For that, the disk is inserted into disk drive to fit on its rod. A disk drive can perform two operations viz. (i) to read, and (ii) to write. Disk drive consists of a *magnetic head* which is known as *read/write head.* This head can read the magnetic impulses and can store a large amount of data. It has the capability of accessing a file also. Since it is a random-access storage medium, it may reach any item and read it. Several kinds of disk drives are in use. These are

1. Floppy drive
2. CD drive
3. CD writer
4. DVD (Digital versatile disk)

CD-ROM and its materials. CD-ROM (Fig. 18.14) is a kind of optical disk, used as storage medium. It is capable of storing a large amount of data and information, data files, text, audio, video, images, animations etc.

Figure 18.14 A CD-ROM is used as memory storage medium

18.10 Specific Electronic, Computer, and Robotic Components; and Their Materials in Automobile Applications

The scenario of electronics, computers and robots in automobile applications are presented below.

- Electronic connectors and interconnection systems (sealed and unsealed) for wiring harness.
- Modular fuse, relay, distribution boxes etc. for electrical connections.
- G-terminator with closed loop crimp height adjustment as a tooling for wiring harness design.
- Printed wiring board for anti-lock braking.
- Sensor connectors for electronic control of engines.
- Electro-optic devices using optical fibre based databus systems for linking the telephone;
- Navigation system, CD (compact disc) and car radio.
- Networking units for safety system, driver information system, communication system, engine
- Systems, power train, and interconnecting systems.
- Touch-screen data-entry systems for car phones, entertainment and traffic management systems.
- Flexible-film-circuits (FFC) and clock-spring lead assemblies for driver air bags and seat belt pre-tensioners.
- Collision-warning systems for driver information.
- Car wireless and antenna systems for communication.

- Sensors and detection systems for parking aids near obstacles and blind spots.
- Electronic clutches with manual gearboxes for heavy trucks.
- Magnetic valves and sensors for new class of commercial engines.
- Intrusion-alarm and immobilization systems for protection against theft.
- Modular hand tools with interchangeable die sets for crimp quality monitoring of reeled terminals and contacts.
- Smart key engine immobilizer for both four wheelers and two wheelers.
- Computer aided design (CAD) of automobile components.
- Computer aided styling (CAS) of car bodies.
- Computer aided manufacturing (CAM) by CNC (computerized numerically controlled) machines.
- Computer aided testing, servicing and repairing
- On-board computer for vehicle's control.
- Computerized multi-dimensional measurement in CAD/CAM interaction.
- Computer aided simulation and modelling.
- Robotized welding of vehicle bodies.
- Robotized production operations for automobile components, and many more.

18.10.1 Recent Advances in Automotive Electronics

Advances in electronics technology have made the automobiles a sophisticated multi-functional vehicle today. An increasing number of vehicles are equipped with electronic equipments for control, communication and entertainment. Various topics related to automotive electronics are as follows.

1. Multiplexing
2. Sensors and actuators
3. Engine and driveline controls
4. Information systems
5. Electronic displays
6. Relay, switching, and interconnections
7. Instrumentation

Latest innovations in relay and switching technology, and others are as follows.

1. Test Tools for electronic diagnostics and service systems such as
   i. CAN/Networks,
   ii. Controller area network for bus and truck applications,
   iii. Aluminiurn based diagnostics and decision support systems for trucks, tractors, semi-trailers and trailers.
2. Microcomputer technology for truck and vehicle applications.
3. Microprocessor controls for truck and bus transmissions, electronic governor for diesel trucks, and timer ICs.
4. Plastic, ceramic and tantalum capacitors, micro and stepper motors.
5. Carbon and metal films, thermistors and varistors.

18.10.2 Multiplexing in Automobiles

Multiplexing reduces the number of wires in wire harnesses while at the same time expanding the signal handling capabilities of electronic systems. Its vast scope includes topics such as

- CAN protocol
- Surface Acoustic Wave (SAW) technique for switch input
- Mux networks
- Thermal analysis of multi-terminal assemblies
- Developing of multiplexing IC'S (integrated circuits)

Some examples of multiplexing are

1. Integrated electronic dashboard for Alfa 75 and Alfa Romeo cars which include specific microcontroller for multiplex wiring, Xerox's distributed real time control system, multiplex data link with token passing, receiver IC for fibre-optic data links, and protocol for in-vehicle communications.
2. Improvement in wire harness products in Ford vehicles, and collision detection (CD) in Chrysler cars.
3. A microprocessor based wiper control system using motors in automotive multiplex systems.

18.10.3 Sensors and Actuators, and Their Materials

The most recent advances in the rapidly developing area of sensors and actuators are as follows.

- Hot wire air flow meter for engine control systems
- Solid-state magnetic sensors
- Electrostatic particle probes
- Optical-fibre gyroscopes for automobiles
- Lean mixture sensor
- Automotive steering sensors
- Intra-cylinder combustion pressure sensor
- Film technology
- Pressure/temperature sensors
- Integrated hybrid pressure sensor using a piezo-resistive thick film sensor element
- Measurement of engine torque with the intra-bearing torque sensor
- Flat response accelerometer
- Liquid level monitoring
- Lead tolerant titania exhaust gas oxygen sensors
- Wide range air-fuel ratio sensor
- Wide range wheel speed sensor
- Poisoning of zirconia exhaust oxygen sensors by silica
- Brushless D.C. motor
- Thick film accelerometer for electronic suspension control

A few types of sensors and actuators are shown in Fig. 18.15.

18.10.4 Engine and Driveline Control

Advancements in various areas to enhance the vehicle performance have led to new applications and better understanding of engine and transmission control systems. In this regard, several developments are as follows.

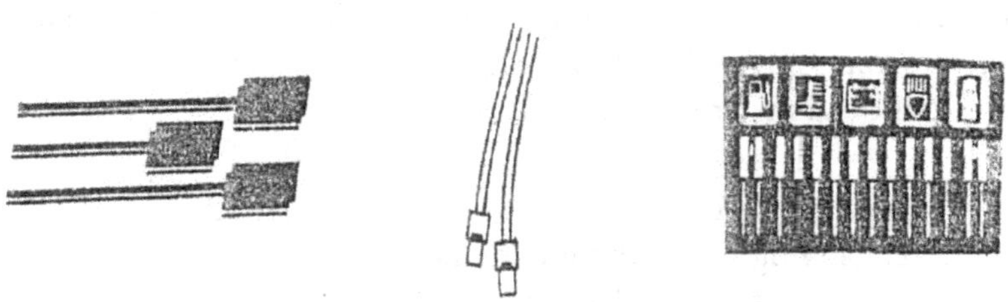

Figure 18.15 Sensors and actuators are backbone of automotive electronics

- Hydraulic coupling unit (HCU) for 4 wheel drive vehicles.
- Adaptive engine controls for fuel consumption and emission reduction.
- Toyota's single-chip microcomputer based engine and transmission control system.
- Soot formation in a multipoint-fuel injected spark ignited engine.
- Knock control of gasoline engines.
- Replacing relays with semiconductor devices.
- In-cylinder measurement of combustion characteristics using ionization sensors.

Current advances in electronic controls of diesel engine are the following.

- Use of *Fuzzy logic* for engine idle speed control.
- Intelligent alcohol fuel sensor.
- Design of diesel smoke feedback control using PI control algorithm.
- Electronic controls for *John Deere* diesel engines.
- Control design for a differential compound engine.
- Turbocharged/intercooled diesel engine with rack-actuated electronic fuel control system.

Smart power ICs provide lower costs and improved reliability in all electronic systems. The latest in this rapidly evolving field in the automotive industry are the following.

i. Smart power technologies for automotive applications.
ii. Combined Hall effect sensors with intelligence and power on a single chip.

18.10.5 Electronic Displays and Information Systems

To meet the increased demand for reliable, attractive and affordable automotive information systems, new technologies continue to evolve. These technical innovations are in the fields of:

1. Navigation systems with map-matching method
2. Fuel gauging by pneumatic methods
3. Lighting system for electronic analog cluster

The increasing application of electronic instrumentation has played a significant role in growth of automotive electronics in recent years. Some of these are the following.

- Guest Host (GH) dot matrix display. .
- Toyota electro multi-vision.
- Heads-up display using a graphic liquid crystal display.
- Vacuum fluorescent display.
- Visibility requirements for automobile CRT displays.
- Voice control in motor vehicles.

Consumers want reliable, attractive, affordable and easy to understand display and instrumentation systems. The display systems are of different types such as

- Visual displays
- LED indicators for automotive telltales
- Auditory displays
- Liquid crystal display (LCD)
- Passive displays
- Emissive displays

The ergonomical aspects of instrumentation desire proper installation, suitable mode of presentation and control for electronic displays. More details on these specialized topics can be referred in some standard text/literature on newly emerging subject 'Automotive Electronics'.

18.10.6 Shape of Things to Come

The vehicle of the very near future will be its own information superhighway. It will include sophisticated diagnostics, intelligent information systems, highly integrated engine control, optoelectronic displays and other advanced features for enhanced performance and security. The trend in the automotive industry is clear: simple mechanical assemblies are being transformed to intelligent electronic systems. Setting the pace, the engineers have been constantly and successfully innovating and integrating newer technologies into automobiles. The future technologies that are likely to play a vital role are the following.

- Computerized motor for starting and other systems.
- Advanced entertainment and security systems

- Satellite navigation to convey longitude - latitude location of the vehicle in remote places.
- Intelligent cruise control which keeps a set speed and maintains a safer distance between the vehicles.
- Sophisticated diagnostics with radio telemetry enabling detection of malfunctions without seeing the vehicle.
- Access to the Internet from an on-hoard computer.
- GPS (Global Positioning System) and automatic phone dialing systems will call for help in case of an accident.
- Interconnection systems for faster signal transmission.
- Digital electronics
- Electronic clutches in smaller automobiles.
- Onboard radar and camera systems to warn of any probable hazards on the road ahead and then to slow down the car automatically.
- Device that 'reads' the white lines on the road and gives an audible warning. If driver fails to pay attention to this warning, the device veers out the car out of its lane.
- A sensor that monitors the driver's line of vision and checks in other directions for any approaching danger.

Some of these have been introduced recently on specific auto-vehicles.

18.10.7 Future Technologies and Designs

The status of electronic systems in autovehicles will be further enriched with many newer technologies and designs. The most likely among them are briefly described below.

18.10.7.1 *Future Cockpit*

The electronic displays will feature more than the conventional dashboard information. The interconnection with electronic control units will become more sophisticated to match the requirements for faster signal transmission, reduced volume and weights, and reduced cost. Examples of such interconnecting systems will be as given below.

1. Board-to-board connector system comprising of
   i. surface mount 1.27 mm, AMP make micro match ribbon cable connectors with anti-fretting corrosion property.
   ii. surface mount flexible printed circuit (FPC) connectors on 0.5 mm contact center line spacing.

Figure 18.16 A hybrid connector is one among various electronic interconnecting systems [Courtesy: AMP]

   iii. single or double row connectors
   iv. flexible etched circuitry.
2. Wire to board connector systems.
3. Hybrid connectors.

These will enhance the driver's access to information. The hybrid interconnections in electronic control units is shown in Fig. 18.16. It will be used in engine, power train, antilock brake, audio and display connections. The terminals and cables will maximize a broad spectrum anticipation.

18.10.8 Optical Technology

The connectors for plastic optical fiber are used extensively in autovehicle applications. This product is used for connecting automotive entertainment and communication devices such as the FM/AM tuner amplifier, CD changer and telephone. In this way, the devices can be arranged in a ring network that allows easy integration of additional devices. A complete connector system includes a light source and detector, mounted in a header and fiber-optic cable assemblies. Their development has involved extensive laboratory testing including accelerated aging testing. Compared to their copper based equivalents, they offer the advantages of decreased weight, smaller size, immunity from electromagnetic interference (EMI), and high intrinsic bandwidth.

18.10.9 Electrical and other Materials in Important Automobile Uses

Table 18.7 presents a list of the preferred and alternative materials for making various automotive components.

Table 18.7 Preferred and alternative materials for making various automotive components

Component	Preferred material	Main advantage	Other materials
• Magnetic poles of generator, alternator and starting motor	Alnico	Hysteresis loss is minimum	Silicon-alloy steel
• Electrical conductor	Copper	High conductivity	Aluminium
• Springs	Phosphor bronze, cold rolled brass, spring steel	Good springing	Nylon, Be-Cu alloy, nimonics, Si-Mn alloy
• Capacitors	Metallised: polypropylene. polycarbonate	Available in film and foil forms	Polyester, polystyrene
• Battery cases	Polyurethane	High strength, resistant to corrosion, chemicals and ultraviolet radiations of sunlight	Nitrile (Buna-N) rubber
• Spark plug electrodes	Ni-Mn alloy	Resistant to high temperature effects, sulphur attack, and leaded petrol attack	Ni-Mn-Si-Pt alloy
• Springs	Phosphor bronze, cold rolled brass, spring steel	Good springing	Nylon, Be-Cu alloy, nimonics, Si-Mn alloy

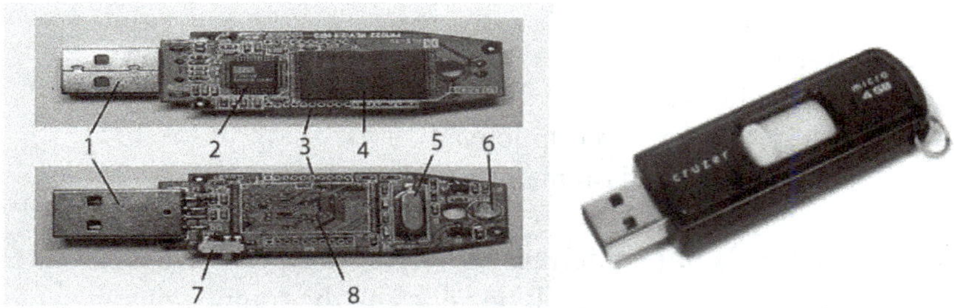

Figure 18.17 Internal parts of a pen drive [1]

18.11 Pen Drives (or Flash Memory) and Its Materials [1]

A Pen Drive (or USB flash drive) is a data storage device that includes flash memory with an integrated Universal Serial Bus (USB) interface. USB flash drives are typically removable and rewritable, and physically much smaller than an optical USB flash drive. Most pen drives weigh less than 30 grams. As of January 2013, pen drives of up to 512 gigabytes (GB) are available. A one-terabyte (TB) pen drive was unveiled at the 2013 Consumer Electronics Show and will be available during 2013 [1]. Storage capacities as large as 2 TB are planned, with steady improvements in size and price per capacity expected. Some allow up to 100,000 write/erase cycles, depending on the exact type of memory chip used, and a 10-year shelf storage time.

A flash drive consists of a small printed circuit board (PCB) carrying the circuit elements and a USB connector, insulated electrically and protected inside a plastic, metal, or rubberized case which can be carried in a pocket or on a key chain. For example, the connector may be protected by a removable cap or by retracting into the body of the drive, although it is not likely to be damaged if unprotected. Most flash drives use a 'standard type-A USB connection' allowing the connection with a port on a personal computer, but drives for other interfaces also exist.

Internals of a typical USB flash drive are shown in Figure 18.17. Its various parts are as follows.

1. USB Standard, Male A-plug
2. USB mass storage controller device
3. Test point

4. Flash memory chip
5. Crystal oscillator
6. LED (Optional)
7. Write-protect switch (Optional)
8. Space for second flash memory

PCB base materials of pen drives. The "base" of a pen drive is a Printed Circuit Board (PCB). It is the PCB on which all the components shown in fig. 18.17 are fitted. Excluding the exotic products using special materials or processes, all printed circuit boards being manufactured today, can be built using the following four materials:

- Laminates
- Copper-clad laminates
- Resin impregnated B-stage cloth (Pre-preg)
- Copper foil

Laminates are manufactured by curing the layers of cloth or paper with thermoset resin under pressure and temperature, to form an integral piece of uniform thickness. The size can be up to 1.2 m in width and 2.4 m in length. Varying cloth weaves (threads per inch or cm), cloth thickness, and resin percentage are used to achieve the desired final thickness and dielectric characteristics.

Dielectric materials for the circuits. There are different dielectrics that can be chosen to provide different insulating values, depending on the requirements of the circuit. Some of these dielectrics are

- polytetrafluoroethylene (Teflon).
- FR-4, FR-1, CEM-1 or CEM-3.

Well known prepreg materials used in the PCB industry are

- FR-2 (phenolic cotton paper)
- FR-3 (cotton paper and epoxy)
- FR-4 (woven glass and epoxy)
- FR-5 (woven glass and epoxy)
- FR-6 (mat glass and polyester)
- G-10 (woven glass and epoxy)
- CEM-1 (cotton paper and epoxy)

- CEM-2 (cotton paper and epoxy)
- CEM-3 (non-woven glass and epoxy)
- CEM-4 (woven glass and epoxy)
- CEM-5 (woven glass and polyester).

Thermal expansion is an important consideration especially with ball grid array (BGA) and naked die technologies, and glass fiber offers the best dimensional stability.

- FR-4 is by far the most common material used today. The board with copper on it is called "copper-clad laminate"[2].
- The conductors used in PCBs are generally copper and silver.

18.12 Remote Control Devices and Materials Used in Them

Remote control handset is an electronics device, most commonly used for a television set, air-conditioner, DVD player and home theater systems etc. for operating the device wirelessly from a short line-of-sight distance. The main technology used in home remote controls is infrared (IR) light. The signal between a remote control handset and the device it controls, consists of pulses of infrared light, which is invisible to the human eye. Typical remote control devices are shown in Fig. 18.18 and the materials used for its various components in Table 18.8.

Figure 18.18 Typical Remote control handsets for A.C., DVD, T.V etc. [3]

Table 18.8 Various components of a typical T.V remote handset and materials used in them

S. No.	Components	Materials
1	Keypad buttons	Silicone rubber, Different types of silicone can be used for different button feel (texture and pressure), Hard plastic.
2	Key board and Membrane key buttons	Membrane keys are different from standard buttons in that the entire keypad is one sheet of keys that resets on top of the face of the remote. The keys are raised areas on the keypad. Generally the keys click, when pressed. Membrane keys are less permeable to environmental contents since there are no areas for liquids to enter from the face of the unit.
3	Integrated circuit	Silicon dioxide (SiO_2), PSG (Phospho silicate glass), Si_3N_4, poly-Si(polycrystalline silicon), mono-Si (monocrystalline silicon), Al, Cu, metal silicides ($TiSi_2$, $MoSi_2$, $TaSi_2$), refractory metals(W, Ti, Mo, Ta) [4].
4	Button contacts	Gold, conductive plastic.
5	Infrared light-emitting diode (LED)	Gallium Arsenide (GaAs) and InSb.

18.13 Hand Held Devices and Materials Used in Them

The hand held devices, also called mobile devices are typically portable computing machines with a touch input or a miniature keyboard. They have installed operating system, and can run various application software. The various hand held devices are mobile phones, smart watches, etc. Some hand held devices are shown in Fig.18.19, and the materials used for different components are listed in Table 18.9.

Figure 18.19 Some Hand held devices [5]

Table 18.9 Various components of typical hand held devices or a smart phone and material used in them

S. No.	Components	Materials
1	Battery (Anode, Cathode, Electrolyte)	Lithium Ion type. carbon (usually graphite), a metal oxide lithium salt. But now a day's, Graphene is the new emerging material in the field of battery technology.
2	Screen	Glass on the screen to make it scratch resistances. The typical glass used is "Gorilla Glass". Gorilla Glass is the registered trademark for an alkali-aluminosilicate sheet toughened glass manufactured by U.S. glassmaker Corning Inc. Engineered for a combination of thinness, lightness, and damage-resistance, it is used primarily as the cover glass for portable electronic devices including mobile phones, portable media players, laptop computer displays, and some television screens. It is manufactured through immersion in a molten alkaline salt bath using ion exchange to produce compressive residual stress at the surface. This prevents cracks from propagating - for a crack to start, it will first have to overcome this compressive stress.

(Continued)

Table 18.9 (Cont.)

S. No.	Components	Materials
3	Display Technologies	Liquid Crystal Display (LCD) or more recently O-LED technologies. The LCD contains liquid crystals which are embedded between the layers of glass for illumination with transistors for an electric charge. The liquid crystalline substances can contain toxic substances such as mercury. Foil display systems are now under development which would make LCD's much more environmentally friendly, especially in small products.
4	Body	Plastic body. The plastic is Polycarbonate/ Acrylonitrile Butadiene Styrene (PC/ABS) [6]. Thermoplastic polyurethanes (TPUs) are the material of choice when it comes to outstanding abrasion resistance and rebound resilience. TPUs cases help to protect mobile phones from accidental damage.
5	Touch Screen Technologies	Touch screen are generally of two types: capacitive and resistive. Capacitive sensors are constructed from many different media, such as copper, Indium tin oxide (ITO) and printed ink. Copper capacitive sensors can be implemented on standard FR4 PCBs as well as on flexible material. ITO allows the capacitive sensor to be up to 90% transparent (for one layer solutions, such as touch phone screens) [7].
6	Integrated Circuits	(i) Silicon monocrystals are the main substrate used for ICs although some III-V compounds of the periodic table such as gallium arsenide are used for specialized applications like LEDs, lasers, solar cells and the highest-speed integrated circuits. (ii) The more advanced ICs also include the wafers up to 300 mm in diameter. Intel, IBM, NEC, and AMD are using ~32 nanometers for their CPU chips. IBM and AMD introduced immersion lithography for their 45 nm processes.

Table 18.9 (Cont.)

S. No.	Components	Materials
		(iii) Copper interconnects where copper wiring replaces aluminum for interconnects. (iv) Low-K dielectric insulators. (v) Silicon on insulator (SOI) (vi) Strained silicon in a process used by IBM known as strained silicon directly on insulator (SSDOI) (vii) Multigate devices such as tri-ga [4].

References

1. http://en.wikipedia.org/wiki/USB_flash_drive
2. http://en.wikipedia.org/wiki/Printed_circuit_board
3. http://www.qunda.com/showpro.asp?lang=en&prid=220
4. http://en.wikipedia.org/wiki/Integrated_circuit#Manufacturing
5. http://www.scottsdaleairparkit.com/2011/05/10/tablet-safety/
6. http://en.wikipedia.org/wiki/Plastic#Common_plastics_and_uses
7. http://en.wikipedia.org/wiki/Capacitive_sensing

19

Recent Advances and Emerging Trends in Electrical and Electronic Materials

19.1 Novel Applications of Functionally Graded Nano, Optoelectronic and Thermoelectric Materials

Abstract: Functionally graded materials are a novel class of materials having unique characteristics. They are graded property materials and are used as medical implants, for thermal protection of space vehicles, as thermoelectric converter for energy conservation etc. Due to their versatility of behaviour, they are now used as nano, optoelectronic and thermoelectric materials also. Future applications demand the materials having extraordinary mechanical, electronical and thermal properties which can sustain different environment conditions and are easily available at reasonable prices. The carbon nanotubes (CNT) reinforced functionally graded composite materials (FGCM) is expected to be the new generation material having a wide range of unexplored potential applications in various technological areas such as aerospace, defence, energy, automobile, medicine, structural and chemical industry. They can be used as gas adsorbents, templates, actuators, catalyst supports, probes, chemical sensors, nanopipes, nanoreactors etc.

This section aims at elaborating the development of such materials by compiling the ongoing researches. In this regard, the research developments of newer materials by different investigators have been presented here. Brief details of the development of CNT reinforced Functionally Graded composite materials, FGM in optoelectronic applications, FG thermoelectric materials have been presented. In these elaborations, it is shown by the respective investigators that these FGMs can be effectively used as MRI scanner cryogenic tubes, tools and dies, solar panels, photodetectors, high temperature joining materials and various structural elements etc.

19.1.1 Introduction to Functionally Graded Materials (FGMs)

Functionally graded materials (FGMs) are a new generation of engineered materials that are gaining interest in recent years. FGMs were initially designed as thermal barrier materials for aerospace structural applications and fusion reactors. FGMs also found applications in structural components operating under extremely high-temperature environment. As an example, FGMs based on ceramic reinforcement in metal matrix are able to withstand high-temperature environments due to better thermal resistance of ceramic constituents, while the metal constituents enhance their mechanical performance and reduce the possibility of catastrophic fracture. FGMs are the composite materials in which the content of reinforcement is gradually varied in some direction to achieve gradient in properties. Due to graded variation in the content of constituent materials, the properties of FGMs undergo appreciable and continuous change from one surface to another, thus eliminating interface problems and diminishing thermal stress concentrations.

There are many areas of application for FGMs. The concept is to make a composite material by varying the microstructure from one material to another material with a specific gradient. This enables the material to have the best of both materials. If it is for thermal or corrosive resistance, or malleability and toughness, both strengths of material may be used to avoid corrosion, fatigue, fracture and stress corrosion cracking.

FGM belongs to novel material category and offers new capabilities to use it at large scale. FGM technology has the potential to drastically re-define the methods used for developing lighter, stronger, and high performance structures with unique and non-traditional properties. FGM-composites often lead to a reduction in weight and costs, and are more environmental friendly. For these reasons the popularity of these composites is increasing in engineering world. However, engineering breakthrough will be required to develop the technology in long-term goals. Technologies expect the FGM kind of advanced materials with revolutionary new capabilities such as in power systems, electronic devices, sensors, science and medical diagnostic instruments.

19.2 CNT Reinforced FGM Composites [1]

CNT reinforced metal or ceramic matrix functional graded composites exhibit improvements in properties such as thermo- mechanical, lightweight, dimensional stability, barrier properties, flame retardancy, heat resistance and electrical conductivity. These special properties desire more challenging technology in future. Compared to metals, these show better higher strength-to- density ratios, higher stiffness-to-density ratios, better fatigue and wear resistance, better elevated temperature properties (higher strength-lower creep rate). In contrast to composite materials, FG-composites have new unique properties like directional mechanical properties, higher temperature capability (lower thermal expansion properties), excellent fatigue and fracture resistance. However, it is difficult to meet stringent dimensional stability requirements during gradation and results in higher cost of the product.

19.2.1 Applications of CNT in FGM

CNT reinforced metal matrix functional graded-composites due to their unique combination of hardness, toughness and strength are universally used in cutting tools, drills, machining of wear resistant materials, mining and geothermal drilling. CNT reinforced functional graded-composite materials have the ability to generate new features and perform new functions that are more efficient than larger structures and machines. Due to functional variation of FG-materials, their physical/chemical properties (e.g. stability, hardness, conductivity, reactivity, optical sensitivity, melting point, etc.) can be manipulated to improve the overall properties of conventional materials. Potential applications of FGM are both diverse and numerous. Some of the current and futuristic applications of FGM in electrical and electronics field are listed as follows [1].

- MRI scanner cryogenic tubes
- Eyeglass frames
- Reflectors
- Solar panels

- Camera housing
- Sonar domes
- CNT reinforced functionally graded piezoelectric actuators.
- As furnace liners and thermal shielding elements in microelectronics.

19.3 FGM in Optoelectronic Devices [2]

Functionally graded materials are perspective materials for modern optoelectronic devices, such as low threshold current edge lasers and tunable photodetectors. Graded layers can also be used as buffers in heteroepitaxy of nitrides. Development and characterization of FGM is a real challenge and needs thorough analysis. The methodology of measuring layers with graded changing properties is difficult. Another problem is the interrelation between different material properties. It is difficult to change one parameter like energetic band-gap, without influencing the others like refractive index.

Functionally graded materials (FGMs) are materials in which some particular physical properties are changed with dimensions. Properties of such materials can be described by material function f(x). In homogenous materials, this function is a constant like in Fig. 19.1a. In case of a junction of two different materials, function f(x) has a stair-shape (Fig. 19.1b). In FGM, this material function should be continuous or quasi-continuous. It means that the particular properties change continuously or quasi-continuously along one direction, like it is shown in Fig. 19.1c. In many cases, FGM can be presented as a composition of several connected thin layers.

Depending on the number of directions the properties change, we can discriminate them as 1- dimensional, 2-dimensional or 3-dimensional FGM. It can be mathematically described for 3-D FGM as,

$$\frac{\mathrm{d}F}{\mathrm{d}x} \neq 0, \quad \frac{\mathrm{d}F}{\mathrm{d}y} \neq 0, \quad \frac{\mathrm{d}F}{\mathrm{d}z} \neq 0 \tag{19.1}$$

where $F(x, y, z)$ is the material function.

19.3.1 Possible Applications of FGM in Optoelectronics

Now-a-days the graded materials are widely used for anti-reflective layers, GRIN (Gradient Index) lenses and other passive elements made from dielectrics, and also for sensors and energy applications. For example, the

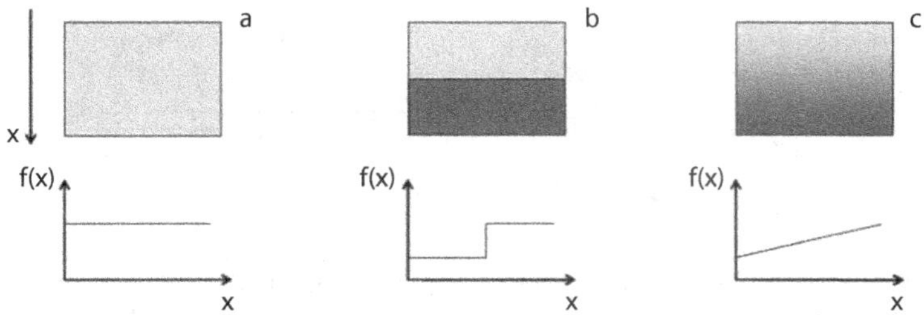

Figure 19.1 Schematic representation of materials function in different structures; homogenous material (a), junction (b), FGM (c) [2]

modulation of refractive index can be obtained in such components through the change in material composition. Another possibility is to apply concept of gradation in semiconductor active devices. In semiconductors, the material function can describe the energetic bandgap, refractive index, carrier concentration, carrier mobility, diffusion length, built-in electric field, and another properties which influence the parameters of optoelectronic devices.

19.3.2 High-Efficient Photodetectors and Solar Cells [2]

The fundamental limitation of the efficiency of homogenous silicon solar cells is the constant energetic band-gap width in bulk material. Because the high-energetic radiation is absorbed in a shallow layer under surface, it is necessary to form electric field in close vicinity to the surface. Generated carriers can effectively be separated in electric field , therefore the diffusion length of carriers should be longer than junction depth.

Another factor which decreases carrier generation efficiency, is the difference of energetic band-gap and absorbed photons energy. By using materials with gradation of energetic band-gap, it is possible to match the absorption edge with band-gap, which improves generation efficiency. The application of cascade of junctions with different energetic band-gap width can be one of the solutions. Another way to overcome this limitation is the use of graded material. The idea of such device is shown in Fig. 19.2.

19.4 Advanced Thermoelectric Materials in Electrical and Electronic Applications

Abstract: Thermoelectric (TE) materials are a novel class of materials having unique characteristics. They are Seebeck and Peltier effect materials and are used as thermocouples, for thermoelectric cooling of microelectronic

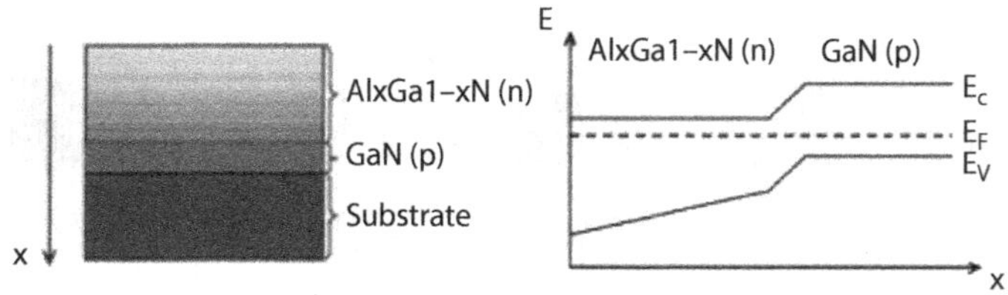

Figure 19.2 Schematic structure of *p-n* photodiode with graded layer electric field, therefore the diffusion length of carriers should be longer than junction depth [2].

products, as thermoelectric converter for energy conservation etc. Due to their versatility of behaviour, they are now used as construction materials for microelectronic equipment, wireless sensors also. This section aims at elaborating the development of such materials by compiling the recent and ongoing researches. In this regard, the research developments of some newer materials by other investigators have been presented here. Brief details of the development of thermoelectric power generator for wearable systems, nano-thermocouple, thermoelectric-microcoolers etc. are presented. In these elaborations, it is shown by the respective investigators that these TE materials can be effectively used in heavily doped semiconductor, thin films, quantum well etc.

19.4.1 Introduction

Thermoelectricity is of strong scientific and technological interest due to its application possibilities ranging from clean energy to photon sensing devices. Recent developments in theoretical studies on the thermoelectric effects as well as the newly discovered thermoelectric materials provide new opportunities for wide applications. One type of these materials is based on the strongly correlated electron system; typical examples are the transition metal oxides, which were not regarded as very promising for thermoelectric applications. The thermoelectric effect is based on the anisotropic Seebeck components in crystals. Upon radiation of heat and/or light on the film surface, a voltage is induced and hence, device which can detect the heat and/or light radiation can be made. In this section some recent progress in this field and special emphasizes on the new applications has been discussed.

Conventional thermocouples based on metal wires are cheap, reliable and widely used for measuring high temperatures. This is the case of furnaces, which are widely used in the microelectronic industry. A thermocouple is a simple electric circuit, formed by two dissimilar conductors joined at both ends (i.e. the junctions). Opening the circuit by cutting one of the wires enables the measurement of a voltage, which is proportional to the difference in temperature at the two junctions (the Seebeck effect). Consequently, the thermocouple can be used to generate a voltage proportional to temperature difference without the need of any external electrical bias.

19.4.2 Thermoelectric Power Generator for Integration in Wearable Microsystems [3]

The functional integration of efficient solid-state TE devices and microelectronic circuits offers many benefits. One is the implementation of local cooling for thermal stabilization of an on-chip reference element or for reducing the leakage current in a critical component such as a photo detector. Another implementation is in TE power generation to enable the operation of a low-power circuit without external electric power source, such as a battery.

The TE effect is very efficient in most materials. The best performance is obtained in the presence of heavily doped semiconductors, such as the bismuth telluride or the silicon germanium. When using semiconductors, the most desirable situation is obtained when the base materials are both n- and p-doped, since this allows the use of same material system for fabrication of the two TE legs between the junctions.

Due to its compatibility with IC technology, polycrystalline SiGe alloys and polycrystalline Si are commonly used in thermopile applications. Their use in microcoolers has shown that their performance is very low when compared to that of tellurium compounds. Tellurium compounds (n-type bismuth telluride, Bi_2Te_3, and p-type antimony telluride, Sb_2Te_3) are well-established room temperature TE materials and are widely employed by the industry in conventional TE generators and coolers.

The techniques explored for the deposition of Bi_2Te_3 (1 μm thick) thin-films are the thermal co-evaporation, the electrochemical deposition, the co-sputtering, the flash evaporation and the metal-organic chemical vapour deposition (MOCVD). Although all these approaches are in principle suitable, the co-evaporation technique is employed in this work to obtain the n-type Bi_2Te_3 and p-type Sb_2Te_3 thin-films. It allows to precisely control the stoichiometry of the deposited thin-film with the lowest costs.

19.4.3 Nano-Thermocouple in Thermoelectric Energy Harvesting [4]

The efficiency of thermoelectric material is often described by a dimensionless number called the figure-of-merit $ZT = \sigma S^2 T/k$, where T is the absolute temperature, σ and k are the electrical and thermal conductivity, respectively, and S is the Seebeck coefficient. Material with high ZT is desired in thermoelectric generator design. In practice, however, it is difficult to increase ZT because increasing S often leads to simultaneous decreasing σ. In bulk thermoelectric materials, Bi_2Te_3 alloys have the highest ZT about 1.0 at 300 K. ZT > 2 can be achieved by low-dimensional semiconductors such as quantum well and quantum wire. The breakthrough of ZT > 2.4 had mainly benefitted from the reduced thermal conductivity. Even with these advances, the applications of low-dimensional thermoelectric materials and their integration with state-of-the-art semiconductor process have not been explored as yet.

For thermoelectric materials with feature size (also called the characteristic length) $a < 100$ nm, the quantum confinement effect and spatial confinement effect have been known to increase the figure-of-merit ZT. The former eliminates some states that the electrons can occupy, since they do not obey the boundary conditions of electronic wave function. The latter reduces the phonon relaxation rate and lowers the thermal transport properties of low-dimensional materials. In quantum wells, the electrons are confined to move in two-dimension so that the electron motion perpendicular to the potential barrier is quantized. This change of the energy band structure and electronic density-of-states (DOS) can increase the asymmetry between the hot/cold electron transport, obtain large transport energy, and increase the number of carriers in the materials, thereby achieving higher Seebeck coefficient.

19.4.4 Low-Cost Micro-Thermoelectric Coolers for Microelectronic Products [5]

The development of a high-efficiency micro-cooler for cooling localized hot-spots is very desirable to many microelectronic products such as power amplifiers, laser diode, and microprocessors. Although conventional cooling devices such as miniature heat pipes, micro-heat dissipation, traditional heat sinks and fans have performed a high cooling performance; their inconveniences in packaging and refilling of the circulating fluid, IC incompatible process and large physical size have kept them from being used in cooling localized tiny hot-spots applications.

Conventional macro-thermoelectric coolers (macro-TECs) were also reported in wide range of applications such as microelectronics and optoelectronics cooling, thermal stabilization and refrigeration/cryosurgery instrument. Two types of miniaturized μ-TECs are reported here [5] for small area hot-spots cooling applications. The basic operation principle of TEC devices is the Peltier effect, which can be observed at the junction of

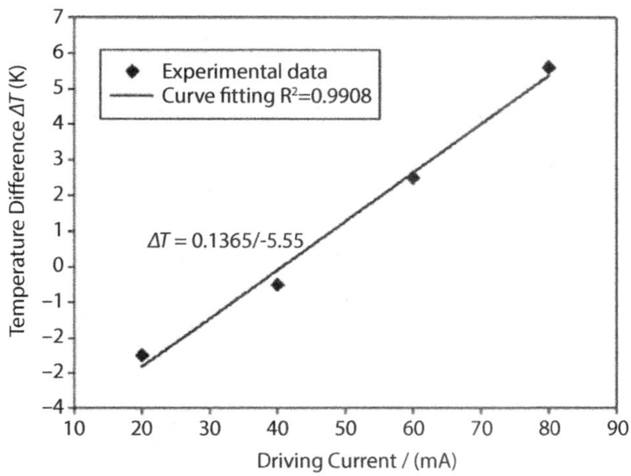

Figure 19.3 Temperature distribution of polysilicon under driving current.

two different metals or n- and p-type semiconductors under appropriate current drive.

Thermoelectric material properties of the n-type- and p-type doped polysilicon (1.5μm thick) and electroplated telluride TE thin films (8μm thick) have been measured. Figure 19.3 presents the temperature distribution of the bridge-type polysilicon μ-TEC under 80 mA driving current. A near linear relationship (R^2 = 99.08%) between the cooling performance of the bridge-type polysilicon μ-TEC and the driving current can be obtained. This research has demonstrated that the bridge-type polysilicon μ-TEC has a better cooling performance than the column-type telluride μ-TECs.

19.4.5 Thermoelectric Water-Cooling Device for Electronic Equipment [6]

With the development of microprocessors, the heat dissipation problems become more and more serious; this raises challenges in electronic cooling. Recently, thermoelectric cooler (TEC) has been applied to electronic cooling with its advantages of sensitive temperature control, quietness, reliability, and small size. A conventional TEC consists of p-type and n-type bulk semiconductor thermoelements connected electrically in series and sandwiched between two ceramic substrates. Thermoelectric cooler is regarded as a potential solution for improving the thermal performances of cooling devices on the package. Air-cooling devices will no longer satisfy the high

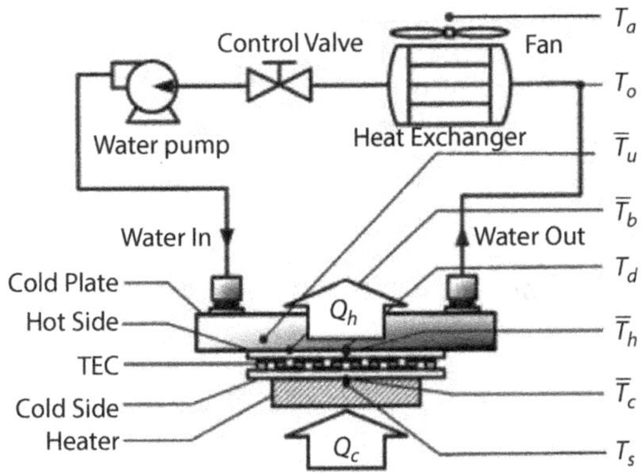

Figure 19.4 Schematic illustration of the thermoelectric water-cooling device.

power requirements of next generation electronic components. With high thermal capacity, liquid cooling techniques had been developed to solve the high power dissipation problem.

Figure 19.4 shows a schematic illustration of the thermoelectric water-cooling device. It consists of a TEC, a cold plate, a heat exchanger, a fan, a water pump, and a control valve. The TEC cold side contacts the heater to absorb the heat dissipation of heater (Q_c), and the TEC hot side touches the cold plate to release the heat from the TEC hot side. The rejected heat (Q_h) from the hot side conducts to the cold plate cooled by water flow. Finally, the rejected heat is carried to the heat exchanger by water flow, and is then transferred to the environment by air flow caused by the fan.

19.4.6 Conclusion

A temperature sensor results if one of the junctions is maintained at a well-known temperature. The thermocouple can also be used as an actuator. The advantages of TE energy conversion is that moving mechanical parts are avoided which enables high system reliability, quiet operation, and it is usually environmentally friendly. It has been found that the tellurium compounds Bi_2Te_3 and Sb_2Te_3 having ZT > 2 are well established TE materials for futuristic uses.

19.5 Frontiers in Electronic Materials Research

Advances in the understanding of electronics have created revolution in all fields of its applications. May it be micro-electronics, micro-electromechanical systems, fluorescent-magnetic devices, nanobiodevices, terahertz devices or a vast field of nanotechnology, everywhere the electronics has made a vital dent. Advancement of any engineering discipline is not possible without the development of materials. Hence continued researches full of vivid novelties are required in electronic materials technologies. Research demands that the development of newer materials, their characterization and appropriate applications be given topmost priority. With this aim, in next sections, the scenario of emerging trends have been focused to present the current and futuristic electronic materials.

19.5.1 Introduction

The development in electronics engineering and electronic materials is currently towards creating the atomic and nano size devices from miniature size. It therefore envisages the synthesis and development of nanomaterials and of other materials in the form of thin films etc. In this respect, the pyroelectric composites and nano ferroelectrics are frontier materials. The search of advanced materials for conventional energy sources and renewable energy applications are also the emerging fields for research. Therefore the research for electrolytes, nanobattery, and electrochemically advantageous materials has been included in this paper. Superconductivity is one of the most desired property in electronic materia. For that, the efforts are being made to produce ultra-thin magnetic tunnel junctions(MTJ's).The frontiers of research also include the development of conducting paper as these play vital role in electronic sector.

19.6 New Pyroelectric Thin Composite Films [7]

Pyroelectricity is a property of polar crystals characterized by the evolution of electric charge when exposed to heat. This property is utilized in various applications such as un-cooled thermal imaging, infrared-red light detectors, thermal switches, thermometers and gas sensors. The number of devices based on pyroelectric materials has increased during the last years due to their application in industries such as airplanes, cars and private home security control. Most of the knowledge today on pyroelectric materials is based on bulk-size single crystals. Pyroelectric polycrystalline thin films are also used in devices but require specific depositing conditions to obtain preferred crystallographic polar orientations. Pyroelectric materials have lower detection sensitivity of infra-red light by few orders of magnitude compared to semiconductor crystals.Their main advantage is the ability to detect the light in a wide wavelengths range, especially in the far infra-red, without any cooling requirements.

There is a high scientific and technological interest in developing new pyroelectric materials capable of better detection limit, higher lateral sensitivity and faster response time. One way of meeting these challenges,suggested by S.Berger and M.Nitzani[7], is to decrease the size and increase the surface density of the detecting crystals to obtain uniform electric response from each of the detecting crystals without any electric interaction between them. Berger *et al.* presented a new thin composite film that consists of nanometer-sized pyroelectric rods grown inside the alumina nano-pores. The rods are single crystals arranged in a uniform crystallographic orientation. Their surface density is about 10^{11} / cm^2. The combination of small size, uniform crystallographic orientation and high surface density results in improved pyroelectric properties from theoretical and experimental view points.

Materials. Thin porous alumina films (about 0.5 μm thick) are formed on the surface of pure aluminum foils (0.5 mm thick) by electrochemical anodization. Prior to anodization, the Al surface is electropolished to a high degree of smoothness using an electrolyte solution of perchloric acid and ethanol (3:7 volume ratio) under a constant potential at 25^{0}C. The anodization is done in an aqueous phosphoric solution (3 % vol.) under a constant current density at 25^{0}C. The anodized samples are inserted into a super-saturated aqueous solution of TriGlycine Sulfate TGS-$(NH_5C_2O)_3H_2SO_4$. The insertion process is done at different temperatures in the range of 25-60^{0}C. The liquid solution is cooled down slowly to enable nucleation and growth of single crystals in the pores. In some experiments a dc voltage is applied on the samples during cooling to enforce the growth of single crystals with preferred polar crystallographic orientation. After cooling, the samples are pulled out of the liquid solution. The crystals on the samples surfaces are removed by delicate wiping. Finally, the samples are dried in desiccators.

19.7 Composite and Nanocomposite Polymer Electrolytes for Electrochemical Energy Sources [8]

Environmental issues and risks of oil shortage, lead to extensive research on new energy sources i.e. solar, wind, and geothermal renewable energies, that are alternatives to power stations based on oil, to produce electricity and, by cogeneration, heat. Whatever the interest of such energy sources, these must be associated in performing the electrochemical energy sources as batteries, in order to store the peak productions of electricity, or to electrolysers allowing pure hydrogen, useful for Proton Exchange Membrane Fuel Cells (PEMFC) to be produced. In most of the developed countries, researches are presently funded to succeed in adapting PEMFC to electric vehicle (EV) use. Both in lithium batteries and PEMFC, one of the main bottlenecks deal with the polymer electrolyte which must provide high performances, high lifespan and above all, safety. In the frame of this paper J.Y.Sanchez *et al.* emphasize the interest of using composite and nanocomposite polymer electrolytes.

Nanocomposite Polymer Electrolytes. Lithium polymer batteries (Li anode) improve the EV autonomy while limiting the battery weight and volume. While Li non-rechargeable batteries which associate liquid electrolyte to a porous polymeric separator are commercialized, their rechargeable version do not meet the requirements of EV in terms of both lifespan and safety. On the other hand, the use of a polymer electrolyte free of added liquid solvent allows several hundred cycles of charge/discharge to be performed with a weak fading , meeting therefore the requirements for an EV application. Batscap company developped an EV i.e. Blue Car based on the lithium polymer battery. The elemental modulus of this battery whose total thickness is about 100 micrometers is shown in Fig.19.5.

In such polymer electrolytes, the polymer behaves both as a "viscous solvent" and as a separator preventing the battery of a short circuiting originating from a contact between the electrodes. The first function implies the salt dissolution, its dissociation, the ion solvation and their displacement. Polyamines, polyethers and polydioxolane are able to dissolve a wide variety of lithium salts and to interact with lithium cations. In addition to these functional properties, the host polymer, amorphous in the operating conditions of batteries, provide mechanical properties compatible with the use of thin films which allows the moderate conductivities (with respect to

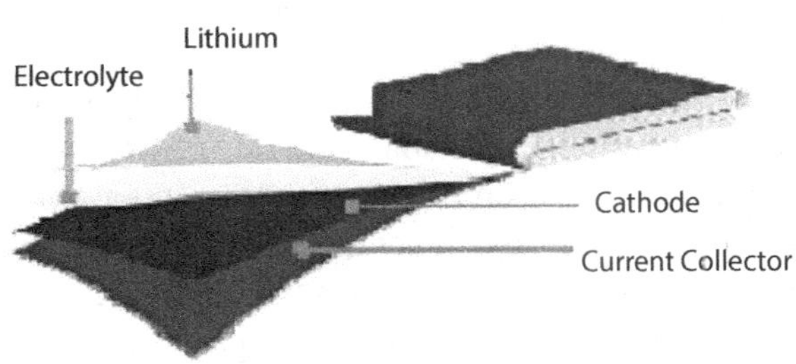

Figure 19.5 A lithium polymer battery of 100 μm thickness [8]

liquid electrolytes) to be compensated in terms of internal resistance. Up to recently, this objective was only achieved by cross-linking the polyether matrix.

It was early proposed to use nanofillers to improve the performances of polymer electrolytes i.e. through the addition of TiO_2, conductivity enhancement and sometimes, improvement of the polymer electrolyte interface with lithium anode. Jean-Yves Sanchez *et al.*[8] approach was to improve the mechanical properties of polymer electrolytes while keeping a high conductivity level. Usual reinforcements by incorporation of woven or unwoven fibres as glass fibres are unsuited to the elaboration of thin films of some tens of micrometers. The authors selected to use highly crystalline cellulosic fibres originating from the mantle of a sea animal i.e. tunicin. These fibres, which have a high 'form factor', form a network by hydrogen bonding which occur at a percolation threshold of some %.

19.8 Novel Nanostructured Materials for A Variety of Renewable Energy Applications [9]

Hydrogen can be generated by a variety of means including the electrolysis of water using the electricity derived from wind power, photovoltaics or by thermo-chemical processing of biomass. Hydrogen can then be reacted with oxygen in fuel cells to generate electricity, combusted in an engine to generate mechanical energy, or simply burned to generate heat. In each of these cases, water is produced in a virtually pollution-free process. Unfortunately, before hydrogen can be employed in the transportation sector, numerous technical hurdles has still to be overcomed. Recent theoretical studies have shown that by complexing the fullerenes with a transition metal (TM), H_2 may be bound with binding energies appropriate for on-board vehicular storage. In an optimal structure, scandium has been predicted to complex with all of the twelve five-membered rings in a fullerene. The $C_{60}[ScH_2(H_2)_4]_{12}$ complex has a reversible hydrogen capacity of 7.0 wt.%.

Recently, hot-wire chemical vapor deposition (HWCVD) has been employed as an economically scalable method for deposition of crystalline tungsten oxide nanorods and nanoparticles. Under optimal synthesis conditions, only crystalline nanostructures with a smallest dimension of $\approx$ 10-50 nm are observed with extensive transmission electron microscopy (TEM) analyses. The incorporation of these particles into porous films led to profound advancement in state-of-the–art electrochromic (EC) technologies. The development of durable inexpensive EC materials could make them suitable for large area window coatings and lead to decreased energy consumption in air conditioning. HWCVD has also been employed to produce crystalline molybdenum oxide nano- rods, particles and tubes at high density. It is also possible to fabricate large area porous films containing the MoO_3 nanostructures. Furthermore, these films have been tested as the negative electrode in lithium-ion batteries, and a surprisingly high reversible capacity as well as dramatically improved charging and discharging insertion.

19.9 Critical Fields in Lithium Niobate Nano Ferroelectrics [10]

An important property like giant polarization with many other properties are commonly observed in ferroelectric materials, which have an wide area of applications. However, the most significant property is their ability to show non-volatile memory, which remains finite when the field is withdrawn. This is clearly noticed in the hysteresis curve of polarization (P) vs external electric field (E), which is non-linear.

The ferroelectric materials such as lithium niobate and lithium tantalate typically show this non-linear hysteresis behaviour, which can be explained by dynamical system analysis. The behaviour of these ferroelectrics are usually explained by domains and the domain wall movements. So, the spatial variation of the domain wall was studied previously by A.K.Bandyopadhyay *et al.* [10] in order to see its effect on domain wall width in the context of Landau-Ginzburg function. By treating it as an eigenvalue problem, the critical values of polarization (Pc) were estimated within the 'zone of stability' through a linear Jacobian transformation, which showed the possibility of a 'giant memory'. The corresponding limits of domain wall width were also found in the case of lithium tantalate and lithium niobate ferroelectric crystal.

19.10 Nanoengineering of Wood Fibres for Conducting Paper [11]

Paper manufacturing is one of the mainstays of the American economic infrastructure, and paper products influence almost every aspect of business and personal life. From currency to tissue, trees and paper influence our lives. Although, it is said that the electronic sector is taking over the paper industry, the present technology cannot meet the need for paper. Therefore, while there is no significant threat from the electronic sector on the paper industry, it is wise to look for new opportunities to develop new paper-based products with extra functionalities.

To fill the many market sector needs, a variety of processes have been developed and introduced for production of pulp, forming sheets, and coating paper after formed into sheets. The coating processes developed thus far include printability improvements, opacity improvements, smoothness, and strength to name a few. These coating processes have in common that they apply a coat to the substrate paper after formation of the sheet from the microfibers. In this work, K.Varahramyan *et al.*[11] demonstrated the enhancement in electrical properties of paper using a systematic layer-by-layer (LbL) nanoassembly of polyelectrolyte thin films on cellulose microfibers prior to papermaking. LbLnanoassembly is a unique method based on sequential deposition of oppositely charged polyelectrolytes or nanoparticles on surfaces of different shapes and sizes.

The autors have used an aqueous dispersion of poly (3,4- thylenedioxythiophene)- poly (styrenesulfonate) (PEDOTPSS) conducting polymer as anionic polyelectrolyte, and poly (allylamine hydrochloride) (PAH) and poly(ethyleneimine) (PEI) as primary cationic polyelectrolytes. By creating alternating layers of these two polyelectrolytes on the surfaces of wood microfibers, the authors have produced a nanocoating that enables the microfibers to exhibit moderate electrical conductivity. Subsequently, they have used these microfibers for the production of handsheets that have a measurable electrical conductivity.

19.11 Effects of E - Waste on Environment and Their Solution by Reclamation of Green Materials From the Waste [12]

ABSTRACT. Use of electronic appliances/devices/equipment has immensely increased, now-a-days. They are turning into waste due to broken pieces, updation of technology and other reasons. These wastes are generally referred to as e-waste. Waste components of electronic appliances comprise of the recoverable green (eco-friendly) raw materials as well as environmentally-unsound materials. Keeping in view the important emphasis on environmental awareness, it is very essential to treat the e-waste materials. The treatment can be imparted by the processes of recovery, recycling and reclamation. Before carrying-out these reclamation processes, their collection and disposal methodologies are also required to be efficiently managed.

This section presents an insight into the sources of e-waste, their impact on environment, and suggests the solution by reclaiming the environmentally-safe (i.e. green) materials and their re-use after recycling. Suggestions for maintaining the healthy -environment with the help of green materials are also incorporated.

19.11.1 Introduction

The eco-friendly materials (i.e. green materials) do not pollute the environment due to following reasons.

i. They are reusable,
ii. They are recyclable, and
iii. They are biodegradable.

That is why they also reduce the

- waste,
- pollution, and
- use of energy

The clean -environment conscious society and the material scientists are well aware of the ill-effects of the waste materials that are non-reusable, non-recyclable and non-biodegradable. Hence, incorporation of green materials in different areas of engineering and technology is envisigated. One of the most important requirements of such materials lies in electronic and allied industries, where the quantity of e-waste is increasing at a very fast rate.

Technical scenario of the globe is changing very fast. The engineering world is dominated by electronic/electrical devices and equipment. The needs and requirements of the modern society cannot be fulfilled now, without the use of electronic gadgets. In fact, the present era can well be claimed to be 'electronic era'. The consumption of electronic items has immensely increased and it is still increasing at a very high rate in every walk of life. May it be industry, trade or commerce, household articles, military, education or entertainment; macro to milli (miniature) size electronic devices have made a deep dent everywhere.

Electronic/electrical devices and equipment cannot last forever. They have a certain life span. Once their life-cycle is over, they are of no use and are thrown away. During operational period, some of their part(s) may get damaged. They may get broken also. Many of these devices get obsolete and are to be removed from the service. They, therefore, accumulate as waste. Thus the electronic wastes (or e-wastes) are broken pieces, used and thrown pieces and undesired electronic/ electrical/ computer components and appliances. These are discarded items which are non-repairable. E-wastes, lying here and there in the office or dust-bin, are the sources of e-pollution. These are undesired and create unhealthy environment.

19.11.2 Major Sources of E-Waste

Electronic waste comes from various sources, the main among them may be categorized as follows.

Primary sources which cannot be re-used. These include:

1. Products used for data processing such as computers, computer peripherals viz. monitors, speakers, keyboards, mice/ mouse, printers.
2. Entertainment electronic items such as TVs, DVDs and CD players.
3. Items used for communications such as mobile phones, landline phones and faxes etc.
4. Household appliances such as vacuum cleaners, microwave ovens, washing machine, air- conditioners and refrigerators etc.
5. Audio-visual components such as VCRs, stereo equipment etc.
6. Office equipment such as photocopier.
7. Automatic systems such as ticket-issuing machines.
8. Medical electronic instruments and appliances.

Secondary sources which can be directly re-used. These include fully functional computers and components that are discarded

- during up-grading
- during re-furbishing
- whose constituent raw materials can be recycled.

19.11.3 Constituent Materials of E-Waste

E-waste normally contains various elements and substances, the most of which is suitable for reclamation. These materials are:

1. Metals
   i. Precious metals like gold (Au), platinum (Pt), silver (Ag), palladium (Pd) etc.
   ii. Useful metals like copper (Cu), aluminium (Al), iron (Fe) etc.
   iii. Toxic metals like lead (Pb), cadmium (Cd), chromium (Cr) etc.
   iv. Hazardous materials like mercury (Hg), radioactive isotopes.
   v. Toxic substances like dioxins, polychlorinated biphenyls (PCBs).
2. Plastics like
   - High-impact polystyrene (HIPS)
   - Acrylonitrile butadiene styrene (ABS)
   - Polycarbonate (PC)
   - Polyphenylene oxide (PPO)
   - Polypenylene ether (PPE) etc.
3. Glasses like
   i. Cathode ray tube (CRT) glasses such as SiO_2.CaO, NaO
   ii. Glasses such as K_2O, ZnO, MgO, PbO, BaO etc. for colouring and protecting from X-rays.
4. Printed wiring board.

19.11.4 Estimation of the Quantity of E-Waste

The worldwide quantity of e-waste amounts to millions of tonnes. On an average, the weight percentage of different types of materials in computers and TVs are as given in Table 19.1.

19.11.5 Problems Created by E-Waste

Electronic wastes, if not treated, create environmental problems. Their disposal,disassembly and uncontrolled burning cause the following major problems.

Table 19.1 Weight percentage of materials that can be reclaimed

Types of materials	*Approximate weight %	
	In Computers	In TVs
Metals		
Precious metals	0.02	27.0
Copper	7.0	5.0
Aluminium	14.0	-
Iron	21.0	-
Lead	6.5	-
Plastics	23.0	15.0
Glass	25.0	48.0
Printed wiring board	-	5.5
Others	Rest	Rest

* Collected from different sources (newspaper reports, information in open domain, etc.) and approximated to round figures.

i. They create toxins, hence toxicity. This is caused due to toxic substances like
   - Hg • Pb • Cd • Cr
ii. They are sources of carcinogens, hence carcinogenicity. Carcinogenic substances in e-wastes are
   - polychlorinated biphenyls (PCBs).
   - dioxins
iii. They cause health problems to all those who are directly involved with the e-waste processing, as they also contain
   - radioactive isotopes.
iv. Printers release dangerous ozone that retards the green-house effect,
v. Used CDs,empty cartridges,toners and replaced parts create disposal problems.

19.11.6 Electronic Waste Reclamation

The e-waste may be reclaimed in many ways such as given below,

1. They may be re-used.
2. They may be recycled to reclaim some useful materials in raw form.

Benefits of reuse of e-waste. There are several social and environment benefits of re-use of e-waste. These are:

i. diminished demand for new products.
ii. saving of a large quantity of water and energy, which otherwise will be consumed in production of new products.
iii. reduced need of packaging
iv. decreased requirement of e-waste in landfills.

19.11.7 Electronic Waste Recovery and Recycling

The process of recovery starts with dismantling of the appliances into metals, glasses, plastics etc. Then for recycling, these materials undergo the following processes one-after another.

1. Feeding into a hopper.
2. Transportation by a conveyor into mechanical separator.
3. Screening and granulation in different stages.

To prevent pollution due to dispersion of metallic dust, the recycling machinery system is kept enclosed with the help of a dust collection system.

The recycling processes has several advantages viz. it

- saves energy
- saves natural resources
- saves landfill space
- reduces pollution

19.11.8 Advanced Methods of Environmental-Friendly Recycling

The recycling methods are primarily targeted at separating the following constituents.

1. Hazardous components/materials, and
2. Valuable (precious) materials/components

Thereafter, the mechanical processes such as given below are executed.

i. Screening
ii. Magnetic separation
iii. Eddy current separation

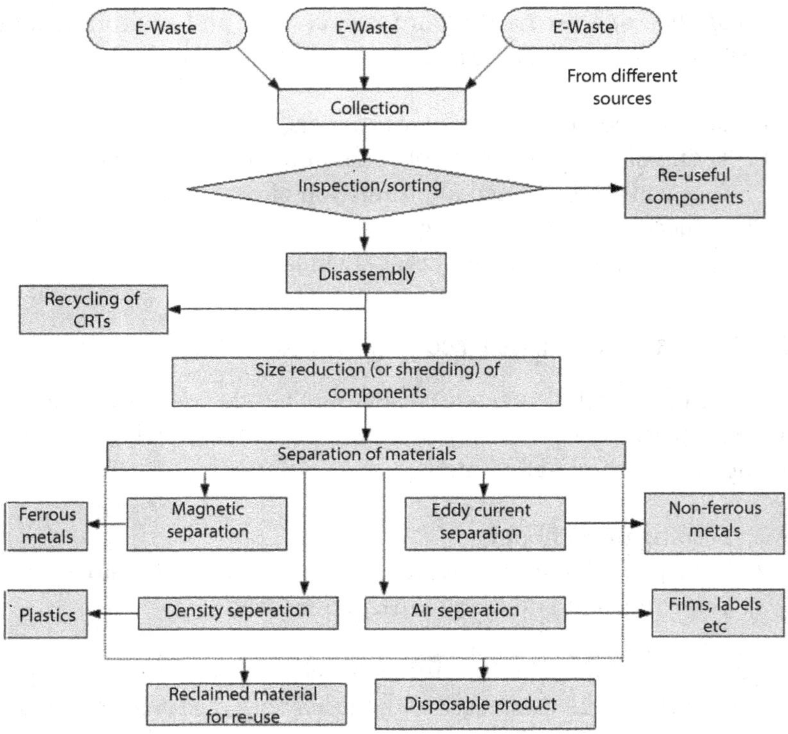

Figure 19.6 Schematic flow chart illustrating the steps involved in recycling from e-waste

iv. Electrostatic separation
v. Shape separation
vi. Jigging

For obtaining maximum separation of materials, the e-waste should be grinded to the size of fine particles, generally below 10 mm size or below 5 mm, if possible.

Various steps involved in recycling from e-waste are shown schematically in Fig. 19.6.

19.11.9 Status of Possible Recovery of Useful Materials from E-Waste in India

India is a fast developing country. It is one of the fastest growing economies of the world. The domestic demand for its consumer durable is also sky-rocketing. Some major details in this regard are the following [12].

- Increase in the sales of domestic household appliances during 1998-2002 was 53.1%.
- Growth in PC ownership per capita during 1993-2000 was 604% as compared to world average of 181%.
- The number of obsolete PCs in 2003 was 1.38 million.

However, the e-waste per capita in India is at a lower level as compared to developed countries. Moreover, in India, the e-waste is sold to second and third hands; therefore the recovery is at a lower rate. There are not too many recovery plants also, hence recovery of useful materials is much less.

19.11.10 Conclusions

To minimize the environmental problems arising due to e-waste, following solutions are suggested here.

1. Rules and regulations must be framed and implemented strictly to improve the recovery, recycling and reuse of materials. This will not only minimize the pollution caused due to e-waste, but will also recover valuable materials.
2. Cost effective electronic waste recycling plant, which should be environmental friendly also, should be installed for component and materials recovery.
3. Advertisement by all possible means should be made for collection of e-waste.
4. Use of ozone-depleting chemicals such as CFCs, chlorine etc. should be eliminated from manufacturing.
5. The components of the appliances should be made of the recyclable plastics.
6. The PCs and peripherals, TVs, CDs etc. should qualify for green-environment throughout their life- cycle.

19.12 Plastics in Electrical and Electronics Applications

19.12.1 Expanding Plastics

Common plastics shrink when cured. *Expanding monomers (EMs) experience zero shrinkage during polymerization.* The polycyclic ring opening monomers are combined with conventional monomers and oligomers to produce expanding plastics. These plastics have enhanced resistance to corrosion and abrasion. Expanding monomers find use in coatings, adhesives, electro-optics, dentistry, medical prostheses, electronics and other industries. EMs may be used to

i. eliminate internal stresses by expanding into micro-cracks in *advanced composites.*
ii. copolymerize with epoxies, acrylics, and urethanes to generate unprecedented properties in *aviation* and *aerospace* plastics applications.
iii. deliverbio-degradablepolymerizedvesiclesin *Pharmaceuticals.*
iv. produce advanced *microchip sealants*, and *conductive links* in electronics.
v. combine with GMA (Glycidyl methacrylate) for superior fillings, dentures, and crowns *in dentistry.*

19.12.2 Conducting Polymers

The advent of *polyacetylene* (PAC) in early seventies raised hopes and interest in conducting polymers. This system, however, is non-conducting, and may be regarded as semiconducting. The conducting fillers such as metals wires and powders, graphite powders and transition metal compounds are incorporated in non-conducting polymers to make them conducting. In a later development, the conductivity of PAC is enhanced by more than ten times when *iodine* is doped into it. The doped PAC possesses conductivity much higher, and almost equal to metallic range. It, however, loses conductivity rapidly when exposed to atmospheric conditions. Another limitation is the migration and loss of dopant from the surface. Conducting polymers can generate electric signals by ion-exchange mechanism. *Conducting electroactive polymers* belong to the class of smart materials and are used as *chemical sensors.*

19.12.3 Polymers in Electronics

Several polymers show excellent *opto-electronic* properties. Many polyimides are used for integrated insulation and cross overs for *microwave integrated circuits. Aviation radome, anti-radar paints*, and *heat shields* are some novel applications of special polymers.

19.12.4 Liquid Crystal Polymers (LCP)

These are made of rigid, rod-type, ordered molecules, which maintain their crystalline order even on melting. Based on the molecular orientation displayed, they are classified as follows.

1. Nematic liquid crystal polymers, and
2. Symectic liquid crystal polymers.

The main difference between these polymers, and the conventional liquid crystals used in electrical display devices, is their molecular weight. *LCPs have much higher molecular weight.*

Different types of LCPs, their examples and advantages. Depending on the packing arrangement of rigid units, there are two types of LCPs.

1. Main chain LCPs (e.g. polyesters, polyamides), and
2. Side chain LCPs (e.g. polyacrylate, polymethacrylate, polysiloxane).

The advantages of these polymers over conventional plastics are their

- high impact strength,
- excellent dimensional stability,
- low thermal expansion,
- better chemical resistance, and
- outstanding processability.

They are used in highly specialised areas viz. telecommunication, fibre optics, chemical and hostile environments. *Xydar* and *Vectra* are the example of commercial LCPs.

19.12.5 Photocurable Polymers

These polymers find applications in *microlithography* for making printing plates, photoresist for microcircuits, video discs and optical fibre coatings etc. In this highly specialized radiation curable system, the *the photoresist* such as *polyvinyl cinammate* and related materials are synthesized by using low temperature interfacial transfer catalysed technique.

19.12.6 Photorefractive Polymers

Photorefractive effect is a light-induced change in optical properties of a material caused by absorption of light and transport of charged photocarriers. This effect is obtained from different photorefractive materials such as high-resistivity semiconductors, multilayer thin films, ferroelectric oxides, and photorefractive polymers etc. The photorefractive materials are useful for, or likely to be useful for the following main applications.

- Optical computers
- Optical control systems
- Holographic optical memory
- Optical image processing
- Phase conjugation, etc.

Photorefractive polymers are high performance photorefractive materials of new era. Compositional flexibility of these materials are similar to electrophotographic polymers and nonlinear optical polymers. Their use is likely to be made in above applications in future. The photoconductivity and optical nonlinearity necessary for photorefractivity are generally provided by combining the sensitizers i.e. photoinduced charge generators, charge transporters, and non-linearly optical (NLO) chromophores. Their examples are the following.

- *Sensitizers*: Charge transfer complexes e.g. PVK/TNF, dyes and dye aggregates, fullerenes (C60)
- *Charge transporters*: Electron donors e.g. hydrazones, aryl amines, carbazoles, conjugated polymers; and electron acceptors e.g. trinitrofluorenone etc.
- *NLO chromophores*: Conjugated molecules e.g. tolanes, stilbenes, styrenes, polyenes, azobenzenes, liquid crystalline mesogens etc.

19.13 Composite Materials for Electronics Applications

Present era is of electronics. It has made dent in every walk of life. Major breakthrough in engineering is attributed to the advancement of electronics. Materials play very important role in such advancements. The development of miniature-sized electronic devices has also become the essential need of today. They also need to be smart enough to exhibit unique and fascinating behaviour and properties. To expose to these scenario, the recent developments taking place in the area of composite materials for various applications in electronics industry have been focused in this section. The details contained in it depict the trends in researches being done by various investigators and their findings.

19.13.1 Magneto-Eletro-Elastic Composite [13]

Magneto-electro-elastic (MEE) composites are multi-functional materials involving the interaction of magnetic, electric and elastic effects. These materials find wide usage in electronics industry. Composite made of $BaTiO_3$ and $CoFe_2O_4$ is one such example, in which the $BaTiO_3$ remains as inclusion constituent in $CoFe_2O_4$ which acts as matrix constituent.

Some main applications of composites having magneto-electric-elastic properties are sensors, transducers, waveguides, phase invertors etc. A ferroelectric and ferrite phase can be combined to produce a coupling of two fields viz. electric field and magnetic field. The electric field can induce magnetization, while a magnetic field may induce electric polarization.

Composites having piezomagnetic + ferrite phase, and piezoelectric + ferroelectric phase show some specific behaviour, when fabricated with varying percentage of their compositions. The specific behaviour may be in the form of yield of maximum magneto-electric signals along with adequate mechanical strength of composite microstructure. In them, the strain is produced by lattice distortion, to induce magnetostriction. This entails creation of stress in these phases.

19.13.2 Magnetic Composites [14]

Magnetic composites are generally made of conducting materials and magnetite (Fe_3O_4), or conducting materials and magneto-rheological (MR) fluids. They exhibit unique electrical performance as well as enhanced magnetic properties. These may be organic or inorganic composites. Polyaniline (PANI) may be a good conducting material due to its chemical stability and good process ability. It can be made of composite with magnetite that finds wide applications in ferrofluids, printing inks, magnetic storage media and magnetic-guided drug delivery etc.

MR fluids are fascinating materials, composed of colloidal particles dispersed in a non-magnetic carrier liquid. These exhibit rapid and reversible changes under external magnetic field. The changes may be in their yield stress and shear viscosity due to the transformation between solid-like and liquid-like states. MR materials are used in applications requiring the control of vibrations, control of force and torque such as in shock absorbers, isolators, dampers and brakes etc. When a magnetic field is applied to MR fluids, the dispersed magnetic particles are polarized and aligned in the direction of magnetic field. It thus forms a solid-like fibrous structure. A conducting polyaniline + nanosized Fe_3O_4 composite, synthesized by a chemical reaction process is an example of magnetic composite.

19.13.3 Ferrite-Silica Hybridized Composite [15]

Miniaturization of devices and their reduced cost are the needs of today's technology. Such needs can be fulfilled by those materials which possess magneto-crystalline anisotropy, magneto-optical properties, magnetostriction, excellent chemical stability and high corrosion resistivity etc. Ferrite composites such as CoZn ferrite + SiO_2, NiZn ferrite + SiO_2, NiZn ferrite + B_2O_3, NiZn ferrite + polymer are such examples. Ferrite composites equipped with above cited properties are useful for applications like magnetic static wave devices, surface acoustic wave transducers, recording media with optical wave guides etc.

Silent features of such composite materials are their high electrical resistivity, low magnetic and dielectric losses which make them useful in absorption of microwaves applications. Similarly, the complex oxides consisting of transition metals like Fe, Co, Ni, Mn are very much suitable as negative temperature coefficient thermistors for temperature measurement and control. CoZn ferrite + SiO_2 composite having general formula (1-x) $Co_{0.5}Zn_{0.5}Fe_2O_4$ + x SiO_2 with x = 0.0 to 0.8 has been prepared by co-precipitation technique. Results of its characterization show that the room temperature resistivity increases from 10^5 ohm-cm to 10^9 ohm-cm when x is increased from 0.0 to 0.8.

19.13.4 Composite Electronic Packaging Material [16]

Metal matrix composites possessing high thermal conductivity, high mechanical strength, low density and an appropriate coefficient of thermal expansion are suitable for electronic packaging purposes. SiC/Al may be one such material. Such materials are useful in applications like base plates and enclosures of power modules, heat sinks and substrates etc.

These composites are metalized by different processes in which a layer of alloy/ metal is plated on the surface of the composite to improve its properties. The improvement may be in terms of an increase or decrease in thermal conductivity, solderability etc. A coating of Ni-P on SiC /Al to improve thermal conductivity makes this composite a very suitable material for electronic packaging application.

19.13.5 Magnetostrictive Material-Shape Memory Piezoelectric Actuator Composite [17]

Ferroelectric materials have multifunctional properties, such as piezoelectricity, large permittivity, nonvolatile charge and electrooptical functions. In order to realize a memory effect of these properties the imprint electrical field has to be controlled. Conventional piezoelectric actuators are driven with a DC voltage to maintain certain positions. However, with the use of imprint electrical field control, a piezoelectric actuator can be operated with a pulse-shaped voltage, so that the conventional piezoelectric actuators can have a memory effect.

In recent years, magnetostrictive materials that have giant magnetostriction (over 1000 ppm) have been produced such as Terfenol-D ($Tb_xD_{y1-x}Fe_2$), and various applications have been investigated. A voltage controllable mechanism has been proposed by using this material with a piezoelectric material and a permanent magnet material. These studies have shown that the shape change induced by the piezoelectric actuator is utilized for change in the permeability of a magnetostrictive material. The permeability change results in a magnetic flux density. This principle has been applied to realize the voltage control for a magnetic flux density without coil; however, to maintain a certain magnetic flux density, continuous voltage supply is required.

In this study, the magnetic permeability of the magnetostrictive material is controllable depending on its strain condition. With a permanent magnet attached to a magnetostrictive–piezoelectric actuator composite, the magnetic flux density can be controlled by the voltage to the piezoelectric actuator. The advantage of this system is a coil-less structure, and voltage operation instead of current operation. The innovative aspect of this study is to apply a shape memory piezoelectric actuator to the system. The shape memory piezoelectric actuator is fabricated with control of the imprint electrical field, which enables the use of a pulse shaped voltage.

By attaching the shape memory piezoelectric to a magnetostrictive material, the magnetic permeability can be operated with a pulsed voltage and the memory effect can be obtained without voltage input. The combination of this composite with a permanent magnet is expected to result in an innovative magnetic actuator with a magnetic memory effect.

19.14 Electrical Behaviour of Ceramics

Ceramics are widely used for insulation purposes in electric power transmission, electrical machines/ equipments/devices etc. Mica, porcelain, glass, micanite, glass bonded mica, asbestos, glass tape etc. are commonly used (ceramic) materials. They show the following characteristics in respect of their electrical behaviour.

Dielectric constant. Ceramics have high values of dielectric constant ε_r. It is 8 for mica, 7 for soda-lime glass, and 6 for porcelain. That is why they are used for insulation purposes in electric motors, alternators, transformers etc. See more details in the chapter on dielectrics.

Dielectric strength of ceramics is high. It is defined as the voltage required per unit thickness to cause a breakdown in material. Dielectric strength of mica is 100 MV/m. This is pretty good strength that makes the mica as a very good insulating material.

Dielectric loss. A good electrical insulator should have a low value of dielectric loss. It is characterized by power factor tan δ, where δ is the loss angle. Value of dielectric loss in mica is only 0.0005 which further supplements its use as a good insulating material.

19.14.1 Applications of Ceramics

Ceramics are widely used materials. They find use in almost all fields of engineering ranging from mechanical, civil and electrical to computer, biomedical and nuclear. Important applications of ceramics in electrical applications are listed below.

i. As firebricks and fireclay for lining of ovens and furnaces
ii. As insulators (dielectrics) in electrical transmission and distribution
iii. As ferroelectric crystals and Piezoelectric crystals to generate ultrasonic waves for non- destructive testing and other purposes
iv. As magnetic materials in electrical machines, devices etc.
v. As powerful superconducting magnets to make magnetic rail for wheel-less train (leviated train)
vi. As ferrites in memory cores of computers.
vii. As miniature capacitors for electronic circuits

viii. As garnet in microwave isolators and gyrators
ix. As components of sonar device that helps to locate a small object in large volume (e.g. searching of a black box from within the ocean)
x. As electronic packaging to protect ICs. Aluminium nitride (AlN), SiC, boron nitride (BN)are used as substrate materials for this purpose

19.15 Giant Magneto-Resistance (GMR) [18]

The giant-magneto resistance is a prototype device which is used as a 'read head' and 'memory–storage cell'. It consists of a sandwich structure composed of alternating layers of ferromagnetic and non-magnetic metals. Depending upon the desired orientation of magnetizations in magnetic layers, the resistance of devices changes from a small to large. The small change occurs in parallel magnetization, while a large change takes place in anti-parallel magnetizations. Recently, the magnetic tunnel junction devices have also been produced in which, the tunnelling current depends on spin orientations of the electrodes.

In order to observe GMR, one has to provide an opportunity to reorient the magnetic moments of ferromagnetic layers relative to one another. In magnetic multilayers, this can be achieved due to the effect of antiferromagnetic interlayer coupling, which is a particular case of interlayer exchange coupling. The interlayer exchange coupling oscillates between ferromagnetic and antiferromagnetic as a function of the thickness of non-magnetic layer. By choosing an appropriate thickness of non-magnetic layer, it is therefore possible to create an antiparallel configuration of ferromagnetic layers and then reorient (align) the moments by an applied magnetic field.

19.16 Ferrorfluids (or Magnetic Fluids) [19]

Although the iron, nickel and cobalt are ferromagnetic materials, yet they are ferromagnetic in solid state only. Contrary to this, the liquids can also be produced which are ferromagnetic in nature. Ferro-fluids are such kind of fluids. A ferrofluid is the fluid media composed of solid magnetic particles of very small size (nanoparticles) which are colloidally dispersed in a liquid carrier such as transformer oil or kerosene. It is highly stable and strongly magnetic liquid.

Ferrofluids are extremely interesting due to interplay between hydrodynamic and magnetic phenomena. Ferrofluids are colloidal suspensions of single-domain superparamagnetic particles,with typical dimensions of about 10 nm, dispersed in a liquid carrier.It responds to a magnetic field. The solid particles are generally stabilized with an attached surfactant layer.A typical ferrofluid may contain by volume 5% magnetic solid, 10% surfactant and 85% carrier [19].In the 1960's, Stephen Pappell at NASA first developed ferrofluids as a method for controlling the fluids in space. Magnets and/or magnetic fields were used to control this magnetic fluid.

19.16.1 Types of Ferrofluids

The ferrofluids may be grouped into following two categories.

1. Surfacted ferrofluid (SFF), if the coating is a surfactant molecule.
2. Ionic ferrofluid (IFF), if it is an electric shell.

Surfacted Ferrofluid (SFF). In this case,

i. The steric repulsion between particles acts as a physical barrier that keeps grains in the solution and stabilizes the colloid.
ii. It is widely used in technological devices.

Schematic arrangement of surfacted ferrofluid grains are shown in Fig.19.7a-b. Its examples are:

- Magnetic particles (usually magnetite, Fe_3O_4) coated with surfactant agents (amphiphilic molecules, as oleic acid and aerosol sodium di-2 ethylhexyl-sulfosuccinate).

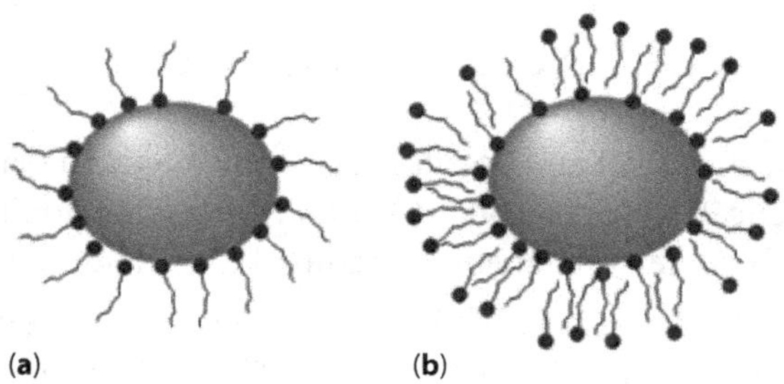

Figure 19.7 Surfacted ferrofluid grains (a) single-layered grains, and (b) double-layered grains [19]

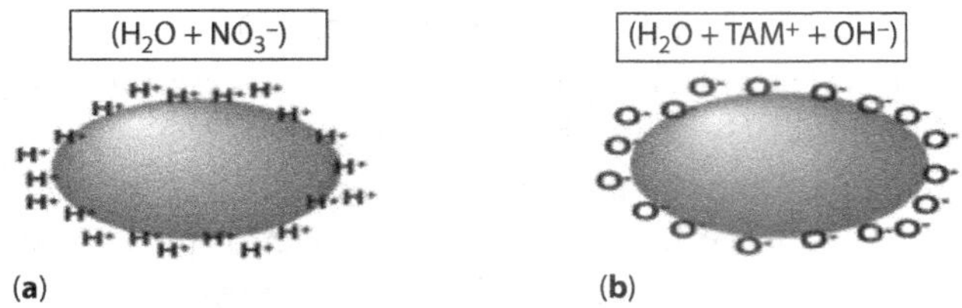

Figure 19.8 Ionic ferrofluids grains [19]

Ionic Ferrofluid (IFF). In this case,

i. The Nanoparticles are electrically charged to keep the colloidal system stable.
ii. An acid-alkaline reaction between particles and the bulk, keeps the surface of them electrically charged.
iii. Acid IFF (pH < 7) have positively charged particles, and alkaline IFF (pH > 7) have negatively charged particles.
iv. They have steric and electrostatic repulsion.

Their examples are

- Magnetic particles (usually maghemite, γ-Fe_2O_3, and different ferrites, MFe_2O_4 where M = Mn, Co, Zn, Cu, Ni)and liquid carrier is water.

Schematic arrangement of ionic ferrofluids are shown in Figs. 19.8.a-b. In it, the counterions and the water molecules are also indicated (a) acid ferrofluid gain, and (b) alkaline ferrofluid, (TAM^+OH is the tetramethyl-ammonium hydroxide)

19.16.2 Properties of Ferrofluids

Various properties of ferrofluids are the following.

i. Ferrofluids are optically isotropic; but in the presence of an external magnetic field, they exhibit induced birefringence.
ii. In them, there is a change in density in proportion to magnetic field strength.
iii. Ferrofluids keep their fluidity even if subjected to strong magnetic fields (10 kG)

iv. Ferrofluids also change their resistance.
v. They have exceptionally high magnetic susceptibility.
vi. The instability is driven by the magnetic field.
vii. When a paramagnetic fluid is subjected to a strong magnetic field, the surface forms a regular pattern of peaks and valleys. Thus they have a normal-field instability.
viii. The forces holding the magnetic fluid in place are proportional to the gradient of the external field and the magnetization value of the fluid.
ix. The magnetization of the ferrofluids responds immediately to the changes in the applied magnetic field; and when the applied field is removed, the moments randomize quickly.

19.17 Information Storage Density

The bubbles, by their presence or absence, may be used for storing the binary information and that the information can be read out. The advantage of using magnetic bubble is mainly, density. With presently available photoengraving techniques, the contiguous structure may lead to a density of 10 million bits per cm^2.This is of course not a random access memory, the information has to be read out serially; the achievable speed may be a few hundred K bit/s. It could be used at an advantage when large blocks of data need to be transferred to the main memory of a computer for processing.

19.17.1 Rare-Earth Aluminates as a Charge Trapping Materials for Nand Flash Memories [20]

Charge trapping based TANOS (TaN–AlO–SiN–SiO–Si) memory is a potential candidate to replace the floating gate technology. It is because of their tolerance to and reduced parasitic coupling with neighboring devices. However, they suffer from critical retention and poor erase performances which are associated with these devices. These memories store charge in electrical defects of dielectric materials, Si_3N_4 for TANOS. The energy of the traps plays an important role in defining the retention capabilities. In an ideal trapping material, traps should have higher trap density to allow wider memory window, and preferentially deep energy traps for better retention. Further, high electron affinity (i.e. a high conduction band offset with blocking and tunnel dielectrics) is also desirable as it helps retention.

High-k (dielectric constant) materials are interesting as a charge trapping layer, as they are found to be trap rich. Moreover, having a k-value higher than that of Si_3N_4, it allows a greater voltage drop at the tunnel oxide which would also improve the programme and erase operations. For these reasons, in recent years, the possible replacement of Si_3N_4 with materials such as HfO_2, HfON, HfAlO, Y_2O_3, ZrO_2 and LaO_3 has been investigated. On the contrary, the rare-earth aluminates like LaAlO, GdAlO, and LuAlO have never been integrated as charge trapping material in TANOS-like stacks.

19.18 Magnetocaloric Materials [21]

Today, everyone wants a readily available cooling. Magnetic refrigeration based on the magnetocaloric effect (MCE) has become a promising competitive technology for the conventional gas-compression/expansion technique in use today. The magnetocaloric effect is the change in temperature of a material as a result of the alignment of its magnetic spins that occurs on exposure to an external magnetic field. The phenomenon forms the basis for magnetic refrigeration. So, the materials having magnetocaloric effect are known as magnetocaloric material.

The magnetocaloric effect (MCE), or adiabatic temperature change (ΔT_{ad}), which is detected as the heating or the cooling of magnetic materials due to a varying magnetic field, was originally discovered in iron by Warburg in 1881.Since the 1930s, the magnetic refrigeration has been a standard technique in low-temperature physics. In 1976, Brown designed the first magnetic refrigerator working at room temperature. The major breakthrough, however, occurred in 1997 when the Ames Laboratory/Astronautics proof of-principle refrigerator showed that the magnetic refrigeration could be competitive with conventional gas-compression cooling. The early prototypes were able to reach high magnetic flux densities in the magnetocaloric material, only if the superconducting magnets were applied. The first "room-temperature magnetic refrigerator" containing permanent magnets was designed and built in 2001 by the same group at Astronautics Corporation.

19.18.1 Magnetocaloric Materials

The discovery of the giant magnetocaloric effect in $Gd_5Si_2Ge_2$and in a few other families of intermetallic compounds were the important developments, both in science of the magnetocaloric effect and in its application to near room temperature cooling. The most promising compounds today are:

- $Gd_5(Si_xGe_{4-x})$ with *x* near 2 (it is worth noting that when x is greater than 2, this family offers excellent conventional MCE materials with Curie temperatures as high as 340 K),
- MnAs and Mn $(As_{1-x}Sb_x)$, which also show a large increase of the MCE when subjected to hydrostatic pressure.
- $FeMn(P_{1-x}As_x)$ and related compounds, and
- La $(Fe_{0.88}Si_{0.12})H_y$, where *y* is 1 or greater.

19.18.2 Working Principle

The MCE is intrinsic to all magnetic materials due to the coupling of magnetic sublattice with the magnetic field, which changes the magnetic part of the entropy of a solid. Just as the compression of a gas, the isothermal magnetizing of a paramagnet or a soft ferromagnet reduces the entropy and, in a reversible process, demagnetizing (which is similar to the expansion of a gas) restores the zero-field magnetic entropy of a system. At constant pressure, the entropy of a magnetic solid which is a function of both: the magnetic field strength and the absolute temperature, is the combined total of the magnetic and lattice contribution.

The MCE can also be expressed by means of isothermal magnetic entropy change, when the magnetic field is applied isothermally.

19.18.3 Applications of Magnetocolorific Materials

The magnetocalorific materials find use in the following applications.

- Refrigerators/air conditioners.
- Food preservation.
- Low temperature electronics.
- Liquefaction of gases, especially natural gases and hydrogen.
- Cryogenic detectors spanning the entire electromagnetic spectrum.
- Storage of cryogenic propulsion fuels for long-term space missions.

A magnetic refrigerator is shown in Fig. 19.9.

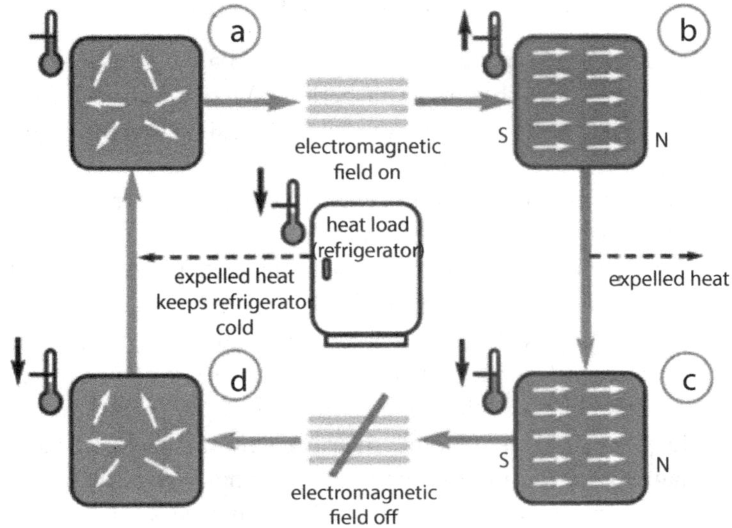

Figure 19.9 Magnetic Refrigerator [22]

19.19 Magneto-Dielectric Materials [23]

Magneto-dielectric materials are magnetically soft materials, made as composites by compressing the magnetic particles in a dielectric material. It provides both, electrical insulation and binding properties. Magnetic properties of the final product depend on the average size of the particles, their shape and the packing density. Mechanical properties are dictated mostly by the type of the binder. Magneto-dielectric materials consist of several families such as (i) spinel, (ii) garnet, and (iii) hexagonal ferrites. These materials have a lot of advantages in the following applications, based on their magnetic and dielectric properties.

- Resonators
- Microwave absorber devices
- Magneto-optical systems
- Memory applications etc.

However, their properties decline at high frequency ranges, majority beyond 10 MHz. Hexagonal ferrites containing a very complicated crystalline structure are magneto-dielectric materials which are able to work in this microwave frequency range.

19.19.1 Applications of Magneto-Dielectrics

- Effective antenna miniaturisation is a challenging problem due to the fact that antenna performance is bound on the fundamental limit of its size. But by using the magneto-dielectrics, this crucial problem is solved.
- Antenna miniaturisation is the most important application of magneto-dielectric materials.
- Materials with magnetic permeability and permittivity >1 aid in antenna miniaturisation.
- High frequency devices i.e. antennas require the use of high refractive index material, materials with a reasonable impedance match to air (close to 1).
- The above parameters are measured by the dielectric and magnetic properties of the materials through its permittivity (ε) and permeability (μ) respectively, as well as the usable bandwidth.
- Magneto-dielectrics have both: the relative permeability and relative permittivity greater than 1, hence can be used to miniaturise the antennas without deteriorating the band width.

Miniaturisation of patch antennas method and materials:

- Miniaturisation in the past has been achieved by structural modifications like the introduction of slots, slits, fractals or using materials of high dielectric constant.
- These modifications lead to a reduced bandwidth.
- An improvement has been made by the introduction of nickel-cobalt-zinc ferrite substrate, the magneto-dielectric material with permittivity 10.3 and permeability 2.3.
- Antennas designed on the above substrate have a lower Q and a wider bandwidth.
- In addition to the above, the use of nickel-cobalt-zinc ferrite substrate leads to a reduction in the excitation of surface waves leading to lesser coupling losses in array antennas.

19.19.2 Different Types of Magento-Dielectric Materials

Magneto-dielectric materials are metamaterials characterised as having real permeability and permittivity greater or less than unity. In order to achieve this, the material must contain an insulator phase which contributes a low dielectric value, and a magnetic phase which contributes a high magnetic value. There are three categories for the incorporation of magnetic phase into the substrate. The inclusion of the following magnetic materials have been observed with increasing difficulty:

Naturally ferromagnetic metal: The direct way of creating a two phase simultaneously high dielectric and magnetic medium is to embed the electrically conducting ferromagnetic metal within an insulating matrix. Ferromagnetic metals such as iron, nickel, cobalt, and their alloys have some of the highest known saturation magnetisation, allowing for the possibility of a very high composite permeability. At the same time, the dielectric constant is increased by the addition of electrically conducting filler.

Ceramic ferrites: Ceramic ferrites have been investigated as an alternative for high conductivity metallic ferromagnets. Ceramic ferrites generally have very high resistivities, allowing for comparatively low loss operation.

Artificial magnetic material: None of the abovementioned methods to fabricate the magneto-dielectric materials have yielded composites that are of use in the microwave region. Ferromagnetic fillers tend to generate large eddy currents when wavelengths approach the scale of the filler, and ceramic ferrites have an inherent resonance that limits their use above 1 GHz. Still another approach exists for creating the materials with magnetic and dielectric properties of permeability and permittivity >1, within microwave frequencies. However, these recently explored materials do not contain any inherent magnetic components. When metals are placed with an oscillating magnetic field, the magnetic dipole moments are created, leading to an effective magnetic permeability.

19.19.3 Merits of Magneto-Dielectric Materials

i. The magnetic moment in an atom is mostly due to electron spin alignment by exchange energy between electrons. Therefore, the ideal ferromagnetic materials have many unpaired electrons that are aligned in parallel or paired in anti-parallel.
ii. Magnetic domains exist to reduce the magneto-static energy, which is proportional to the square of the magnetic field create by the aligned atomic moment.
iii. Ferromagnetic materials have a non-unity permeability dependent upon the field strength and previous magnetic history.

iv. A non-magnetic conductor can exhibit non-unity permeability at high frequencies, however when a sizeable magnetic response is needed at low frequencies the ferromagnetic materials must be selected.
v. Desired ferromagnetic materials exhibit sufficient magnetic ordering of the solid's atom in the absence of magnetic field.
vi. A single domain block of material can reduce its magneto-static energy through splitting into smaller domains by inserting Bloch walls.
vii. Bloch walls introduce energy and a block of material that will divide into multiple domains until there is equilibrium between the reduction in magneto-static energy and the energy required to create new Bloch walls.

19.20 Biomimetics and Biomimetic Materials in Electrical and Electronics Applications

19.20.1 Nanotechnology Used to Harness the Power of Fireflies [24]

Firefly light is one of nature's best examples of bioluminescence, Fig.19.10. Its light is extremely bright and efficient. Fireflies produce light through a chemical reaction between luciferin and it's counterpart, the enzyme luciferase. To develop the same effect in laboratory through an experiment, enzyme is attached to the nanorod's surface; luciferin which is added later, serves as

Figure 19.10 A firefly is nature's best example of bio-luminescence

the fuel [24]. The energy that is released when the fuel and its the enzyme interact, is transferred to the nanorods, causing them to glow. The process is called Bioluminescence Resonance Energy Transfer (BRET). The trick for increasing the efficiency of the system is to decrease the distance between the enzyme and the surface of the rod, and to optimize the rod's architecture.

The scientists at Syracuse University Dr. Mathew Maye, Assistant Professor of Chemistry in SU's College of Arts and Science; and Rebeka Alam, a Chemistry Ph.D. candidate says, "We designed a way to chemically attach, genetically manipulated luciferase enzymes directly to the surface of the nanorod." Maye's collaborators at Connecticut College provided the genetically manipulated luciferase enzyme. The nanorods are composed of an outer shell of cadmium sulfide and an inner core of cadmium seleneide. Both are semiconductor metals. Manipulating the size of the core and the length of the rod, alters the color of the light that is produced. The colors produced in the laboratory are not possible for fireflies. The nanorods glow green, orange, and red. Fireflies naturally emit a yellowish glow. The efficiency of the system is measured on a BRET scale. The researchers found their most efficient rods (BRET scale of 44) occurred for a special rod architecture (called rod-in-rod) that emitted light in the near-infrared light range. Infrared light has longer wavelengths than visible light and is invisible to the eye. Infrared illumination is important for such things as night vision goggles, telescopes, cameras, and medical imaging.

19.20.2 Biomimetic Coating for Electric Transmission [25]

Energy is one of the most important issues in this century. To conserve energy in high voltage transmission lines, such as by preventing leaking currents, dry band arcing and flashover caused by contamination of the insulator surface; a self-cleaning surface is proposed. This is where the lotus leaf has provided inspiration. Despite growing in the muddy waterways of Asia, the leaves and flowers of the lotus plant remain clean. Its water repelling characteristics result from its microscale surface structures and nanoscale waxy protrusions. When rain hits the leaves of the lotus plant, it simply beads up. Because of the combination of nanoscale and microscale structures, the water droplets can only make contact with about 3% of the surface. Such properties have already prompted the development of 'self-cleaning' coatings on glass, hard disks drives and magnetic tape.

However, to mimic the lotus coating effect for outdoor applications, degradation over time - especially UV irradiation degradation – poses the most severe problem. The team at Georgia Institute of Technology (GIT) USA, [25] has addressed this issue by using UV stable materials – such as a combination of silicone, fluorocarbons and inorganics like titanium dioxide and silicon dioxide. The prototype has demonstrated the impressive weathering resistance and the surface has higher contact angle measurements and a lower hysteresis loss than the natural analogue. This indicates that the novel coatings are hydrophobic and could have a better self-recovery mechanism than the lotus leaf. However, additional research is needed to understand the effect of surface structure on superhydrophobicity. The previous work on self-cleaning surfaces for transmission lines has been theoretical. In addition to its self-cleaning nature, the biomimetic surface has other benefits also such as anti-corrosivity, biocompatibility, antistiction and transparency. These properties could be exploited in a number of applications, such as

- implantable medical devices
- micro-electromechanical systems (MEMs)
- space suits.

19.20.3 Optical Biomimetics: Materials and Applications [26]

Optical biomimetics is the study of natural systems to inspire the novel solutions to the problems in optical technologies. It is used to characterise and mimic the naturally occurring optical effects. Beginning with an overview of natural photonic structures, it goes on to discuss optical applications of biomolecules such as retinylidene and bacteriorhodopsin, polarisation effects in natural photonic structures and their applications, and biomimetic nanostructures for anti-reflection (AR) devices. Control of iridescence in natural photonic structures is explored through the case of butterfly scales, alongside a consideration of nanostructure fabrication using natural synthesis. The investigation into silk optical materials is followed by a final discussion of the control of florescence in natural photonic structures.

19.20.4 Display Technology Inspired by Butterfly

The photonic display technology is a good technological achievement, developed on account of the inspiration of butterfly. As shown in Figs.19.11.a-b, their details are summarized below.

i. The scales of these butterflies consist of microstructures such as ridges, cross-ribs, ridge lamellae, that have been responsible for colouration.

Figure 19.11 Display technology inspired by Butterfly [27]

ii. Morpho butterfly wings contain microstructures that create its coloring effect through structural colouration rather than pigmentation.
iii. Incident light waves are reflected at specific wavelengths to create vibrant colours due to multilayer interference, diffraction, thin film interference and scattering properties.
iv. The photonic microstructure of butterfly wings can be replicated through biomorphic mineralisation to yield similar properties.
v. The same principles behind the colouration of soap bubbles apply to butterfly wings. The colour of butterfly wings is due to multiple instances of constructive interference from structures.
vi. The photonic microstructures can be replicated using metal oxides or metal alkoxides such as titanium sulphates ($TiSO_4$), zirconium oxide (ZrO_2), and aluminium oxide(Al_2O_3).

References

1. Gururaja Udupal, S. Shrikantha Rao, K. V. Gangadharan, "Future applications of Carbon Nanotube reinforced Functionally Graded Composite Materials", IEEE-International Conference on Advances in Engineering, Science and Management (ICAESM -2012) March 30–31, 2012.
2. Mateusz Wosko, Bogdan Paszkiewicz, Tomas Piasecki, Adam Szyszka, Regine Paszkiewicz, Marex Tlaczala "Application of functionally graded materials in optoelectronic devices" *Optica Applicata* Vol. 35, No. 3, 2005.
3. Joao Paulo Carmo, Luis Miguel Goncalves, Reinoud F. Wolffenbuttel, José Higino Correia "A planar thermoelectric power generator for integration in wearable microsystems" *Sensors and Actuators A* 161 (2010) 199–204.

4. S.M. Yanga, M. Cong, T. Lee "Application of quantum well-like thermocouple to thermoelectric energy harvester by BiCMOS process" *Sensors and Actuators A* 166 (2011) 117–124.
5. I-Yu Huang, Jr-Ching Lin, Kun-Dian She, Ming-Chan Li, Jiann-Heng Chen, Jin-Shun Kuo "Development of low-cost micro-thermoelectric coolers utilizing MEMS technology" *Sensors and Actuators A* 148 (2008) 176–185.
6. Hsiang-Shen Huang, Ying-Che Weng, Yu-Wei Chang, Sih-li Chen, Ming-Tsun Ke "Thermoelectric water cooling device applied to electronic equipment" *International communications in heat and mass Transfer* 37 (2010) 140–146.
7. S. Berger and M. Nitzani, '*New Pyroelectric Thin Composites Films*', Faculty of Materials Science and Engineering, Technion, Haifa, 32000, Israel.
8. Jean-Yves Sanchez, Fannie Alloin, France Chabert, Alessandra d'Apprea, Alain Dufresne, Nadia El Kissi, Wim Thielemans, Cristina Iojoiu, ' Composite and Nanocomposite Polymer Electrolytes For Electrochemical Energy Sources,' LEPMI , BP 75, LGP2, BP 65, Laboratoire de Rhéologie, BP 95, F38402 St Martin d'Hères Cedex.
9. C. Dillon, E.S. Whitney, C. Curtis, C. Engtrakul, M. Davis, Y. Zhao, Y-H. Kim, S.B. Zhang, S-H. Lee, R. Deshpande, P.A. Parilla, K.M. Jones, B. To and A.H. Mahan 'Novel Nanostructured Materials for a Variety of Renewable Energy Applications' *National Renewable Energy Laboratory*, 1617 Cole Blvd. Golden, Colorado 80401
10. A.K. Bandyopadhya, L. Vu-Quoc , P.C. Ray and V. Gopalan 'Critical Fields in Lithium Niobate nano Ferroelectrics'. Govt. College of Engineering & Ceramic Technology, WBUT, Kolkata- 700010, India; 0LT, WBUT, Salt Lake, Kolkata-700098, India ; Dept. of Materials Science and Engineering, Pennsylvania State University, U.S.A.
11. Kody Varahramyan, Mangilal Agarwal and Yuri Lvov '*Nanoengineering of Wood Fibers for Conducting Paper Institute for Micromanufacturing*', Louisiana Tech University, Ruston, LA 71272.
12. Deepali Sinha-Khetriwal Philipp Kraeuchi and Markus Schwaninger, 'A Comparison Of Electronics Waste Recycling In Switzerland and in India: A Review" *Environmental Impact Assessment Review* 25 (2005) 492–504.
13. G.C. Siha and H.Y. Yu, 'Volume fraction effect of magneto electroelastic composite on enhancement and impediment of crack growth', *Composite Structures* 68(2005) 1–11.
14. J.H. Kim, *et al.*, 'Magnetic composites of conducting polyaniline/nano-sized magnetite and their magnetorheology', *Materials Letters* (2008), doi: 10.1016/j.matlet.2008.01.067.
15. M.U. Islam, *et al.*, 'Electrical transport properties of CoZn ferrite-SiO_2 composites prepared by co-precipitation technique', *Materials Chemistry and Physics* (2008), doi :10.1016/j.matchemphys.2007.12.021.

16. H.F. Zhao, *et al.*, 'Thermal conductive properties of Ni-P electroless plated SiC_p/Al composite electronic packaging material', *Surface & Coating Technology* 202(2008) 2540–2544.
17. Tomoya Ozaki *et al.*, 'Magnetic flux memory effect using a magnetostrictive material-shape memory piezoelectric actuator composite', *Sensors and Actuators A* 154 (2009) 69–72.
18. E. Y. Tsymbal, and D. G. Pettifor , "Perspectives of Giant Magnetoresistance", *Solid State Physics*, ed. by H. Ehrenreich and F. Spaepen, Vol. 56 (Academic Press, 2001) pp.113–237.
19. C. Scherer, and A.M. Figueiredo Neto, "Ferrofluids: properties and applications", *Braz. J. Phys.* vol. 35 no. 3a São Paulo Sept. 2005
20. A. Suhane *et al.*, 'Rare-earth aluminates as a charge trapping materials for NAND flash memories: Integration and electrical evaluation', *Solid-State Electronics* 65–66 (2011) 177–183.
21. http://mafija.fmf.uni-lj.si/seminar/files/2009_2010/MAGNETOCALORIC effect.pdf(page:3, 4,5).
22. http://newenergyandfuel.com/http:/newenergyandfuel/com/2009/05/25/progress-update-on-magnetic-refrigeration/
23. Hua Su, Xiaoli Tang, Huaiwu Zhang, Yulan Jing, "Low-Loss Magneto-Dielectric Materials: Approaches and Developments", *Journal of Electronic Materials*, Vol. 43, No. 2, 2014, Doi: 10.1007/S11664-013-2831-5.
24. Judy Holmes, "Syracuse University researchers use nanotechnology to harness the power of fireflies", published in *Nano Letters*, Jun 14, 2012.
25. Susan Essien Etok, "Biomimetic coating for electric transmission", *Materials World Magazine*, 01 Dec 2006.
26. M.C.J. Large, "Optical Biomimetics" *Materials and Applications*, A volume in Woodhead Publishing Series in Electronic and Optical Materials, 2012.
27. http://goodcleantech.pcmag.com/science/281558-breakthrough-qualcomm-imod-display-imitates-butterfly-wings-uses-less-energy-than-lcd.

Appendix I: SI Prefixes of Multiples and Submultiples

Factor	Symbol	Prefix	Factor	Symbol	Prefix
• 10^{-1}	d	deci	• 10^{-9}	n	nano
• 10^{+1}	da	deka	• 10^{+9}	G	giga
• 10^{-2}	c	centi	• 10^{-12}	p	pico
• 10^{+2}	h	hecto	• 10^{+12}	T	tera
• 10^{-3}	m	milli	• 10^{-15}	f	femto
• 10^{+3}	k	kilo	• 10^{+15}	P	peta
• 10^{-6}	μ	micro	• 10^{-18}	a	atto
• 10^{+6}	M	mega	• 10^{+18}	E	exa

Appendix II: Greek Alphabet

Name	Symbol	Analogous English sound	Name	Symbol	Analogous English sound
• Alpha	α	a	• Xi	Ξ	X
• Beta	β	b		ξ	x
• Gamma	Γ	G	• Omicron	ο	o
	γ	g	• Pi	Π	P
• Delta	Δ	D		π	p
	δ	d	• Rho	ρ	r
• Epsilon	ε	e	• Sigma	Σ	S
• Zeta	ζ	z		σ	s
• Eta	η	e	• Tau	τ	t
• Theta	θ	Th	• Upsilon	υ	u
• Iota	ι	i	• Phi	Φ	Ph
• Kappa	κ	k		φ	ph
• Lambda	Λ	L	• Chi	χ	ch
	ι	l	• Psi	Ψ	Ps
• Mu	μ	m		ψ	ps
• Nu	ν	n	• Omega	Ω	O
				ω	o

Appendix III: Conventions to be Followed While Using SI UNIT

1. Full stop, dot, dash or plural is not used while writing unit symbols, e.g.,
   write 10 mm and not 10 m.m.
   write 5 kg and not 5 kgs
   write 20 MN and not 20 M-N
2. Prefix symbol is used in continuation (without gap) with the unit symbols, e.g.,
   write MW and not M W.
3. Two symbols should be separated by one space. For example, for milli second
   write m s and not ms (millisecond)
4. For temperature,
   write K and not ^{0}K
5. Write proper names in small letters when used as a word, and their symbols with a capital letters. For example
   write coulomb and not Coulomb
   write T (tesla) and not t or Tesla
6. Use single prefix instead of double prefixes, e. g., for terahertz
   write THz and not MMHz
7. Prefix should be attached to numerator instead of denominator,e.g.,
   write MWb m^{-2} and not Wb m m^{-2}

8. Group three digits together on either side of the decimal point. Do not group four digit numbers in this way. For example
   write 20 465 and not 20465
   write 1398 and not 139 8 or 1 398
   write 4.061 872 and not 4.0618 72

9. Write 10 MV and not 10^7 V;
   write 1.602×10^{-19} and not 1602×10^{-22}

Appendix IV: Physical Constants

Quantity	Symbol used in this book	SI uint
Acceleration due to gravity	g	9.807 m/s^2
Atomic mass unit	(amu)	1.660×10^{-27} kg = 10^{-3} N
Avogadro number	N_A	6.023×10^{23}/mol
		= 6.023×10^{23} particles/mol
Bohr magneton	β	9.273×10^{-24} Am^2
Boltzmann's constant	k	1.381×10^{-23} J/K
		= 8.614×10^{-5} eV K^{-1}
Charge of electron	e	1.602×10^{-19} C
Electron rest mass	m_e	9.109×10^{-31} kg
Faraday constant	F	9.649×10^{7} C/mol
Gas constant	R	8.314 J/mol K
Molar gas volume at STP	-	2.241×10^{-2} m^3/g-mol
		= 22.4 m^3/kg-mol
Permeability of vacuum	μ_0	$4\pi \times 10^{-7}$ H/m (henry/meter)
Permittivity of vacuum	ε_0	8.854×10^{-12} F/m
Planck constant	h	6.626×10^{-34} J s
Proton mass	m_p	1.672×10^{-27} kg
Speed of light in vacuum	c_0	2.998×10^{8} m/s
		= 3×10^{8} m/s

Appendix V: Conversion Factors

Quantity	Unit and Symbol	Converted Value
• Length	1 Å	$=10^{-10}$ m = 10^{-8} cm = 10^{-4} μm = 0.1 nm
• Density	1 kg/m^3	$=10^{-3}$ g/cm^3 = 10^{-3} kg/litre = 1 g/litre
• Force	1 N	= 0.102 kgf* = 10^5 dyne = 1 kg-m/s^2
		= 0.2248 lbf
	1 kgf*	= 9.807 N ≈10 N = 2.2 × 10^{-3} kip
• Pressure	1 mm of Hg at 0^{0}C	= 133.3 Pa = 1 torr = 1.333 mbar
Intensity	1 atm	= 101.325 kPa = 760 mm of Hg at 0^{0}C
	1 Pa	= 1 N/m^2 = 10^{-6} N/mm^2 = 10^{-5} bar
		= 1.45 x 10^{-4} psi
	1 ksi	= 1000 psi = 6.895 MPa
• Work	1 J	= 1 N-m = 0.2389 cal
• Power	1 kW	= 1000 W = 1000 J/s = 3.6 MJ/h
		=0.9478 Btu/s
	1 hp	=746 W = 10.69 kcal/min =550 ft-lb/s
• Energy	1 eV	= 1.602 x 10^{-19} J
	1 J	$=10^7$ erg = 1 W s
	1 kWh	=3.6 MJ 3412 Btu = 860 kcal
	1 kcal	= 1000 cal = 4187 J = 3.967 Btu

Quantity	Unit and Symbol	Converted Value
• Temperature	0^0C	= 273.15 K
	0 K	= – 273.15^{0}C
	t^{0}C	= (1.8 × t + 32) ^{0}F
• Magnetic		
- field strength	1 A/m	= 0.01257 oersted
- flux	1 Wb	= 10^8 maxwell
- flux density	1 (trsla) T	= 1Wb/m^2 = 10^4 gauss
• Thermal Conductivity	1 W/(m-K)	= 0.00237 cal/(cm-s-^{0}C)
• Electrical		
- conductivity	1 (siemen) S/m	= 1mho/m = 1/(ohm m)
- teststivity	1 ohm m	=10^2 ohm cm
• Quantity of electricity	1 Ah	=3600 C = 3.6 kC
• Frequency	1 Hz	=1/s = 1c/s
• Stress, strength	1 Pa	=10 dynes/cm^2 = 1.0197 × 10^{-2} g/cm^2

ABBREVIATIONS

Ft = foot/feet
yd = yard
ml = millilitre
Lb = pound
psi = pound/sq. inch
ksi = kip per sq inch
Btu = British thermal unit
Ah = ampere-hour
s = second
kip = kilo pound
W = watt
Wb = weber

GLOSSARY OF TERMINOLOGIES

Anode is an electrode which supplies electrons to an external circuit.

A-15 Superconductors are intermetallic compounds such as Nb_3Sn, Nb_3Al, V_3Ga having extremely low critical temperature. Their structure *is* cubic crystalline, which are also known as beta-tungsten (β-W) structure.

Ageing of Mineral Insulating Oils It refers to degradation of physical as well as electrical properties of oils with time due to ingress of moisture, oxygen and heat. Ageing of oils leads to deterioration of their insulating properties such as loss of electric strength, change in viscosity, formation of sludge etc.

Asbestos is a fibrous silicate mineral. It is used as insulator in the sheet, powder and wool forms.

Bipolar Solids are those which conduct current by means of two charge carrying particles of opposite sign (+ and -). For example, the semiconductors conduct current by negative electrons and positive holes both, so they are bipolar solids.

Black Varnish It is a resinuous solution of bitumen and asphalt in oil. It is of black colour due to the presence of black coloured bitumen as base material. Black varnish is less hygroscopic and does not age quickly, but has poor heat resistance. It is used to manufacture varnished cloth for use as impregnent in winding the armature coil.

Bubble Theory is one of the theories of breakdown for liquid insulation (other theory is colloidal theory). It explains that the breakdown occurs due to formation of vapour bubble in liquid. The formation of bubbles may be caused due to gas pockets on the surface of electrodes, irregular electrode surfaces, change in temperature and pressure etc. This theory suggests that the *bubble breakdown strength (i.e.* electric field) E_b in a gas bubble which is immersed in a liquid of permittivity ε_L, is given by

$$E_b = \frac{3E_0}{\varepsilon_L + 2}$$

where E_0 is field in the liquid in absence of bubble.

Bucky Ball is a large carbon crystal made-up of 60 carbon atoms. Known as C 60, this uniquely beautiful structure has 20 hexagonal and 12 pentagonal planes. The C 60 crystal of 0.7 nm diameter is completely a bad conductor. Its likely uses are in making superconductors, lubricants, catalyst in chemical reactions etc.

Cathode is an electrode which receives electrons from an external circuit.

Cellophane is a film form product of regenerated cellulose, obtained by precipitating the natural cellulose in an acid.

Ceramic Firing refers to high-temperature treatment of agglomerated materials such as cermets (ceramic + metal) and dispersion- strengthened materials such as Cu, Ag etc. in Mo and W carbides. The dispersion-strengthened materials are used to make electrical contacts.

Closed Packed Structure refers to a crystal structure whose APE (atomic packing efficiency) has maximum possible value. Maximum APE is 0.74 for FCC structure such as those of Au, Ag, Cu, Al etc.

Coaxial Connectors. Connectors are used to provide interconnection between various components of an electronic equipment/system or a connection with printed circuit board (PCB). Several configurations of connectors are employed, among which coaxial is one type. In it, the braid shielding is crimped to the connector. The centre insulation is usually made of teflon, rubber, ceramic, or mica-filled bakelite; and the connector bodies of copper, brass or zinc. The male and female pin contacts are made of brass and beryllium-copper respectively.

Continuity Equation refers to a fundamental law that governs the flow of charge. This equation/law considers that the charge can neither be created nor destroyed in processes like generation, recombination, drift and diffusion.

Copper Braid Transmission of a signal from one place to another is accomplished by various means such as transmission lines, cables, waveguides, and optical fibres. A transmission cable is composed of a conductor, insulation around it, shield, and a jacket. Shields are used to isolate the transmitted signals from the electromagnetic interference. Different types of shields are used for this purpose among which the braid shield is an important type. It offers good flexibility. It is made of groups of fine wires woven around the centre conductor. The wires are generally made of copper, hence the name 'copper braid'.

Devitrification In some materials such as glass, the solid state is achieved when the molten glass is supercooled. The molecular arrangement in solid state corresponds to that in the frozen liquid state. This results in formation of an amorphous material such as glass. It is due to high viscosity of liquid, consequent upon which the crystal could not grow for lack of time. However, such glassy materials may crystallize on annealing. Crystallization in this way is called *devitrification.* Quartz is one such example that vitrifies on annealing.

Dry ice is the name given to solid CO_2. It is used to freeze foods, meat, ice cream, and is used as a coolant also.

Electric Stress When experienced by a dielectric or an electrical insulating material, the electric field intensity is often referred to as electric stress.

Electrode is a component that serves as an electrical pole.

Electrohydrodynamic (EHD) Motion Whenever conduction in an insulating fluid (polar and nonpolar liquids) *is* accompanied with a significant charge injection, the *convection motion* occurs. This is called as EHD motion.

Electro-optic Effect *is* the behaviour of a material in which its optical isotropic nature changes to anisotropic nature on application of an electric field. This effect is seen in $LiNb0_3$, $LiTiO_3$, benzene etc.

Epitaxy and Epitaxial Growth The meaning of *epitaxy* is 'arranged layer' from Greek literature. The lattice structure of a newly grown layer is an exact extension of the substrate crystal structure (of semiconductor). Source of silicon to be grown is in gaseous form. This gas also contains impurity atoms which are precipitated alongwith silicon, to form an expitaxial layer. Formation of such layer is one process in the sequence of several processes to fabricate an Integration Circuit (IC).

Equi-cohesive Temperature refers to a temperature at which the strengths of grains (crystals) and grain boundaries are equal.

Extrusion is a manufacturing process to manufacture the pipes, tubes, rods etc. through a *die,* by deforming the hot metal plastically.

Fluorescence is a type of luminescence that occurs in a device immediately after its excitation by electricity. The delay time between absorption of energy and re-emission of visible light is much less than *one second.* Fluorescent lamps are the example of such phenomenon.

Galena (or Galenite) is a semiconductor compound (PbS) of lead and sulphur. This crystal was used earlier in radio sets.

Hall Angle It is the non-vanishing angle made by the total electric field in a semiconducting sample, with the x-axis along which the carriers drift.

Hall Generator is a Hall Effect based measuring device which is used to measure high amplitude a.c. and impulse currents, and very high direct currents. Such high currents are used in testing of cables, lightening arresters, circuit breakers etc. Hall generators employ semiconductor materials in their construction. It is because the Hall coefficient of semiconductors is high as compared to metals whose Hall coefficient is very small.

Hardness is a surface property of materials by virtue of which they resist permanent deformation, indentation, scratch, mechanical wear etc. on the surface. Diamond is hardest known material. It is rated as 10 on Moh's hardness scale and 2600 on Brinell hardness scale.

Heteropolar Bonds are formed due to transfer of electrons between the atoms. These are also called ionic bonds. NaC1, MgO, Al_2O_3 etc. are such examples.

Homogenization is a heat induced diffusion to produce uniformity in a solid. This is performed on electrical materials to achieve isotropy.

Homopolar Bonds are those bonds in which the atoms share electrons with each other. This is also known as covalent bond. Si and Ge are such examples.

Inclusions refer to impurities present in the form of particles, within the material.

Induced Dipole Moments The induced dipole moment is the difference of 'orbital dipole moment under an applied magnetic flux' and 'orbital magnetic dipole moment in the absence of any magnetic field'. It is related to magnetic dipole moment *M* per unit volume by

$$M = \mu_{m\ induced} \times N$$

where *N* is the number of atoms per unit volume. The direction of induced dipole moment is opposite to the direction of applied magnetic field.

Ion is an atom which carries a charge due to the addition or removal of electrons while keeping its nucleus unchanged. Ion is electrically non-neutral, and its properties are different from its atom. Relative size of an ion *is* smaller than the size of its atom.

Iron Group of Elements Those transition elements in which the 3d states are partly empty are called 'iron group of elements'. Their atomic numbers are from 21 to 29. Fe, Ni, Co, Gd are such examples. Due to incompletely filled inner states, these elements influence the magnetic properties of materials considerably.

Luminescence is a phenomenon in which the light is emitted by re- radiation of photons, after initial activation. Some materials are capable of absorbing energy and then re-emitting the visible light. Depending upon the source of production, this can be cathode luminescence, photo luminescence, electro luminescence, chemical luminescence and others. The picture viewed on television screen is an example of luminescence.

Magnetic Annealing Annealing is a heat treatment process which is performed to improve ductility, enhance machinability, refining the grains, and softening the metals. When the process is performed in absence of magnetic field, it is known as mechanical annealing (or simply annealing). And if the process is performed in presence of an applied magnetic field, it is called 'magnetic annealing'. Its treatment alters the mechanical properties of materials. For example, the demagnetizing force and magnetization of Alnico increases considerably on magnetic annealing.

Magnetostatic Energy and Isotropic Exchange Energy Magnetostatic energy is the energy of dipole moments of a material, whereas the *isotropic exchange energy* is the energy of interaction between the dipoles. Due to this energy, the dipoles line-up together.

Masking is the process of repeating oxidation-photolithography-doping-epitaxial growth-metallization; a number of times on wafer surface during making of an integrated circuit (IC).

Metallization This is a process of IC-fabrication in which the interconnections of components are formed on the chip. To accomplish this process, aluminium is vapourized so as to get a thin layer on the slice. Then a layer of photoresist is spread over it, to etch everything else except aluminium.

Metal-film Resistors are manufactured by depositing a film of metal (generally Cr, Ni, Cr-Ni) on high grade ceramic rod. The end caps are force-fitted on the ceramic rod alongwith the specially coated axial *leads*. The 'leads' are welded to ensure good mechanical and electrical contact. The ceramic rod is then helically grooved to achieve desired value of resistance. Such resistors are generally used in oscilloscopes, oscillators, and measuring bridges etc.

Nernst-Einstein Equation relates electrical conductivity of the ionic crystals with their diffusivity *D*. It is related as

$$\frac{\sigma}{D} = \frac{ne^2 z}{kT}$$

where n is number of diffusing ions per unit volume, z is valency, e is electronic charge, T is temperature, and k is Boltzmann's constant.

Noble Metal is one that is nonreactive. It does not corrode. Gold is a noble metal. Platinum and silver also fall in this category.

Oxifer is a special kind of ferrite which is obtained by firing a mixture of two different sulphates.

Paschen's Law It explains that the spark breakdown voltage V_b of a uniform field gap is a function of the product of gas pressure *p* and gap length *d* for a particular gas and electrode material. Thus V_b = fn (p x d). This law helps in determining the sparking potential for breakdown at the gaps. For air, the value of *(p.d.)* =0.55 and V_{min} = 352 volt.

Phase Coherency indicates two such phases whose structures are matching to each other.

Phase Diagram Metals exist in different phases such as α-phase, β-phase, ϒ-phase etc. Two or more metals are mixed together to form an alloy. The plots showing relations between different phases in equilibrium versus composition at different pressure and temperature are called phase diagrams. Depending upon the number of metal components involved, the phase diagrams may be unary phase diagram, binary phase diagram, ternary phase diagram etc.

Phosphorescence refers to a delayed luminescence for certain period, after excitation. The delay period *is* more than 'one' second.

Photolithography This is a process of planar technology to fabricate an IC in which, a desired pattern of doping is transferred onto the silicon wafer. The entire process involves *(i)* coating of SiO_2 layer with a film of photosensitive emulsion called *photoresist (ii)* blackening the portions to be doped on a transparent sheet, placing this sheet on photoresist layer and then exposing to ultravoilet light *(iii)* removing the mask and dipping the wafer in trichlorothylene, and *(iv) etching* the SiO_2 with hydrofluoric (HF) solution.

Preferred Orientation refers to a non-random alignment of crystals in a solid,

Printed Circuit Boards (PCBs) are built on a variety of substrates/laminates. The type of laminate depends on operating requirements (temperature, humidity, vibration etc.) and the cost. Laminates are generally made of copper cladded phenolic resin, glass reinforced silicon resin, epoxy, polyester etc.

Relaxation Time is the time required for the time-dependent component of strain or a process to reach 1/e of its final value. Here *e* is exponential factor.

Reluctivity It is the reciprocal of permeability and is expressed as $1/\mu_r$ for a medium. Its value is little less than *one* for diamagnetic materials, little more than *one* for paramagnetic materials, and a small fraction of *one* for ferromagnetic materials. For steel it is 1/2000.

Ruby Crystal $Al_2O_3.Cr_2O_3$ is a crystal of aluminium oxide in which some aluminium atoms are replaced by chromium atoms. Cr^{+3} ions are active material in ruby rod which is an essential part in ruby laser. A ruby crystal containing about 0.05 percent chromium is of pink colour.

Scale is a layer of oxide on the metal surface. It is caused due to corrosive environment, and is detrimental to electrical conduction.

Silver Tarnishing is a type of dry corrosion that occurs when the silver reacts with sulphur. It can be prevented by adding aluminium or silicon in silver. The alloying elements provide protective surface layers of Al_2O_3 and SiO_2.

Sintering is a fabrication process in which the agglomeration of powdered metal is done by application of pressure and temperature. This process is employed to prepare the intricate and tiny sized products such as miniature magnets.

Skin Effect When the thin laminated cores (sheet) of apparatuses like transformers are subjected to alternating current, the alternating magnetic flux should normally get distributed over the entire cross-section of the sheet. But normally it is not so. The flux is forced outwards by the outer skin of magnetic sheet. This behaviour of sheet is called 'skin effect'. This is an undesired effect and *is* caused due to demagnetizing effect of eddy current.

Soft and Hard Directions Ferromagnetic materials exhibit directional nature in their behaviour. They get magnetized differently in different directions when subjected to an applied magnetic field. Some magnetic materials offer high permeability along certain directions and are easily magnetised. Such directions are called *soft directions*. But these materials are difficult to magnetise in some other directions which are called *hard directions*. These directions are different in different materials.

Spinels are the compounds having general formula XY_2O_4, where X and Y are the cations, and oxygen is the anion. The cations X are in the four tetrahedral voids and cations Y in the eight octahedral voids for every four FCC packed oxygen anions.

Splat Cooling refers to extremely fast cooling rate, generally exceeding one million degrees per second, and is used in the production of *metallic glass* by transforming the liquid state directly into a solid state, without undergoing through the intermediate state.

Stark Effect is similar to Zeeman effect but in this case, the light source is placed in strong electric field instead of a magnetic field as in Zeeman effect.

Steatite is a kind of porcelain and belongs to ceramic family of insulators. It possesses low dielectric loss, and is suitable for low voltage high frequency applications. It is used for switches, spacers, bushings and resistor shaft purposes.

Superlattice Alloys. These are alloys of platinum formed with iron or cobalt. Since structure of Pt is of FCC type, of Fe is BCC and Co is HCP; therefore the alloys formed by them are of special characters. When heated for hardening and then cooled, they form a superlattice of FCC and BCC, or FCC and HCP configurations. These are used to make hard magnetic materials.

Superplasticity is the behaviour of material in which it deforms by many hundred percent without necking. High purity aluminium and glass pulled to very long fibres are the examples of superplastic materials.

Super-Refractory Materials are very high temperature resisting refractories (a kind of ceramics) such as beryllia (BeO), Zirconia (ZrO_2); nitrides and carbides *e.g.* AlN, boron nitride (BN), hafnium carbide (HfC). Melting point of HfC is 4200 K.

Thermistors are temperature measuring devices, usually made of semiconducting materials. They have a negative temperature coefficient of resistance. That is why their resistance decreases with increase in temperature. Due to their high sensitivity of temperature changes, they are used in temperature measurement, control and compensation of precision systems. Thermistors can operate in a temperature range of -100°C to 300°C. Commercial thermistors are made from sintered mixture of Mn_2O_3, NiO_2 and Co_2O_3.

Thermocouple is a temperature measuring instrument used in ovens and furnaces. This is made of thermoelectric materials.

Transducer is a device that converts energy from one form to another. For example, a loudspeaker converts electrical energy into sound energy while an electric heater converts electrical energy into heat energy. Components of transducers are generally made of electrical, electronics and optical materials.

Unipolar Solids are those which conduct current by means of mobile charges of one sign (either + or -) only. For example, the metals conduct current by free electrons (of —ve sign) only, therefore they are unipolar solids.

Variable Inductors are electrical devices that are used in tuning, timing, and calibration circuits for varying the value of inductance. They use a variable magnetic core of adjustable inductance.

Villari Effect. It is the converse of magnetostriction effect. It states that the longitudinal deformation in a material under the influence of mechanical load, induces a change in its permeability in the direction of applied strain.

Vitreous Enamel Capacitors are manufactured by spraying vitreous enamel on the metal plates that are stacked and fired at a temperature high enough to vitrify the glaze. These capacitors have superior high frequency characteristics and can be used at high temperatures (150-200°C). They can also withstand high humid conditions.

Vulcanisation refers to a process through which the rubber (a linear chain compound) is cross-linked using sulphur, in the presence of heat and pressure.

Zone Refining means purifying a material from impurity contents present in it. Principle of phase separation is employed for zone refinement. Ultrapure silicon single crystal is produced by this method.

REFERENCES

The following books have provided the guidelines to prepare the manuscript of this book.

A.J. Dekker, *Electrical Engineering Materials,* Prentice-Hall of India Private Limited, New Delhi, 2009.

Adir Bar-Lev, *Semiconductor and Electronic Devices,* Prentice Hall, 1993.

B.W. Williams, *Power Electronics: Devices, Drivers, Applications and Passive Components,* Palgrave Macmillan Press, 1992.

Ben G. Streetman and Sanjay Banerjee, *Solid State Electronic Devices,* PHI Learning Private Limited, New Delhi, 2006.

Jacob Millman and C.C. Halkias, *Integrated Electronics: Analog and Digital Circuits and Systems,* Tata-McGraw Hill Pub. Co. Ltd., New Delhi, 1991.

K.H. Jurgen Buschow *et. al.*, Editors-in-chief, *Encyclopedia of Materials: Science and Technology,* Vol. 4, Elsevier, 2011.

K.M. Gupta, *Electrical and Electronics Engineering Materials,* Umesh Publications, Delhi, 2006.

K.M. Gupta, *Engineering Materials: Research, Applications and Advances,* CRC Press, Taylor & Francis Group, 2014.

Robert F. Pierret, *Modular Series on Solid State Devices-Semiconductor Fundamentals,* Addison-Wesley, 1996.

Robert M. Rose, L.A. Shepard and John Wulff, *Electronic Properties: Volume IV,* John Wiley & Sons, 1966.

W. Bolton, *Electrical and Magnetic Properties of Materials,* Harlow: Longman Scientific & Technical, England, 1992.

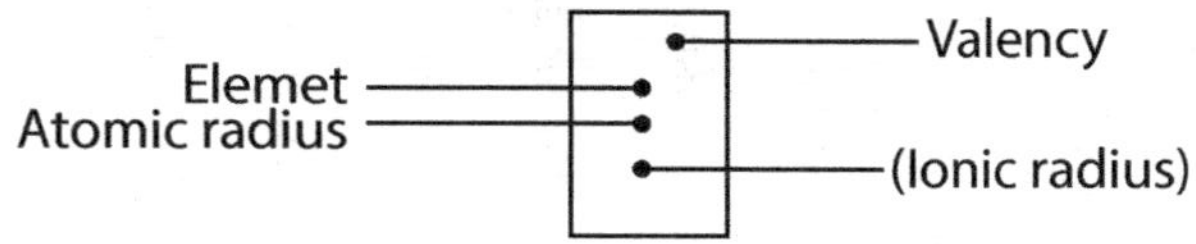

H +1 0.46 (=0)																	He 1.76 –
Li 1+ 1.52 (0.68)	Be 2+ 1.12 (0.34)											B 3+ 0.46 (0.25)	C 4+ 0.71& 0.77 (0.20)	N 0.71 –	O 2– 0.68 (1.32)	F 1– 0.64 (1.33)	Ne 1.58 –
Na 1+ 1.86 (0.98)	Mg 2+ 1.59 (0.78)											Aj 3+ 1.43 (0.57)	Si 4+ 1.18 (0.38)	P 5+ 1.10 (0.30)	S 2– 1.04 (1.74)	Cl 1– 0.90 (1.81)	Ar 1.88 –
K 1+ 2.31 (1.33)	Ca 2+ 1.97 (0.94)	Sc 3+ 1.61 (0.68)	Ti 4+ 1.46 (0.60)	V 3+ & 5+ 1.32 (0.65& 0.40)	Cr 3+ 1.25 (0.64)	Mn 2+ 1.12 (0.91)	Fe 2+ & 3+ 1.24 (0.83 & 0.67)	Co 2+ 1.25 (0.82)	Ni 2+ 1.25 (0.78)	Cu +1 1.28 (0.96)	Zn 2+ 1.31 (0.83)	Ga 3+ 1.22 (0.62)	Ge 4+ 1.22 (0.44)	As 5+ 1.25 (0.40)	Se 2– 1.14 (1.91)	Br 1– 1.13 (1.96)	Kr 2.00 –
Rb 1+ 2.44 (1.48)	Sr 2+ 2.15 (1.27)	Y 3+ 1.79 (1.06)	Zr 4+ 1.58 (0.87)	Nb 5+ 1.43 (0.69)	Mo 4+ 1.36 (0.68)	Te 1.34 –	Ru 4+ 1.35 (0.65)	Rh 3+ 1.34 (0.68)	Pd 1.38 (0.86)	Ag +1 1.44 (2.13)	Cd 2+ 1.49 (1.03)	Jn 3+ 1.63 (0.92)	Sn 4+ 1.51 (0.74)	Sb 5+ 1.45 (0.90)	Te 2– 1.43 (2.11)	I 1– 1.35 (2.20)	Xe 2.17 –
Ca 1+ 2.62 (0.65)	Ba 2+ 2.17 (1.43)	La 3+ 1.69 (1.03)	Hf 4+ 1.59 (1.84)	Ta 5+ 1.43 (0.68)	W 4+ 1.37 (0.68)	Re 1.36 –	Os 4+ 1.36 (0.67)	Ir 4+ 1.36 (0.66)	Pt 1.39 (0.80)	Au +1 1.44 (1.37)	Hg 2+ 1.55 (1.12)	Ti 3+ 1.70 (1.05)	Pb 2+ & 4+ 1.75 (1.32 & 0.84)	Bi 1.56 –	Po 1.7 –	At	Rn
Fr 1+ (1.75)	Ra 2+ (1.48)	Ac 3+ (1.13)		Rf	Ha	Unh	Uns	Uno	Une								

Ce 3+ (1.02)	Pr 3+ (0.99)	Nd 3+ (0.98)	Pm 3+ (0.97)	Sm 3+ (0.96)	Eu 3+ (0.95)	Gd 3+ (0.94)	Tb 3+ (0.92)	Dy 3+ (0.91)	Ho 3+ (0.90)	Er 3+ (0.89)	Tm 3+ (0.88)	Yb 3+ (0.87)	Lu 3+ (0.86)
Th 3+ (1.08)	Pa (1.04)	U 4+ (1.025)	Np (1.01)	Pu (1.00)	Am (0.98)	Cm (0.97)	Bk (0.96)	Cf (0.95)	Es	Fm	Md	No	Lr

Atomic and Ionic radii and Valency of Elements (Values in Å), 1 Å = 10^{-10}m

Answers to Numerical Questions

Chapter 1

NA

Chapter 2

1. (*a*) 13.6 eV (*b*) – 27.2 eV (*c*) –13.6 eV (*d*) 13.6 eV
2. $v = 2.25 \times 10^6$ m/s, *En*= 6700 T Hz, $f = 6.7 \times 10^{15}$Hz
3. (*a*) 15 Phosphorus (*b*) 46 Palladium
4. (*a*) 2.19×10^6m/s (*b*) 0.0529 m (*c*) $1.519 \times 10^{-16}s$
5. 6.57×10^{-34}Js
6. 6600 Å
7. 4.03×10^8
8. 4.56×10^{14}/s, 1.52×10^6/m
9. 4.05Å, 1.43Å
10. 4380 kg/m^3
11. 2.878 Å, Lead, 2.966×10^{19}
12. (*a*) 1.5 (*b*) 2 (*c*) 58; SC or BCC, BCC, SC or BCC
13. 6000 kg/m^3
14. 5.324
15. --
16. (*a*) 6 (*b*) 90.72×10^{-30} m^3 (*c*) 7177 kg/m^3
17. 2.3 Å, 2.3 Å
18. --
19. (418)
20. --
21. 3.49×10^9 atoms/m
22. (110)

23. $6.24 \times 10^{12}/mm^2$
24. 2.5×10^9atoms/m, 7.21×10^{18} atoms/m^2, plane (111)
25. 1.732 Å
26. (*a*) 3.25 Å (*b*) 3.98 Å (*c*) 5.64 Å
27. $1.4 \times 10^{19}/m^2$, $10^{19}/m^2$, $8.8 \times 10^{18}/m^2$

Chapter 3

1. 20°
2. 3.5 Å
3. 2.58 Å, 1.824 Å, 1.289 Å
4. 2.752 Å
5. 50x to 675x
6. yes, no, yes

Chapter 4

1. (a) 1.16×10^5m/s (b) 4.33×10^{-3} m^2/V s (c) 0.433 m/s
2. 1.016 m
3. 400 mm, 4.13×10^{-8} ohm m
4. 2.5 mm
5. 0.451 ohm, 2.6 times
6. 4.34×10^{-6} ohm cm
7. 6.76×10^{-3} m^2/V s, 6.76×10^{-3} m/s
8. (a) $6.62 \times 10^{28}/m^3$ (b) 5.3×10^{-15} s
9. 160 V/m
10. 1.39 mm
11. 0.14 m^2/Vs
12. --
13. (a) 0.7 m/s (b) 7×10^{-3} mWs (c) 4×10^{-14} s
14. (a) 5.8×10^{-8} ohm m (b) 222%
15. 1.72×10^{-8} ohm m
16. 1.5×10^8 J
17. 432 and 5.25 W/m°
18. 2.3 eV, potassium

Chapter 5

NA

Chapter 6

1. 0.43 ohm m
2. (a) 3000 ohm m (b) $6 \times 10^{25}/m^3$
3. $225 \times 10^9/m^3$, EF is located at 0.407 eV above E_t towards E_c
4. 2.563/ohm m
5. 2.0/ohm m, p-type
6. 0.015 eV
7. 2.26 eV
8. (a) 8.93×10^{-4} ohm m (b) $-625 \times 10^{-6} m^3/C$ (c) $-62.562.5$ μV
9. n-type
10. $D_h = 0.00091\ m^2/s$, $D_e = 0.0044\ m^2/s$

Chapter 7

NA

Chapter 8

NA

Chapter 9

1. 1.0001548
2. (a) 3.57 (b) 0.037 (c) 0.1323 (d) 87.8°
3. 360 000 volt
4. (a) 95.5 W (b) 1.59×10^{-10} farad (c) 1.81 W
5. Increase of 0.0117 farad
6. 1.21×10^{-7} F
7. 5.4×10^{-22} m
8. 4500 cm^2, 1.6×10^4 V
9. 8×10^4 V
10. $2.2 \times 10^{-7}\ C/m^2$

Chapter 10

1. 857 V
2. 0.02 mm

Chapter 11

1. 0 .25 Wb/m^2, –1.2 A/m
2. 10.14 A/m, 0.098 tesla
3. 0.251 tesla, – 1.2 A/m
4. 1 MA/m
5. (a) 534.85 W (b) 366.14 W
6. 85.7%
7. 85.7%
8. 366.1 W, 534.8 W
9. 2.59 MJ, 471.9 cm^2
10. 1.257 Wb/m^2, 5 A/m
11. 2.2 Bohr magnetons/atom
12. 83.33 kWh
13. 1.283 kW
14. (a) 3.5×10^4 A/m (b) 0.44 tesla
15. (a) 1 T (b) 3×10^5 A/m (c) 1.2 T (d) $\approx 7 \times 10^6$ J/m (e) $\approx 1.4 \times 10^5$ TA/m
16. 3 W
17. 6.3 W
18. 20 W, 100 W
19. (a) 1.25 T (b) 5 A/m
20. (a) 1.25 T (b) 1400 A/m (c) 5000 A/m

Chapter 12

NA

Chapter 13

1. 0.06624 A/m, –0.0758 A/m
2. 1.71×10^8 A/m^2
3. 18.15 K, 24.66 A/m, 23.34 A/m
4. (a) 7.19 K (b) 38.99 nm

Chapter 14

NA

Chapter 15

NA

Chapter 16

NA

Chapter 17

1. (a) 6.62×10^{-19} J (b) 4.13 eV
2. 5.9×10^{-6} eV
3. (a) 3×10^{8} Hz, 1.25×10^{-6} eV, (b) 3×10^{11} Hz, 1.25×10^{-3} eV, (c) 6×10^{14} Hz, 1.25 eV, (d) 3×10^{18} Hz, 1.25×10^{4} eV
4. (a) 9 mW (b) 2.57×10^{-3} J/s (c) 2.81×10^{16} photons/s
5. 2.26 eV

Chapter 18

NA

Chapter 19

NA

Answers to Objective Questions

Chapter 1

1. d 2. a 3. c 4. c 5. b

Chapter 2

1. a	2. a	3. a	4. a	5. d
6. b	7. a	8. a	9. d	10. d
11. c	12. a	13. a		

Chapter 3

NA

Chapter 4

1. a	2. c	3. b	4. F
5. T	6. F	7. T	8. T
9. (loosely)	10. (zero)	11. (Joule's law)	12. (high)
13. (Seebeck)	14. d	15. a	16. d

Chapter 5

1. b 2. c 3. d 4. b 5. a 6. d

Chapter 6

1. c	2. a	3. a	4. d	5. b
6. d	7. c	8. b	9. b	

Chapter 7

1. b	2. d	3. a	4. d	5. a	6. d

Chapter 8

1. c	2. d	3. c	4. c	5. a
6. d	7. b	8. b	9. b	10. c

Chapter 9

1. c	2. b	3. d	4. b	5. c	6. b	7. c
8. d	9. a	10. c	11. a	12. a	13. c	14. d

Chapter 10

1. d	2. a	3. a	4. d

Chapter 11

1. b	2. c	3. (reducing)	4. a	5. d
6. a	7. a	8. d	9. d	10. b
11. a	12. b	13. b	14. a	

Chapter 12

1. c	2. b	3. b	4. a	5. d	6. a
7. b	8. a	9. b	10. a	11. (F)	

Chapter 13

1. b	2. a	3. c	4. d	5. a
6. (LTS)	7. (London's)	8. (Less)	9.(MRI)	
10. (wheel-less)	11. d	12. a		

Chapter 14

NA

Chapter 15

NA

Chapter 16

NA

Chapter 17

1. c 2. d 3. c 4. d

Chapter 18

NA

Chapter 19

NA

Index

Also of Interest

Check out these published volumes in the Advanced Materials Series

Advanced Materials for Agriculture, Food and Environmental Safety
Edited by Ashutosh Tiwari and Mikael Syväjärvi
Published 2014. ISBN: 978-1-118-77343-7

Advanced Biomaterials and Biodevices
Edited by Ashutosh Tiwari and Anis N. Nordin
Published 2014. ISBN 978-1-118-77363-5

Biosensors Nanotechnology
Edited by Ashutosh Tiwari and Anthony P. F. Turner
Published 2014. ISBN 978-1-118-77351-2

Advanced Sensor and Detection Materials
Edited by Ashutosh Tiwari and Mustafa M. Demir
Published 2014. ISBN 978-1-118-77348-2

Advanced Healthcare Materials
Edited by Ashutosh Tiwari
Published 2014. ISBN 978-1-118-77359-8

Advanced Energy Materials
Edited by Ashutosh Tiwari and Sergiy Valyukh
Published 2014. ISBN 978-1-118-68629-4

Advanced Carbon Materials and Technology
Edited by Ashutosh Tiwari and S.K. Shukla
Published 2014. ISBN 978-1-118-68623-2

Responsive Materials and Methods
State-of-the-Art Stimuli-Responsive Materials and Their Applications
Edited by Ashutosh Tiwari and Hisatoshi Kobayashi
Published 2013. ISBN 978-1-118-68622-5